Microwave Excited Plasmas

Plasma Technology

Advisory Editor: L. Holland

Vol. 1. Plasma Etching in Semiconductor Fabrication (R. A. Morgan)
Vol. 2. High Vacuum Production in the Microelectronics Industry (P. Duval)
Vol. 3. Plasma Polymerization Processes (H. Biederman and Y. Osada)
Vol. 4. Microwave Excited Plasmas (edited by M. Moisan and J. Pelletier)

Plasma Technology, 4

Microwave Excited Plasmas

Edited by

Michel Moisan

Groupe de Physique des Plasmas
Université de Montréal
Montréal, Québec (Canada)

Jacques Pelletier

Centre National de la Recherche Scientifique
Laboratoire de Physique et Chimie des Procédés Plasma
Grenoble, France

ELSEVIER

Amsterdam · Lausanne · New York · Oxford · Shannon · Singapore · Tokyo

ELSEVIER SCIENCE B.V.
Sara Burgerhartstraat 25
P.O. Box 211, 1000 AE Amsterdam, The Netherlands

First edition 1992
Second impression 1999

Library of Congress Cataloging-in-Publication Data

Microwave excited Plasmas / edited by Michel Moisan, Jacques Pelletier.

p. cm. - (Plasma Technology: 4)
Includes bibliographical references and index.
ISBN 0-444-88815-2 (alk. paper)

1. Microwave plasmas.   I. Moisan, Michel.   II. Pelletier, Jacques.

QC718.5.M5M53   1992
621.044-dc20   92-30694
                  CIP

ISBN: 0 444 88815 2

♾ The paper used in this publication meets the requirements of ANSI/NISO Z39.48-1992 (Permanence of Paper).

Transferred to digital printing 2006

PREFACE

In the fifties, it was considered good taste to distinguish between the "new" physics of plasmas and the "old" physics of gas discharges. The latter, with its emphasis on the dark and bright areas near the electrodes, the fluctuating cathode spots and the ramified and unpredictable structure of lightning, was considered reminiscent of zoology rather than an exact science. This, despite its admitted success in such practical applications as illumination, electrical breakdown, high tension, etc. By contrast, the new physics was considered to serve a "higher" course: nuclear fusion and its new ideas like the topology of magnetic field lines, charged particle orbits, MHD theory, plasma waves and instabilities, gas kinetic methods, etc. were firmly rooted into the established science of classical electrodynamics.

This argumentation proved useful in that it helped establishing a new community who spoke the same language. However, it gave no credit to the historic development of the new theories from the gas discharge studies of, in particular, I. Langmuir around 1930. After that, the two disciplines remained intertwined as the most practical way to study fusion and astrophysical plasmas in the laboratory was by looking at gas discharge experiments.

In this interplay between plasmas and discharges many interesting phenomena came to the fore. Most interesting perhaps are the high-frequency and ultra high-frequency discharges, which have been in use from the early days of the studies of plasmas. With discharges at around 20 MHz and, one could make a large volume (some dm^3) of pure plasmas of known composition in a controllable way, using capacative and self inductive coupling. This was also done at ultra-high frequencies, either in a cavity or by radiating low pressure gas with antennas. An authoritive review of the properties of all these discharge techniques was given by S.C. Brown of M.I.T. in a classical article in the Handbuch der Physik, which appeared in 1956.

However, since the seventies, the field of high frequency discharges has grown considerably. One has created discharges at surfaces, in electron cyclotrons and at multipolar magnetic confinement configurations. The many new applications are notably in mechanical surface processing and, especially, in microelectronics.

Now it is appropriate to review all these new developements, in which the French (and Quebecian) school has played a very important rôle, in the form of a book. I am very pleased to announce this very complete volume that has been prepared by research teams at Montréal, Grenoble and Orsay and which, I am confident, will remain the central source of reference for a long time to come.

Orsay, July 30, 1992   J.L. Delcroix
Amsterdam, September 14, 1992

PREFACE

Dans les années cinquante, il était de bon ton dans les milieux scientifiques d'opposer la toute nouvelle physique des plasmas, à la "vieille" physique des décharges dans les gaz. Celle-là était considérée comme un peu "zoologique", avec par exemple ses espaces sombres ou lumineux près des électrodes, ses taches cathodiques fluctuantes, et la structure ramifiée et imprévisible de la foudre. On lui reconnaissait certes un intérêt utilitaire avec des applications bien ancrées dans le génie électrique (éclairagisme, claquage, haute tension, ...). Mais cela n'avait rien à voir avec la "nouvelle" physique des plasmas qui était poussée par une application majestueuse, la fusion nucléaire contrôlée, et qui développait à partir de l'électrodynamique classique des domaines scientifiques nouveaux (topologie des champs magnétiques, orbites et invariants adiabatiques, magnétoplasmadynamique, ondes et instabilités diverses, théorie cinétique des gaz coulombiens, amortissement de Landau, ...). Cette argumentation était utile pour ceux qui, comme l'auteur de cette préface, cherchaient à obtenir des moyens pour travailler dans ce nouveau domaine. Mais elle était évidemment simpliste: historiquement d'abord, car la physique des plasmas était née dans les décharges étudiées vers 1930 notamment par I. Langmuir; conceptuellement ensuite car il y a continuité entre les gaz ionisés des décharges, et les plasmas de la fusion et de l'astrophysique; utilitairement enfin, car pour étudier les plasmas au laboratoire, il faut les créer et les techniques des décharges dans les gaz fournissent pour cela le moyen le plus simple.

Dans cette dualité plasmas-décharges, il y a d'ailleurs des zones frontières très intéressantes, des décharges qui sont des sources de plasma propre et qui se prêtent bien à des études de physique des plasmas; la colonne positive des décharges luminescentes et surtout les décharges en hautes fréquences et en hyper-fréquences en font partie. Dès le début de la physique des plasmas, on a beaucoup utilisé les décharges HF. En fonctionnant aux environs de 20 MHz, avec un couplage capacitif ou selfique, on pouvait créer dans des volumes assez grands (jusqu'à quelques dm^3) des plasmas propres (pas d'électrodes dans le plasma), de propriétés assez bien connues et reproductibles. On pouvait aussi fonctionner dans le domaine des hyper-fréquences, soit dans une cavité, soit en irradiant une enceinte contenant du gaz à basse pression par des cornets. La synthèse des propriétés de toutes ces décharges a été présentée par S.C. Brown du M.I.T. dans un article classique du Handbuch der Physik paru en 1956.

Mais depuis les années soixante-dix, ce domaine des décharges haute-fréquence s'est enrichi et diversifié: on a vu apparaître des décharges à onde de surface, à résonance cyclotronique électronique, à confinement magnétique multipolaire. De nombreuses applications nouvelles sont également apparues notamment pour les traitements de surface en mécanique et surtout en microélectronique.

Il était temps de faire le point de tous ces nouveaux développements. Le rôle de l'école française (et québécoise) y ayant été très important, je suis heureux de présenter aujourd'hui l'ouvrage très complet préparé par les équipes de Montréal, Grenoble et Orsay. Je pense que cela sera pour un temps que je leur souhaite long, l'ouvrage de référence dans ce domaine.

J.L. Delcroix

Orsay, le 30 juillet 1992

ACKNOWLEDGMENTS

This book is the result of a close collaboration between our two laboratories that started thanks to the France-Québec scientific exchange program. This collaboration explains the strong emphasis throughout the book on surface-wave sustained plasmas (Montréal) and on microwave plasmas confined by multipolar magnetic fields including DECR (Grenoble). Recent interest in these two types of plasmas has been responsible for a renewed interest in microwave sustained plasmas and resulted in the idea of taking the opportunity of our collaboration to review and analyze microwave excited plasmas.

We would like to thank first of all the authors who contributed to the various chapters of this collective work. It is their participation that has allowed us to cover the many different aspects of microwave plasmas so thoroughly. Besides their scientific contribution, we would also praise their patience and confidence during the preparation of this book. We are deeply indebted to the editing team, Lison Néel, Gaston Sauvé, Jean-Eudes Samuel, Nicole Bourhy, Jacques Cocagne and Fabrice Fabiano. Lison not only typed the numerous versions of the text but also made sure that all editorial details and conventions were obeyed. We wish her good luck in her last year of law studies. During the last six months, she was assisted by Nicole. Gaston devoted many hours to reading and checking all the texts, coordinating the book into its final camera-ready form. We strongly acknowledge the work of Jacques, Fabrice and Sam in drawing the many figures. We also need to thank all those who acted as reviewer and helped improve the text and its contents: P. Lorrain, M. Wertheimer, G. Lister, C. Scott, T. Mantei, J. Waymouth, F. Dias, Yu. Lebedev, A. Sola and A. Gicquel. We also want to express our appreciation to C. Barbeau and L. St-Onge for proofreading the final version of the text. Finally, we acknowledge the support at various levels of J. Hubert, J.-R. Derome, G. Beaudet, R. Cochrane, J. Brebner, F. Senez, J. Labrie, D. Lefebvre and, last but not least, of our wifes, Marie-Claude and Marguerite.

Michel Moisan
Jacques Pelletier

June 1992

AFFILIATION

Yves Arnal - Laboratoire de physique et chimie des procédés plasma, Université Joseph Fourier (Grenoble 1), Unité CNRS, CNET-Grenoble, B.P. 98, 38243 Meylan Cedex, France.

Rudolf Burke - Laboratoire de physique et chimie des procédés plasma, Université Joseph Fourier (Grenoble 1), Unité CNRS, CNET-Grenoble, B.P. 98, 38243 Meylan Cedex, France.

Carlos Matos Ferreira - Centro de Electrodinâmica da Universidade Técnica de Lisboa, Instituto Superior Técnico, 1096, Lisboa Codex, Portugal.

Richard A. Gottscho - Bell Labs, 600 Mountain Avenue, Murray Hill, N.J. 07974, U.S.A.

John Heidenreich - IBM T.J. Watson Research Center, P.O. Box 218, Yorktown Heights, N.Y. 10598, U.S.A.

Tudor W. Johnston - Institut national de la recherche scientifique (INRS) - Énergie, B.P.1020, Varennes, J3X 1S2, Québec.

Joëlle Margot - Groupe de physique des plasmas, Département de physique, Université de Montréal, B.P. 6128, Succursale A, Montréal, H3C 3J7, Québec.

Gilles Matthieussent - Laboratoire de physique des gaz et des plasmas, Université Paris XI, Unité de recherche associée 73 du CNRS, Bâtiment 212, 91405 Orsay Cedex, France.

Michel Moisan - Groupe de physique des plasmas, Département de physique, Université de Montréal, B.P. 6128, Succursale A, Montréal, H3C 3J7, Québec.

Jindrich Musil - Institute of Physics, Academy of Sciences of Czecho-Slovakia, Na Slovance 2, 180-40, Prague 8, Czecho-Slovakia.

Jurij Paraszczak - IBM T.J. Watson Research Center, P.O. Box 218, Yorktown Heights, N.Y. 10598, U.S.A.

Jacques Pelletier - Laboratoire de physique et chimie des procédés plasma, Université Joseph Fourier (Grenoble 1), Unité CNRS, CNET-Grenoble, B.P. 98, 38243 Meylan Cedex, France.

Michel Pichot - Centre national d'études des télécommunications, CNET-Grenoble, B.P. 98, 38243 Meylan Cedex, France.

Claude Pomot - Laboratoire de physique et chimie des procédés plasma, Université Joseph Fourier (Grenoble 1), Unité CNRS, CNET-Grenoble, B.P. 98, 38243 Meylan Cedex, France.

Gaston Sauvé - Groupe de physique des plasmas, Département de physique, Université de Montréal, B.P. 6128, Succursale A, Montréal, H3C 3J7, Québec.

Zenon Zakrzewski - Academy of Sciences of Poland, IMP-PAN, ul. Fiszera 14, 80952 Gdansk, Poland.

CONTENTS

10. Discharges confined by multipolar magnetic fields
R. Burke and J. Pelletier

11. Ambipolar diffusion model of multipolar plasmas
G. Matthieussent and J. Pelletier

12. Homogeneity in multipolar discharges: the role of primary electrons
J. Pelletier and G. Matthieussent

13. High frequency sustained multipolar plasmas
C. Pomot and J. Pelletier

14. Distributed electron cyclotron resonance (DECR) plasmas
M. Pichot and J. Pelletier

15. Applications of microwave plasmas in microcircuit fabrication

J. Paraszczak and J. Heidenreich

CHAPTER 1

INTRODUCTION *

The objective of this work is to deal with questions on microwave discharges that are asked both by the user, who may not be overly concerned about plasma science, and by the plasma expert, who may wish to redirect his interest towards plasma applications, such as materials processing. Hence, whenever feasible, this text aims at providing at least partial answers to the questions that are bound to arise when trying to choose the best plasma for a specific application: it is the purpose of this book to propose a synthesis on microwave discharges, laying out the corresponding physical references but without developing too much plasma theory. This approach is worthwhile in view of the great variety of available microwave plasma sources: it is not enough to put the essential questions, as it is equally important to supply useful answers. Selecting a particular microwave discharge and, eventually, the configuration of its magnetic field confinement, should be the outcome of a careful analysis. This analysis should rely on corroborated and concrete physical considerations, and not on subjective opinions or judgments.

The trend in novel plasma applications is towards reactors of ever increasing size. It is, however, a challenge to excite microwave plasmas over diameters that are larger than the vacuum wavelength. Also, providing, when required, an adequate magnetic field confinement for large diameter discharges should minimize use of costly and clumsy equipment. Another problem is the insertion of substrates or other objects at locations in the plasma where the microwave field is important. This can modify, hinder, or even frustrate the very excitation of the plasma as well as damage the substrate. Hence, for certain applications, one needs to spatially separate the plasma production region (the active zone) from its utilization zone (the reactor). Depending on conditions, a microwave-field free reactor may still contain plasma (charged particles) or there may be very little electrons and ions left, most of the active species by then being neutral atoms, molecules and radicals. Another consideration is the increasingly stringent requirement for processes that are uniform over large dimensions. This can be realized in different ways. For example, when charged particles are needed, uniformity can be achieved in some instances by plasma confinement whereas, when neutral species are driving the process, it is more appropriate to operate in the so-called flowing afterglow of the discharge.

---

* **Presented by M. Moisan and J. Pelletier**

The useful volume that the plasma will be able to fill outside the active zone depends on the characteristic diffusion length of the charged species. In weakly ionized plasmas, dealt with here, this length is usually about one order of magnitude larger than the ion-neutral mean free path. Practically speaking, this leads to the distinction between two pressure regimes. The first one, termed "high pressure", is when the ion-neutral mean free path is much smaller than the characteristic dimension of the plasma reactor. The plasma is then highly collisional (ion-neutral collisions) and one cannot benefit from the action of charged particles outside the region of plasma production. The second regime corresponds to "low pressure", i.e. the ion-neutral mean free path is distinctly larger than the characteristic dimension of the reactor. The plasma is then slightly collisional and may also be used as such outside the production zone. For most gases, it may be assumed that the transition from one regime to the other occurs at a pressure of about one tenth of a pascal (1 µbar or roughly one mtorr). At that pressure, the mean free path is a few centimeter long.

A crude classification permits one to put into the category of "high pressure" plasmas most discharges excited in the absence of a static magnetic field B_0. Among them, one can mention the resonant cavity excited plasmas and the plasmas sustained by electromagnetic surface waves. Conversely, in the presence of a static magnetic field, since it reduces the charged particle losses, one can sustain lower pressure microwave plasmas than with $B_0 = 0$, whatever the means of imposing the microwave field on the discharge. In this "low pressure" plasma category, one finds the special case of electron cyclotron resonance (ECR) sustained discharges.

When considering plasma processes, as already mentioned, one has to distinguish those dominated by the ion flux from those where the flux of neutral (atom or radical) species prevails. In the former limiting case, the process is usually the most efficient within the active zone of the discharge (where the power transfer from the electromagnetic field to the plasma takes place), whereas, in the latter limiting case, it occurs far outside this active zone, in what is called the flowing afterglow. In a flowing afterglow far enough from the active zone, most species are uncharged and thus little control is possible on their flux, in contrast to what can be done within the plasma, where the ion flux and the ion energy can be electrically or magnetically modified. As a matter of fact, in surface treatment applications, it is often essential to be able to independently control the flux and the energy of the charged species impinging upon the substrate. While impossible to achieve in a classical 13.56 MHz parallel plate discharge, such an independent control can be implemented in most other high frequency (HF) discharges by biasing the substrate surface with respect to the plasma potential. The selection of a biasing method, namely a constant or alternating voltage and, in the latter case, the choice of its frequency, are interrelated, and are a function of both the substrate type (conducting or insulating) and the desired application.

Having completed our introductory considerations, let us now examine the structure and the content of the book. Schematically, it is divided into four parts of unequal length. These are:

- Physical principles of microwave plasma generation (Chap. 2).
- Microwave discharges in the absence of a static magnetic field (Chaps. 3-5).
- Microwave discharges in the presence of a static magnetic field: classical axial confinement including ECR (Chaps. 6-8) and multipolar magnetic confinement including Distributed ECR (Chaps. 9-14).
- Applications of microwave plasmas in microcircuit fabrication (Chap. 15).

The book thus starts by considering the physical principles of microwave plasma generation. The first part of Chap. 2 is devoted to the basic physical phenomena occurring in microwave discharges, most of the material presented being in fact applicable to HF discharges in general. Important points are the interaction between the electromagnetic (EM) field and the charged particles of the discharge, and the discharge maintenance process under steady-state conditions. Further on, the microwave discharge is analyzed as a load to the microwave field applicator, these two elements together with the impedance matching network forming the plasma source. Finally, a classification of microwave discharges is proposed.

The first question to be discussed, when considering a specific application, concerns the discharge optimal excitation frequency: zero frequency, i.e. direct current (DC), radio-frequency (RF), or microwave. Put differently, which are the plasma parameters that are foremost influenced by the choice of the excitation frequency? The third chapter endeavors replying to that question by indicating how the electron energy distribution function (EEDF) evolves when the discharge frequency proceeds from DC to RF ($\approx$ 10 - 300 MHz) on to microwaves ($\geq$ 300 MHz). Changes in the EEDF cause the steady state relative concentrations of excited atoms, radicals and ions to shift as a function of the applied frequency. The appropriate frequency depends on the application under consideration: as a rule, one has to promote the generation of either charged particles, on the one hand, or of neutral atoms, molecules, or radicals (eventually excited species), on the other hand. Determining the optimum species concentrations require a thorough knowledge of the part played by those species in the intended application and, further, of the cross-section energy thresholds leading to the production of those active species. General rules on frequency optimization are given at the end of the chapter to serve as guidelines.

Surely, the absolute answer to the variety of applications is a total flexibility in the choice of frequency for the HF discharges. In practice, however, such a reply is confronted with other requirements, such as those imposed by the EM spectrum regulations and by the domain of application as well as with challenges concerning the availability and the cost of implementing such a frequency flexibility in plasma processing equipment. Though there exist microwave power generator and equipment at other frequencies than 2450 MHz, the latter frequency is nowadays the one most commonly found in plasma applications. A series of reasons can be set forward to explain this situation: 1) 2450 MHz is a frequency authorized in all countries by the International Telecommunication Union (ITU) for Industrial, Scientific and Medical (ISM) applications (see Appendix 1.1), in contrast with, for example, 915 MHz (authorized in the USA but not in Europe and Japan) and 433 MHz (not authorized in the USA and Japan); 2) microwave gas discharges at 2450 MHz have been produced and investigated in many laboratories as early as the 1950s, mostly employing available diathermy equipment: these investigations have shown that efficient microwave power coupling could be obtained that yielded high density plasmas; 3) more recently, the advent of low cost 2450 MHz magnetrons for domestic use (microwave oven: 500 - 1500 watts) and to a lesser extent for industrial purposes (e.g. vulcanization, drying of various materials: 3 to 6 kW per unit with sometimes 10 to 20 units in a production line) has opened the way to today's vendors of plasma processing equipment.

Over the last few years, people have been trying to get the most out of 2450 MHz microwave discharges for given applications such as elemental analysis, materials processing, ion sources, lasers, to name only a few. This has led to a profusion of discharge excitation schemes, a large number of researchers trying to design their own system to fit their particular needs. The selection of the ideal excitation and reactor structures indeed depends on the kind of application considered. An example will illustrate this statement. A DC excitation should be avoided if the plasma is to be used for the deposition of insulating layers. Likewise, an excitation structure comprising dielectric materials to allow HF power access to the discharge has to be ruled out when deposition of metal layers is to be achieved. In addition, the stringent requirements of certain plasma processes mandate a precise control of the kind and scale of contamination emanating from the reactor walls, and more specifically from the excitation structure. The selection of metal or dielectric walls, the exact chemical composition and the operating temperature of the walls affect the exuded contamination one way or the other. A certain number of the various microwave plasma sources now available are described in Chaps. 4 and 5 for those not requiring the presence of an external static magnetic field and in Chaps. 7, 13 and 14, for those using either axial or multipolar magnetic fields. Chapter 4 on "microwave plasma sources using discharges sustained within microwave circuits", besides describing in general realizations of plasma sources, constitutes an introduction to the modelling of HF discharges. In Chap. 5, we consider a special case of HF discharges, the traveling-wave discharge (TWD), to which belongs the surface-wave sustained plasma. The latter, besides possessing sound physical characteristics, can well serve a didactic aim concerning the basis of HF field applicators and the various parameters of HF produced plasma columns: this plasma is, to our knowledge, the most fully and extensively modeled HF discharge.

We then proceed to the excitation structures of the "low-pressure" plasmas in the presence of a static magnetic field. Chapter 6 describes the fundamentals of magnetically assisted microwave discharges. The operation and properties of such discharges form the subject of Chap. 7. Chapter 8 proposes a new theoretical approach, based on the surface wave plasma formalism, to illustrate the power transfer mechanisms and identify the physical parameters in ECR discharge operation.

As previously pointed out, one major advantage of plasma excitation at low pressures, because of the large velocities and large mean free paths of the charged particles, lies in the possibility of using the various plasma components outside their zone of production. However, an important question to be dealt with concerns the possibility of obtaining homogeneous plasmas or achieving uniform treatments in this diffusion zone. In boundless plasmas, it is easily granted that the density of the expanding plasma decreases as a function of the distance from the plasma source, this decrease being enhanced by ion recombination. The expansion of the plasma generally produces a plasma beam whose characteristics mainly depend on geometry and initial (spatial) conditions. In other words, the plasma evolution from the source is mainly determined by the upstream conditions. In bounded plasmas, the above behavior is in no way modified, provided that ions and electrons are assumed to recombine totally on the walls (no reflection). Hence, an expanding plasma, bounded or not, is inhomogeneous by nature, and this applies to the collisional plasma (ambipolar diffusion regime) and, to a lesser extent, to the noncollisional plasma (free fall regime). An attractive solution to modify boundary conditions and in particular to minimize plasma loss on walls is to confine the plasma inside a magnetic field tube. When the discharge tube is linear, a positive gradient of B_0 (i.e. a magnetic mirror) is needed at each extremity to minimize plasma losses. When closed onto itself, the tube takes the characteristic torus form found in tokamaks. However, in both cases, one runs into the prohibitive cost of the devices that produce large volumes of magnetic field. Besides, the plasma thus obtained is strongly anisotropic, since B_0 affects charged particles differently in the parallel and in the transverse direction of the field. This proves to be a handicap for a number of surface treatment processes: for this kind of application, it is advisable to work with an isotropic plasma and then to use the ion sheath that surrounds the substrate to direct charged particles in the direction perpendicular to its surface. At the present time, the only example of magnetic confinement yielding isotropic, field free plasmas with satisfactory homogeneity, is the multipolar magnetic confinement (Chap. 9), originally developed with discharges excited by filaments.

Chapter 10 describes the experimental aspects of these multipolar discharges sustained by a hot cathode. It starts with a description of the existing reactors and the various types of magnetic configurations that were tested. We particularly stress the properties of this plasma and show how its confinement contrasts with that of the so-called primary electrons, i.e. those electrons coming directly from the cathode and trapped by the confining multipolar field. The modeling of multipolar discharges under ambipolar diffusion regime (Chap. 11) corroborates the mediocre plasma confinement observed experimentally in such a case. The main conclusion of this theoretical chapter is that the magnetic confinement results in a perceptible improvement of the plasma density profile, but that the confinement by itself does not account for the observed uniformity. To reach a better agreement between model and experiment, one has to take into account the ionization provided at the periphery of the reactor by the primary electrons trapped in the multipolar magnetic structure (Chap. 12).

Although the hot cathode multipolar discharges generate plasmas which present all the characteristics required for controlling the interaction of the plasma with any surface to be treated, the use of hot filaments to emit electrons generally introduces drawbacks such as thermal effects, a short up time and considerable contamination. A straightforward solution to this problem is to combine multipolar magnetic confinement with any of the available RF or microwave plasma sources (Chap. 13). In the first investigated configuration of multipolar microwave plasmas, a single plasma source of the surface wave-type is used and localized at one end of the confinement structure. Under "low pressure" conditions, the diffusion plasma at the center of the reactor is then homogeneous. Considering that the power supplied to a given EM field applicator can generally not exceed some safety value, the plasma density is thus also limited. A simplistic solution for increasing the flux of plasma emerging into the reactor is to increase the number of such plasma sources connected to the reactor. However, this leads to higher costs and bulkiness. The distributed electron cyclotron resonance (DECR) scheme elegantly solves these problems. This technique, described in Chap. 14, uses a multipolar magnetic field to achieve and localize ECR conditions for plasma production at the reactor periphery, the microwave power being individually supplied at a low level to several excitation antennas.

Our panorama of microwave excited plasmas is completed by Chap. 15 devoted to the fabrication of microcircuits, a very attractive application of these plasmas. This chapter illustrates the wide scope of plasma conditions required by this application. The contrasting examples selected should show the capability of microwave plasmas not only to cover the totality of expressed needs in that particular field but in many others. For example, the ions and reactive neutral species, indispensable for the synergetic effects in etching and deposition processes, can be used in metallurgical treatment, and for materials processing in general. They also have the ability to dissociate molecules and excite atoms as required in analytical chemistry where the information on the constituent concentrations is obtained through optical spectroscopy or mass spectrometry. Finally, microwave plasmas can supply the photons for laser and lighting applications. It is noteworthy that microwave plasmas cover an impressive pressure range of eight orders of magnitude, from 10^{-3} Pa (10^{-5} torr) to above atmospheric pressure. The versatility of microwave plasmas, their moderate cost, and their ease of implementation particularly appeal to the industrial entrepreneur, confronting the laboratory scientist with increasingly stringent requirements.

8

APPENDIX 1.1

**FREQUENCY BANDS FOR INDUSTRIAL,
SCIENTIFIC AND MEDICAL APPLICATIONS**

The International Telecommunication Union (ITU) defines the rules for the assignment and use of frequencies. As for the so-called Industrial, Scientific and Medical (ISM) frequencies, ITU states that they should be utilized for applications of radio-frequency energy that concern:

> "the operation of equipment or appliances designed to generate and use locally radio frequency energy for industrial, scientific, medical, domestic or similar purposes, excluding applications in the field of telecommunications... Administrations shall take all practicable and necessary steps to ensure that radiation from equipement used for ISM applications is minimal and that, outside the bands designated for use by this equipment, radiation from such equipment is at a level that does not cause harmful interference to a radiocommunication service or any other safety service".[1]

For the allocation of frequencies, the world has been divided into three Regions. Region 1 includes Africa and Europe. Turkey, Mongolia and USSR[†] territory including Siberia also belong to Region 1. Region 2 includes America (with Greenland) whereas Region 3 includes Oceania and Asia (with Iran). All the frequency bands designated by the ITU for ISM applications are listed in Table 1.1. Except where otherwise stated (e.g. 433 and 915 MHz frequencies), these frequencies may be used in all countries. However, besides these bands designated for worldwide use, other frequency bands, allowed by national legislations, may be used in some countries. They are summarized in Table 1.2.[2]

References

[1] Radio Regulations, International Telecommunication Union (ITU), General Secretariat (Geneva), vol. 1, Edition of 1990.

[2] Internal report, Comité Consultatif International des Radiocommunications (CCIR), unpublished (1989).

[†] At the time of publication, the ITU had not yet decided to modify the Regions to account for the various countries emanating from the former USSR.

Table 1.1. Frequency bands designated for ISM applications above 1 MHz.

Center frequency	Frequency band	Validity
MHz	MHz	
6.78	6.765 - 6.795	*
13.56	13.553 - 13.567	world-wide
27.12	26.957 - 27.283	world-wide
40.68	40.66 - 40.70	world-wide
433.92	433.05 - 434.79	Region 1 **
915.00	902.00 - 928.00	Region 2
GHz	GHz	
2.45	2.40 - 2.50	world-wide
5.80	5.725 - 5.875	world-wide
24.125	24.00 - 24.25	world-wide
61.25	61.00 - 61.50	*
122.50	122.00 - 123.00	*
245.00	244.00 - 246.00	*

*   With special authorization.

** Special authorization required except in Germany, Austria, Lichtenstein, Portugal, Switzerland and Yugoslavia.

Table 1.2. Additional frequency bands alloted by specific legislations to ISM applications above 1 MHz.

Center frequency	Frequency band	Country
MHz	MHz	
1.76	-------	U S S R
2.64	-------	U S S R
5.28	-------	U S S R
81.36	-------	U S S R
152.50	-------	U S S R
460.00	-------	U S S R
896.00	-------	United Kingdom
922.00	918.00 - 926.00	Australia *
GHz	GHz	
-------	≥ 1.00	United Kingdom **
2.375	-------	U S S R
3.39	3.37 - 3.41	Netherlands *
22.125	-------	U S S R
42.30	-------	U S S R
46.20	-------	U S S R
48.40	-------	U S S R

*   Alloted by the ITU.

** Absence of legislation for ISM applications above 1 GHz.

CHAPTER 2

PHYSICAL PRINCIPLES OF MICROWAVE PLASMA GENERATION [*]

2.1. Introduction

A microwave plasma source can be realized, at the engineering level, in many different ways. However, there are some general considerations that are common to all: the plasma results from a gas discharge sustained by an electromagnetic field and the conditions for its maintenance are essentially determined by the charged particle and energy loss mechanisms. In order to understand the operation and characteristics of any microwave plasma source, it is necessary to have some knowledge of these processes. The present chapter is intended to provide such a general background.

The term microwave is used in our presentation to denominate the frequency domain above 300 MHz. This is a convenient delimitation since these frequencies correspond to free space wavelengths (one meter and less) which are comparable to the dimensions of the various plasma chambers and discharge vessels encountered in practice.

The validity of the considerations brought up in this chapter in many cases apply as well to radio frequency (RF) sustained discharges. This is why, on many occasions, we use the term high frequency (HF) discharges to designate both the RF and the microwave discharges.

2.1.1. Contents of the chapter. We end the present introduction with a short historical survey of research in the field of microwave discharges. Then, in Sec. 2.2, we give a concise review of well known concepts and relations concerning the maintenance processes in HF discharges sustained at reduced gas pressures, in the absence of an external magnetic field.

The discharge is a major constituent of the *plasma source*, which further includes the HF field applicator and the impedance matching network. The discharge interacts strongly with the electromagnetic field imposed by the applicator, influencing in this way the electrodynamic characteristics of the source as a whole. The source can be represented by an equivalent circuit, which is useful in analyzing impedance matching. These various aspects of the plasma source are the subject of Sec. 2.3. The fourth and final section of the chapter concerns the classification of microwave discharges.

---

[*] **Presented by C.M. Ferreira, M. Moisan and Z. Zakrzewski**

2.1.2. Brief historical review. The extensive research program on radar during World War II provided the microwave technology that opened new possibilities in various fields, including microwave discharges. As a result, the late 1940s and early 1950s were a period of intense theoretical and experimental investigations of microwave discharges. The leading role was played by MIT scientists such as W.P. Allis, S.C. Brown, D.J. Rose and A.D. MacDonald. In only a few years, they accumulated a wealth of interesting results on the nature of the physical processes governing the initial stage (breakdown) and the maintenance of the microwave discharge. A review of these achievements has been presented by Golant.[1] The main outcome of the work done in those, and in the few following years, was a full theoretical description of the microwave breakdown in gases and its experimental verification. MacDonald's monograph[2] contains a complete summary of the state of the knowledge on microwave breakdown, which did not evolve much further since that time. However, the understanding of the physics of steady-state microwave discharges remained far behind, because of the complexity of the processes involved.

The interest in microwave discharges was renewed in the late 1960s. It happened because of the increasing number of applications of microwave plasmas in materials processing, analytical chemistry, spectroscopy, laser engineering, and rocket propulsion, to name only a few.

One of the obstacles on the way to wider applications of microwave discharges had been the relatively small volume of the created plasma. This is because the discharges sustained within microwave circuits (cavities, waveguides) necessarily have dimensions smaller than or comparable to the free space wavelength, i.e. a few centimeters at 2.45 GHz. A breakthrough occurred in the 1970s when it was shown that microwave discharges could be sustained by propagating electromagnetic waves, for example, by using a slow wave field applicator along the discharge tube[3] or by having surface waves propagating along the plasma column itself.[4] These discharges form a completely new category of microwave discharges, as reflected in our classification (Sec. 2.4) which distinguishes between conventional and traveling wave discharges.

In the literature, to our knowledge, there is no full up to date summary of the state of the art in microwave discharge physics and techniques, but rather presentations limited to various topics. For example, the review by MacDonald and Tetenbaum[5] mainly concerns the problems of microwave breakdown. Similarly, the works by Bell,[6] and by Lebedev and Polak[7] deal with some aspects of microwave plasmas related to applications in plasma chemistry. Moisan et al.,[8] Marec et al.,[9] Ferreira,[10] and Moisan and Zakrzewski[11] concentrate on surface-wave discharges. The recent book by Batenin et al.[12] is primarily devoted to the physics and techniques of atmospheric pressure microwave plasma sources. The present book endeavors to supply the required knowledge for analyzing all microwave plasma sources, suggesting by the same token how to classify them. It dwells on the surface-wave and DECR plasma sources, which are newcomers in the field.

2.2. Physical phenomena in microwave discharges

2.2.1. Interaction between an electromagnetic field and a charged particle.

We consider a harmonic electromagnetic field, that is, a field whose components are sinusoidal functions of time, t. In this case, it is useful and customary to represent the field as complex vectors (also called phasors) $\mathbf{E} \exp j\omega t$ and $\mathbf{H} \exp j\omega t$, where $\mathbf{E}$ and $\mathbf{H}$ are the complex amplitudes of the electric and magnetic fields respectively, $j = \sqrt{-1}$, and ω is the field angular frequency. The true harmonic fields are given by the real part of the corresponding phasors, that is, $E(t) = \mathrm{Re}\,(\mathbf{E} \exp j\omega t)$, $H(t) = \mathrm{Re}\,(\mathbf{H} \exp j\omega t)$, where the operator Re means "real part of the expression in brackets". Note that vectors are designated by bold face letters while tensors will be represented by letters underlined with a tilde. With harmonic functions of time, we use italic letters to designate their real values and upright letters for their complex amplitudes.

Consider now a volume in which a plasma and an electromagnetic field coexist. Until Chap. 6, we shall assume that no external static magnetic field is imposed on this volume. In such a case, the plasma is an isotropic medium and its properties can thus be represented by a scalar permittivity. Within this so-called *dielectric description*, the complex amplitudes $\mathbf{E}$ and $\mathbf{H}$ obey Maxwell's equations

$$\nabla \times \mathbf{E} = -j\omega\mu_0\mathbf{H} \tag{2.1}$$

$$\nabla \times \mathbf{H} = j\omega\varepsilon_p\varepsilon_0\mathbf{E} \tag{2.2}$$

$$\nabla \cdot (\varepsilon_p\varepsilon_0\mathbf{E}) = 0 \tag{2.3}$$

$$\nabla \cdot (\mu_0\mathbf{H}) = 0. \tag{2.4}$$

In the above, the symbols $\cdot$ and x indicate a scalar and a vector product respectively. The permeability of the plasma has been assumed equal to μ_0, that of free space. Customarily, the plasma permittivity is expressed relatively to ε_0, the permittivity of free space. This relative permittivity is a dimensionless complex number $\varepsilon_p = \varepsilon_r + j\varepsilon_i$ whose real and imaginary parts describe the reactive (energy storage) and resistive (energy dissipation) properties of the plasma respectively.

To express the electromagnetic properties of a plasma, one can also rely on the *charges in free space description*, which rests on the complex electric conductivity σ. It is fully equivalent to use either ε_p or σ, as these parameters are related by

$$\sigma \equiv \sigma_r + j\sigma_i = [-\varepsilon_i + j(\varepsilon_r - 1)]\ \omega\varepsilon_0. \tag{2.5}$$

The charged particles within the discharge are accelerated by the electric component of the electromagnetic field. The force due to the magnetic field component can generally be neglected. This is because it is of the order of eEw/c where w, the individual (or microscopic) velocity of the particle, is much less than c, the speed of light in free space, e being the electron charge. At microwave frequencies the ions, because of their large mass, remain practically unaffected by the field and can be considered as motionless. As a consequence, the hydrodynamic description of the response of this plasma to an electromagnetic field is limited to that of a fluid of electrons of average velocity v.

In a simplified description of this response, one can ignore the thermal motion of the electrons and use the equation of motion for a single particle

$$m_e \frac{d\boldsymbol{v}}{dt} = -e\boldsymbol{E}(t) - m_e \nu \boldsymbol{v}, \tag{2.6}$$

where m_e denotes the electron mass. This equation takes into account both the force exerted on the electron by the external field, namely $-e\boldsymbol{E}$, and a "friction" force represented by the term $-m_e\nu\boldsymbol{v}$ where ν is an average electron-neutral collision frequency for momentum transfer. This "friction" force, usually referred to as the Langevin force, represents the effects of electron collisions with the background of heavy particles.

The differential eqn. (2.6) can easily be solved using phasors. Since the steady-state solution of eqn. (2.6) is also a sinusoidal function of time, we can write $v = \text{Re}\,(\mathbf{v}\,\exp j\omega t)$ where $\mathbf{v}$ is the complex amplitude of the velocity. Then, the operator d/dt acting on the phasor $\mathbf{v}\,\exp j\omega t$ simply becomes a factor $j\omega$ and we obtain an algebraic equation

$$j\omega m_e\mathbf{v} = -e\mathbf{E} - m_e\nu\mathbf{v}, \tag{2.7}$$

which yields, solving for $\mathbf{v}$,

$$\mathbf{v} = -\frac{e\mathbf{E}}{m_e}\frac{1}{\nu + j\omega}. \tag{2.8}$$

The same procedure may be applied to obtain the electron displacement $r(t)$ in the HF field which is given by the real part of the phasor $\mathbf{r}\,\exp j\omega t$. One has d/dt $(\mathbf{r}\,\exp j\omega t) = j\omega\,\mathbf{r}\,\exp j\omega t = \mathbf{v}\,\exp j\omega t$, so that $j\omega\mathbf{r} = \mathbf{v}$. Substituting this result in eqn. (2.8) and solving for $\mathbf{r}$ yields

$$\mathbf{r} = \frac{e\mathbf{E}}{m_e\,\omega\,(\omega - j\nu)}. \tag{2.9}$$

The magnitude of the electron displacement is conveniently expressed by the quantity r_{rms}, the root mean square (rms) value of $r(t) = |\mathbf{r}(t)|$, where the symbol $|\;|$ represents the magnitude of a vector quantity. One has therefore $r_{rms}^2 = \mathbf{r}\cdot\mathbf{r}^*/2$, where $*$ means "complex conjugate", so that we obtain from eqn. (2.9)

$$r_{rms} = \frac{eE_{rms}}{m_e\,\omega\left(\nu^2 + \omega^2\right)^{1/2}}, \tag{2.10}$$

where $E_{rms} = \left(\mathbf{E}\cdot\mathbf{E}^*/2\right)^{1/2}$ is the rms electric field intensity.[†] At microwave frequencies, r_{rms} is most generally much smaller than the dimensions of the plasma vessel, and thus one can ignore the impacts of the electrons on the walls of the container in the course of their oscillations.

From eqn. (2.8), one can calculate the electron *conduction current density*

$$\mathbf{J} = -\,en\mathbf{v}, \tag{2.11}$$

where n is the electron number density. The current density can also be written in the form

$$\mathbf{J} = \sigma\mathbf{E}, \tag{2.12}$$

[†] The root mean square of a quantity $A(t)$ is the square root of the time-averaged value of $A^2(t)$, that is, $A_{rms} = \left(\overline{A^2(t)}\right)^{1/2}$, where the bar means the time average. The time-averaged value of the product of two scalar sinusoidal quantities, say $A(t)$ and $B(t)$, of the same frequency is identical to $\mathrm{Re}(AB^*)/2$, where A and B represent the respective complex amplitudes. This is also true for scalar or vector products of vector quantities. Note that when $B(t) = A(t)$, we have $A_{rms}^2 = \overline{A^2(t)} = AA^*/2$ since the product AA^* is pure real.

where σ is the complex electric conductivity mentioned earlier. We obtain for σ the following expression

$$\sigma = \frac{ne^2}{m_e}\frac{1}{\nu + j\omega}\,,\tag{2.13}$$

and since, from eqn. (2.5), $\varepsilon_p = 1 - j\sigma/\omega\varepsilon_0$, we readily arrive at the corresponding relative plasma permittivity

$$\varepsilon_p = 1 - \left(\frac{\omega_{pe}}{\omega}\right)^2 \frac{1}{1 - j\frac{\nu}{\omega}}\,,\tag{2.14}$$

where $\omega_{pe} \equiv (ne^2/m_e\varepsilon_0)^{1/2}$ is the so-called *electron plasma angular frequency*.[†]

In the absence of collisions (i.e. $\nu = 0$), $\sigma = j\sigma_i$ is an imaginary number and $\varepsilon = \varepsilon_r$ is real: the plasma behaves like a lossless medium, the only motion being the *ordered oscillations*[††] of the electrons, induced by the field. On the average, there is no energy transfer from the electromagnetic field to the electrons and from the latter to the background gas: in this case, the medium is entirely passive. Collisions introduce disorder into this motion and constitute the mechanism by which power can flow from the electromagnetic field to the electron fluid and from the latter, to the gas background. To elucidate this important point, we shall now discuss the energy gain and the power transfer in a discharge.

2.2.2. Energy gain and power transfer in a discharge. The kinetic energy associated with the electron oscillatory motion can easily be found from eqn. (2.8). The true time-dependent electron velocity is represented by the real part of the phasor **v** exp $j\omega t$, that is

$$v(t) = -\frac{e}{m_e}\,\text{Re}\left(\frac{\mathbf{E}}{\nu + j\omega}\,\exp j\omega t\right).\tag{2.15}$$

The velocity component along a given axis, say the x-axis, is therefore of the form

---

[†] This frequency is a fundamental parameter in plasma physics. It represents a proper frequency of electron oscillations with respect to immobile ions. Such oscillations produce deviations from charge neutrality over some small scale of length, the Debye length, and are thus accompanied by the buildup of a space charge field. The Debye length, λ_D, is given by the expression $\lambda_D = (\varepsilon_0 k_B T_e/ne^2)^{1/2}$, where k_B is the Boltzmann constant and T_e is the electron temperature. This may also be written as $\lambda_D = (k_B T_e/m_e)^{1/2}\omega_{pe}^{-1}$. Under this form, one sees that λ_D is of the order of the distance traveled by a thermal electron during one period of oscillation.

[††] The ordered oscillation of the electrons in the HF field is to be opposed to their random thermal motion.

$$v_x(t) = \frac{e}{m_e} \frac{E_x}{(v^2+\omega^2)^{1/2}} \cos(\omega t + \psi_x),$$ (2.16)

where E_x is the intensity of the electric field component along the same axis and ψ_x is a phase angle.[†] The kinetic energy associated with this velocity component, $u_x(t) = m_e\, v_x^2(t)/2$, is therefore

$$u_x(t) = \frac{1}{2}\frac{e^2}{m_e}\frac{E_x^2}{v^2+\omega^2}\cos^2(\omega t + \psi_x) = \frac{1}{4}\frac{e^2}{m_e}\frac{E_x^2}{v^2+\omega^2}\left[1 + \cos(2\omega t + 2\psi_x)\right],$$ (2.17)

whose average value over a period is $\bar{u}_x = e^2\, E_x^2 / 4\, m_e\,(v^2+\omega^2)$. The total time-averaged kinetic energy in 3-dimension space is therefore $\bar{u} = e^2\, E_p^2 / 4 m_e\,(v^2+\omega^2)$ where $E_p^2 = 2\, E_{rms}^2$[††] is the sum of the squared field intensities along three orthogonal axes (for example, $E_p^2 = E_x^2 + E_y^2 + E_z^2$ in ordinary cartesian coordinates). The expression above for $\bar{u}$ could also have been directly obtained from eqn. (2.8) using the relations given in the footnote following eqn. (2.10). In fact, one has $\bar{u} = m_e\, \mathbf{v}(t)\cdot\mathbf{v}(t)/2 = m_e\,\mathbf{v}\cdot\mathbf{v}^*/4 = e^2\, E_{rms}^2 / 2m_e\,(v^2+\omega^2)$.

The mean kinetic energy $\bar{u}$ over a period of oscillation is usually much smaller than the energy required for the electrons to produce excitation and ionization of the molecules upon collisions with them, thus, to sustain a discharge. However, due to elastic collisions with the background molecules, the electrons can continuously pick up energy, as we now demonstrate.

As the force exerted by the field on an electron is $-e\mathbf{E}(t)$, the work done by this force by unit time is simply $-e\mathbf{E}(t)\cdot\mathbf{v}(t)$. The average of this quantity over a period of oscillation, termed the *mean power transfer per electron* θ_A, is easily obtained taking into account eqn. (2.8), yielding

$$\theta_A \equiv \overline{-e\mathbf{E}(t)\cdot\mathbf{v}(t)} = -e\,\mathrm{Re}\left(\frac{\mathbf{v}\cdot\mathbf{E}^*}{2}\right) = \frac{e^2}{m_e}\frac{v}{v^2+\omega^2}\,E_{rms}^2 = 2v\bar{u}.$$ (2.18)

[†] To be general, we assume this phase angle to be different for each axis since the phase of each field component can also be different. This occurs, for example, when considering the electric field components in the electron cyclotron resonance effect (Sec. 8.2.1). Along each axis, the phase of the velocity with respect to that of the electric field is equal to $\pi/2 \le \arctan(-\omega/v) \le \pi$; thus, in a low collision plasma ($v/\omega \ll 1$), it is close to $\pi/2$ whereas it tends to π when there are many collisions over one period of the HF field.

[††] Note that the rms field intensity along any axis, say the x-axis, is $E_{xrms} = E_x/\sqrt{2}$. Therefore $E_{rms} = E_p/\sqrt{2}$, where E_{rms} is the root mean square of the total field intensity defined before.

In the absence of collisions ($v = 0$), the electron velocity is 90° out of phase with the field, and there results no average power transfer from the field to the electron motion, thus $\theta_A = 0$. In the presence of collisions, however, $\theta_A \neq 0$, and the random energy of the electrons increases steadily until the energy is sufficient to excite or ionize a molecule. When this happens, the electron is usually left with only a small amount of energy, and the process of energy buildup starts over again. In energy space, the trajectory of an average electron as a function of time is some rising curve (most of the time) with a small ripple superimposed on it which represents the oscillating energy of the ordered motion. When excitation or ionization of the molecule occurs, the electron loses most of its energy. Such processes are depicted in Fig. 2.1.

In what preceded, we have considered the motion and properties of a *single electron* in the HF field. A more complete description of the electrons in the plasma is obtained by considering them as a fluid (*hydrodynamic model*) of average velocity **v**, as we will do in the coming section. A still more refined description consists in taking into account in a statistical way (*kinetic model*), the microscopic motion of the particles subjected to the applied field and to collisions of all types, yielding the electron velocity distribution function (EVDF).† This distribution results from the partial randomization of the electron motion due to collisions with the discharge constituents. It is characterized by a mean energy much larger than that obtained from the field on each collision. This distribution always has a few fast electrons with enough energy to excite and ionize the gas molecules; these electrons form what is usually called the *tail of the distribution*. They play a fundamental role in the discharge sustaining mechanism, as we shall see later.

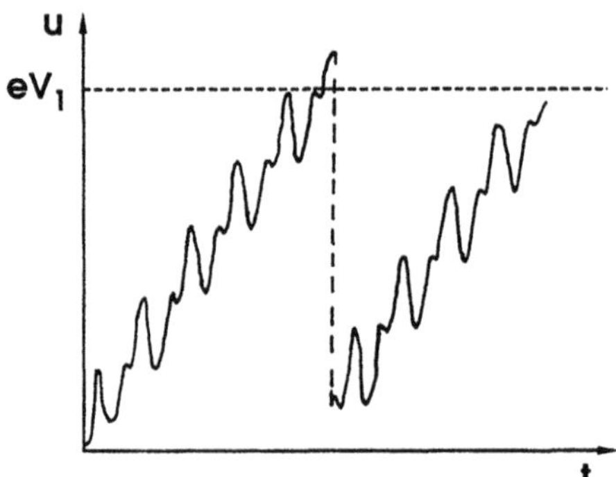

Fig. 2.1. Schematic illustration of the step-like increase of the electron energy u in the HF field as a result of elastic collisions with heavy particles. The large sudden drop corresponds to an inelastic collision where the threshold energy is eV_1.

---

† See Sec. 2.2.5 for a definition of EVDF and Chap. 3 for an introduction to the kinetic model.

2.2.3. Power and charged particle balance in the active zone. The domain over which the interaction between the electromagnetic field and the plasma takes place is called the *active zone of the discharge*. It follows from the above that two key phenomena related to the discharge mechanism occur in the active zone; i) the first of these is the *energy transfer* from the HF field to the electrons. *In a steady-state discharge*, it is known as the *power balance per electron*: the power gained per electron from the field adjusts itself so as to balance the power lost per electron through electron collisions of all kinds; ii) the second phenomenon concerns the *charged particle balance*. Under steady-state conditions, the rate of creation of electron-ion pairs obtained at the expense of the HF power equals the rate of loss within the volume (volume recombination, attachment) and on the discharge vessel wall on which these particles arrive either by diffusion or by direct impact (free-fall regime), and recombine. In the steady-state, the equation for charge conservation (continuity equation) is of the form

$$\nabla \cdot (n\mathbf{v}) = S_p, \tag{2.19}$$

where S_p represents the net rate at which the charged particles are created (that is, the difference between the rates of creation and loss in the volume), whereas the particle removal from the volume toward the wall due to gradients of electron density and of average energy is taken into account by the term $\nabla \cdot (n\mathbf{v})$. Let us observe that a set of boundary conditions is needed to solve the field equation and the balance equation (2.19). These are imposed by the design chosen for the discharge setup, namely the plasma vessel and the way the electromagnetic field is applied to the plasma. We shall show in Sec. 2.2.9 that eqn. (2.19) can lead to the determination of the average electron energy in a steady-state discharge.

To better understand the power balance and the charged particle balance mechanisms in the discharge, we must dwell on the electron-heavy particle collision processes which, in turn, requires to introduce the electron energy distribution function (EEDF). The hydrodynamic approach, based on the continuity equation (2.19), is then complemented with the information provided by the EEDF, as we shall see in the following sections.

2.2.4. Electron-heavy particle collisions. Any interaction between electrons and heavy particles can conveniently be described in terms of a collisional process. Collisional phenomena in ionized gases have found ample coverage in the literature (e.g. Massey and Burhop,[13] McDaniel,[14] Golant et al.[15]). In what follows, we review some aspects of electron collisions which are relevant to the mechanism of microwave discharges.

Let us first consider the scattering of electrons upon collisions with the heavy particles. The collisions causing electron scattering may be divided into the two following classes:

Elastic collision. The electron interacts with a heavy particle in a way that only leads to an exchange of kinetic energy, the internal state of the heavy particle remaining unmodified.

Inelastic collision. Part of the incoming electron's energy is converted to change the internal state of the heavy particle (e.g. excitation of electronic states, ionization, rotational or vibrational excitation of molecules). Occasionally, internal energy may also be converted into kinetic energy; in this case, the collision is termed *superelastic collision* or *collision of the second kind.*

The probabilities of both elastic and inelastic collisions are usually defined in terms of collision cross-sections, or expressed as collision frequencies. To define these parameters, we consider a uniform beam of electrons moving with initial velocity $\mathbf{w}$ toward a fixed heavy particle, as depicted in Fig. 2.2. Let I denote the incident flux, i.e. the number of electrons in the incident beam crossing a unit area normal to $\mathbf{w}$ in unit time. Due to interaction with the heavy particles, a particular electron of the beam will acquire a different velocity $\mathbf{w}'$, the direction of which relative to $\mathbf{w}$ can be characterized by the scattering angles χ (polar angle) and ϕ (azimuthal angle). The *differential cross-section* q_s for the scattering, elastically or inelastically, of the electrons into a small element of solid angle $d\Omega = \sin \chi \, d\chi \, d\phi$ is so defined that $I \, q_s \, d\Omega$ represents the

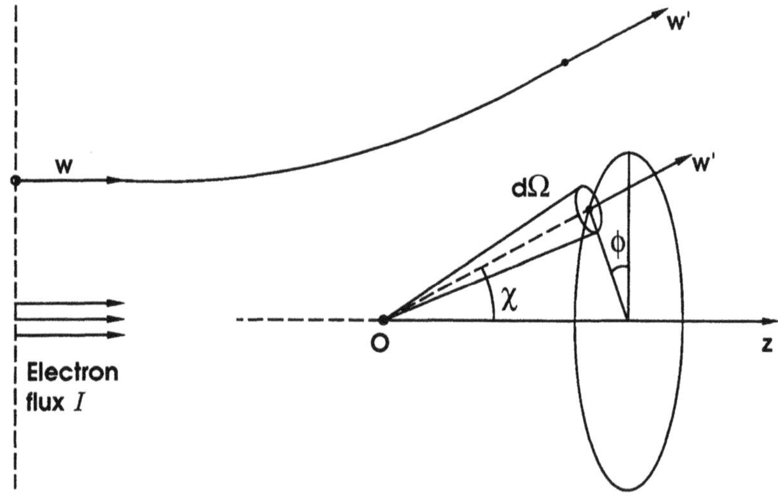

Fig. 2.2. Scattering of a beam of electrons by a heavy particle at rest at point O.

number of electrons in the incoming flux I which are scattered into $d\Omega$ per unit time. In general, $q_s = q_s(\Omega, w)$ where Ω stands for both χ and ϕ. This differential cross-section being fully determined by the specific interaction law between the electron and the heavy particle, it can, in principle, be obtained from calculations.

From the above, it follows that the number of electrons scattered from the incident beam per one fixed heavy particle, per unit time, through all angles, is given by $IQ_s(w)$, where

$$Q_s(w) \cong \int_\Omega q_s(\Omega, w)\, d\Omega = \int_0^{2\pi} d\phi \int_0^\pi q_s(\Omega, w)\, \sin \chi\, d\chi. \tag{2.20}$$

Equation (2.20) defines the total scattering collision cross-section $Q_s(w)$, often referred to simply as the *total collision cross-section.*

Consider now a beam of electrons incident on a slab of gas of cross-sectional area $d\mathcal{S}$ normal to the beam, and of thickness dx. The number of heavy particles (density N) in the slab is $N\, d\mathcal{S}\, dx$. Since these particles are randomly distributed, the number of scattered electrons per unit time in all directions is given by $IQ_s(w)\, (N\, d\mathcal{S}\, dx)$. Dividing this quantity by $I\, d\mathcal{S}$, i.e. the number of electrons incident on the slab per unit time, one obtains the probability that a given projectile will suffer scattering within the distance dx. This probability is therefore $\mathcal{P}_s\, dx = N\, Q_s(w)\, dx$. Due to the scattering, the incident flux is reduced by an amount $dI = -\,IN\, Q_s(w)\, dx$ as the beam crosses the slab of gas.

It is customary to define the probability of collision $\mathcal{P}_s(w)$ as the mean number of collisions experienced by a projectile in a unit path length through the gas, which is

$$\mathcal{P}_s(w) = N\, Q_s(w). \tag{2.21}$$

This probability of collision (actually a density of probability) has the dimensions of the reciprocal of a length.

Let now $dt = dx/w$ be the time that an electron of the beam takes to travel across the slab of gas. The probability that a given electron will suffer scattering within the time interval dt is $N\, Q_s(w)w\, dt$. The quantity

$$\nu_s(w) = N\, Q_s(w)\, w \tag{2.22}$$

is thus the probability per unit time that an electron suffers a collision, or the *collision frequency* as is usually known.

The reciprocal of $v_s(w)$ represents the mean free time between two successive collisions, or simply the *collision time*.

The mean distance traveled by an electron between two successive collisions is called the *mean free path*, λ_s. For an electron of velocity w experiencing elastic collisions with the background gas molecules, it is simply given by

$$\lambda_s(w) = \frac{w}{v_s(w)} = \frac{1}{N\,Q_s(w)}\,.$$
(2.23)

It can be shown that λ_s is the mean depth of penetration into a finite slab of gas before scattering. When averaged over the EEDF, the values of $v_s(w)$ and $\lambda_s(w)$ are called the average collision frequency and average mean free path respectively, and we will identify them as $<v_s(w)>$ and $<\lambda_s(w)>$.

It should be noted at this point that any inelastic collision leading to the excitation or ionization of the heavy particles is characterized by the existence of a threshold energy value, say eV_s, that the electron must exceed for the process to be possible. The corresponding cross-section $Q_s(w)$ and collision frequency $v_s(w)$ are, therefore, identical to zero for all electron energies $m_e w^2/2 \leq eV_s$.

Up to now, we have considered collisions as a single process characterized by a single collision frequency. However, it is desirable to distinguish among various kinds of collisional processes and to characterize them separately. The above formalism can be applied to each (say, k-th) kind of process, which is then described in terms of $\mathcal{P}_{sk}(w)$, $Q_{sk}(w)$ and $v_{sk}(w)$. The *overall collision frequency* $v_s(w)$ is the sum of component frequencies, and so are the overall collision probability $\mathcal{P}_s(w)$ and cross-section $Q_s(w)$; thus

$$v_s(w) = \sum_k v_{sk}(w) = Nw \sum_k Q_{sk}(w) = N \sum_k \mathcal{P}_{sk}(w)\,.$$
(2.24)

As we have shown, the total collision cross-section $Q_s(w)$ and the corresponding parameters $\mathcal{P}_s(w)$ and $v_s(w)$ are directly related to the attenuation of a beam of particles due to scattering. They do not however contain any information concerning the crucial questions of energy and momentum transfer, in particular for elastic collisions. To see this, consider the head-on elastic collision of an electron with a heavy particle of mass M after which the electron is scattered backwards. In such a collision, the electron will lose approximately twice its initial momentum, since its velocity reverses sign after the collision, keeping nearly the same absolute value since $m_e \ll M$. On the other hand, little momentum exchange occurs when the electron is scattered at small angles relative to its initial direction. In order to describe the momentum exchange, we must therefore take into account the properly weighted influence of the scattering angle χ in a collision. It may be shown (see, for example, ref. 15) that the angle dependence of changes in electron energy and momentum in elastic collisions is characterized by a common factor $(1 - \cos \chi)$: an electron scattered through an angle χ will lose approximately $2(1 - \cos \chi)\, m_e/M$ of its initial energy, and the momentum transferred in the forward direction[†] to the heavy particle is $m_e w\,(1 - \cos \chi)$. A beam of electrons with flux I incident on the heavy particle will therefore transfer a forward momentum in unit time given by $Im_e w \int_\Omega (1-\cos \chi)\, q_{el}(\Omega, w)\, d\Omega$, where $q_{el}(\Omega, w)$ is the differential cross-section for elastic collisions and $Im_e w$ is the momentum flux carried by the incident beam. The integral term of this expression, namely the quantity

$$Q_c(w) = \int_\Omega (1 - \cos \chi)\, q_{el}(\Omega, w)\, d\Omega, \qquad (2.25)$$

represents the *momentum transfer cross-section*. It is a weighted average of the differential cross-section for elastic collisions in which backward scattering $(\chi = \pi)$ counts double, right angle scattering $(\chi = \pi/2)$ has the weight one, and forward scattering $(\chi = 0)$ is not counted. By analogy with eqn. (2.22), one defines the *collision frequency for momentum transfer* as

$$v_c(w) = N\, Q_c(w)\, w. \qquad (2.26)$$

This quantity plays a very important role in the electron transport theory. It is often referred to simply as the collision frequency, which may be confusing (see eqn. (2.22)).

For $q_{el}(\Omega, w)$ independent of Ω, as is usually the case for sufficiently slow electrons, one has $Q_c = Q_{el}$, hence $v_c = v_{el}$, since on the right-hand side of eqn. (2.25), the integral of the $\cos \chi$ term vanishes in this case. With the exception of helium, as a rule, the value of v_c is very close to that of v_{el} at electron energies below a few eV. At some energies above 10 eV, v_c can be lower than v_{el} by as much as 25% in neon and 50% in argon.[16] Physically, this means that the electron is then more likely to be scattered forward than backward.

---

† Note that in general there is also some amount of momentum transferred perpendicularly to the direction of the incident particle.

The function $v_c(u)/p_0$ is plotted in Fig. 2.3 for the case of helium and argon, as examples. Here, $p_0 = p\ 273/T$ is the *"reduced" gas pressure*, where p is the pressure in torr and T the temperature in Kelvin.

We also need to consider collisions in which electrons are either created or lost. The main types of such collisions occurring in a weakly ionized gases are the following:

Ionization by electron impact. We have already referred to this process as a type of inelastic collision. Here, it is important to stress that a new electron is created. The probability of ionization per unit time by an electron with velocity w (such that $m_e w^2/2 > eV_i$, the threshold energy for ionization or, in short, the ionization energy) is given by

$$v_i(w) = N\ Q_i(w)\ w, \qquad\qquad (2.27)$$

where Q_i is the ionization cross-section. Since v_i is a function of w, the determination of the average ionization frequency requires the knowledge of the electron velocity distribution function.

Electron attachment. The electron attaches to an atom or a molecule, causing it to acquire a negative charge. With regard to the discharge mechanism, attachment is equivalent to the removal of an electron from the active zone: the large mass of the resulting negative ion prevents it to absorb any meaningful energy from the

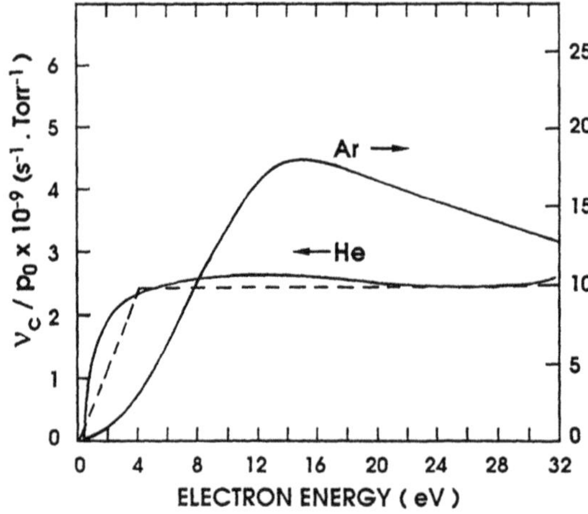

Fig. 2.3. Value of electron-heavy particle collision frequency v_c for momentum transfer in argon and helium where p_0 is the corresponding reduced ($0°$ C , 1 torr) gas pressure (after MacDonald and Tetenbaum[5]). The dashed line shows a possible approximation for v_c/p_0 in the case of helium.

electromagnetic field. The collision frequency for attachment depends on the electron affinity of the given gas, being large for the so-called "attaching" gases like oxygen, chlorine and fluorine.[16]

Electron-ion collisional recombination. A possible outcome of an electron-ion collision is the capture of an electron by the ion which results in a neutral particle. Such a process is called volume or collisional recombination (to distinguish it from wall recombination) and its occurrence is significant at low electron energies, typically below 1 eV. In plasmas with low average electron energy, recombination can strongly affect the charged particle balance. It is customary to characterize volume recombination not by a collision frequency but by a recombination coefficient α_r, defined through the rate of removal of charged particles by that process, namely

$$\frac{\partial n}{\partial t} = -\alpha_r \, n \, n_i \approx -\alpha_r \, n^2, \tag{2.28}$$

where n_i is the density of ions, and generally $n \approx n_i$.

All parameters describing the interaction between electrons and heavy particles depend on the electron velocity, from which stems the crucial influence of the electron velocity distribution function on the discharge processes. This function is the solution of the Boltzmann equation, to be discussed in more detail in Chap. 3. In the following, we deal only with what is necessary to understand the present considerations.

2.2.5. Electron velocity and energy distribution functions. The electron *velocity* distribution function (EVDF) $f(\mathbf{w})$ describes the probability density of finding electrons in a given interval of velocities $\mathbf{w}, \mathbf{w} + \mathbf{dw}$. In the presence of a harmonic electric field, $\mathbf{E} \exp(j\omega t)$, it depends on the electron velocity and on the angle χ between the velocity direction and that of the electric field.[†] Using a first order spherical harmonic expansion in velocity space and a first order Fourier expansion in time, one obtains

$$f(\mathbf{w}) = f_0(w) + \frac{\mathbf{w}}{w} \cdot \mathbf{f}_1 \exp(j\omega t) = f_0(w) + f_1(w) \cos \chi \, \exp(j\omega t). \tag{2.29}$$

The isotropic component $f_0(w)$ is independent of time. It determines the average value of the electron energy, and serves to calculate the average value of any quantity

---

† For the sake of simplicity, we assume here that the field is linearly polarized. This means that it is always directed along the same axis during the oscillation. In this case, the components of $\mathbf{E}$ along the coordinate axis are all in phase and, thus, $E_p = (E_x^2 + E_y^2 + E_z^2)^{1/2}$ as defined in Sec. 2.2.2, is now the field amplitude.

depending only on the absolute value of the electron velocity and not on its direction (e.g. kinetic energy). We shall show later that this isotropic component is directly related to the electron *energy* distribution function (EEDF). The component $f_1(w)$, called the anisotropic term, determines the average value of the electron velocity under the influence of the electric field E; in the present description of the time dependence, it is a complex quantity. In a harmonic field, the two components of the distribution function expansion are related by

$$f_1(w) = \frac{eE/m_e}{v_c(w) + j\omega} \frac{\partial f_0(w)}{\partial w} .$$

(2.30)

The approximation (2.29) is valid only under certain conditions which need to be better specified. In particular, the isotropic component of $f(w)$ can be modulated in time in some cases, in which instance the Fourier expansion must be extended beyond $f_0(w)$ to higher orders in order to account for this modulation. Let us examine under which conditions these higher order terms can be neglected, so that the isotropic component is represented only by $f_0(w)$.

The relaxation of the electron energy $u = m_e w^2/2$ through electron-neutral collisions is described by the characteristic frequency for energy transfer v_u

$$v_u(u) = (2m_e/M) \, v_c(u) + \sum_k v_k(u),$$

(2.31)

where M is the atomic (or molecular) mass, and $v_k(u)$ is the electron-neutral collision frequency for atomic or molecular excitation (including ionization) to energy level k. It can be shown[17] that the time evolution of the isotropic component is determined by the relative values of ω and v_u, the characteristic frequency for energy transfer: when $\omega << v_u$, there are so many collisions during one field period that the energy accumulated by the electron between two successive collisions is closely related to the instantaneous value of the HF field intensity. When this is the case, the isotropic component is modulated over its entire energy range (at twice the frequency ω, since it depends on the square of the electric field).[17] The amplitude of this modulation decreases with increasing ω and it becomes negligible at all energies u for which $v_u(u) < \omega$. For atomic gases, this implies that it is the bulk of the distribution that first stops oscillating when ω increases, because v_u (eqn. 2.31) is smaller there than in the tail, where inelastic collisions occur. With neon, the calculations[18] show that the modulation of f_0 has almost completely stopped for $\omega/p_0 \geq \pi \times 10^8$ s^{-1} torr^{-1} (p_0 is the "reduced" gas pressure in torr). This means that at 200 mtorr, for example, the EVDF should not be oscillating at all with the HF field for frequencies above 10 MHz. A very similar situation is expected with argon since the values of $v_u \simeq \sum_k v_k$ for these two gases are comparable. With molecular gases, the EVDF modulation is predicted to persist to higher frequencies, typically up to 40 MHz in H_2 and CO at 200 mtorr,[17,19] which results from additional energy losses due to vibrational excitation. This means that the modulation is likely to influence most of the EVDF energy range at 13.56 MHz, with some modulation in the tail possibly remaining up to 40 MHz.

In gas discharge physics, it is customary to use the EEDF, $F_0(u)$, instead of $f_0(w)$, where $F_0(u)\,u^{1/2}\,du = f_0(w)\,4\pi\,w^2\,dw$; thus, $F_0(u)\,u^{1/2}$ represents the probability density of finding electrons in the energy interval u, $u + du$, which corresponds to the absolute velocity interval w, $w + dw$.   Recalling that $u = m_e w^2/2$, we have $F_0(u) = 4\sqrt{2}\,\pi\,m_e^{-3/2}\,f_0(w)$.

Assuming that the EEDF is independent of space and time, one obtains the homogeneous Boltzmann equation[20]

$$- \frac{2}{3}\frac{d}{du}\left[u^{3/2}\,v_c(u)\,u_c\,\frac{dF_0(u)}{du} \right] = S_0(F_0) \tag{2.32}$$

where

$$u_c = \frac{(eE_{rms})^2}{m_e}\,\frac{1}{v_c^2(u) + \omega^2} \tag{2.33}$$

represents the average (over a period) *energy transferred per collision* from the HF field to electrons having a random energy u: thus, $v_c\,u_c$ is the *power transferred per electron*. S_0 is the collisional term which accounts for electron-neutral collisions (both elastic and inelastic) and electron-electron collisions. Equation (2.32) can be regarded as a continuity equation in the electron energy space. The term on the left-hand side represents the divergence of the electron flux in this space, while the collisional term $S_0(F_0)$ on the right represents the net rate of appearance of electrons per unit energy interval as a result of collisions.

2.2.6. Transport properties of the plasma in an HF field. The transport of mass, charge and momentum in gases due to directed particle fluxes are described by specific macroscopic coefficients. In this section, we briefly review these transport coefficients in weakly ionized gases, as they play a significant part in the maintenance of microwave discharges.

2.2.6.1. Charged particle diffusion coefficients and constant electric field (DC) mobilities. Diffusion is one of the principal mechanisms of charged particle removal from the plasma. It occurs whenever there exists a nonuniform spatial

distribution of particle density or average energy, and it results in a flow of particles in the direction opposing that gradient. The proportionality factor linking particle flux $n\mathbf{v}$ and density gradient is called the *diffusion coefficient*.

In the case of extremely low electron densities, there is no interaction between charged particles, so that one can define a *free diffusion coefficient* for each such type of particle. For electrons, it is

$$D_e = \frac{4\pi}{3} \int_0^\infty \frac{w^4}{\nu_c(w)} \, f_0(w) \, dw. \tag{2.34}$$

An electron moves in the direction opposite to that of a DC electric field, but collisions with heavy particles tend to impede this motion. Nevertheless, the electron drifts at some constant velocity in the specified direction. The absolute value of the proportionality factor between drift velocity and field intensity is called the *electron mobility*, μ_e, and we have

$$\mu_e = -\frac{4\pi e}{3m_e} \int_0^\infty \frac{w^3}{\nu_c(w)} \frac{\partial f_0(w)}{\partial w} \, dw. \tag{2.35}$$

The diffusion coefficient and the mobility of ions, respectively D_i and μ_i, are phenomenologically defined in a similar way. The velocity distribution of ions may often be assumed to be Maxwellian, and their temperature equal to that of the neutrals.

In the case of high electron densities, encountered in typical laboratory microwave discharges, the electrons and ions interact strongly, manifested by the occurrence of a space charge field, and a collective diffusive motion of electrons and ions, called ambipolar diffusion, takes place. This interaction increases the mean ion transport velocity, while considerably slowing down the electron diffusion (compared with free diffusion conditions). The resulting *ambipolar diffusion coefficient*

$$D_a = \frac{\mu_e D_i + \mu_i D_e}{\mu_e + \mu_i} \approx \frac{\mu_i}{\mu_e} D_e \tag{2.36}$$

is thus always much smaller than D_e.

In the intermediate range of electron density values, the diffusion coefficient is density dependent. Thus, it always varies spatially within any bounded plasma. To cope with this problem, an effective electron diffusion coefficient D_s can be introduced.

Figure 2.4 shows the theoretical dependence of D_s on the electron density (reminder: λ_D is proportional to $n^{-1/2}$, where λ_D is the Debye length). In steady-state microwave discharges, the diffusion is ambipolar in character in most practical cases (Sec. 2.2.9).

The process in which the directed motion of particles is caused by a temperature gradient is called thermal diffusion.

2.2.6.2. HF electric conductivity and permittivity. As we indicated in Sec. 2.2.1, the electric conductivity σ relates the current density and the electric field intensity through $\mathbf{J} = \sigma\mathbf{E}$, where σ is a scalar quantity in plasmas with no external magnetic field. Since the response of the ions to the field can be ignored in most microwave discharges, the current density is determined only by the electron motion, and we can write

$$\mathbf{J} = - \, en\mathbf{v}, \tag{2.37}$$

where $\mathbf{v}$, in our formalism, is a complex quantity. The direction of this average electron velocity is that of the E-field and of the anisotropy $\mathbf{f}_1(w)$. It may be shown that

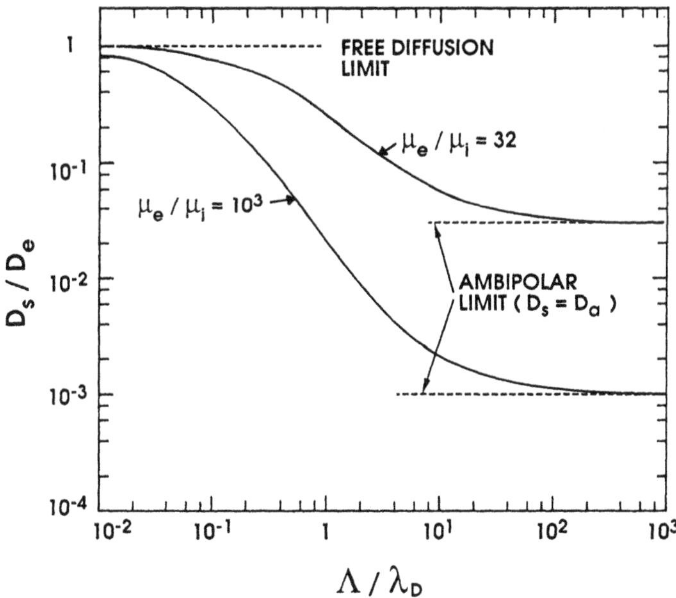

Fig. 2.4. Calculated effective electron diffusion coefficient D_s normalized to the free diffusion coefficient D_e as a function of the characteristic diffusion length Λ normalized to the Debye length λ_D, for two electron-to-ion mobility ratios (from ref. 21).

$$v = \frac{4\pi}{3} \int_0^\infty w^3\, f_1(w)\, dw. \tag{2.38}$$

Using eqns. (2.38) and (2.30), one may express the electric conductivity directly in terms of $f_0(w)$, namely

$$\sigma = \sigma_r + j\,\sigma_i = \frac{-4\pi e^2}{3m_e}\, n \int_0^\infty \frac{w^3}{v_c(w) + j\omega} \frac{\partial f_0(w)}{\partial w}\, dw. \tag{2.39}$$

Using the connection between σ and ε_p (eqn. (2.5)), we can write the relative permittivity of the plasma

$$\varepsilon_p = 1 - j\frac{\sigma}{\omega \varepsilon_0} = 1 + \frac{4\pi e^2}{3m_e \omega \varepsilon_0}\, n \int_0^\infty \frac{w^3}{\omega - j v_c(w)} \frac{\partial f_0(w)}{\partial w}\, dw. \tag{2.40}$$

In the case of a velocity independent collision frequency, i.e. $v_c(w) = \text{const.}$, eqns. (2.39) and (2.40) reduce to

$$\sigma = \sigma_r + j\,\sigma_i = \frac{ne^2}{m_e} \frac{1}{v_c + j\omega}, \tag{2.41}$$

and

$$\varepsilon_p = 1 - \frac{\omega_{pe}^2}{\omega\,(\omega - j v_c)}. \tag{2.42}$$

Note that these expressions are the same as eqns. (2.13) and (2.14) respectively, which were derived in Sec. 2.2.1 using the equation of motion for a single particle. Equation (2.41) is the so-called *Lorentz expression for conductivity*. It does not hold in general for real gases, where v_c depends on electron energy. Nevertheless, the possibility of using the Lorentz conductivity is surely tempting since its simplicity facilitates the analysis of the energy transfer. For that purpose, one strives to define an effective electron-neutral collision frequency (see, for example, Ginzburg,[22] and Heald and Wharton[23]) that would directly replace v_c in eqns. (2.41) and (2.42). This is relatively easy in the limiting cases of small and large values of ω where expressions (2.39) and (2.40) simplify. The corresponding appropriate effective collision frequencies are

$$v_{eff} = -\frac{4\pi}{3} \int_0^\infty v_c(w) \frac{\partial f_0(w)}{\partial w} w^3 \, dw \qquad\qquad \text{for } \omega^2 \gg v_c^2, \qquad\qquad (2.43)$$

and

$$v'_{eff} = \left[-\frac{4\pi}{3} \int_0^\infty v_c^{-1}(w) \frac{\partial f_0(w)}{\partial w} w^3 \, dw \right]^{-1} \qquad\qquad \text{for } \omega^2 \ll v_c^2 . \qquad\qquad (2.44)$$

The two limiting values above are often referred to as the microwave and DC effective collision frequencies respectively. This terminology must be used with care, since, in atmospheric pressure plasmas for example, the "DC condition" $\omega^2 \ll v_c^2$ can be satisfied even at microwave frequencies.

2.2.7. Energy stored and dissipated in the active zone, and the related flow of microwave power into it. It is worthwhile to review the main relations concerning energy storage and dissipation within the active zone of a discharge. As we shall see in Sec. 2.3.3, these are needed when constructing an equivalent lumped-element circuit of the discharge setup.

Consider the active zone contained within a closed surface S. Poynting's theorem, derived from Maxwell's equations, expresses the relation between the electromagnetic energy flowing inwards across S, and that stored and dissipated within the enclosed volume V

$$-\frac{1}{2} \int_S (\mathbf{E} \times \mathbf{H}^*) \cdot \mathbf{dS} = j\,\omega \left(\int_V \mu_0 H_{rms}^2 dV - \int_V \varepsilon_r\, \varepsilon_0\, E_{rms}^2 dV \right) - \omega \int_V \varepsilon_i\, \varepsilon_0\, E_{rms}^2 dV \qquad (2.45)$$
$$\equiv P_A + j\, 2\omega(W_M - W_P) ,$$

where $\varepsilon_r + j\varepsilon_i = \varepsilon_p$. The physical meanings of the reactive components W_M and W_P and of the dissipative term P_A defined by eqn. (2.45) will now be discussed.

The reactive part of the power flux, $2\omega(W_M - W_P)$, is related to energy storage processes within the active zone, namely

$$W_M = \frac{1}{2} \int_V \mu_0\, H_{rms}^2 dV \qquad\qquad (2.46)$$

which denotes the average energy stored in the magnetic field over one period, while

$$W_P = \frac{1}{2} \int_{\mathcal{V}} \varepsilon_r \, \varepsilon_0 \, E_{rms}^2 \, d\mathcal{V} \tag{2.47}$$

is the average energy stored in the electric field in a dielectric medium of permittivity $\varepsilon_r \, \varepsilon_0$. Instead of using this "dielectric representation", as already discussed, one could equivalently describe the plasma as consisting of electrons moving in free space. In the following, we assume that v_c is independent of the electron velocity w and that the HF field is uniform in the active zone, these approximations leading to simple and physically meaningful expressions. Then W_P represents the combined effect of the E-field energy stored in free space

$$W_E = \frac{1}{2} \int_{\mathcal{V}} \varepsilon_0 \, E_{rms}^2 \, d\mathcal{V}, \tag{2.48}$$

with the total average kinetic energy related to the ordered motion of electrons under the influence of the HF electric field, which is by definition

$$W_K = \bar{u} \int_{\mathcal{V}} n \, d\mathcal{V}, \tag{2.49}$$

where $\bar{u} = e^2 E_{rms}^2 / 2 m_e (v_c^2 + \omega^2)$ (see Sec. 2.2.2). It can be easily shown that

$$W_P = W_E - W_K. \tag{2.50}$$

The total energy stored in the enclosed volume $\mathcal{V}$ is thus

$$W \equiv W_M + W_E + W_K = W_M + 2W_E - W_P. \tag{2.51}$$

The resistive part P_A of the power flux is a Joule-like loss term, and it denotes the power deposited in the plasma. Using eqns. (2.5) and (2.41) together with the definition of P_A in eqn. (2.45), one obtains

$$P_A \equiv \int_{\mathcal{V}} \sigma_r E_{rms}^2 \, d\mathcal{V} = \frac{e^2}{m_e v_c} E_e^2 \int_{\mathcal{V}} n \, d\mathcal{V}, \tag{2.52}$$

where the integral denotes the total number of electrons in the enclosed volume and E_e is the *effective field intensity*, defined by

$$E_e^2 = E_{rms}^2 \frac{v_c^2}{\omega^2 + v_c^2}.$$ (2.53)

Note that P_A and W_K are related by the simple formula

$$P_A = 2v_c W_K$$ (2.54)

A similar formula (eqn. (2.18)) but referring only to a single electron was obtained in Sec. 2.2.2, using the single particle motion analysis. Here, P_A refers to all electrons contained in $\mathbb{V}$ so that $P_A = \theta_A \int_{\mathbb{V}} n \, d\mathbb{V}$.

The use of E_e in eqn. (2.52) allows one to express the power transfer independently of the field frequency ω, in a form identical to that for the DC field. In the latter case, we have

$$P_A = \mu_e e E_{DC}^2 \int_{\mathbb{V}} n \, d\mathbb{V},$$ (2.55)

where $\mu_e = e/m_e v_c$ is the electron mobility for $\omega = 0$.

The particular physical state of the system for which the power flux entering $\mathbb{V}$ is purely real, is defined as a *resonance*: all the power flowing into the closed surface $\mathbb{S}$ is absorbed in the plasma, and there is no reactive component in the power flux. As seen from eqn. (2.45), the resonance occurs when

$$W_P - W_M = 0,$$ (2.56)

which also means that

$$W_K = W_E - W_M \quad \text{and} \quad W = 2W_E.$$ (2.57)

2.2.8. Gas breakdown and steady-state discharge in an HF field. Although we have chosen to concentrate on steady-state discharges in this book, we must give at least a brief physical picture of the discharge's initial stage, the so-called *breakdown phenomenon*. This term refers to the sudden appearance of an electric current in a gas when subjected to a sufficiently intense electric field.

To illustrate the breakdown phenomenon and the ensuing steady-state discharge, we consider a diffusion controlled discharge sustained in a cylindrical vessel by a uniform HF field. All the charged particles are assumed to be created in the gas volume by electron impact ionization, and to vanish by recombination at the tube wall, where they arrive by diffusion. The continuity equation is then

$$\frac{\partial n}{\partial t} = \nabla^2(Dn) + <v_i>n, \tag{2.58}$$

where $<v_i>$ is the energy averaged electron impact ionization frequency; D is the diffusion coefficient which we have seen generally depends on the electron density n, and thus on position. However, in the two limiting cases of low and high n, D is density independent (see Fig. 2.4).

In the low density range, the electrons diffuse freely at a rate corresponding to the free diffusion coefficient D_e, while at high n values, the ambipolar diffusion coefficient D_a applies. For these two cases, we can solve eqn. (2.58) by separation of variables, that is, by factoring the density as $n(r,t) = n(r)\,\mathbb{N}(t)$, which yields

$$\frac{d\mathbb{N}}{dt} = (<v_i> - \mho)\mathbb{N} \tag{2.59}$$

and

$$\nabla^2 n = -\frac{\mho\, n}{D}, \tag{2.60}$$

where $\mho$ is a constant and D stands for D_e or D_a, depending on the density range. The solution to eqn. (2.59) is

$$\mathbb{N} = \mathbb{N}(0)\,\exp(<v_i> - \mho)t. \tag{2.61}$$

Solving eqn. (2.60) leads to the spatial profile $n(r)/n(0)$ for a given shape of the discharge vessel. The boundary conditions for eqn. (2.60) are imposed both by the vessel dimensions and by the electron density at the wall. The latter depends on the particle mean free path.[14] However, for gas pressures exceeding some tens of mtorr, the assumption that n becomes zero at the wall is realistic. Note that the ratio $\mho/D$ is the eigenvalue of eqn. (2.60) and it depends on the mentioned boundary conditions. Whatever the discharge vessel configuration, this eigenvalue can be expressed in a general way as

$$\frac{\mho}{D} = \frac{1}{\Lambda^2}, \tag{2.62}$$

which defines the *characteristic diffusion length* Λ, to be discussed below.

We are now ready to characterize the breakdown and the steady-state of the discharge. At very low N values, eqn. (2.59) becomes

$$\frac{dN}{dt} = (<v_i> - \frac{D_e}{\Lambda^2})\, N. \tag{2.63}$$

From the above follows the so-called Townsend *criterion for breakdown*, namely

$$<v_i> \gtrsim \frac{D_e}{\Lambda^2}. \tag{2.64}$$

It reflects the fact that the production rate need be only slightly greater than the loss rate, namely $dN/dt \gtrsim 0$, for the electron density to increase very rapidly (eqn. 2.61), resulting in breakdown. As N grows with time, the diffusion coefficient of electrons changes from D_e to D_s, an effective diffusion coefficient (Fig. 2.4). Once the density has become large enough so that $\lambda_D << \Lambda$, typically when

$$N\,\Lambda^2 > 10^8 \text{ cm}^{-1}, \tag{2.65}$$

the diffusion becomes ambipolar, characterized by D_a. Condition (2.65) is always met in "real" steady-state microwave discharges.

Under steady-state conditions, we have $\mathcal{C} = <v_i>$ (eqn. 2.59), so that

$$<v_i> = \frac{D_a}{\Lambda^2}. \tag{2.66}$$

This condition describes the *steady-state charged particle balance*. It is a fundamental equation in plasma modeling.

For the particular and rather common case of a cylindrical discharge vessel (inner radius R_1, length ℓ), the solution of eqn. (2.60) yields the electron density[†] profile

$$n(r,z) = n_0\, J_0(x_0 r/R_1)\, \cos(\pi z/\ell), \tag{2.67}$$

[†] Since from now on the electron density no longer depends on time, we come back to our initial notation $n(r,z)$. Note that the absolute value of $n(r,z)$ is not known since it is referred to n_0, hence the term profile.

where r, z are the radial and axial cylindrical coordinates respectively, z = 0 being the middle of the cylinder length; $n_0 = n(0,0)$, and $x_0 = 2.405$ is the first zero of $J_0(x)$, the Bessel function of the first kind and of order zero. The characteristic diffusion length is then given by

$$1/\Lambda^2 = (x_0/R_1)^2 + (\pi/\ell)^2. \tag{2.68}$$

For long cylinders, $\ell^2 >> R_1^2$ and $\Lambda \approx \Lambda_\infty = R_1/2.405$.

It could be shown, though it seems obvious, that an increasing ionization implies an increasing electric field intensity. Since we always have $D_e >> D_a$ (recall eqn. 2.36), it follows from conditions (2.64) and (2.66) that the electric field intensity for breakdown is always larger than that necessary to maintain the corresponding steady-state discharge. An example of this is shown in Fig. 2.5.

For a review of detailed theoretical and experimental studies of gas breakdown in HF fields, see MacDonald.[2]

2.2.9. Maintenance processes in a steady-state discharge at reduced gas pressure. This section concerns steady-state HF discharges sustained in a pressure

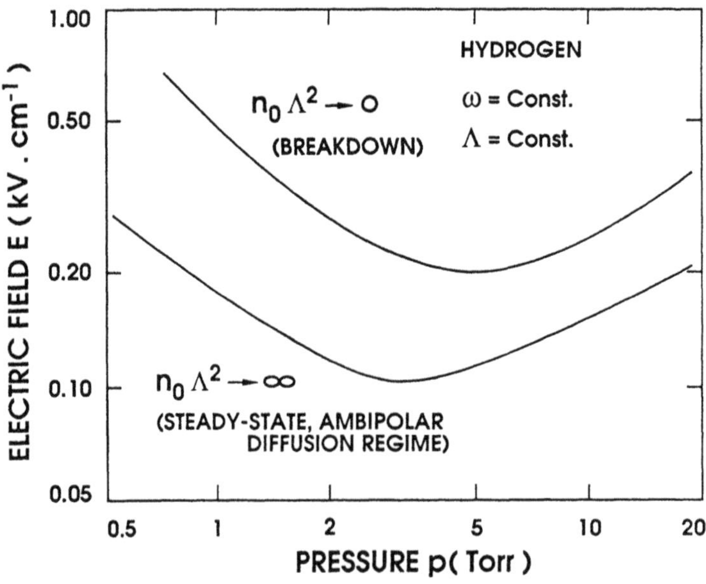

Fig. 2.5. Measured dependence of the breakdown and maintenance field intensity on the gas pressure in a hydrogen discharge at 3.2 GHz (after Brown[16], copywright © 1966 John Wiley & Sons, Inc., reprinted by permission).

range including the ambipolar diffusion regime, on the high pressure side, and the free-fall regime, on the low pressure side; it deals mainly with the charged particle balance and the power balance per electron, two relations that are fundamental for discharge modeling.

Let us assume, as a first approximation, that the EEDF does not vary within the active zone of the discharge. Thus, all quantities obtained from an averaging over the EEDF, such as the electron rate coefficients and transport parameters, are uniformly distributed in space. Also, we consider that the neutral particles are ionized by electron impact. The charged particles vanish due to recombination on the tube wall, where they arrive by ambipolar diffusion (high pressure limit) or by direct impact (low pressure limit). The power acquired by the electrons from the electromagnetic field is transferred to the heavy particles through collisions of various kinds. This power is finally removed from the dischage to the surrounding environment, mainly through optical radiation and via particle impacts on the vessel wall.

2.2.9.1. Charged particle balance. The simple formulation presented in the previous section for a steady-state discharge is not completely correct at gas pressures which are low enough that the ion mean free path yields $\lambda_{in} \gtrsim \Lambda$, where Λ characterizes the dimensions of the discharge vessel (see Sec. 2.2.8). Under such circumstances, the ions make practically no collisions with the gas atoms or molecules; they are simply accelerated by the space-charge field towards the wall, onto which they fall freely, that is without experiencing any significant retardation in the course of their motion. For this reason, this situation is known as the *free-fall regime*. However a general theory, valid at all pressures, that is for all values of λ_{in}/Λ, has been developed by Self and Ewald.[24] As later shown by various authors,[21,25] the equation for the charged particle balance found from this theory can be expressed in the form

$$<v_i> = \frac{D_{se}}{\Lambda^2},$$

(2.69)

where D_{se} represents a new effective diffusion coefficient. This equation has the same form as eqn. (2.66) but with D_{se} in place of D_a. At sufficiently high pressures where $\lambda_{in} << \Lambda$ (classical ambipolar diffusion regime), D_{se} tends to D_a and we again encounter eqn. (2.66). On the other hand, we find $D_{se} << D_a$ when $\lambda_{in} >> \Lambda$ (free-fall-regime). Figure 2.6 shows the variation of D_{se}/D_a as a function of Λ/λ_{in} for both cylindrical and planar geometries, for the case of argon.

The radial profile of n obtained from the theory of Self and Ewald[24] in a cylindrical geometry is illustrated in Fig. 2.7, for the free-fall and the classical ambipolar diffusion

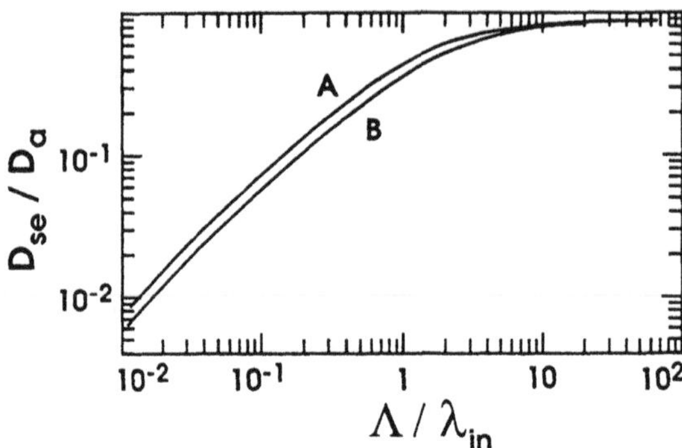

Fig. 2.6. Calculated effective diffusion coefficient D_{se}, related to Self and Ewald's theory,[24] normalized to the ambipolar diffusion coefficient D_a as a function of the diffusion length Λ normalized to the ion mean free path λ_{in}. (a) Cylindrical configuration and (b) planar configuration.

limits, and for an intermediate situation. In the ambipolar diffusion limit, the radial profile of n is well represented by the function $J_0(r/\Lambda)$, as found from the simple formulation of the previous section. However, for $\Lambda \lesssim \lambda_{in}$, the profile differs considerably, n reaching at the plasma boundary[†] some finite value that depends on the value of Λ/λ_{in}. Self and Ewald showed that this boundary appears naturally in their

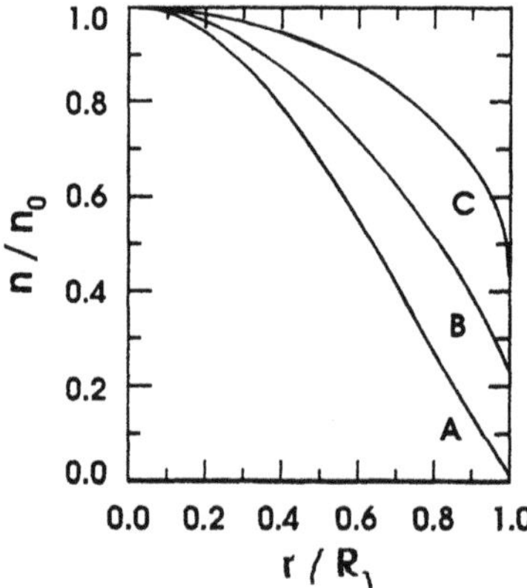

Fig. 2.7. Calculated radial profile of electron density related to Self and Ewald's theory[24] as a function of radial position, for a cylindrical tube of inner radius R_1 ($n_0 \equiv n(r = 0)$). (a) $\Lambda/\lambda_{in} \gg 1$, ambipolar regime, (b) $\Lambda/\lambda_{in} \approx 1$, intermediate regime, and (c) $\Lambda/\lambda_{in} \ll 1$, free-fall regime.

[†] Note that this boundary does not coincide exactly with the position of the wall since a very thin space-charge *sheath* forms between the neutral plasma and the wall. The thickness of this sheath is just a few times the Debye length λ_D; in most microwave plasmas, it is much smaller than the vessel dimensions.

theory at the location in space where the mean ion velocity equals the isothermal ion sound speed $[e(u_k + T_i)/m_i]^{1/2}$; m_i is the ion mass, $u_k \equiv D_e/\mu_e$ is called the *electron characteristic energy*,[†] and T_i is the ion temperature, both expressed in eV. This conclusion is consistent with Bohm's criterion[26] for the formation of a space charge sheath between the plasma and the wall.

Equation (2.69) for the charged particle balance has a very simple physical meaning, namely it merely states that, under steady-state conditions, the average ionization frequency $<v_i>$ must exactly balance the average frequency D_{se}/Λ^2 at which charged particles are lost by diffusion to the wall. It yields a number of important consequences concerning the discharge operation; we now discuss these. i) In practice, for a given gas, $<v_i>$ is a function of the EEDF which is obtained from the Boltzmann equation (2.32); inspection of eqns. (2.32) and (2.33) reveals that the EEDF is, in turn, a function of the amplitude of the electric field E_{rms} in the plasma. Hence, for a given gas, $<v_i>$ is ultimately a function of the amplitude of the electric field. We can therefore conclude that eqn. (2.69) serves to determine, in an implicit way, the value of this amplitude as a function of the gas density N (D_{se} is proportional to N^{-1} whereas $<v_i>$ is proportional to N) and of the geometry and dimensions of the vessel (through Λ). These considerations will be further developed in Chap. 3. ii) The fundamental point to be retained from the above is that the field amplitude in the plasma under steady-state conditions is determined not by the HF field applicator and power level delivered to the plasma, but rather by the charged particle balance requirement. *Whatever the external conditions, the electric field amplitude in the plasma adjusts itself in such a way that the resulting average ionization rate exactly compensates the average loss rate of charged particles.* Note that this statement remains valid even in more complex situations in which the charged particle balance cannot be expressed by a single-term equation as in (2.69). This is the case, for example, when both electron-ion recombination processes and ionization of excited states (also termed step-wise ionization) play important roles in the charged particle balance.

In the above, we have assumed that the field in the plasma is uniform. However, this field must satisfy Maxwell's equations and at the same time appropriate boundary conditions at the plasma-wall interface. Full compliance with these conditions and with the charged particle balance equation generally implies that the field is nonuniform in the plasma, which contradicts our assumption of a uniform field and of a uniform EEDF.

[†] This parameter can be used to define an approximate or effective value for the ion mean free path λ_{in}, an important quantity already mentioned above in this section. Noting that in general $u_k >> T_i$, from the above we have that the ion velocity at the plasma boundary is of the order of $(eu_k/m_i)^{1/2}$. Then we can consider $(eu_k/m_i)^{1/2} v_{in}^{-1}$, where v_{in} is the ion-neutral collision frequency for momentum transfer, as being an estimation of the ion mean free path λ_{in}. See also Sec. 3.2.1.

This assumption nevertheless remains valid as far as one concentrates only on determining the average field amplitude over the plasma volume, which follows from the charged particle balance. Clearly E_{rms}, the rms value of the field intensity in the plasma, is then a cross-section average value. Under steady-state conditions, it is usually termed the *maintenance field*, for reasons that are now obvious. Maxwell's equations and the boundary conditions are to be used to determine the spatial profile of the field when needed.

2.2.9.2. Power balance per electron. The power balance per electron expresses a relation between the maintenance field, E_{rms}, and the power lost by an average electron as a result of collisions of all kinds, θ_L. However, this is not a further independent equation, since the electron Boltzmann equation implicitly contains all this information. To make it explicit, one has to multiply eqn. (2.32) by the electron energy u and then integrate over all energies. The result can be written in the form

$$\sigma_r \, E_{rms}^2 \equiv \theta_A \, n = \theta_L \, n, \tag{2.70}$$

where σ_r is the real component of the complex electric conductivity as defined by eqn. (2.39). Equation (2.70), which is a moment of the Boltzmann equation, is thus automatically satisfied once the latter equation is solved. Equation (2.70) expresses the fact that the average power absorbed from the field $\sigma_r \, E_{rms}^2 \equiv n \, \theta_A$ (where θ_A is thus the power absorbed from the field per electron) exactly balances the average power $n \, \theta_L$ lost in collisions with the gas molecules, per unit volume. It can be shown that θ_L is given by

$$\theta_L = \frac{2m_e}{M} <v_c(u)u> + \sum_k <v_k(u)> eV_k + <v_i(u)> eV_i, \tag{2.71}$$

where the terms on the right-hand side represent, in order, the average power lost in elastic collisions, in excitation of all atomic or molecular states k, and in ionization; v_k represents the excitation frequencies, and eV_k the corresponding excitation threshold energies. We note that the actual value of $\theta = \theta_A = \theta_L$ can be calculated in two ways: i) it can be determined from $\theta_A \equiv n^{-1} \sigma_r \, E_{rms}^2$, once the Boltzmann equation has been solved for a given E_{rms}. The value of $n^{-1} \sigma_r$ is obtained from eqn. (2.39); ii) alternatively, one could also use eqn. (2.71) to calculate θ_L once the Boltzmann equation has been solved. Given the excitation and ionization cross-sections for the gas considered, it suffices to calculate the energy averaged quantities in the bracketts $<>$ on the right-hand side of eqn. (2.71). Both procedures i) and ii) are equivalent since $\theta = \theta_A = \theta_L$ under steady-state conditions. In fact, verifying the equality

$\theta_A = \theta_L$ is a method often used to check the calculations when numerically solving the Boltzmann equation.

In summary, the power balance equation per electron, eqn. (2.70) relates θ to E_{rms} while the charged particle balance determines the value of E_{rms}, the maintenance field, for a given gas density N, characteristic diffusion length Λ and angular frequency ω, as we have seen. By combining these two equations, one ultimately obtains θ as a function of N, Λ and ω for a steady-state discharge. In other words, the steady-state operation of a microwave discharge for given N, Λ and ω requires well defined values of both the magnitude of the field within the plasma and the mean absorbed-lost power per electron, $\theta = \theta_A = \theta_L$. For this reason, the functions $E_{rms}(N, \Lambda, \omega)$ and $\theta(N, \Lambda, \omega)$ are usually referred to as *discharge characteristics.*

In microwave discharges, it is much easier to measure θ than E_{rms} whereas it is the opposite with the positive column of DC discharges. Equation (2.70) allows one to make the connection between these two parameters when comparing these discharges.

2.2.9.3. Power balance equation over a given plasma volume. Equation (2.70) can be integrated over the plasma volume to yield the power balance equation

$$P_A = P_L, \tag{2.72}$$

where $P_A = \theta_A \int_V n \, d\mathcal{V}$ † is the total power absorbed from the field, already defined (see eqns. (2.45) and (2.52)) while

$$P_L = \theta_L \int_V n \, d\mathcal{V} \tag{2.73}$$

is the total power lost by the electrons in collisions with the gas atoms or molecules within the plasma volume. Given the power P_A absorbed in the plasma and provided θ is known, eqn. (2.73) serves to determine $\bar{n}$, the average value of electron density over the volume considered. Recall that the charged particle balance yields the spatial profile $n(r,z) / \bar{n}$ but that $\bar{n}$ remains to be determined (see, for example, eqn. (2.67)).

---

† In general, $P_A = \int_V \theta_A n \, d\mathcal{V}$ but, starting in Sec. 2.2.7, we have assumed for simplicity E_{rms} to be uniform in the plasma. Also, in general, $P_L = \int_V \theta_L n \, d\mathcal{V}$ but we have assumed θ_L to be independent of n.

In concluding, we remind that throughout most of the chapter, we have considered charged particles to be mainly lost by diffusion to the walls. This has allowed us to assume in this section θ to be independent of the electron density. However, this is not valid when electron-electron collisions and/or step-wise ionization processes need to be taken into account; this subject is discussed in more detail in Secs. 3.4 and 3.5 respectively. In the simple case treated here, this leads to the electron density being proportional to the amount of power deposited in the plasma: the result of any increase in the absorbed power density is then an increase in the average electron density value and *not in the maintenance field amplitude*. This point needs to be emphasized since it seems to have escaped the attention of many researchers.

To relate the above results with the coming section, we note that for a given power P_I delivered at the source input, the power P_A actually absorbed depends on the overall electrodynamic characteristics of the plasma source, and its value must be determined separately for any specific design of the plasma source. We discuss this matter below and further in Chaps. 4 and 5.

2.3. The microwave discharge as a constituent of the plasma source

2.3.1. Microwave plasma source system. A microwave plasma source is a device in which a microwave discharge is employed to yield a plasma of the desired physical properties. To this end, the source design and operation must satisfy three main objectives: i) filling the discharge tube with the required gas composition, pressure and, eventually, flow rate; ii) providing the required electromagnetic field distribution and intensity to initiate and sustain the discharge; iii) ensuring an efficient power transfer from the microwave feed line to the plasma. The first objective is attained by using either permanently sealed plasma vessels, or vessels connected to a vacuum apparatus. We now concentrate on the remaining two tasks.

To understand how the practical realization of the source affects the physical processes in the discharge, one must remember some basic facts: i) the design of the field applicator and the plasma vessel impose the boundary conditions for the electromagnetic field and for the charged particle and energy fluxes; ii) the maintenance processes within the discharge and the source characteristics as an entity are strongly coupled because of the electric field-plasma interaction within the active zone.

A schematic diagram of a typical plasma source configuration showing all the parts essential to perform the specified tasks is shown in Fig. 2.8. As a rule, the actual system includes some additional elements. These are directional couplers and a

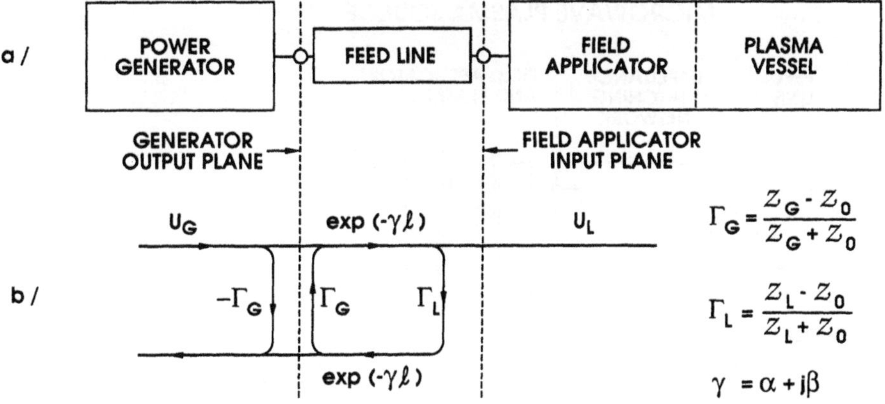

$$\Gamma_G = \frac{Z_G - Z_0}{Z_G + Z_0}$$

$$\Gamma_L = \frac{Z_L - Z_0}{Z_L + Z_0}$$

$$\gamma = \alpha + j\beta$$

Fig. 2.8. (a) The essential parts of a typical HF sustained plasma source. (b) Flow chart showing the voltages U along the transmission line where Γ_G and Γ_L are the reflection coefficients at the input and output of the feed line of length ℓ.

power meter, needed to determine the power delivered to the plasma, and a circulator between the generator and the source, to improve its operating stability (Fig. 2.9). When the source's tuning ability proves insufficient to achieve efficient power transfer to the plasma, a tuner must be added. Not all these elements are absolutely required in a given system but, if they are present, they must be connected in a definite order (Fig. 2.9) into the feed line, and as close to the applicator input as possible. To learn more about the various aspects of power feeding and measurement in microwave discharges, see refs. 27-30.

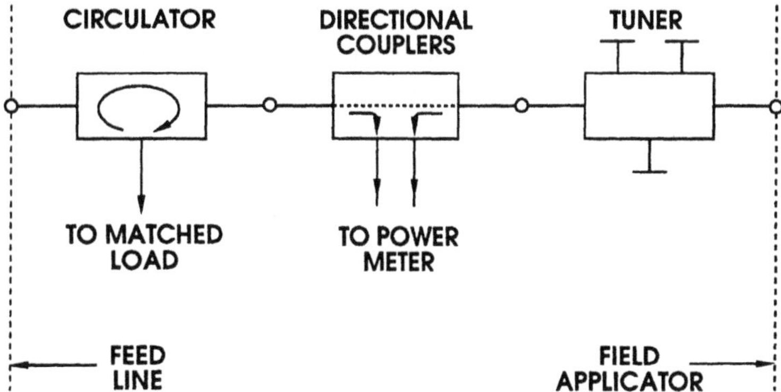

Fig. 2.9. Set of auxiliary elements for an HF plasma source. The insertion sequence of the elements must be observed.

MICROWAVE PLASMA SOURCE

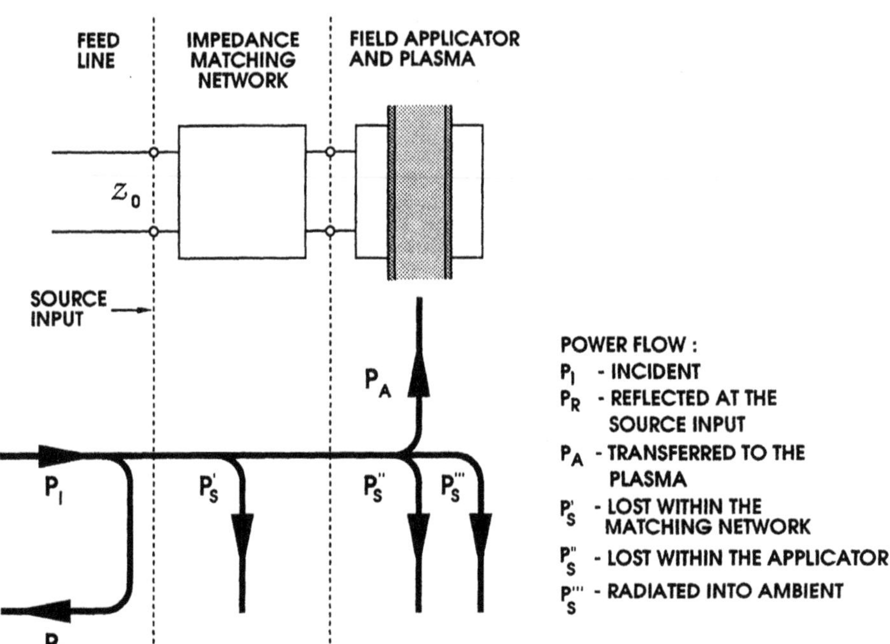

Fig. 2.10. Simplified graph of the power flow in a microwave plasma source.

In the following section, we discuss the efficiency of power transfer from the feed line to the plasma. We shall show that when losses within the source's component elements can be neglected, this efficiency is characterized by a single overall parameter η relating the power absorbed in the plasma to that delivered at the input.

2.3.2. Efficiency of power transfer from the HF generator to the plasma.

The plasma source's overall efficiency is defined as the fraction of power delivered by the generator at the source input which is absorbed by the plasma; a simplified graph of the power flow in the source is shown in Fig. 2.10. Besides reflection at the line-source interface, there are two other possible loss mechanisms capable of diverting HF power from the discharge, namely losses in the dielectric and metal components of the plasma source (P'_S, P''_S), and HF radiation into the surrounding space (P'''_S). However, the cumulative power loss related to P'_S, P''_S and P'''_S can be made negligibly small by a proper source design. When this is the case, reflection at the source input is the only factor determining the overall efficiency η, so that

$$\eta \equiv P_A/P_I \approx (P_I - P_R)/P_I, \tag{2.74}$$

where P_I and P_R are the incident and reflected wave power at the input respectively, and P_A is the power absorbed in the plasma. Note that η can be measured directly in the power feed line using standard reflectometer techniques.

Clearly, then, optimum power transfer requires that the reflected power be minimal. This can be accomplished by using an appropriate impedance matching network ahead of the field applicator (Fig. 2.10). This may either be an *intrinsic tuning* means, so-named because it is incorporated into the applicator device, or it may be a separate tuning element.

2.3.3. **The equivalent circuit representation.** The equivalent circuit description

is used to represent power transfer phenomena taking place within the plasma source; the various circuit elements correspond to electromagnetic energy storage or dissipation processes occurring in various parts of the system. This method serves to analyze the HF characteristics of the plasma source without having to solve a full set of electromagnetic field equations. The impedance (admittance) concept, the field-impedance relations and the equivalent circuit representation of microwave networks are well covered in the literature. The interested reader may find details in books on electromagnetic field theory and circuit analysis.[31-35]

Let us again refer to Fig. 2.10. Clearly, the conventional formalism of microwave circuit theory can be readily applied to construct the equivalent circuit of any matching network. We need to show how this formalism can be extended to describe the field applicator enclosing the plasma. To this end, the power loss in the applicator, excluding power dissipation within the plasma, is assumed to be negligible. The voltage and current of amplitudes U and I respectively at the input of the field applicator can be defined so that their product gives the power entering the applicator, namely

$$UI^{*} = - \int_{S} (\mathbf{E} \times \mathbf{H}^{*}) \cdot \mathbf{dS} \tag{2.75}$$

where S denotes integration over the input area. The minus sign results from the fact that we are considering an inward flow of power, while the vector $\mathbf{dS}$ is directed outward. From the definition of admittance $\mathcal{Y}$ (I = $\mathcal{Y}$ U, thus I* = $\mathcal{Y}^{*}$U*) and from eqn. (2.45), it follows that

$$\frac{1}{2}\mathcal{Y}^{*} UU^{*} = P_A + j\, 2\omega(W_M - W_P). \tag{2.76}$$

From this, one obtains the *input admittance* of the single-port network, representing the field applicator loaded with the plasma,

$$Y \equiv G + j\,B = \frac{2P_A - j\,4\omega(W_M - W_P)}{UU^*}. \tag{2.77}$$

Remembering that $W_P = W_E - W_K$, we can write it as

$$Y = Y_D + j\,B_0, \tag{2.78}$$

where

$$j\,B_0 = j\,\frac{4\omega(W_E - W_M)}{UU^*} \tag{2.79}$$

represents the input admittance of the same network in the absence of the discharge ($W_K = 0$), and

$$Y_D = G_D + j\,B_D = \frac{2P_A - j\,4\omega W_K}{UU^*} \tag{2.80}$$

accounts for the presence of the discharge in the network.

The HF power P_A delivered to the plasma (eqn. 2.52) goes entirely to the electron gas. Then, provided we assume that the value of v_c does not vary with the electron energy, P_A equals the time averaged kinetic energy gain of all electrons[36] and we have (eqn. (2.54))

$$P_A = 2v_c W_K. \tag{2.81}$$

Therefore, eqn. (2.80) becomes

$$Y_D = G_D + j\,B_D = \frac{2P_A}{UU^*}\left(1 - j\,\frac{\omega}{v_c}\right) = \frac{4\omega W_K}{UU^*}\left(\frac{v_c}{\omega} - j\right). \tag{2.82}$$

As one may see from eqn. (2.80), G_D represents the power loss, and it is proportional to P_A. Note that $B_D < 0$, thus, the influence of the discharge on the properties of the microwave network, as far as reactive energy is concerned, is that of an inductance.

The equivalent circuit of the plasma source is shown in Fig. 2.11. The impedance matching network is represented here by a linear (i.e. independent of input power), lossless two-port circuit. On the other hand, that part of the network input impedance (eqn. (2.78)) depicted by Y_D is both lossy and nonlinear because G_D and B_D depict energy absorbed from the electromagnetic field by the electrons, and then either spent in collisions with heavy particles or stored as kinetic energy. In Figs. 2.10 and 2.11a, the field applicator and the impedance matching network are represented as separate units. This need not, and sometimes cannot, be the case. For example, when the field applicator is equipped with intrinsic tuning means, a single two-port network representation is more appropriate.

The source's input impedance Z_L constitutes the load impedance for the feed line as shown in Fig. 2.11b. In order to relate this impedance to the power transfer efficiency η, we resort to standard methods of transmission line analysis.

2.3.4. Impedance matching for minimum reflection and maximum power transfer. The quantity characterizing both the impedance mismatch in a transmission line terminated by the impedance Z_L and the wave reflection level within it (Fig. 2.8) is the reflection coefficient

$$\Gamma_L = \frac{Z_L - Z_0}{Z_L + Z_0} = \frac{z_L - 1}{z_L + 1}. \tag{2.83}$$

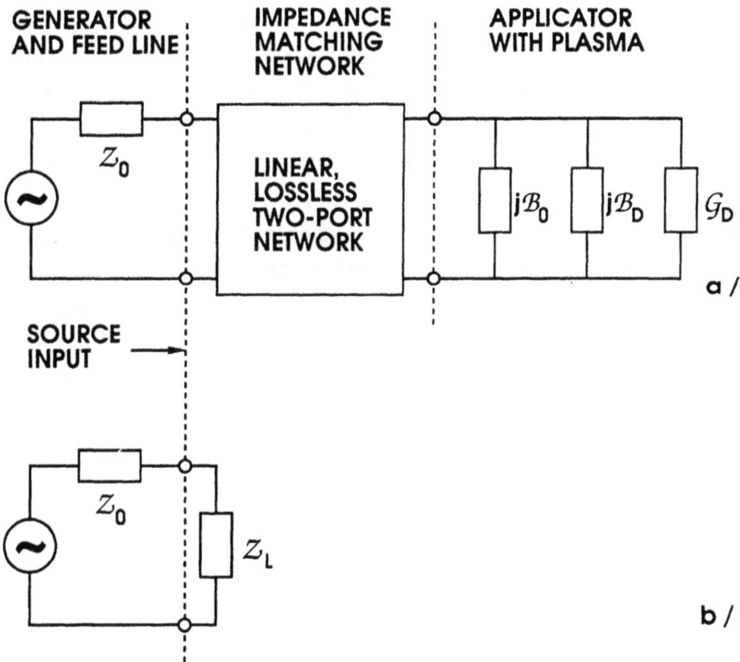

Fig. 2.11. (a) Equivalent circuit of a microwave plasma source where jB_0 represents the input admittance of the applicator in the absence of the discharge while $G_D + jB_D$ accounts for its presence. The term G_D relates to the power loss. (b) The impedance Z_L is the plasma source's impedance and it constitutes the load for the feed line of characteristic impedance Z_0.

Lower case symbols denote impedances normalized with respect to Z_0, the characteristic impedance of the feed line (a real quantity), namely

$$z \equiv Z/Z_0 \equiv \mathcal{R}/Z_0 + j\ X/Z_0 \equiv r + j\ x. \tag{2.84}$$

By definition of Γ_L, we have

$$|\Gamma_L|^2 = P_R/P_I, \tag{2.85}$$

which through eqn. (2.74) defines the *efficiency of the power transfer*

$$\eta \approx 1 - |\Gamma_L|^2. \tag{2.86}$$

The Voltage Standing Wave Ratio (VSWR) is directly related to $|\Gamma_L|$ and to the amount of reflected power as shown in Fig. 2.12.

We are now ready to discuss the conditions for impedance matching in the plasma source. A variable load impedance will receive the maximum power from the generator when its value is adjusted to equal the complex conjugate of the generator's output impedance (*conjugate impedance matching*). On the other hand, it is evident from eqns. (2.83) and (2.85) that only when $\Gamma_L = 0$, is the reflected wave in the feed line eliminated. This requires that the feed line be terminated with an impedance Z_0 (*image impedance matching*). One therefore generally faces a dilemma: whether to get maximum power delivery to the plasma, or to avoid undesired effects associated with a reflected wave in the feed line, namely power loss and possible unstable operation of the plasma source[30]. This dilemma is usually resolved in favor of image impedance matching, as two additional factors tip the scale: the tuning for $\Gamma_L = 0$ is easy to monitor and, in most practical plasma source setups, a circulator (with input and output impedances Z_0) is inserted between generator and applicator input, mainly to avoid instabilities of operation. In such a configuration: i) the generator is always terminated with Z_0; ii) the impedance seen from the applicator input toward the generator is also Z_0 (resulting in zero reflection hence no loss in the feed line). As a result, then only impedance matching on an image-impedance basis is possible and the conditions for impedance matching at the launcher input are

$$\mathcal{R}_L = Z_0, \quad X_L = 0, \quad \text{i.e.} \quad r_L = 1, \quad x_L = 0. \tag{2.87}$$

To fulfill the above conditions, at least two of the reactances appearing in the equivalent circuit must be variable. In practical terms, this means that the source must be equipped with no less than two independent tuning means.

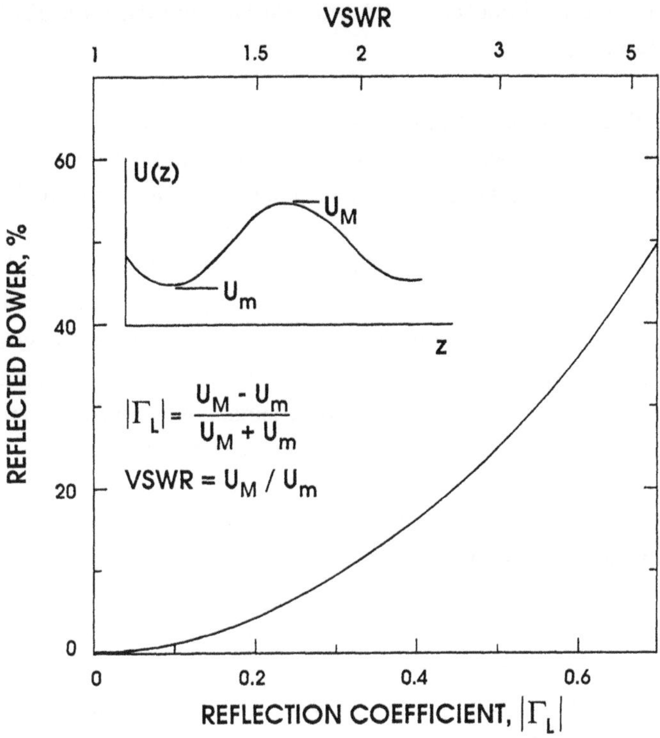

Fig. 2.12. Graphs connecting the standing wave pattern to the reflection coefficient $|\Gamma|$ and to the Voltage Standing Wave Ratio (VSWR), and indicating the corresponding percentage of power reflected at the line-load interface. $U(z)$ represents the voltage distribution along the line (after Hubert et al. [30]).

2.4. Classification of RF and microwave discharges

A wide variety of RF and microwave plasma setups have been described in the literature but no one seems to have succeeded in proposing a widely accepted classification of these discharges. Various criteria of classification can be chosen (see, for example, Kunkel[37]). We chose to base our classification on the way the electromagnetic field is imposed on the plasma but, in contrast to earlier such classifications (see, for example, Marec et al.[9]), our implies that all RF and microwave discharges fall into one or the other of the following two broad categories.

The first category consists of discharges sustained within HF circuits. This generally implies that the phase difference between field oscillations at any two points within the discharge is so small that the field intensity throughout the discharge is essentially the same at all times. Since in this case the active zone of the discharge is largely confined within the field applicator, we propose for them the term *discharges with a localized active zone.*

The second category includes all the remaining discharge types, more specifically HF discharges with a spatial extent which is large compared with the wavelength. These discharges are sustained by the electric field of a wave propagating along them, and are thus labeled *traveling-wave discharges.*[11,38] Their active zone extends in the direction of wave propagation, and the time difference between field variations at different points along that direction cannot be neglected in comparison with the traveling wave wavelength. In such a situation, the field applicator may either need to extend fully along the discharge tube[3], or it may enclose only a short section of the tube.[4,39] In the latter case, corresponding to surface-wave discharges, the power is carried outward from the localized applicator, a wave launcher, by a wave that propagates along the plasma-dielectric interface. As is illustrated throughout this book, the distinction between these two discharge categories also corresponds to two different methods of analyzing them.

The ideas and relations presented in the present chapter can be put to work in other chapters to provide a complete, although approximate, analysis of microwave discharges of either category above. Strictly, a self-consistent formulation must be used to fully account for the interaction between microwave field and plasma. It should comprise a set of equations including Maxwell's equations, and those describing the discharge maintenance processes. It must also include proper boundary conditions for the electromagnetic field within the complete source setup, and those for the fluxes of charged particles and power at the surfaces surrounding the discharge area. The problem formulated in such a self-consistent way obviously eludes any one simple treatment. To render it more tractable, it is convenient to perform the analysis in two separate operations.

The first operation deals with the discharge maintenance processes, which can be analyzed by considering the charged particle continuity, transport and energy balance equations; it yields, under given discharge conditions, the relations between the plasma parameters of interest[†] and the power lost by the electrons through collisions with heavy particles. This stage of the analysis is performed in a similar manner for both discharge categories.

[†] For a reduced pressure discharge, these parameters are essentially the EEDF, the maintenance field intensity, and the spatial distribution of electron density.

The second operation, dealing with power transfer from the electromagnetic field to the plasma, is performed in a way which is specific to either category. In the case of discharges with a localized active zone, it is convenient to construct equivalent circuits in which the energy storage and dissipation processes are directly represented by lumped admittance (or impedance) elements. On the other hand, when dealing with traveling wave discharges, it is more adequate to use dispersion relations and the resulting phase and attenuation coefficients when analyzing the local interaction between the plasma column and the traveling wave field. However, in analyzing the power transfer from the feed line to the wave power flux at the launcher exit, an equivalent circuit description may be used. Processes occurring in the wave excitation region are then represented by the lumped elements which account for the wave power carried away from this region and for field energy storage and dissipation. From such a circuit, one can determine the electrodynamic characteristics of the wave launcher.

References

[1]   V.E. Golant, Uspiekhi Fiz. Nauk **65**, 39 (1958).

[2]   A.D. MacDonald, *Microwave Breakdown in Gases* (Wiley, New York, 1966).

[3]   R.G. Bosisio, C.F. Weissfloch and M.R. Wertheimer, J. Microwave Power **7**, 325 (1972).

[4]   M. Moisan, C. Beaudry and P. Leprince, Phys. Lett. **50A**, 125 (1974).

[5]   A.D. MacDonald and S.J. Tetenbaum, in *Gaseous Electronics*, M.N. Hirsh and H.J. Oskam eds. (Academic Press, New York, 1978) Chap. 3.

[6]   A.T. Bell, in *Techniques and Applications of Plasma Chemistry*, J.R. Hollahan and A.T. Bell eds. (Wiley, New York, 1974) Chaps. 1 and 10.

[7]   Yu. A. Lebedev and L.S. Polak, High Energy Chem. **13**, 331 (1979).

[8]   M. Moisan, C.M. Ferreira, Y. Hajlaoui, D. Henry, J. Hubert, R. Pantel, A. Ricard and Z. Zakrzewski, Revue Phys. Appl. **17**, 707 (1982).

[9]   J. Marec, E. Bloyet, M. Chaker, P. Leprince and P. Nghiem, in *Electrical Breakdown and Discharges in Gases*, E.E. Kunhardt and L.H. Luessen eds. (Plenum, New York, 1982) p. 347 (part B).

[10]   C.M. Ferreira, in *Radiative Processes in Discharge Plasmas*, J.M. Proud and L.H. Luessen eds. (Plenum, New York, 1986) p. 431.

[11]   M. Moisan and Z. Zakrzewski, in *Radiative Processes in Discharge Plasmas*, J.M. Proud and L.H. Luessen eds. (Plenum, New York, 1986) p. 381.

[12]   V.M. Batenin, I.I. Klimovski, G.V. Lisov and V.N. Troiski, *Microwave Plasma Generators: Physics, Techniques, Applications* (Energy-Atomizdat, Moscow, 1988), in Russian.

[13]   H.S.W. Massey and E.H.S. Burhop, *Electronic and Ionic Impact Phenomena* (Clarendon Press, Oxford, 1952).

[14]   E.W. McDaniel, *Collision Phenomena in Ionized Gases* (Wiley, New York, 1964).

[15]   V.E. Golant , E.P. Zhilinsky and I.E. Sakharov, *Fundamentals of Plasma Physics* (Wiley, New York, 1980).

[16]   S.C. Brown, *Introduction to Electrical Discharges in Gases* (Wiley, New York, 1966).

[17]   R. Winkler, J. Wilhelm and A. Hess, Ann. Phys. **42**, 537 (1985).

[18]   R. Winkler, H. Deutsch, J. Wilhelm and Ch. Wilke, Beitr. Plasma Physik **24**, 285 (1984).

[19]   R. Winkler, M. Capitelli, M. Dilonardo, C. Gorse and J. Wilhelm, Plasma Chem. Plasma Proces. **6**, 437 (1986).

[20]   W.P. Allis, *Motions of Ions and Electrons*, Handbuch der Physik **21**, 383 (1956) (Springer Verlag, Berlin).

[21]   Ch. H. Muller III and A.V. Phelps, J. Appl. Phys. **51**, 6141 (1980). For more detail see A.V. Phelps, J. Res. National Inst. of Stds. and Technology **95**, 407 (1990).

[22]   V.L. Ginzburg, *The Propagation of Electromagnetic Waves in Plasma* (Pergamon, New York, 1974).

[23]   M.A. Heald and C.B. Wharton, *Plasma Diagnostics with Microwaves* (Wiley, New York, 1965).

[24]   S.A. Self and H.N. Ewald, Phys. Fluids **9**, 2486 (1966).

[25]   C.M. Ferreira and J. Loureiro, J. Phys. D: Appl. Phy. **17**, 1175 (1984).

[26]   D. Bohm, in *The Characteristics of Electrical Discharges in Magnetic Fields*, A. Guthrie and R.K. Wakerling eds. (McGraw Hill, New York, 1949) Chap. 3.

[27]   B.A.H.G. Jutte and J. Agterdenbos, Spectrochim. Acta **34B**, 133 (1979).

[28]   M. Outred, Spectrochim. Acta **35B**, 447 (1980).

[29]   M. Outred and C.B. Hammond, J. Phys. D: Appl. Phys. **13**, 1609 (1980).

[30]   J. Hubert, M. Moisan and Z. Zakrzewski, Spectrochim. Acta **41B**, 205.(1986).

[31]   E.L. Ginzton, *Microwave Measurements* (McGraw Hill, New York, 1957).

[32]   R.N. Ghose, *Microwave Circuit Theory and Analysis* (McGraw-Hill, New York, 1963).

[33]   S. Ramo, J.R. Whinnery and T.V. Duzer, *Fields and Waves in Communication Electronics* (Wiley, New York, 1965).

[34]   R.E. Collin, *Field Theory of Guided Waves* (McGraw Hill, New York, 1960).

[35]   P. Lorrain, D.P. Corson and F. Lorrain, *Electromagnetic Fields and Waves* (Freemann, New York, 1988).

[36]   D. Quémada, *Ondes dans les plasmas* (Hermann, Paris, 1968).

[37]   W.B. Kunkel, *Plasma Physics in Theory and Application*, W. B. Kunkel ed. (McGraw Hill, New York, 1966).

[38]   Z. Zakrzewski, J. Phys. D: Appl. Phys. **16**, 171 (1983).

[39]   M. Moisan and Z. Zakrzewski, J. Phys. D: Appl. Phys. **24**, 1025 (1991).

CHAPTER 3

KINETIC MODELING OF MICROWAVE DISCHARGES: INFLUENCE OF THE DISCHARGE STIMULATING FREQUENCY [*]

3.1. Introduction

It is now generally accepted that the frequency $\omega/2\pi$ at which a high frequency (HF) discharge is sustained has considerable influence on the properties of the plasma. For example, the electron density obtained for a given HF power density deposited into the plasma is usually higher at microwave than at radio frequencies (RF). Along the same line, it has been shown that the plasma stimulating frequency can be adjusted to improve the efficiency of plasma processing, for example, in the etching and deposition of thin polymer films.[1] The analysis developed in the literature for explaining such results calls on the dependence of the electron energy distribution function (EEDF) upon ω. Reviewing this analysis is the leading object of the chapter, which ultimately aims at drawing guidelines for the frequency optimization of plasma processes. An indirect but important aspect of this review is that it introduces the reader to the kinetic modeling of microwave discharges.

We must from the beginning clearly delimit the physical context to which the present analysis applies. This context must be borne in mind when attempting to predict, for example, the effects on a given plasma process of converting from a RF capacitive discharge at 13.56 MHz to a microwave induced plasma (MIP) at 2.45 GHz. Clearly, considering such a move implies making other changes in the system besides frequency. The following remarks are thus in order:

(i) Most MIP can be generated using a microwave field applicator located outside the discharge vessel which usually consists of a low loss dielectric material such as fused silica. This method of field application allows a substrate exposed to the MIP to be biased independently of the microwave power, which is rarely possible for RF capacitive discharges where the parallel plates are in direct contact with the discharge. In the latter case, the powered electrode holding the substrate is said to be self-biased. The corresponding potential drop across the sheath before the substrate depends on the RF power applied to the discharge and it cannot be adjusted independently of the bulk plasma parameters such as, for example, plasma density. An independent biasing is advantageous in providing the means for adjusting the flux and the energy of

---

[*] **Presented by C.M. Ferreira and M. Moisan**

ions bombarding the substrate during reactive ion etching (RIE) or deposition processes. The model that we present ignores any biasing effect and thus, when using it to compare results from RF and microwave plasma reactors, one implicitly assumes that the substrate is at the floating potential in both cases.

(ii) The coupling scheme of RF power, not only for RF capacitive discharges but for RF discharges in general, is different from that employing microwaves, preventing the use of the same reactor in both situations. To our knowledge, the only case where the RF and microwave fields are applied in the same way to the discharge is that of surface-wave produced plasmas. Modifications to the reactor dimensions not only affect the charged particles' characteristic diffusion lengths but also influence the hydrodynamics of the gas flow which often plays a significant role in plasma processing.

(iii) The power density absorbed into the plasma is generally an operator-dependent parameter. A priori, one would think that it suffices to keep it constant when changing frequency to observe nothing else but the influence of the discharge excitation frequency upon plasma parameters. Generally this is not that simple since the electron density may also vary with frequency, precisely as a result of changes in the EEDF with ω.[1] Such a density variation with ω can induce two side effects: i) a further variation of the EEDF as a function of electron density. This occurs with discharges where the charged particles' losses are dominated by volume recombination. It also shows up with diffusion controlled plasmas when changes in the degree of ionization, which determines the transition regime from dominating electron-neutral collisions to dominating electron-electron collisions, modify the EEDF; ii) a variation in $\dot{n}_k$, the density of given species produced per second, when they result from electron impacts: $\dot{n}_k$ then increases with the electron density (see Sec. 3.4). Thus, when varying frequency under constant power density (a condition which is experimentally more easily achieved than working at a constant electron density), one must take into account the above two side effects when interpreting the results. Our analysis will consider both constant electron density and constant power density operations.

3.2. Changes in the EEDF as a function of the field frequency

As we have indicated in Sec. 2.2.5, the EEDF is nearly stationary when $\omega > \nu_u$, where ν_u is the characteristic frequency for electron energy relaxation (eqn. (2.31)). This condition is usually fulfilled at low to moderate gas pressures for frequencies above a few tens of MHz. The stationary EEDF F_0 then satisfies the homogeneous Boltzmann equation (see eqn. (2.32))

$$-\frac{2}{3}\frac{d}{du}\left[u^{3/2}\,v_c(u)\,u_c\,\frac{dF_0(u)}{du}\right] = S_0\,(F_0)\,, \tag{3.1}$$

where $S_0(F_0)$ is the collisional term that accounts for electron-neutral collisions (both elastic and inelastic) and electron-electron collisions. The quantity

$$v_c\,u_c = \frac{e^2}{m_e}\,\frac{v_c(u)}{v_c^2(u)+\omega^2}\,E_{rms}^2 \tag{3.2}$$

represents the mean power transferred from the field to electrons of energy u. The dependence of F_0 on ω is entirely due to the presence of this quantity since no other term in eqn. (3.1) depends on ω. We see from eqn. (3.2) that $v_c u_c$ depends on both u and ω, which means that electrons with different energies absorb different powers from the field. The power transfer for a given ω is maximum for electrons whose collision frequency $v_c(u)$ is equal to ω, since $v_c = \omega$ corresponds to the maximum of the ratio $v_c / (v_c^2 + \omega^2)$. When ω is varied, the maximum power transfer will therefore occur for electrons of different energies, which results in changes in the shape of $F_0(u)$. These changes obviously depend on the exact way that v_c varies with u, which is a property of the gas under consideration.

When one considers a collisional term that includes all types of collisions, it can be shown that eqn. (3.1) determines $F_0(u)$ as a function of the set of independent parameters E_{rms}/N, ω/N, n/N and n_k/N. The degree of ionization, n/N, appears as a parameter through the electron-electron collision contribution to $S_0(F_0)$. The relative concentration of species in the k-th excited state, n_k/N, comes from that part of $S_0(F_0)$ describing electron collisions with excited species. If and when the above two types of collisions can be neglected, that is, provided only electron-ground state atom (molecule) collisions are of importance, $F_0(u)$ depends only on the two parameters E_{rms}/N and ω/N. This is the simplest situation one can consider and we shall pay special attention to it in the remaining of Sec. 3.2 and in Sec. 3.3. The effects of electron-electron collisions and of electron collisions with excited atoms (molecules) will be discussed in Secs. 3.4 and 3.5 respectively.

3.2.1. Basic discharge model. Consider, thus, the case where only the collisions of electrons with ground state atoms (molecules) are taken into account. In this case, $F_0(u)$ is only a function of E_{rms}/N and ω/N. As $\omega \to 0$, the shape of $F_0(u)$ evolves toward that corresponding to a DC electric field of magnitude E_{rms}. Though mathematically such a frequency evolution may be followed in a continuous way, the solutions become physically meaningless in the range $0 < \omega < v_u$ where eqn. (3.1) does not apply. The case $\omega = 0$ is however correctly described by eqn. (3.1). On the other hand, for sufficiently large ω, one reaches a situation in which $\omega \gg v_c(u)$ for all electrons. In this case, one can neglect $v_c^2(u)$ as compared to ω^2 in eqn. (3.2) and it can be shown that $F_0(u)$ then becomes a unique function of the ratio E_{rms}/ω. This limit is usually designated as the *microwave case* since it is reached, in practice, at frequencies of a few hundred MHz, at low to moderate gas pressures.

Equation (3.1) provides $F_0(u)$ as a function of E_{rms}/N and ω/N and one thus needs to determine E_{rms} as a function of N. This can be done by introducing the solution of eqn. (3.1) in a set of hydrodynamic-type continuity and momentum transport equations for electrons and ions. When the field is uniform, resulting in a spatially constant ionization rate $<v_i>$ (where $< >$ denotes averaging on the EEDF), and when the charged particles are predominantly lost on the wall (ambipolar diffusion or ion free fall), the solution to these hydrodynamic-type equations can be expressed in the form

$$<v_i> = \frac{D_{se}}{\Lambda^2} , \tag{3.3}$$

as discussed in Sec. 2.2.9. Recall that D_{se} represents an effective diffusion coefficient and Λ is the characteristic diffusion length for the discharge vessel (see Secs. 2.2.8 and 2.2.9). For further consideration, we add that for a given geometry, the ratio D_{se}/D_a, where $D_a \simeq \mu_i u_k$ is the ambipolar diffusion coefficient (see eqn. (2.36)), can be expressed [2] as a unique function of Λ/λ_{in}, where

$$\lambda_{in} = \mu_i \left(\frac{3 m_i u_k}{e}\right)^{1/2} = \frac{1}{v_{in}} \left(\frac{3 e u_k}{m_i}\right)^{1/2} \tag{3.4}$$

is an *effective ion mean free path*.[3] Herein, m_i is the ion mass, $\mu_i = e/m_i v_{in}$ is the ion DC mobility, and v_{in} is the ion-neutral collision frequency for momentum transfer. The quantity $u_k \equiv D_e/\mu_e$ is the electron characteristic energy introduced in Sec. 2.2.9.[†]

For a given gas, eqn. (3.3) is a relationship between the variables E_{rms}/N, ω/N and $N\Lambda$. To see this, we rewrite eqn. (3.3) in the form

$$C_i = \left(\frac{D_{se}}{D_a}\right)\frac{D_a N}{(N\Lambda)^2} \simeq \left(\frac{D_{se}}{D_a}\right)\frac{\mu_i N}{(N\Lambda)^2} u_k, \tag{3.5}$$

where $C_i = <v_i>/N$ is the electron *rate coefficient* for ionization; this rate coefficient is, by definition, given by $C_i = <Q_i(w)w>$ where $Q_i(w)$ is the ionization cross-section by electron impact at a given velocity w. The ratio D_{se}/D_a on the right-hand side of

---

[†] Note that throughout this book u_k is expressed in volt whereas u, the energy of electrons, is in true energy units unless otherwise indicated.

eqn. (3.5), as mentioned, is a unique function of the ratio Λ/λ_{in}. Since $\lambda_{in} \propto v_{in}^{-1}u_k^{1/2} \propto N^{-1}\,u_k^{1/2}$, we see that D_{se}/D_a is a function of the combination of the parameters $N\Lambda/u_k^{1/2}$. As for the product $\mu_i N$ on the right-hand side of eqn. (3.5), it is independent of N since $\mu_i \propto N^{-1}$ and we can conclude that the right-hand side of eqn. (3.5) is a function of u_k and $N\Lambda$. Finally, the electron parameters C_i and u_k being functions of the EEDF alone, they in turn depend on the combined parameters E_{rms}/N and ω/N as previously discussed. The full set of parameters involved in eqn. (3.5) is therefore E_{rms}/N, ω/N and $N\Lambda$ (or equivalently, the $p_0 R$ product in cylindrical geometry, where p_0 is the reduced gas pressure and R is the plasma radius). For this reason, eqn. (3.3) constitutes an implicit function of the form $\mathcal{F}(E_{rms}/N,\ \omega/N,\ N\Lambda) = 0$.

This is a fundamental result. It means that, for a steady-state discharge in a given vessel (setting Λ) at a given pressure p_0 (setting N) sustained by a HF field of frequency ω, the electric field intensity necessarily adjusts itself so as to satisfy eqn. (3.3). *In other words, E_{rms} must be such that the EEDF provides an exact balance between the rate of ionization and the rate of charged particle diffusion loss.* For this reason, the quantity E_{rms} in the plasma is usally referred to as the maintenance field,[†] as pointed out in Sec. 2.2.9.

Figures 3.1a and 3.1b show the EEDF as calculated from the above model for an argon discharge in a long cylindrical tube, at two $p_0 R$ values. In each figure, results are presented for the DC (curve A) and microwave ($\omega >> v_c(u)$) limits (curve H). Also shown is the Maxwellian distribution whose temperature T_e is such that eqn. (3.3) is satisfied (curve M). Since in the latter case C_i is a unique function of T_e, and that $u_k = T_e$ (with T_e expressed in volt), then eqn. (3.3) directly relates T_e to $p_0 R$. As seen from the figures, the shape of the distribution is quite different is these three cases. In particular, the relative number of electrons in the body and in the *tail* (the latter corresponding to the energy range above the lowest energy threshold for excitation which, for argon, is $eV_1 = 11.55$ eV) considerably varies from one case to another, implying that the electron rate coefficients for excitation and ionization are also significantly different. Consider, for example, the rate coefficient for excitation of the Ar $^3P_2$ metastable state whose energy threshold is eV_1. We have

$$C_1 = <Q_1(w)w> \equiv \int_{V_1}^{\infty} Q_1(u)\left(\frac{2u}{m_e}\right)^{1/2}F_0(u)\,u^{1/2}du, \tag{3.6}$$

[†] The concept of maintenance field can be generalized to situations where the field is nonuniform. In such situations, the maintenance field may refer either to the field intensity in some particular point (for example, at the axis in the case of a cylindrical geometry) or to some volume averaged value of the intensity. Detailed calculations of the maintenance field taking into account the field nonuniformity have been reported for surface-wave discharges.[4]

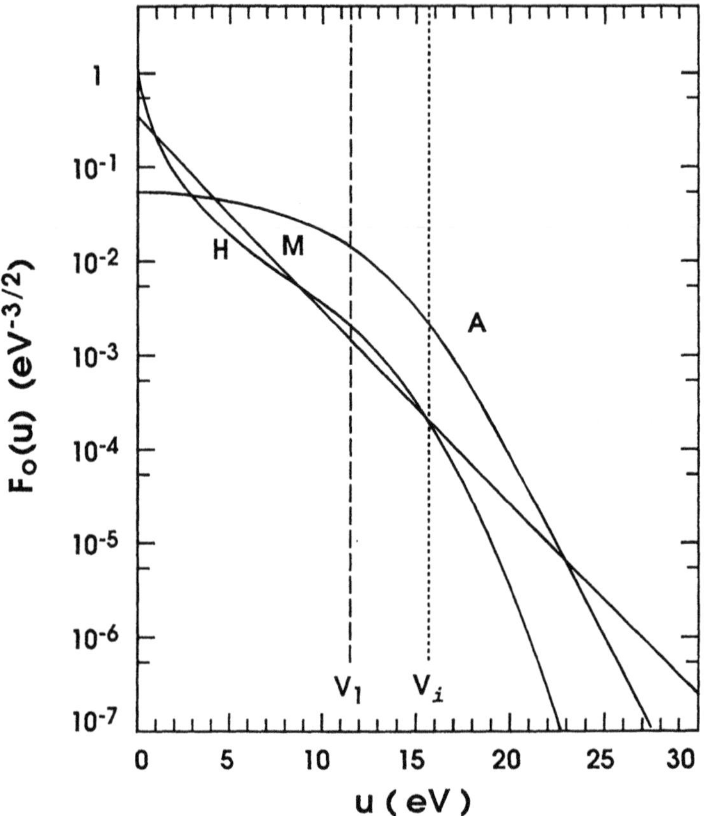

Fig. 3.1a). Electron energy distribution functions $F_0(u)$ calculated from a self-consistent model for an argon discharge in a long cylindrical tube, under ambipolar diffusion conditions, for $p_0R = 0.15$ torr-cm in three limiting cases. Curve A: no electron-electron collisions and $\omega \to 0$ (DC case); curve H: no electron-electron collisions and $\omega \to \infty$ (microwave case); curve M: dominating electron-electron collisions, any ω value (Maxwellian distribution). The vertical dashed and dotted lines correspond to the first excitation level and ionization level threshold energies respectively. The distributions are normalized such that $\int_0^\infty F_0(u)\,\sqrt{u}\,du = 1$ (from ref. 1).

where Q_1† is the excitation cross-section of Ar $^3P_2$. Since the relative number of electrons in the range $u \geq 11.55$ eV is larger for $\omega = 0$ than in the microwave case, C_1 is then larger for $\omega = 0$ than for a microwave sustaining field. The same is true for C_i, the ionization rate coefficient, as well as for the rate coefficient for excitation of any other state (whose energy threshold lies in-between eV_1 and eV_i). An important consequence of these facts is that the mean power absorbed or dissipated per electron

---

† Note that $Q_1(u < V_1) = 0$. Therefore, the energy averaging in eqn. (3.6) only concerns the tail.

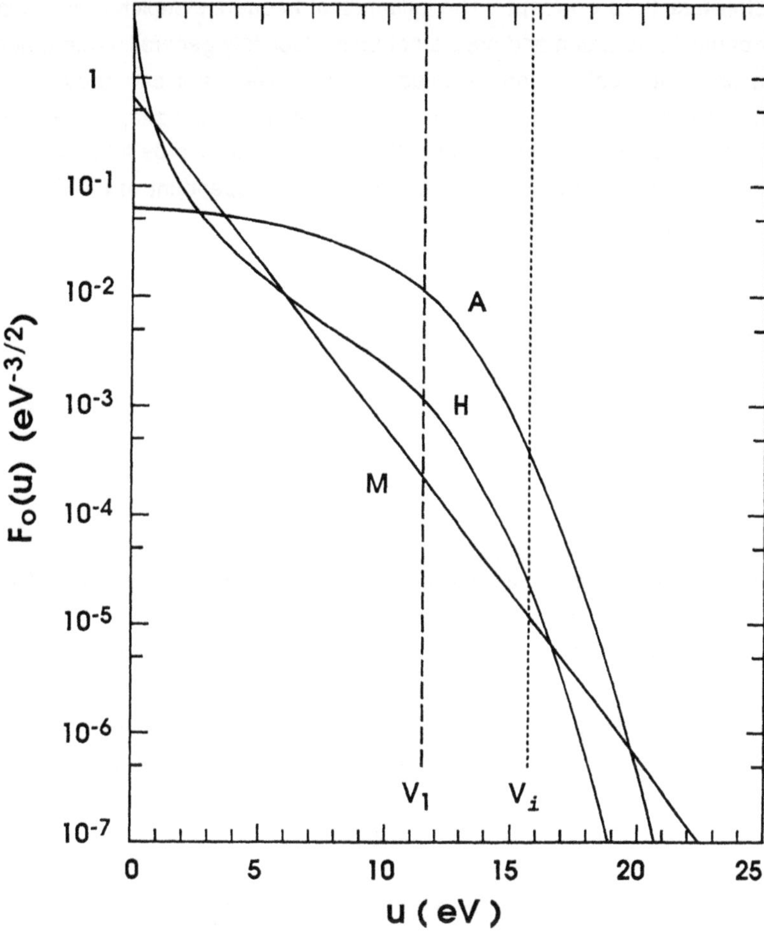

Fig. 3.1b). Same as in Fig. 3.1a) but for $p_0R = 1$ torr-cm (from ref. 1).

is smaller in the microwave case than in the DC case for fixed p_0R, as shown below in Sec. 3.2.3 (see Fig. 3.8).

One may wonder why this is so. The reason is that the changes in the body of $F_0(u)$ make that $u_k^{MW} < u_k^{DC}$ (see Fig. 3.16), which means that the rate of escape of electrons to the wall is smaller in the microwave (MW) case. Therefore, the rate of ionization necessary to sustain the discharge also becomes smaller, which requires a smaller fraction of electrons with energy $u \geq eV_i$. In the present case, this is achieved through a decrease in the whole tail of the distribution function which makes all excitation rate coefficients decrease.

3.2.2. Effective field concept. The influence of frequency depicted from Fig. 3.1 is however particular to argon and should not be immediately generalized to other gases. Only the same full calculations applied to each given gas can provide the right evolution. Nevertheless, as already pointed out, changes in the shape of the EEDF will always be induced by ω when v_c varies with energy u. This can be further illustrated by considering the ideal case where v_c is, in contrast, independent of u. One can then write eqn. (3.2) in the form

$$v_c u_c = \frac{(eE_e)^2}{m_e v_c} , \tag{3.7}$$

where

$$E_e = E_{rms} \frac{v_c}{\left(v_c^2 + \omega^2\right)^{1/2}} \tag{3.8}$$

is the effective field intensity, as defined in Sec. 2.2.7 (see eqn. (2.53)). The power transfer expressed by eqn. (3.7) has the same form as that obtained in a DC field of magnitude E_e. Therefore the EEDF remains unchanged for any value $\omega \geq 0$, provided E_e/N is kept constant. In other words, when v_c is independent of u, the EEDF only depends on E_e/N, not explicitly on ω. As a consequence, it follows from eqn. (3.3) that DC discharges and HF discharges operated under the same conditions but at any frequency for the latter must then have exactly the same properties, namely the same effective field, rate coefficients and state populations.

Thus one would like to extend, even partially, the effective field concept to situations where v_c is a function of u. For a given gas, one can define an effective field

$$E_e = E_{rms} \frac{v_{ce}}{\left(v_{ce}^2 + \omega^2\right)^{1/2}} , \tag{3.9}$$

where v_{ce} is the momentum-transfer collision frequency for electrons of a given energy in that gas. The choice of this energy is a priori arbitrary and, in practice, one usually chooses an energy value representative of the bulk electrons. The quantity v_{ce} so defined for a given gas is simply proportional to the gas density N, that is, $v_{ce} = $ constant times N. It is therefore fully equivalent to use either N or v_{ce} as a parameter. In particular, one can represent $F_0(u)$ as a function of the pair of independent parameters E_e/N and ω/v_{ce} in place of the pair E_{rms}/N and ω/N. Note that this is only a change of representation which facilitates the interpretation of the results, not an approximation to calculate $F_0(u)$ from eqn. (3.1) that would assume $v_c(u) = v_{ce}$.

Indeed, when solving eqn. (3.1) for a given gas, the exact variation law of $v_c(u)$ must always be taken into account.

In summary, the only case where $F_0(u)$ is independent of ω/v_{ce} (or ω/N) and, thus, only depends on E_e/N, is the ideal case where $v_c(u)$ = constant, as discussed above. In all real cases, a dependence on ω/v_{ce} exists to an extent that depends on the particular variation law of $v_c(u)$.

Helium is to our knowledge the gas where the dependence of $F_0(u)$ on ω/v_{ce} is the smallest and, thus, the effective field concept of eqn. (3.8) most usefully applies. This can be seen from Fig. 3.2 where calculations of the EEDF in helium[5] are shown for $E_e/N = 10^{-16}$ V cm^2, with $v_{ce}/N = 6.8 \times 10^{-8}$ cm^3 s^{-1} (this corresponds to $v_{ce}/p_0 \approx 2.4 \times 10^9$ s^{-1} torr^{-1}; compare this value with that of $v_c(u)/p_0$ in Fig. 2.3), for values of ω/v_{ce} ranging from zero (DC) to much larger than unity (microwave case).

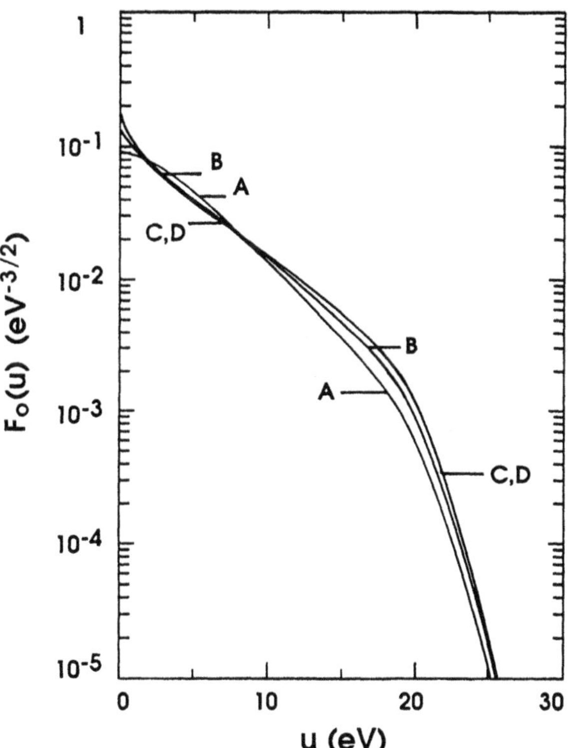

Fig. 3.2. Electron energy distribution functions, calculated from a self-consistent model as in Fig. 3.1, for an helium discharge at $E_e/N = 10^{-16}$ V cm^2 and for the following values of ω/v_{ce}: 0(A); 1.47(B); 14.7(C); 147(D) where $v_{ce}/N = 6.8 \times 10^{-8}$ cm^3 s^{-1} (from ref. 5).

The EEDF remains almost unchanged, which is quite different from the situation in argon.

Figure 3.3 shows similar calculations in oxygen[6] for $E_e/N = 10^{-15}$ V cm^2 where now $v_{ce}/N = 1.3 \times 10^{-7}$ cm^3 s^{-1}. The changes induced by ω are larger than in He but comparatively small with respect to those in Ar.

Large changes as for Ar can also be expected for the heavy rare gases Kr and Xe where the momentum transfer cross-section for electrons, $Q_c(w)$, also exhibits a pronounced Ramsauer minimum at low energies.[7]

3.2.3. Power balance per electron. As pointed out in Sec. 2.2.9, the electron mean power balance equation is simply obtained by multiplying eqn. (3.1) by u and then integrating over all energies. The result can be expressed in the form (see eqns. (2.70) and (2.71))

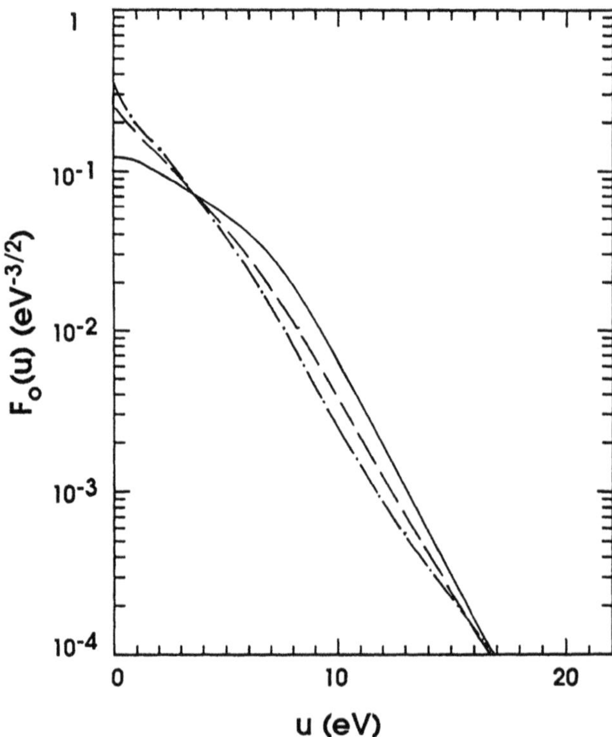

Fig. 3.3. Electron energy distribution functions, calculated from a self-consistent model as in Fig. 3.1, for an oxygen discharge at $E_e/N = 10^{-15}$ V cm^2 and for the following values of $\omega/v_{ce} = 0$ (full curve); 1 (dashed curve); >> 1 (chain curve). Here $v_{ce}/N = 1.3 \times 10^{-7}$ cm^3 s^{-1} (from ref. 6).

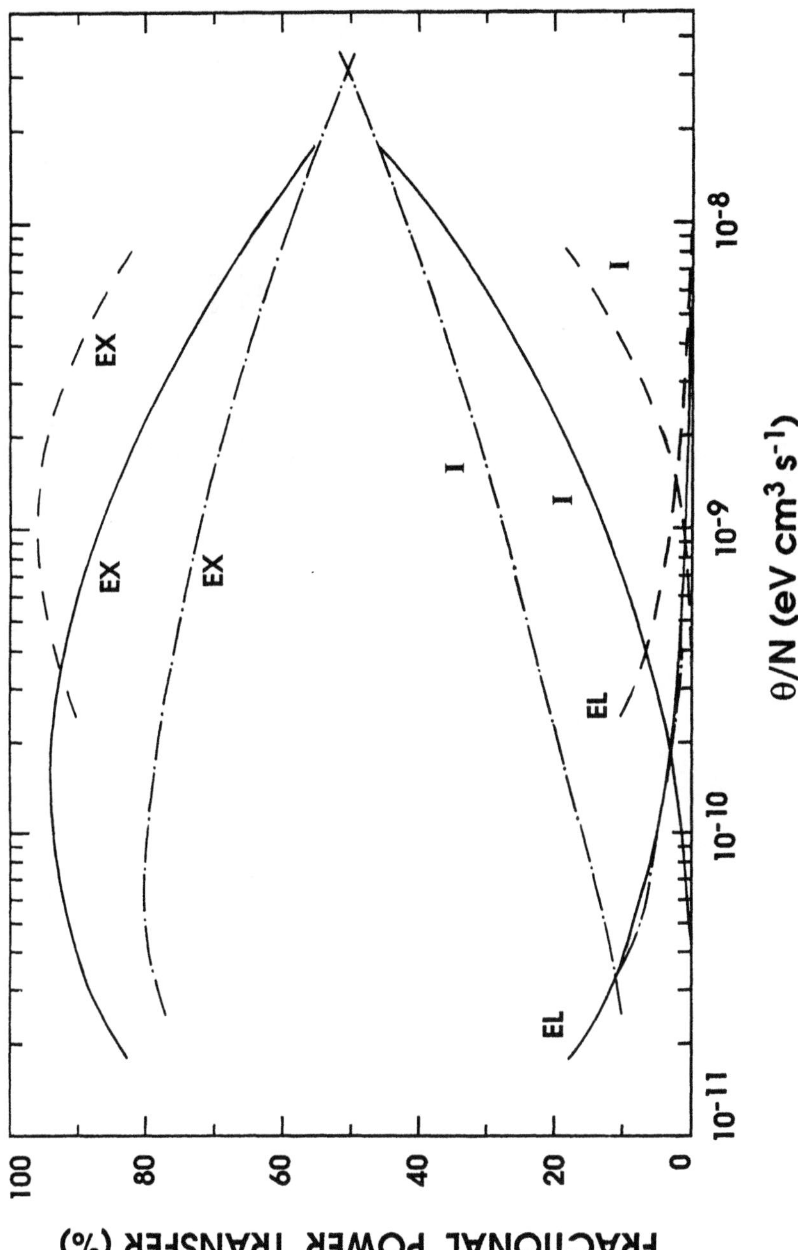

Fig. 3.4. Mean fractional power transferred by electrons to argon atoms in elastic collisions (EL), excitation (EX) and ionization (I) as a function of θ/N, for: $\omega = 0$ (dashed curves); $\omega \gg \nu_{ce}$ (solid curves); Maxwellian EEDF (chain curves) (from ref. 8).

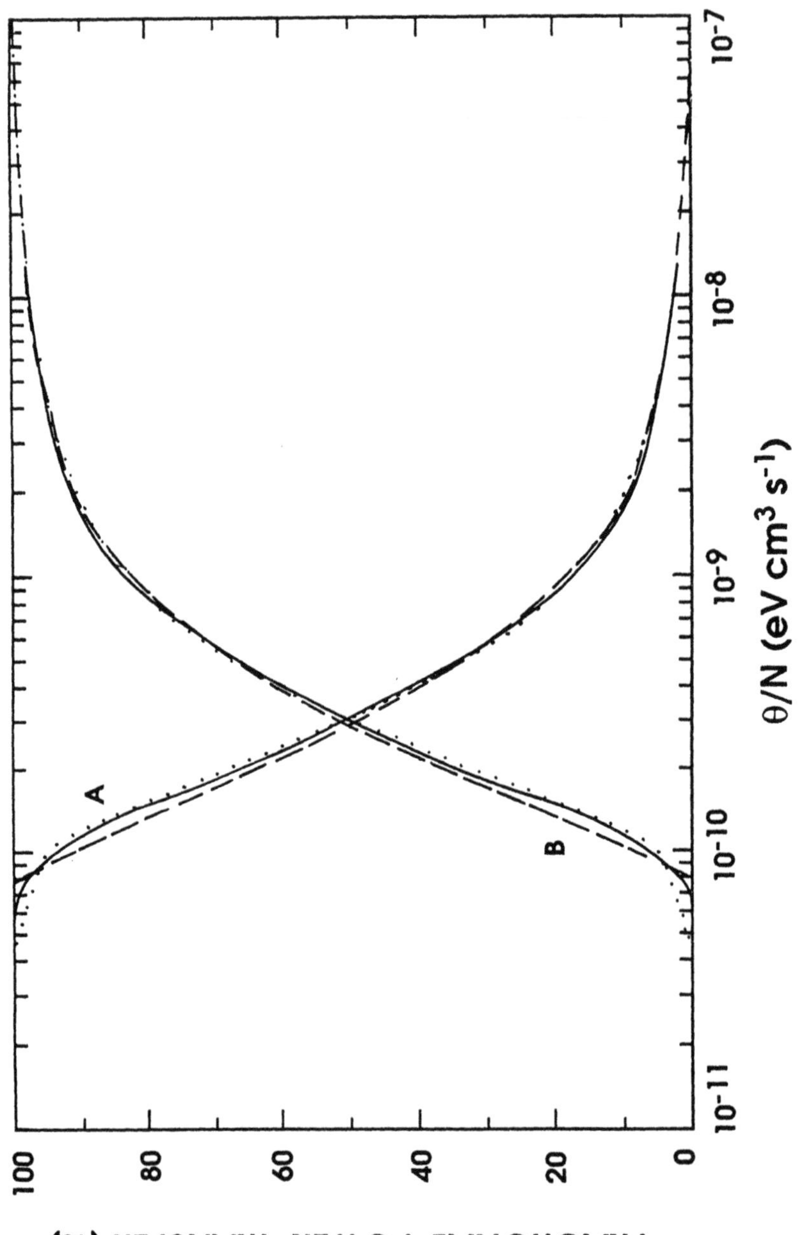

Fig. 3.5. Mean fractional power transferred by electrons to helium atoms in elastic (A) and inelastic (B) collisions as a function of θ/N for the following values of $\omega/\nu_{ce} = 0$ (full curves); 1.47 (dashed curves); 147 (dotted curves) (from ref. 5).

$$\theta_A \equiv n^{-1}\sigma_r E_{rms}^2 = \theta_L, \tag{3.10}$$

where θ_A is the mean absorbed power per electron,

$$\sigma_r = -\frac{2}{3}\frac{e^2}{m_e}\, n \int_0^\infty \frac{v_c(u)}{v_c^2(u) + \omega^2}\, u^{3/2}\, \frac{dF_0(u)}{du}\, du \tag{3.11}$$

is the real part of the complex electric conductivity,[†] and

$$\theta_L \equiv \frac{2m_e}{M} < v_c(u)u > + \sum_n < v_n(u) > eV_n + < v_i(u) > eV_i \tag{3.12}$$

is the mean power lost per electron in collisions of all kinds.[††]

We note that eqn. (3.10) is a local equation since it is obtained from the homogeneous Boltzmann equation, which results when the spatially dependent term $\nabla_r F_0$ is neglected. This means that, at a given point, $F_0(u)$ is assumed to be determined only by the local value of E_{rms} and the local electron collision frequencies (local approximation). Since, for example, E_{rms} generally varies with position, the quantity $\theta = \theta_A = \theta_L$ will also vary with position. Besides E_{rms}, θ depends on other quantities that can also vary with position as can be seen by inspection of eqn. (3.12). It shows that the ratio θ/N is a functional of $F_0(u)$ (through the energy averaging) and is, eventually, also a function of the relative populations in excited states, n_k/N, if and when electron collisions with such states need to be considered (note that $v_n \propto n_k$ when v_n refers to a collision with an atom in the k-state). In reality, θ/N is a function of the same parameters as $F_0(u)$, namely E_{rms}/N, ω/N (or E_e/N, ω/v_{ce}), n/N and n_k/N, in general. In the simple case of interest in this section, it is assumed that the field is uniform and that electron-electron collisions and collisions with excited states can be neglected. In this case, θ/N is a unique function of E_{rms}/N and ω/N and is therefore spatially constant.

Some useful information related to eqn. (3.10) is, for example, the way that the absorbed power θ_A is dissipated via the various collisional channels. It is customary to present this information as in Figs. 3.4 to 3.7 where the percentage of electron energy

---

[†] Note that eqn. (3.11) for σ_r is the same as eqn. (2.39), apart from the fact that we used here $F_0(u)$ instead of $f_0(w)$. See Sec. 2.2.5.

[††] The sum over the integer n on the right-hand side of eqn. (3.12) accounts for all excitations from the ground state to the k-states ($n = k$) and, eventually, also for electron induced transitions k - k', in which case $V_n = V_{k'} - V_k$.

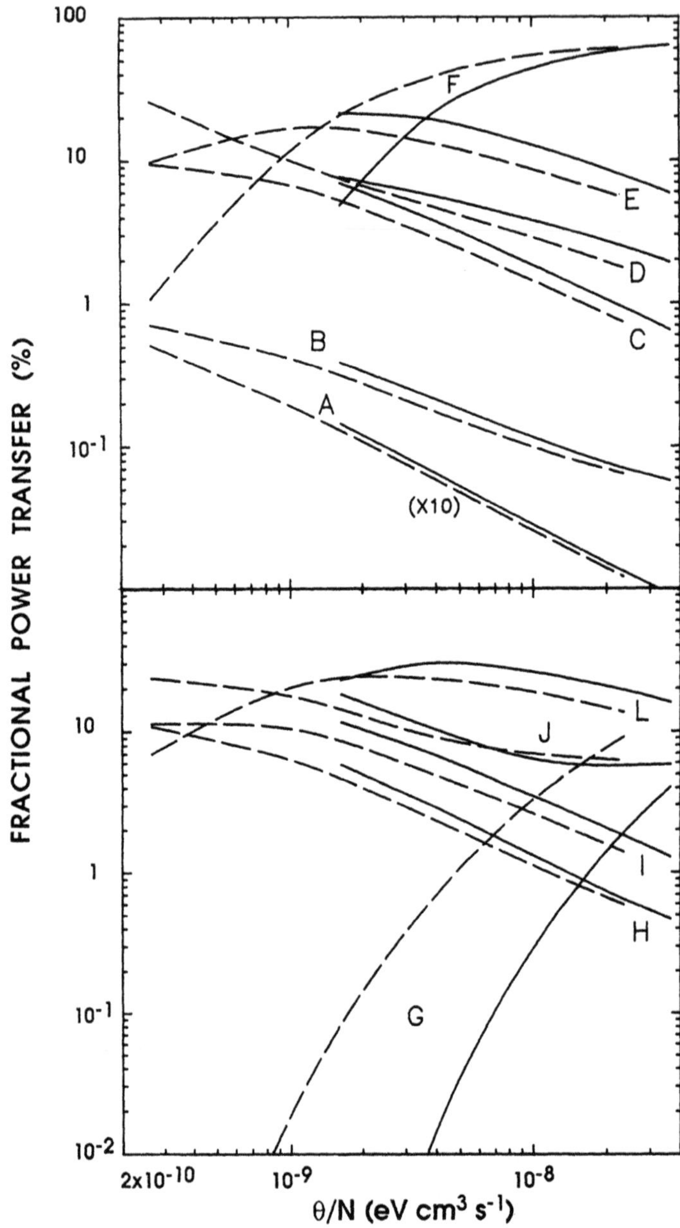

Fig. 3.6. Mean fractional power transfer by electrons in DC (full curves) and microwave ($\omega \gg \nu_{ce}$; dashed curves) oxygen discharges as a function of θ/N. The curve labels correspond to the following processes: A. rotational excitation; B. elastic collisions; C. excitation of $O_2(b^1\Sigma)$; D. vibrational excitation; E. excitation of states with energy thresholds of 4.45 eV, 9.97 eV and 14.7 eV; F. 8.4 eV dissociation channel; G. ionization; H. reexcitation of $O_2(a^1\Delta)$; J. excitation of oxygen atoms from $O(^3P)$; L. 6.0 eV dissociation channel (from ref. 9).

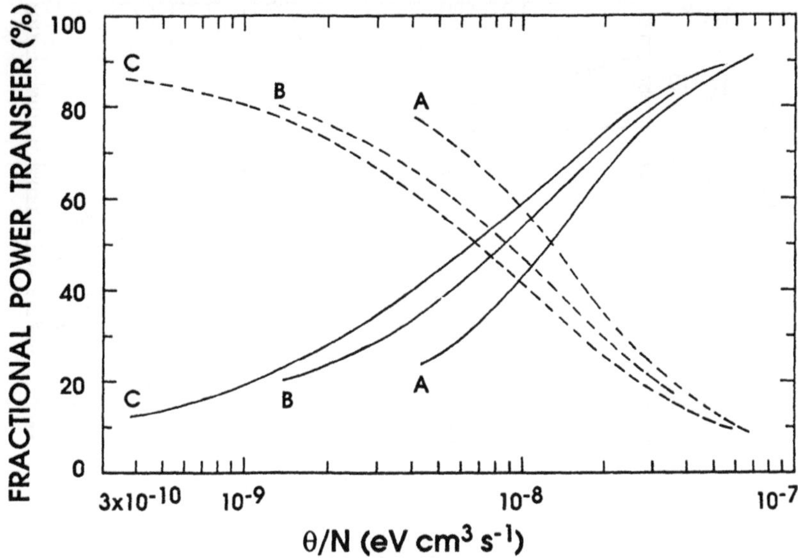

Fig. 3.7. Mean fractional power transferred by electrons to N_2 molecules in electronic excitation plus ionization (full curves) and vibrational excitation (dashed curves) as a function of θ/N, for the following values of $\omega/\nu_{ce} = 0(A)$; $0.83(B)$; $\gg 1(C)$. A vibrational temperature of $4 \times 10^3 K$ was taken in the calculations (from ref. 10).

losses via various types of collisions is plotted against θ/N, using ω/ν_{ce} as a parameter, for Ar,[8] He,[5] O_2,[9] and N_2.[10] With the exception of He, for which the parameter ω/ν_{ce} has negligibly small effects, we note as a general trend that the microwave ($\omega \gg \nu_{ce}$) percentage energy loss curves are always shifted towards lower θ/N values as compared to those for the DC case ($\omega = 0$). This is a consequence of the changes in the shape of $F_0(u)$ as ω varies.

It is however more immediately useful to analyze the effects of ω on the power transfer directly as a function of the p_0R product. Equations (3.1) and (3.10) are not sufficient for this purpose and they must be complemented with a discharge model describing the balance of charged particles, for example, as in Sec. 3.2.1. Then, using the EEDF obtained for a given p_0R and a given ω (see Figs. 3.1a and 3.1b), one can calculate from eqn. (3.10) the dependence of θ/p on p_0R (Fig. 3.8) as well as the percentage of electron energy losses according to specific collisional channels as a function of p_0R (Fig. 3.9).

We note from Fig. 3.8 that for a fixed gas pressure, the value of θ decreases as the frequency is increased from zero (case A) to the microwave range (case H). Physically, this means that less power is needed to create an electron-ion pair in a microwave

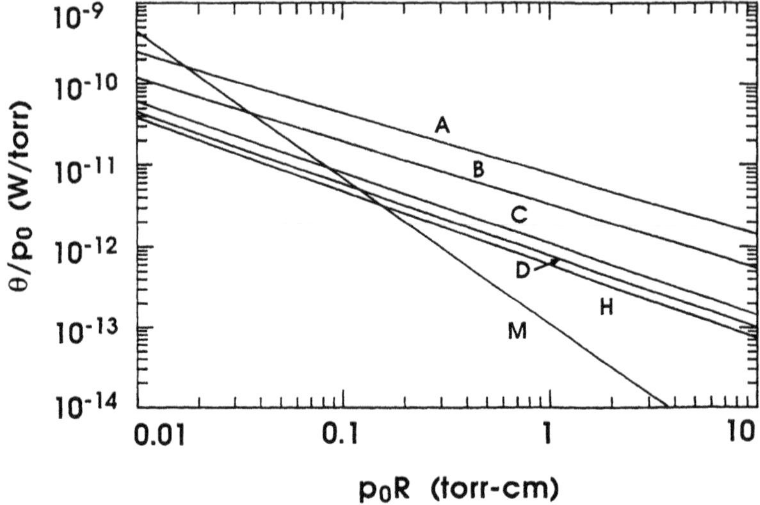

Fig. 3.8. Calculated plot of θ/p_0 as a function of p_0R for cylindrical argon discharges, at the following values of ω/ν_{ce}: 0(A); 0.15(B); 0.80(C); 1.5(D); $\gg$ 1(H). Curve M is for a Maxwellian EEDF (from ref. 8).

discharge than in a DC one. Therefore, for a fixed power density absorbed in the plasma, a higher electron density will be achieved in the microwave case.

Figure 3.9 shows the percentage of electron energy losses via a given type of collision (elastic, excitation or ionization). We note that excitation constitutes by far the dominant energy loss mechanism. For a given p_0R, we further observe that these percentages are very nearly the same for DC (case A) and microwave conditions (case H). This results from the frequency shifting of the percentage energy loss vs. θ/N curves observed in Fig. 3.4, which compensates for the decrease of θ with increasing frequency. An important conclusion is then that, for a fixed power density, the total power *per unit volume* transferred into excited states is very nearly the same under DC and microwave conditions, while it is less for a Maxwellian EEDF. However, when considering the power transferred *per electron*, it is the largest in the DC case since θ is then the largest.

The important difference noted between the results obtained assuming a Maxwellian EEDF (case M) and those for the DC and microwave cases will tend to decrease when we will take into account electron-electron collisions (Sec. 3.4). Then curves A and H tend toward curve M as electron-electron collisions become important, as is the case when increasing the degree of ionization.

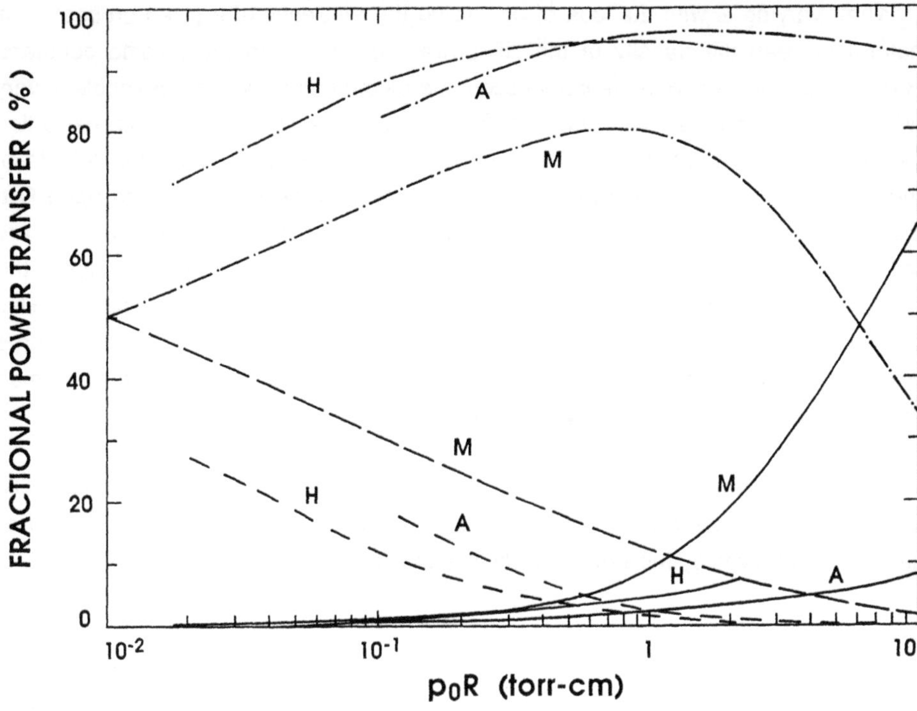

Fig. 3.9. Calculated fraction of the power θ per electron transferred to argon atoms as a result of elastic collisions (full curves), excitation (chain curves) and ionization (dashed curves), as a function of the product p_0R. Curves A, H and M refer to three limiting cases: DC, microwave and a Maxwellian EEDF respectively (from ref. 1).

3.3. Rate of creation of active species as a function of the plasma stimulating frequency in argon-dominated gas mixtures

Argon is the major constituent of many gas mixtures currently used for plasma chemistry and plasma processing applications. One example is the plasma polymer deposition using a monomer molecule in an argon-based HF discharge. Experimental results in that field have been obtained as a function of ω.[1] To interpret these and similar data, it is instructive to extend the study of the previous section for a pure argon discharge to calculate the excitation rate coefficients of atoms or molecules with lower excitation energy thresholds than those of argon, in an argon controlled discharge. Such an approach, using the EEDF calculations from the model for an argon discharge, is strictly speaking only valid if the molecular additives can be assumed to be at the trace level.

When varying ω with the idea of optimizing the efficiency of a given process, it is natural to keep the density of the HF power absorbed into the plasma constant. However, as mentioned in the introduction, the electron density can nevertheless vary with ω, which interferes with determining the true influence of the plasma excitation frequency on the process. In the following, we shall examine both the effects of frequency upon $C_k = \langle Q_k(w)w \rangle$ (constant electron density operation) and upon $C_k n$ (constant power density operation). Both these quantities in the above situations are related to the number of species produced per second, $\dot{n}_k = (C_k N)n$. The higher the density of species produced, the more efficient is the corresponding process. This study is developed as a function of the process threshold energy eV_k, for the three limiting cases A, H and M previously described. These calculations were performed using the EEDF for pure argon given in Fig. 3.1.

In order to estimate to what extent our results on C_k and $C_k n$ depend on the specific shape of the process cross-section, we considered two very distinct types of cross-sections, both given by the same analytical expression

$$Q_k(u) = Q_0 \, (V_k/u)^{c_1} \, (1 - V_k/u)^{c_2} \tag{3.13}$$

for different values c_1 and c_2, Q_0 being a constant. Note that the cross-section maximum is attained for energy $u_{max} = [(c_1 + c_2)/c_1]V_k$. We consider $c_1 = c_2 = 1$ to represent optically forbidden transitions (which usually have their maximum at about twice the threshold energy) and $c_1 = 1$, $c_2 = 4$ to represent allowed transitions (whose cross-section maximum is generally located at 5 to 10 times the threshold energy).[11]

3.3.1. Constant electron density operation. Figure 3.10 presents the excitation coefficient $C_k(V_k)$ as calculated for an argon controlled discharge with the above cross-section for optically allowed transitions. Figure 3.10a shows, for the three limiting cases A, H and M, the ratios of the corresponding coefficients C_k at $p_0 R = 0.15$ torr-cm. We note the following: i) C_k in the DC case is always larger than in the microwave limit, whatever the threshold energy of the process; ii) when the discharge evolves from the microwave limit where electron-electron collisions are ignored to the Maxwellian case where electron-electron collisions dominate (and frequency effect is lost), the Maxwellian rate coefficient exceeds the microwave one for energy close to or larger than eV_1; iii) when considering the three limiting cases in Fig. 3.10 together, we note that it is the DC limit that leads to the largest coefficient for all energies from zero to approximately 17 eV (the argon ionization threshold energy is 15.76 eV), above which the largest of the three coefficients is that for the Maxwellian distribution.

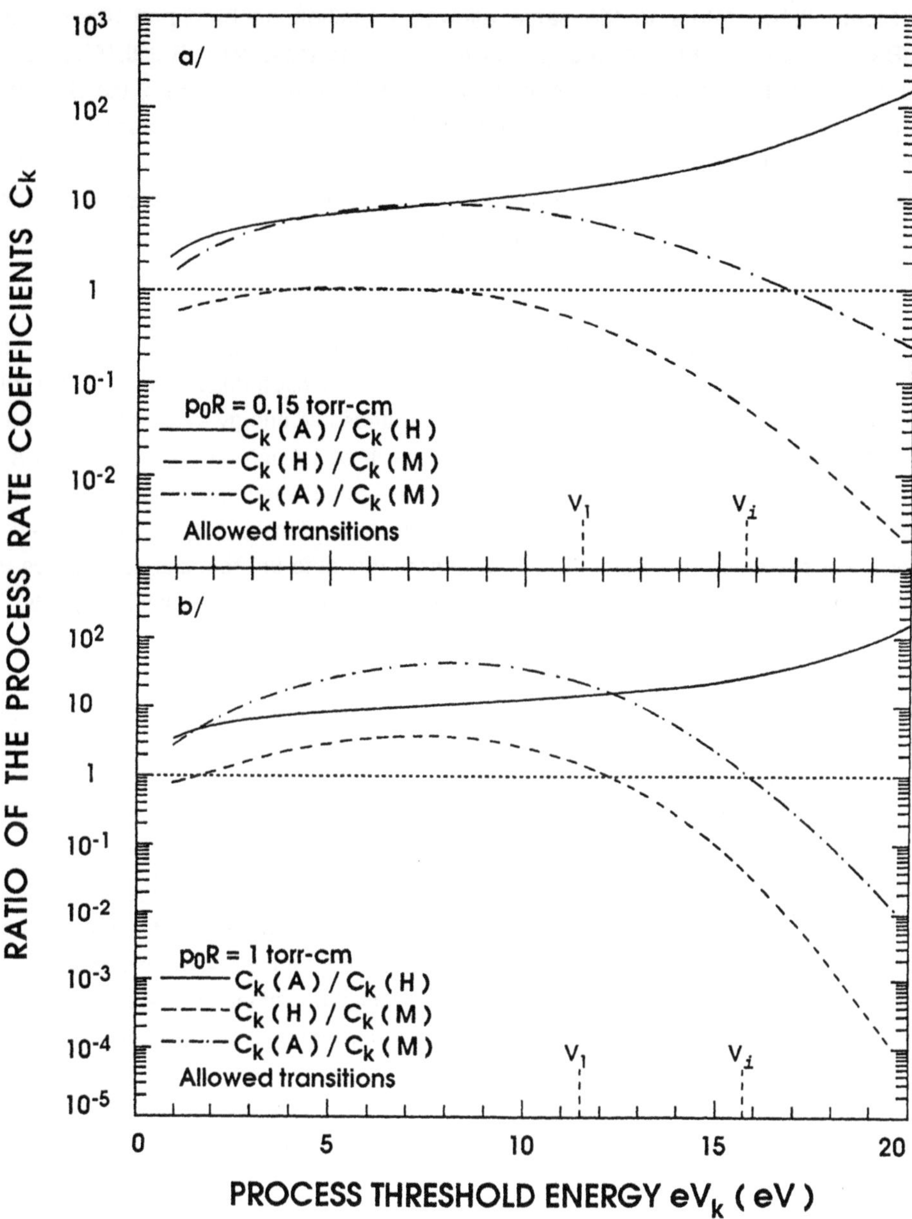

Fig. 3.10. Ratio of the process rate coefficients C_k calculated with the EEDF corresponding to three limiting situations, namely DC (curve A), microwave (curve H) and Maxwellian EEDF (curve M) cases, as a function of the process threshold energy eV_k. The assumed process cross-section simulates an optically allowed transition (a) $p_0R = 0.15$ torr-cm and (b) $p_0R = 1$ torr-cm (from ref. 1).

Figure 3.10b refers to the same allowed transition as in Fig. 3.10a but at $p_0R = 1$ torr-cm. The results are qualitatively similar to those for $p_0R = 0.15$ torr-cm, but the larger the p_0R-product, the more important is the electron rate of the DC limit compared to the Maxwellian one, for energies below the ionization threshold. The results for the forbidden transition (not shown) do not significantly differ from those for the allowed transition at the same p_0R value, suggesting that the exact shape of the process cross-section, in contrast with the process energy threshold, does not really matter.[1]

3.3.2. Constant absorbed power density operation. Figure 3.11 compares the density of species produced per second by direct electron impact on the atom (or molecule) in the ground state, as a function of the threshold energy eV_k, under constant absorbed power density conditions, for the same three limiting cases A, H and M as before. These results were obtained with the above assumed cross-section for optically allowed transitions. The value of the electron density n appearing in these calculations is adjusted so that the power density (equal to θn) remains constant as θ varies. The values of θ used in these three limiting cases are those shown in Fig. 3.8.

Figures 3.11a and 3.11b refer to $p_0R = 0.15$ and 1 torr-cm respectively. Determining the limiting case that is the most energy efficient in Fig. 3.11a requires a division of the energy range into three distinct intervals (recall that $eV_1 = 11.55$ eV and $eV_i = 15.76$ eV in argon): i) from zero to about 3 eV: the best situation is that for the Maxwellian EEDF; ii) from 3 eV to about 9 eV: the microwave case leads; iii) above 9 eV: the Maxwellian distribution is again the most efficient one. Figure 3.11b shows that at higher p_0R values, except for a narrow range of threshold energies around 7.5 eV, the Maxwellian is the most efficient case everywhere. The results for a forbidden transition (not shown) are qualitatively the same as those for an allowed transition at the same p_0R value, proving once more that the exact shape of the cross-section for the process does not play a significant role, at least in an argon-based discharge.

In conclusion, the energy efficiency of a given process depends not only on the applied field frequency but also on the process threshold energy compared to the ionization energy of the gas controlling the discharge.

3.3.3. Guidelines for frequency optimization of plasma processing. From the previous analysis, we can set out some indications about the frequency optimization of plasma processing. For this purpose, we need to distinguish two operating conditions: constant electron density and constant power density, and we are going to consider three possible scenarios for the evolution of the discharge characteristics as frequency f is raised from DC to microwaves. These scenarios are

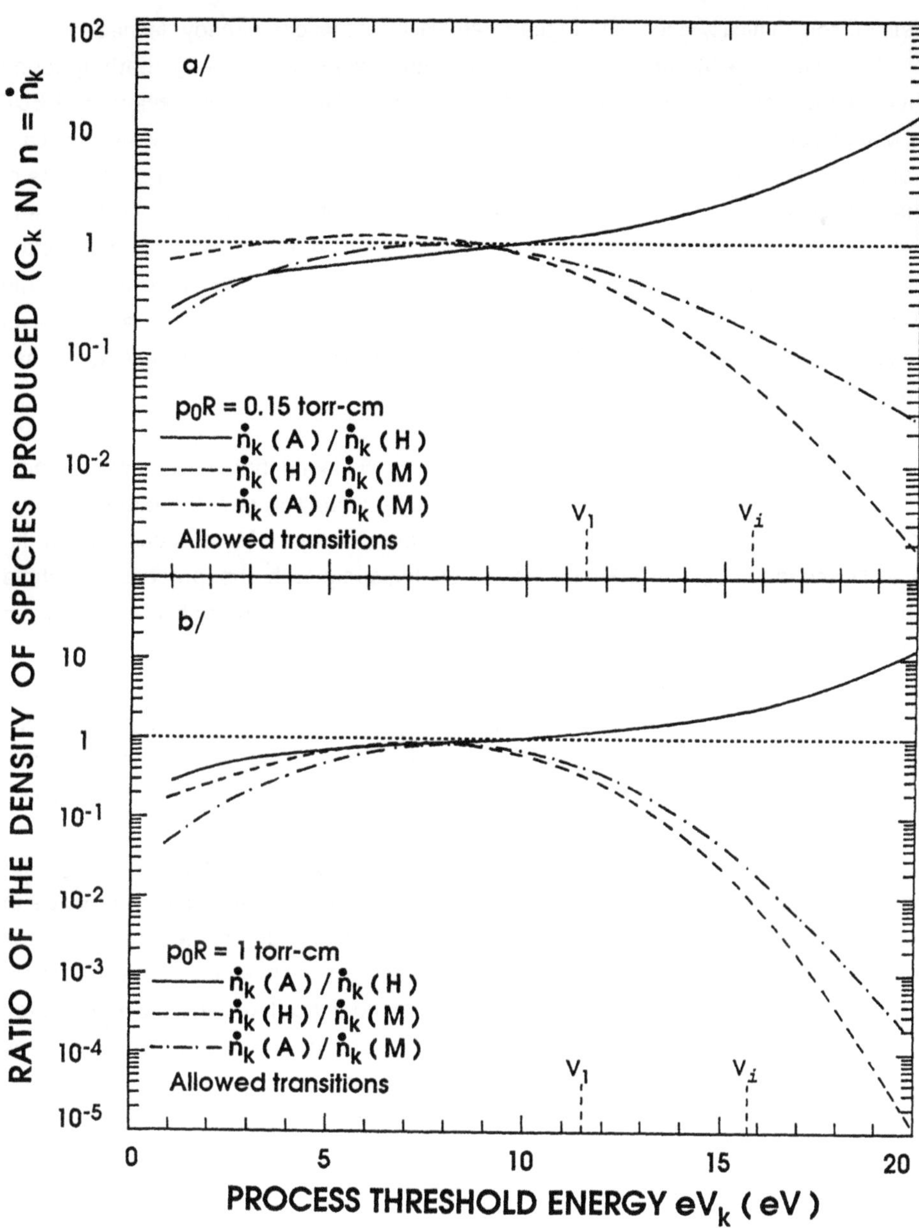

Fig. 3.11. Ratio of the densities of species produced per second, $(C_kN)n$, calculated with the EEDF corresponding to three limiting situations, namely DC (curve A), microwaves (curve H) and Maxwellian EEDF (curve M) cases, as a function of the process threshold energy eV_k, under constant power density conditions. The process is assumed to result from electron impact on the atom (molecule) in the ground state and the cross-section used simulates an optically allowed transition: (a) $p_0R = 0.15$ torr-cm and (b) $p_0R = 1$ torr-cm (from ref. 1).

defined schematically according to the three limiting situations already defined: case A or DC case ($\omega \to 0$) and case H or microwave case ($\nu_{ce}/\omega \to 0$), both ignoring electron-electron collisions, and case M, whatever the field frequency, when the EEDF is Maxwellian. The possible scenarios for $0.15 \le p_0 R \le 1$ torr-cm, as f varies typically from 13.56 to 2450 MHz, are: i) A $\to$ H: this situation is governed by the decrease of the parameter ν_{ce}/ω; ii) A $\to$ M: it means that the ionization degree increases with frequency to such an extent that it reaches a level above which the electron-electron collisions are dominating, yielding then a Maxwellian EEDF. In such a case, the parameter ν_{ce}/ω is meaningless; iii) A $\to$ H $\to$ M: it corresponds to a degree of ionization becoming large enough to yield a Maxwellian EEDF only once ν_{ce}/ω has decreased to be close to or smaller than unity.

Considering operation at *constant electron density*, we found that, for a process energy threshold below that for ionization of the discharge controlling gas, the rate coefficient for plasma processing is always the largest in the DC case. More specifically in the A $\to$ H scenario, $C_k(A)/C_k(H)$ increases continuously with the process threshold energy whereas, in the A $\to$ M scenario, $C_k(A)/C_k(M)$ goes through a maximum at about half the ionization energy of the controlling gas.

Considering operation at *constant power density*, we have markedly different results. The maximum of efficiency for the production of active species now strongly depends on the scenario envisaged: i) A $\to$ H scenario: $\dot{n}_k$, the density of species produced per second, is lower in the DC case for processes with a threshold energy below about half the ionization energy eV_i of the discharge controlling gas, whereas it is larger for processes with threshold energies larger than eV_i; ii) A $\to$ M scenario: the value of $\dot{n}_k$ is almost always the largest when the EEDF is Maxwellian, whatever the process energy threshold; iii) A $\to$ H $\to$ M scenario, where the microwave case is reached, say, between 13 and 2450 MHz: the results depend critically on the process energy threshold: for low threshold energies ($\le eV_i/2$), one gets a maximum of species production per watt at large enough frequencies; for large threshold energies, there is a minimum of efficiency at the frequency for which the microwave case is reached.

3.4. Effects of electron-electron collisions

The analysis of the previous sections revealed that the shape of the EEDF has a strong influence on all discharge properties. Indeed, the results obtained in argon are markedly different when one considers either the EEDF of case A ($\omega = 0$, no electron-electron (e-e) collisions), or that of case H ($\omega \gg \nu_{ce}$, no e-e collisions), or a Maxwellian distribution. An accurate calculation of the EEDF from eqn. (3.1) is therefore necessary to get reliable predictions.

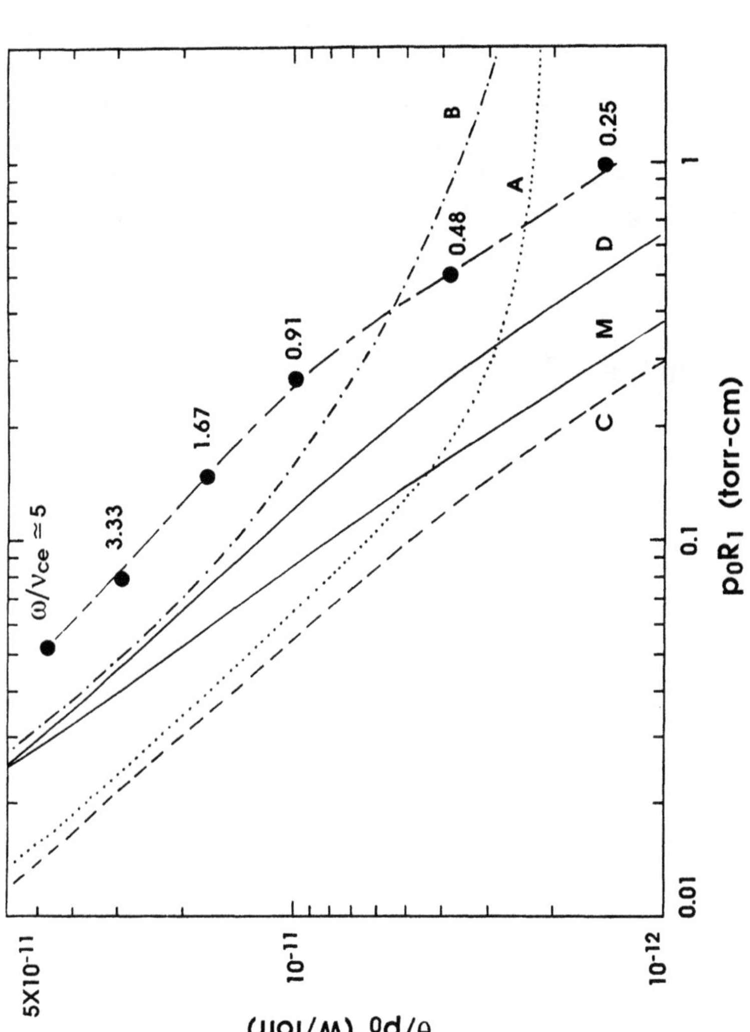

Fig. 3.12. Characteristics of θ/p_0 as a function of p_0R_1, where p_0 is the argon gas pressure reduced to 300 K, for a surface-wave discharge operated at 200 MHz in a cylindrical tube with $R_1 = 1.3$ cm where $n \simeq 1 \times 10^{11}$ cm⁻³. Data points from surface-wave experiments conducted in Montréal (ref. 1). Curves are calculations for the following cases: only direct ionisation without e-e collisions (A) and with such collisions (B); including multi-step ionization but without e-e collisions (C) and with such collisions (D); Maxwellian EEDF and direct ionization only (M) (from ref. 14).

A crucial test of the model is, for example, the comparison between calculated and measured values of θ/N vs. NR for a given gas discharge. Measurements of θ in surface-wave-sustained discharges have been performed over the last ten years or so for various gases, over a very wide range of frequency, pressure and tube diameters. Figures 3.12 to 3.14, for example, show a comparison of measured and calculated values of θ/N as a function of NR in argon, for the following values of frequency and tube inner radius R_1: 200 MHz, 1.3 cm (Fig. 3.12); 900 MHz, 1.3 cm (Fig. 3.13); 2.45 GHz, 0.15 cm (Fig. 3.14). Let us concentrate for the moment on the curves labelled A shown on these figures, which are obtained from the model described in Sec. 3.2.1, with $F_0(u)$ determined from eqn. (3.1), neglecting e-e collisions. Note that these calculations are at fixed ω, which means that ω/ν_{ce} decreases along the curve as the pressure increases. It is clear from these figures that curve A strongly differs both in

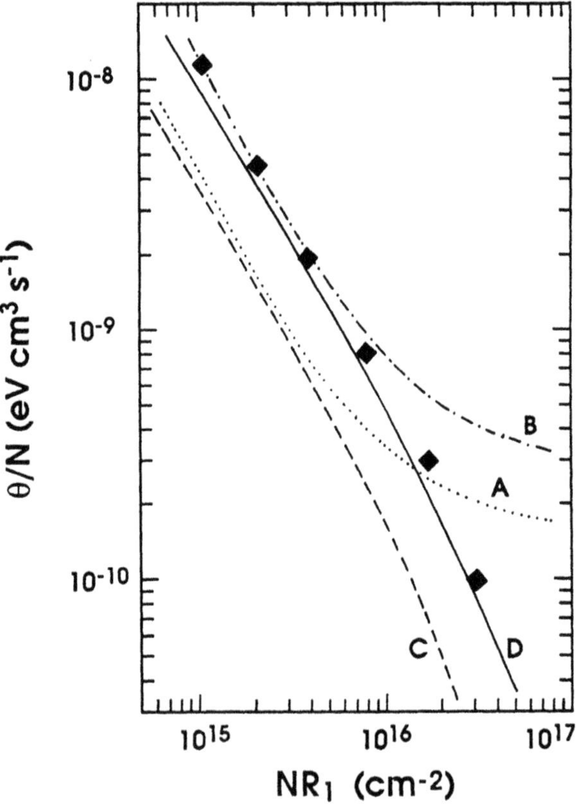

Fig. 3.13. Characteristics of θ/N as a function of NR_1, for a microwave discharge operated at 900 MHz in a cylindrical tube with $R_1 = 1.3$ cm where $n \simeq 2 \times 10^{11}$ cm^{-3}. Curves are calculations as explained in caption to Fig. 3.12. Data points are from experiments on surface-wave discharges (Montréal and Orsay groups) (from ref.14).

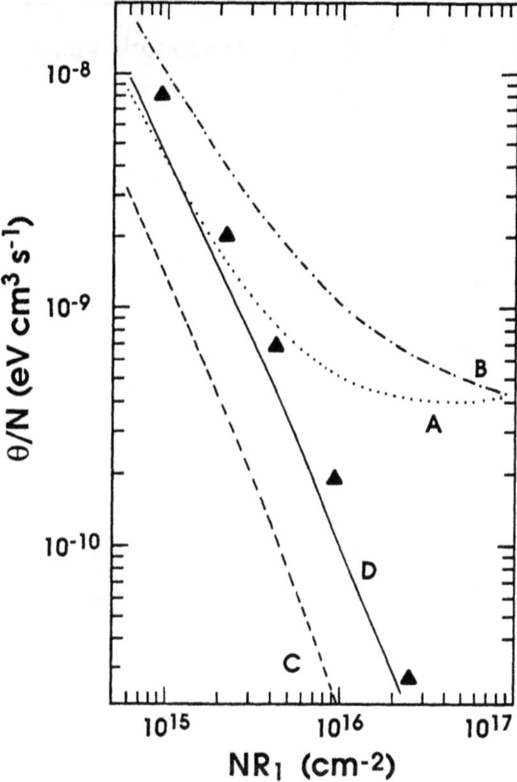

Fig. 3.14. As in caption to Fig. 3.13 but at 2.45 GHz, $R_1 = 0.15$ cm and $n \simeq 2 \times 10^{13}$ cm^{-3} (from ref. 14).

shape and in magnitude from the behavior of the experimental data, indicating that the initial model used is too crude.

One of the sources of so large an inaccuracy is precisely the neglect of e-e collisions as we now demonstrate. At the beginning of Sec. 3.2, we stressed that $F_0(u)$ becomes a function of the ionization degree n/N when the e-e collision term is retained in eqn. (3.1). Figure 3.15 shows the effect of these collisions on the EEDF in argon for $E_e/N = 3 \times 10^{-16}$ V cm^2, $n/N = 10^{-4}$, for both the DC ($\omega = 0$) and microwave ($\omega \gg \nu_{ce}$) cases. We see that the EEDF is indeed strongly affected, especially in the microwave case, when e-e collisions are taken into account; one of the important effects in this case is a strong reduction in the peak of the EEDF at $u = 0$. It is easy to understand on physical grounds why this is so. In argon, low energy electrons cannot efficiently absorb energy from a microwave field because their collision frequency with the neutrals is very small in the range $u/e \leq 0.5$ eV (see Fig. 2.3) due to the presence of a Ramsauer minimum in the momentum transfer cross section[7] (remind that the

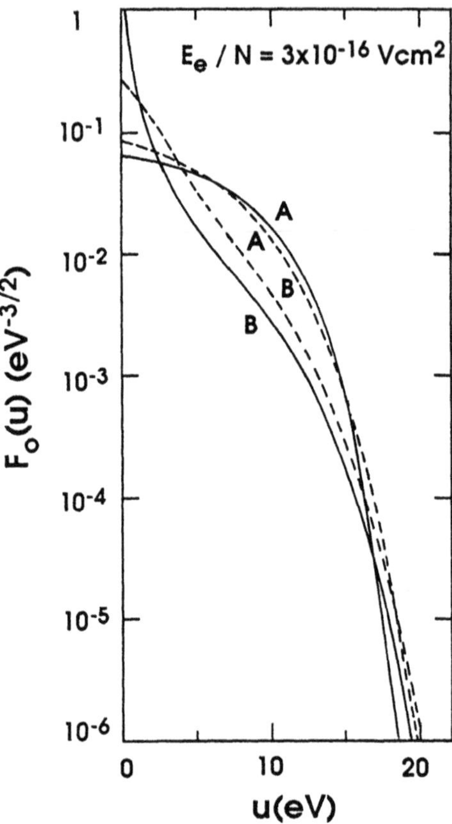

Fig. 3.15. Calculated EEDF in argon for $E_e/N = 3 \times 10^{-16}$ V cm^2, for $\omega/\nu_{ce} = 0$ (A) and $\omega/\nu_{ce} = 10$ (B), at $n/N = 0$ (full curves) and $n/N = 10^{-4}$ (broken curves) (from ref. 14).

power transferred from the field to the electrons, $\nu_c u_c$, is zero when $\nu_c(u) = 0$, as seen from eqn. (3.2)). Therefore, electrons tend to accumulate at zero energy, which causes a strong peak near $u = 0$. However, these low energy electrons can interact with other electrons of higher energy and, in so doing, climb up in energy space. For this reason, when e-e collisions are considered in the calculation, they counteract the accumulation of electrons near $u = 0$ and, thus, drastically contribute in reducing the peak of the EEDF at $u = 0$.[12-14]

The changes induced by e-e collisions on the EEDF have a huge influence on the electron characteristic energy, u_k, a key parameter that determines, as we have seen, the ambipolar diffusion loss rate (recall that $D_a \simeq \mu_i u_k$). Figure 3.16 shows the variation of the ratio u_k/T_e as a function of n/N, for three values of E_e/N, in the DC and microwave cases. In the former case, u_k/T_e decreases as n/N increases whereas, in the

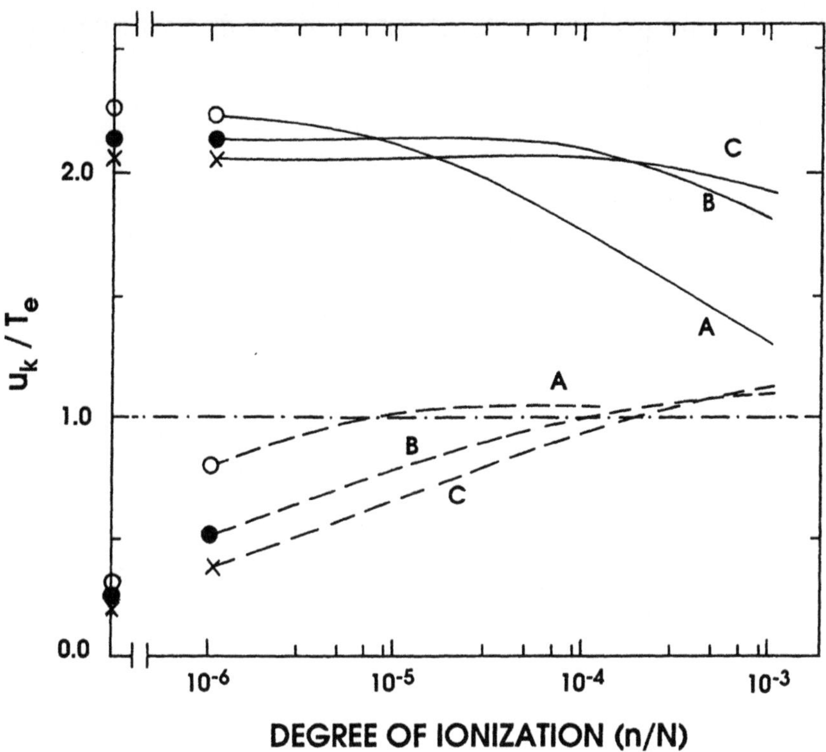

Fig. 3.16. Characteristic energy u_k normalized to the electron temperature T_e as a function of the degree of ionization, n/N, in argon, for the DC (full curves) and microwave (ω/ν_{ce} = 10; broken curves) limiting cases, and the following values of E_e/N in 10^{-17} V cm²: 10 (A); 50 (B); 100 (C) (from ref.14).

latter case, it is the opposite. The change in u_k/T_e in the microwave case is quite large even at relatively small n/N values and we see that u_k becomes of the order of T_e as n/N approaches 10^{-5} or 10^{-4}, depending on E_e/N. Remembering that $u_k = T_e$ for a Maxwellian distribution, this indicates that the body of the EEDF in the microwave case nearly Maxwellizes at moderate values of n/N. We see however, from Fig. 3.16, that u_k systematically, though slightly, overshoots the Maxwellian limit (u_k/T_e = 1) as n/N increases, which indicates that Maxwellization is not yet fully achieved in the range of n/N considered. In the DC case, u_k remains considerably larger than T_e, even for relatively high degrees of ionization. Therefore, as can be seen from Fig. 3.15, the body of the EEDF still deviates considerably from that of a Maxwellian, which would correspond to a straight line in that figure.

The ionization rate coefficient C_i is also somewhat affected by increased e-e collisions but to a lesser extent than u_k, as can be seen from Fig. 3.17. The effects are

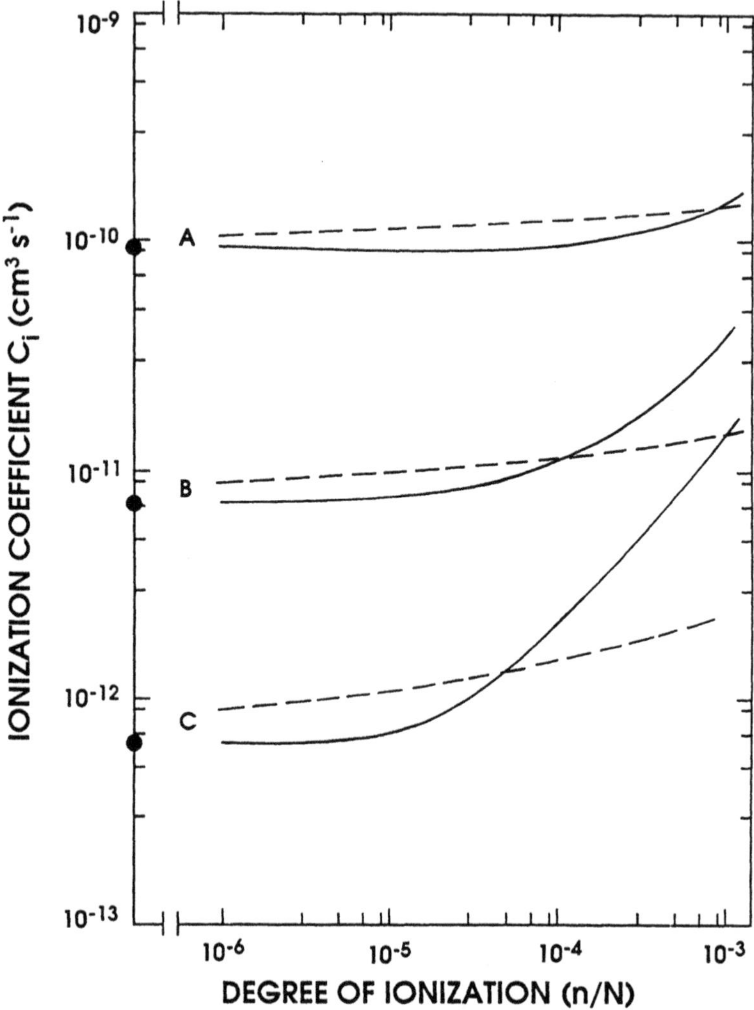

Fig. 3.17. Electron ionization rate coefficient C_i in argon as a function of the degree of ionization, for the DC (full curves) and microwave ($\omega = 10\,\nu_{ce}$, broken curves) cases, at the following values of E_e/N in 10^{-17} V cm^2: 100 (A); 50 (B); 30 (C) (from ref. 14).

larger at low E_e/N values (higher pressures) for, then, the high-energy tail of the EEDF is largely depleted and e-e collisions can strongly contribute to enhance it.

The changes induced by e-e collisions on C_i and u_k have a large effect on the maintenance field as determined from eqn. (3.3), thus, also on the mean absorbed and lost power per electron, θ, as illustrated in Fig. 3.18. One sees from this figure that e-e collisions considerably increase the value of θ/N in the microwave case (due to the

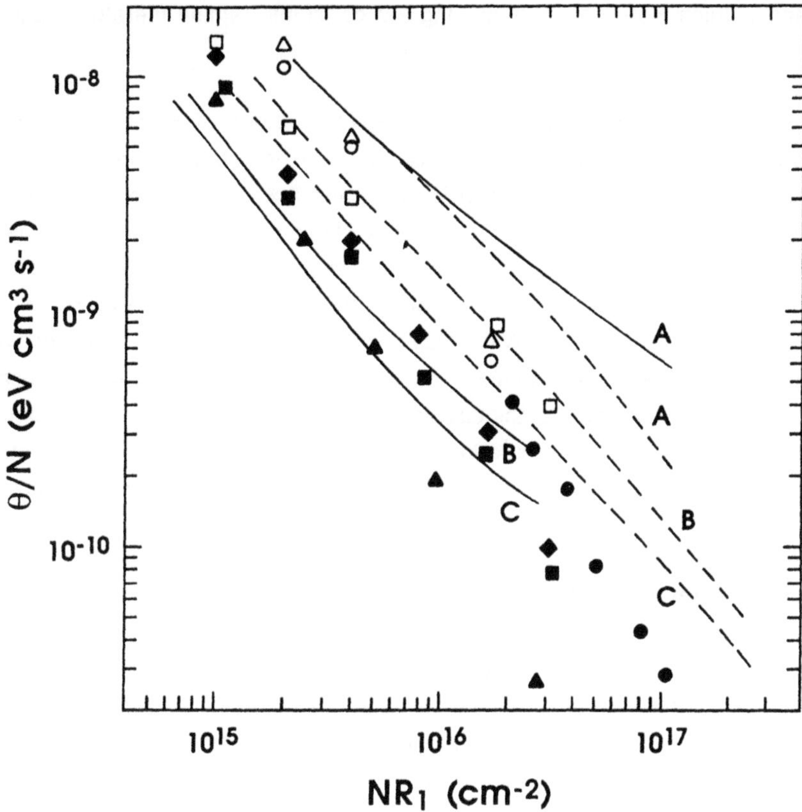

Fig. 3.18. Mean absorbed power θ per electron, at unit gas density N, as a function of NR_1 in argon, for $\omega/\nu_{ce} = 0$ (A), 0.8 (B), and 10 (C), at $n/N = 0$ (full curves) and $n/N = 10^{-4}$ (broken curves). Data points are from experiments on surface-wave discharges (Orsay and Montréal groups) for various values of tube radius R_1 and wave frequency: closed triangles-0.15 cm, 2.45 GHz; closed squares-1.3 cm, 1625 MHz; closed diamonds-1.3 cm, 900 MHz; closed circles-3.8 cm, 390 MHz; open squares-3.25 cm, 200 MHz; open circles-1.3 cm, 50.5 MHz; open triangles-1.3 cm, 25.8 MHz (from ref. 14).

resulting increase in u_k, thus, in the diffusion loss rate), but tend to decrease θ/N in DC discharges (due to a resulting decrease in u_k), especially at the higher pressures (lower E_e/N). The global effect of e-e collisions on the set of characteristics of θ/N as a function of NR_1 at constant ω/ν_{ce} is thus to bring all the curves corresponding to different ω/ν_{ce} values closer together, as shown in Fig. 3.18. This could be expected since all these curves must collapse into a single curve when the Maxwellian limit is reached at sufficiently high n/N values. Nevertheless, one sees from Fig. 3.18 that the θ/N curve for the microwave case still lies considerably below that for DC conditions, even though we have considered a relatively high degree of ionization, namely $n/N = 10^{-4}$. The general conclusions drawn in Sec. 3.2.3 on the influence of the plasma stimulating frequency, which were obtained ignoring e-e collisions, remain therefore qualitatively valid.

We now return to the comparison of predicted values of θ with measurements performed at a fixed frequency, i.e. the situation depicted in Figs. 3.12 to 3.14. Assuming that $n = 10^{11}$ cm^{-3}, 2×10^{11} cm^{-3}, and 2×10^{13} cm^{-3} respectively in the discharges considered in these three figures (these are typical values of the mean electron density in these discharges for the corresponding operating conditions) and taking into account e-e collisions, one obtains the curves labelled B on these figures. Note that now both ω/ν_{ce} and n/N decrease along curve B as pressure increases. The agreement with experiment is clearly improved at low pressures, except in the smallest radius tube, but the behavior of the calculated curves considerably deviates from that of experimental data as the pressure increases. This is due to the neglect of stepwise ionization processes, that is, of ionization from excited states, as demonstrated in the next section.

3.5. Multi-step ionization. Collisional-radiative models

An important source of inaccuracy in the model of Sec. 3.2.1 is the assumption that direct ionization from the ground state prevails. This is essentially correct at very low pressure (high maintenance fields), since then the EEDF has a rich tail of high-energy electrons capable of directly ionizing ground state atoms. However, as the pressure is increased, the high-energy tail becomes dramatically depleted and multi-step ionization processes take over the discharge sustaining mechanisms. In general, multi-step ionization mainly proceeds via highly populated excited states, such as metastables and resonance levels. In many circumstances, ionization upon collisions between excited atoms may also significantly contribute to the total ionization rate.

In order to account for ionization from excited atoms, it is necessary to extend the model of Sec. 3.2.1 to include rate balance equations for the relevant excited species. These are equations accounting for the principal collisional and radiative processes that create or destroy a given species, whose solution yields the species' concentration. A complete model must therefore include the Boltzmann equation (3.1), a charged particle balance equation taking into account all the ionization and loss processes, and the rate balance equations for the relevant excited species. Such a model is referred to as a *collisional-radiative model*.

Collisional-radiative (C-R) models have been developed in recent years for microwave discharges in various gases, such as Ar, He and O_2. In this section, we briefly describe the basic content of these models and discuss some results obtained.

In a C-R model for argon, [6,14,15] one must account for ionization from the four levels of the $3p^5 4s$ configuration, that is, metastable states $^3P_0$ and $^3P_2$ and resonance states $^1P_1$ and $^3P_1$, which are usually highly populated. The principal collisional and radiative processes that determine the populations of these states are listed in Table 3.1, where $Ar(^1S_0)$ is the ground state of the atom, $Ar(p_k)$, with $k = 1 - 10$, denotes a level of the $3p^5 4p$ configuration, and $Ar(p)$ denotes the levels of the $3p^5 4s$ configuration, with $p = 1$, 2, 3 and 4 corresponding to $^3P_2$, $^3P_1$, $^3P_0$ and $^1P_1$ respectively. The rate coefficients for these processes appear in Table 3.1. The model also accounts for metastable diffusion to the wall (rate D_p/Λ^2, where D_p is the diffusion coefficient for atoms in the energy level p) and emprisonment of the resonance radiation, which is dealt with through an *escape factor* g_p. Details on all these processes and their corresponding rates can be found in ref. 14 where this model was first formulated for a DC discharge.

Table 3.1. List of elementary processes considered in the collisional-radiative model for argon. Ar^* and Ar^+ indicate an excited and an ionized argon atom respectively and e^- is an electron (refs. 14 and 15).

No.	Process	Rate
1(a)	$Ar(^1S_0) + e^- \rightarrow Ar(p) + e^-$	C_{0p}
1(b)	$Ar(^1S_0) + e^- \rightarrow Ar^* + e^-$	[total rate for
	$Ar^* \rightarrow ... \rightarrow Ar(p) + h\nu$	1(a)+1(b)]
2	$Ar(p) + e^- \rightarrow Ar(q) + e^-$	C_{pq}
3	$Ar(p) + e^- \rightarrow Ar(p_k) + e^-$	C_{pp_k}
4	$Ar(p_k) \rightarrow Ar(p) + h\nu$	$A_{p_k p}$
5	$Ar(p) + e^- \rightarrow Ar^+ + 2e^-$	C_{p+}
6	$Ar(p) + Ar(q) \rightarrow Ar(^1S_0) + Ar^+ + e^-$	K_p
7	$Ar(p) \underset{\leftarrow}{\overset{\rightarrow}{}} Ar(^1S_0) + h\nu$, for $p = 2,4$	$\nu_p = A_{p0}\, g_p$
8	$Ar(p) + \text{wall} \rightarrow Ar(^1S_0)$, for $p = 1,3$	$\nu_p = D_p/\Lambda^2$

The ionization-diffusion loss balance for charged particles in argon is described as in eqn. (3.3) but replacing $<\nu_i>$ with an effective ionization rate per electron, ν_{it}, given by

$$n\nu_{it} = n\, N\, C_{0+} + \sum_{p=1}^{4} n\, n_p\, C_{p+} + \frac{1}{2} K_p \left(\sum_{p=1}^{4} n_p \right)^2 . \tag{3.14}$$

Here, the terms on the right-hand side account, in order, for direct ionization (whose rate coefficient is now designated C_{0+}), ionization of the atoms in the $3p^54s$ levels by electron impact, and ionization via the energy pooling reactions (reactions 6 in Table 3.1).

Table 3.2. **List of elementary processes considered in the collisional-radiative model for helium (ref. 16).**

Process	Rate
$He(q) + e^- \rightleftarrows He(p) + e^-$ $q \equiv (n, L, s) < p \equiv (n', L', s')$	C_{qp}, C_{pq}
$He(p) + e^- \rightarrow He^+ + e^- + e^-$ $p \equiv (n, L, s)$	C_{p+}
$He(q) \rightarrow He(p) + h\nu$ $q \equiv (n, L, s) > p \equiv (n, L \pm 1, s)$	$A_{qp}\, g_{pq}$ (escape factor $g_{pq} \neq 1$ for $p = 1^1S$ and the following transitions $2^3P \rightarrow 2^3S$; $2^1P \rightarrow 2^1S$; $3^3P \rightarrow 2^3S$; $3^3D \rightarrow 2^3P$)
$He(q) + He(1^1S) \rightleftarrows He(p) + He(1^1S)$ $q \equiv (n, L, s) > p \equiv (n, L', s) \geq 3^3P$	K_{qp}, K_{pq}
$He(q) + He(p) \rightarrow He(1^1S) + He^+ + e^-$ $\quad\quad\quad\quad \rightarrow He_2^+ + e^-$ $q, p = 2^3S, 2^1S, 2^3P$	$K_p\, \chi$ $K_p\, (1 - \chi)$ with $\chi = 0.30$
$He(p) + He(1^1S) \rightarrow He_2^+ + e^-$ $p = (n, L, s) \geq 3^3D$	S_{p+}
$He^+ + 2He(1^1S) \rightarrow He_2^+ + He(1^1S)$	η
$He_2^+ + e^- \rightarrow He(2^3S) + He(1^1S)$	α_r
$He(p) + wall \rightarrow He(1^1S)$ $p = 2^1S, 2^3S$	D_p/Λ^2

Curve D in Figs. 3.12 to 3.14 show the results of this model with respect to the parameter θ. Note that these results take into account the effects of e-e collisions as well. The agreement with experiment is now excellent at 900 MHz and 2.45 GHz. At 200 MHz, the shape of the calculated curve is the same as for the experimental data but a disagreement in magnitude by a factor of about 2 still persists. This can be due either to experimental uncertainties in the measurements of θ or to inaccuracies in some data used in the C-R model. Also shown in these figures are the results obtained from this C-R model when e-e collisions are ignored (curves C). The general conclusion to be drawn from all these results is that both e-e collisions and multi-step ionization need to be considered in order to correctly explain the observations.

A C-R model for microwave discharges in He was also recently developed by Alves et al.[16] Table 3.2 summarizes the basic processes considered.[†] This C-R model accounts for ionization by electron impact from the ground state and the excited states $p \equiv n^{2s+1}L$, with $2 \leq n \leq 6$; $s = 0,1$; $L = $ S, P, D, F, G, H, I, and also for Penning reactions (rate coefficient K_p) between all pairs of 2^1S, 2^3S and 2^3P states. It also accounts for the creation and destruction kinetics of He_2^+ molecular ions, which are mainly formed by associative ionization, namely $He(p) + He \rightarrow He_2^+ + e^-$ (rate coefficient S_{p+}), and for the dissociative recombination of electrons with He_2^+ (recombination coefficient α_r) which is known to become important at pressures above 5 to 10 torr. The creation-loss balance for electrons in helium can also be expressed by an equation similar to eqn. (3.3) but with $<v_i>$ replaced by an effective ionization rate per electron, v_{it}, given in this case by

$$n v_{it} = n\,N\,C_{0+} + \sum_p n\,n_p\,C_{p+} + \frac{1}{2}K_p\left(\sum_{p=2^1S,2^3S,2^3P} n_p\right)^2 + \sum_p S_{p+}\,n_p\,N - \alpha_r\,n\,n_m, \qquad (3.15)$$

where, the notation is similar to that used for argon, and n_m denotes the number density of molecular ions He_2^+.

The characteristics of θ/N as a function of NR calculated from this C-R model for helium are compared in Fig. 3.19 with measurements performed in surface-wave discharges[17-19] for an operating frequency of 2.45 GHz. The calculations are relatively insensitive to the values of the electron density in the range 5×10^{11} to 5×10^{12} cm^{-3} as seen from the figure. Also shown is a calculation taking only into account direct excitations and ionization. The latter curve considerably deviates from the experimental data as the pressure increases, which again shows that a C-R model is necessary to explain the observations. Finally, we wish to point out that for helium, the

---

[†] In Table 3.2, we use the letters n, L, s to designate the usual quantum numbers.

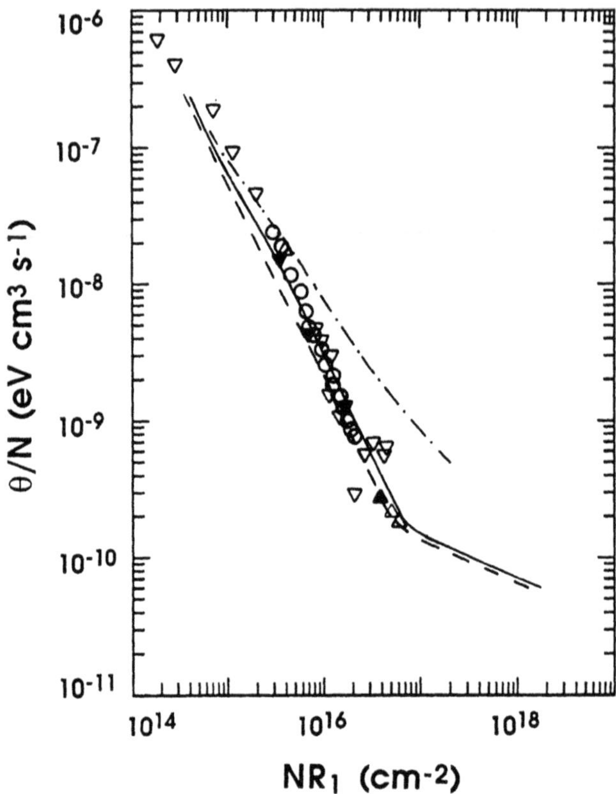

Fig. 3.19. Characteristics of θ/N vs. NR_1 in helium. Curves are calculated for $\omega/2\pi = 2.45$ GHz, $R_1 = 0.3$ cm and the following cases: C-R model with $n = 5 \times 10^{11}$ cm^{-3} (solid) and $n = 5 \times 10^{12}$ cm^{-3} (dashed); simple model with excitation and ionization from the ground state only (chain). Data points are from experiment for $\omega/2\pi = 2.45$ GHz and $R_1 = 0.3$ cm (0); 0.4 cm (∇); and 0.8 cm (Δ) (from ref. 16)

results for 2.45 GHz may be extrapolated to other frequencies, even to DC conditions, without much error. This is because the EEDF in He is quite insensitive to the changes in ω, as we have seen in Sec. 3.2.

The formulation of a C-R model for a discharge in oxygen faces a number of additional complications. First of all, oxygen is an electronegative gas which makes that large amounts of negative ions are usually present in an oxygen discharge. At low pressures, the dominant negative ion is O$^-$ which is formed by dissociative attachment, namely $e^- + O_2 \rightarrow O^- + O$; its density can be comparable to, or even higher than, the electron density. Therefore, the plasma contains two types of negative species, electrons and O$^-$ ions, with quite different masses, and positive ions O$_2^+$. An additional complication is that electron collisions dissociate very efficiently O$_2$(X) molecules (see

Fig. 3.6) which leads to the presence of large amounts of $O(^3P)$ dissociated atoms. Large concentrations of $O_2(a^1\Delta)$ metastables are also usually present. These two species, $O(^3P)$ and $O_2(a^1\Delta)$, play an important role in the discharge kinetics since they are the principal detaching species, which remove the electron from the O^- ions. At the end, one faces a very complex kinetic scheme involving three types of charged species and a neutral background which is, in fact, a mixture of three species, namely $O_2(X)$, $O_2(a^1\Delta)$, and $O(^3P)$.

Tables 3.3 and 3.4 list the basic electron and heavy particle kinetic processes that have been considered in recent models for low pressure discharges in oxygen.[9,20]

Table 3.3. List of elementary processes considered in the collisional-radiative model for oxygen (refs. 9 and 20).

Electron processes
Molecular oxygen
1) $e^- + O_2(X, v) \rightleftarrows e^- + O_2(X, v')$
2) $e^- + O_2(X, v = 0) \rightleftarrows e^- + O_2(a^1\Delta)$
3) $e^- + O_2(X, v = 0) \rightarrow e^- + O_2(b^1\Sigma)$
4) $e^- + O_2(X, v = 0) \rightarrow e^- + O_2(4.5\ eV)$
5) $e^- + O_2(X, v = 0) \rightarrow e^- + O_2(6.0\ eV)$
6) $e^- + O_2(X, v = 0) \rightarrow e^- + O_2(8.4\ eV)$
7) $e^- + O_2(X, v = 0) \rightarrow e^- + O_2(9.97\ eV)$
8) $e^- + O_2(X, v = 0) \rightarrow e^- + e^- + O_2^+$
9) $e^- + O_2(X, v = 0) \rightarrow e^- + O_2(14.7\ eV)$
10) $e^- + O_2(a^1\Delta) \rightarrow e^- + O_2(b^1\Sigma)$
11) $e^- + O_2(a^1\Delta) \rightarrow e^- + e^- + O_2^+$
Atomic oxygen
12) $e^- + O(^3P) \rightarrow e^- + O(^1D)$
13) $e^- + O(^3P) \rightarrow e^- + O(^1S)$
14) $e^- + O(^3P) \rightarrow e^- + O(^3S)$
15) $e^- + O(^3P) \rightarrow e^- + e^- + O^+$

Table 3.4. List of elementary processes considered in the collisional-radiative model for oxygen to determine the populations of the heavy species $O_2(X^3\Sigma)$, $O_2(a^1\Delta)$, O, and of O^-. T is the gas temperature in kelvin (refs. 9 and 20).

No.	Reaction	Rate
1	$O_2(X^3\Sigma) + e^- \rightarrow O^- + O$	$K_1 = f(E/N)$
2	$O_2(a^1\Delta) + e^- \rightarrow O^- + O$	$K_2 = f(E/N)$
3	$O^- + O \rightarrow O_2(X^3\Sigma) + e^-$	$K_3 = 1.4 \times 10^{-10}\ cm^3\ s^{-1}$
4	$O^- + O_2(a^1\Delta) \rightarrow O_3$	$K_4 = 3 \times 10^{-10}\ cm^3\ s^{-1}$
5	$O_2(X^3\Sigma) + e^- \rightarrow O + O + e^-$	$K_5 = f(E/N)$
6	$O_2(a^1\Delta) + e^- \rightarrow O + O + e^-$	$K_6 = f(E/N)$
7	$O + wall \rightarrow 1/2\ O_2(X^3\Sigma)$	$K_7 = 200\ \sqrt{T/300}\ s^{-1}$
8	$O_2(X^3\Sigma) + e^- \rightarrow O_2(a^1\Delta) + e^-$	$K_8 = f(E/N)$
9	$O_2(a^1\Delta) + e^- \rightarrow O_2(X^3\Sigma) + e^-$	$K_9 = f(E/N)$
10	$O_2(a^1\Delta) + e^- \rightarrow O_2(b^1\Sigma) + e^-$	$K_{10} = f(E/N)$
11	$O_2(a^1\Delta) + wall \rightarrow O_2(X^3\Sigma)$	$K_{11} = 0.4\ s^{-1}$
12	$O_2(b^1\Sigma) + e^- \rightarrow O_2(a^1\Delta) + e^-$	$K_{12} = f(E/N)$
13	$O_2(X^3\Sigma) + e^- \rightarrow O_2(b^1\Sigma) + e^-$	$K_{13} = f(E/N)$
14	$O_2(b^1\Sigma) + e^- \rightarrow O_2(X^3\Sigma) + e^-$	$K_{14} = f(E/N)$
15	$O_2(b^1\Sigma) + wall \rightarrow O_2(X^3\Sigma)$	$K_{15} = 400\ s^{-1}$
16	$O + O + O_2(X^3\Sigma) \rightarrow O_3 + O$	$K_{16} = 2.1 \times 10^{-34}\ e^{345/T}\ cm^6\ s^{-1}$
17	$O + O_2(X^3\Sigma) + O_2(X^3\Sigma) \rightarrow O_3 + O_2(X^3\Sigma)$	$K_{17} = 6.4 \times 10^{-35}\ e^{663/T}\ cm^6\ s^{-1}$
18	$O_3 + O \rightarrow O_2(a^1\Delta) + O_2(X^3\Sigma)$	$K_{18} = 1 \times 10^{-11}\ e^{-2300/T}\ cm^6\ s^{-1}$
19	$O_3 + O \rightarrow 2O_2(X^3\Sigma)$	$K_{19} = 1.8 \times 10^{-11}\ e^{-2300/T}\ cm^6\ s^{-1}$
20	$O_2(b^1\Sigma) + O_3 \rightarrow 2O_2(X^3\Sigma) + O$	$K_{20} = 1.5 \times 10^{-11}\ cm^3\ s^{-1}$
21	$O_2(a^1\Delta) + O_3 \rightarrow 2O_2 + O$	$K_{21} = 5.2 \times 10^{-11}\ e^{-2840/T}\ cm^3\ s^{-1}$
22	$O_3 + e^- \rightarrow O + O_2 + e^-$	$K_{22} = 5K_5\ cm^3\ s^{-1}$

Due to the presence of negative ions, the charged particle creation-loss balance is not as simple as that expressed by eqn. (3.3) with D_{se} determined from the theory of Self and Ewald, as in Sec. 3.2.1. However, it has recently been shown[21] that a similar equation is also valid in this case but with an effective diffusion coefficient whose meaning is now quite different: D_{se} depends in this case on parameters such as the positive and negative ion mobilities, and the attachment and detachment rates. Physically, this means that D_{se} depends on the negative ion concentration. The important point to make here is that eqn. (3.3), with D_{se} so reinterpreted, constitutes again an indicial equation determining the maintenance field, and thus θ, for given pressure and geometry, for the same reasons as before.

Figure 3.20 shows a comparison of calculated[9] and measured values of θ/N as a function of NR_1 for oxygen discharges operated at 390 MHz, in tubes of various

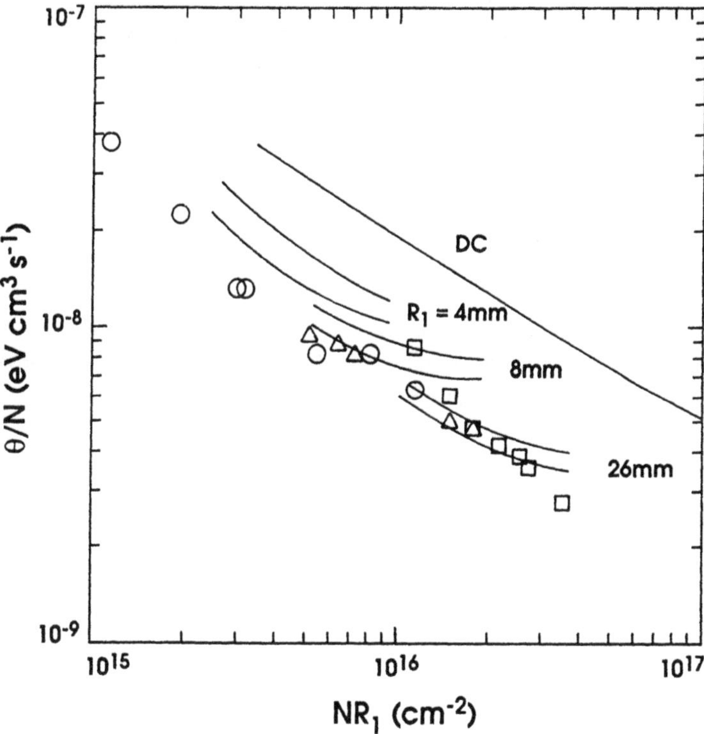

Fig. 3.20. Characteristics of θ/N vs. NR_1 in oxygen discharges. Curves $R_1 = 4$ mm, 8 mm and 26 mm where R_1 is the discharge tube radius, are calculated for $\omega/2\pi = 390$ MHz. Upper and lower curves correspond to $n = 2.10^{10}$ cm^{-3} and $n = 10^{11}$ cm^{-3} respectively. Data points are from surface-wave discharge experiments with tube radii of 4 mm (circles), 8 mm (triangles) and 26 mm (squares). Also shown is a calculated characteristic for $\omega = 0$ (DC) (from refs. 9, 22).

diameters. The experimental data is from surface-wave experiments.[22] The agreement of predictions with measurements is also very satisfactory in this case. Also shown in this figure is a power characteristic calculated for a DC discharge ($\omega = 0$). We note that, for a fixed pressure, θ is lower in the microwave case than in the DC one, as in argon.

3.6. Conclusion

The analysis presented in this chapter should help one understanding the role of the plasma stimulating frequency upon discharge parameters and plasma processing. It assumes the EEDF to be time independent, corresponding typically to operating conditions such that $(\omega/2\pi)/p \geq 20$ MHz torr^{-1}. We have shown that the plasma excitation frequency has in general a large effect on the shape of the EEDF and, thus, on the electron transport parameters and collision rates. One of the parameters that usually exhibits a large dependence on ω is the mean absorbed or lost power per electron under steady-state conditions, θ. With the exception of helium, in all the cases investigated so far, θ decreases, for fixed pressure, as ω varies from zero (DC) up to the microwave range. This behavior is in fact a quite general law valid for all gases in which the electron-neutral collision frequency for momentum transfer, $v_c(u)$, grows rapidly with u in the bulk of the EEDF. Such a law simply stems from eqn. (3.2) for the power transfer between an electric field and electrons.

As far as comparing the above model with experimental results is concerned, we have stressed in Sec. 3.1 the necessity of making sure that all other external parameters (so-called discharge conditions and power density) are kept constant when varying ω. For the time being, the only such experiments in the domain 13.56 - 2450 MHz seem to have been performed with surface-wave produced plasmas. Some of these results have been presented in this chapter. To provide the reader with more data showing the interest of frequency optimization, we end with the following facts: i) with respect to charged particle production, in argon gas at reduced pressure, depending on discharge conditions, 2 to 7 times less power per ion-electron pair is required at 2450 MHz than at radio-frequencies;[1] ii) as far as the rate of plasma processes is concerned, the plasma polymer deposition from hydrocarbon and fluorocarbon monomers, the growth of amorphous silicon and the etching of polyimide indicate a possible frequency optimization that leads to typically a factor of 5 to 10 in increased efficiency.[1] These experiments point out that the optimum process efficiency is not necessarily attained at either 13.56 or 2450 MHz: under the operating conditions of the above experiments, the polymer deposition is more efficient at frequencies above typically 100 MHz and its quality satisfactory below 400 MHz; the amorphous silicon deposition in a RF capacitive discharge is the most efficient around 70 MHz; the etching of polyimide in an O_2 - CF_4 discharge is optimum close to 50 MHz. These various results imply that some flexibility in choosing the operating frequency of the plasma reactor is desirable to achieve the optimum process efficiency.

Depending on the situation considered, a higher process efficiency may translate into more rapid processing and lower cost. On the other hand, the technological problems associated with plasma processing at other than ISM authorized frequencies (see Appendix 1.1) and, for the time being, the related scarcity and high cost of equipment, are obviously factors which should not be overlooked.

References

[1]   M. Moisan, C. Barbeau, R. Claude, C.M. Ferreira, J. Margot, J. Paraszczak, A.B. Sá, G. Sauvé and M.R. Wertheimer, J. Vac. Sci. Technol. **B9**, 8 (1991).

[2]   Ch. H. Muller III and A.V. Phelps, J. Appl. Phys. **51**, 6141 (1980).

[3]   J. Ingold, Phys. Fluids **15**, 75 (1972).

[4]   A.B. Sá, C.M. Ferreira, S. Pasquiers, C. Boisse-Laporte, P. Leprince and J. Marec, J. Appl. Phys. **70**, 4147 (1991).

[5]   L.L. Alves and C.M. Ferreira, J. Phys. D: Appl. Phys. **24**, 581 (1991).

[6]   C.M. Ferreira, L.L. Alves, M. Pinheiro and A.B. Sá, IEEE Trans. Plasma Sci. **19**, 229 (1991).

[7]   S.C. Brown, *Basic Data of Plasma Physics* (MIT Press, Cambridge, Mass., 1961).

[8]   C.M. Ferreira and J. Loureiro, J. Phys. D: Appl. Phys. **17**, 1175 (1984).

[9]   M. Pinheiro, G. Gousset and C.M. Ferreira, to be published.

[10] C.M. Ferreira and J. Loureiro, J. Phys. D: Appl. Phys. **22**, 76 (1989).

[11] H.S.W. Massey and E.H.S. Burhop, *Electronic and Ionic Impact Phenomena* (Clarendon, Oxford, 1952).

[12] E.V. Karoulina and Yu. A. Lebedev, J. Phys. D: Appl. Phys. **21**, 411 (1988).

[13] U. Kortshagen, H. Schlüter and A. Shivarova, J. Phys. D: Appl. Phys. **24**, 1571 (1991).

[14] P.A. Sá, J. Loureiro and C.M. Ferreira, J. Phys. D: Appl. Phys. **25**, 960 (1992).

[15] C.M. Ferreira, J. Loureiro and A. Ricard, J. Appl. Phys. **57**, 82 (1985).

[16] L.L. Alves, G. Gousset and C.M. Ferreira, J. Phys. D: Appl. Phys. submitted (1992).

[17] S. Daviaud, C. Boisse-Laporte, P. Leprince and J. Marec, J. Phys. D: Appl. Phys. **22**, 770 (1989).

[18] S. Daviaud, Thèse de Doctorat, Université Paris-Sud, Orsay, France (1989).

[19] M. M'Harzi, Thèse de 3e Cycle, Université Paris-Sud, Orsay, France (1987).

[20] G. Gousset, C.M. Ferreira, M. Pinheiro, P.A. Sá, M. Touzeau, M. Vialle and J. Loureiro, J. Phys. D: Appl. Phys. **24**, 290 (1991).

[21] C.M. Ferreira, G. Gousset and M. Touzeau, J. Phys. D: Appl. Phys. **21**, 1403 (1988).

[22] A. Granier, S. Pasquiers, C. Boisse-Laporte, R. Darchicourt, P. Leprince and J. Marec, J. Phys. D: Appl. Phys. **22**, 1487 (1989).

CHAPTER 4

PLASMAS SUSTAINED WITHIN MICROWAVE CIRCUITS [*]

4.1. Introduction

The first microwave discharges emerged as adaptations of RF discharges to a domain of higher frequencies. This can be seen by recalling that RF discharges were originally all sustained either between electrodes (e.g. parallel plates, two-ring setups) or within a coil, that were respectively capacitive or inductive elements of the RF circuit. Clearly, such lumped-circuits were of no practical use at microwave frequencies. One obvious step was thus to turn to the microwave counterparts of RF circuits such as sections of coaxial lines and waveguides, or resonant cavities, and form microwave circuits within which discharges could be sustained. In such discharges, in most practical cases, the requirement of small phase differences between the field at any two points within the active zone is met and the discharges belong to the *localized active zone* category (Sec. 2.4). Within this category, the discharges may be divided into specific groups according, for example, to the structure of the field applicator.[1] The delimitations that we have adopted is presented below.

The first, and by far the most common group is that of *discharges sustained within resonant cavities*. These discharges can generally be operated at gas pressures ranging from few mtorrs to above atmospheric pressure. Due to the limited volume of their active zone, their use has been mainly restricted to plasma sources providing the excitation of atomic and polyatomic species for emission spectroscopy studies, an application where they have some distinct advantages[2] over RF excited plasmas. These resonant-cavity plasma sources are extensively described in the literature; as a matter of fact, during the last decade, they have been part of almost all reviews dedicated to specific applications of HF discharges (see, for example, Zander and Hieftje,[2] Matousek et al.,[3] Goode and Baughman[4]).

The second group also consists of *resonantly sustained discharges*, but the field applicator is not a resonant structure. The resonance is achieved by exploiting the eigen modes of a bounded plasma, for example, the Tonks-Dattner modes.[5] Such discharges can be sustained only at comparatively low pressure, the upper pressure limit depending on the particular resonance mode.[6] They have been investigated both theoretically and experimentally,[7-9] but do not seem to have found applications yet.

---

[*] **Presented by Z. Zakrzewski, M. Moisan and G. Sauvé**

The third group is limited to *discharges sustained inside a waveguide*, where the discharge tube is placed parallel to the electric field. These found applications mainly in gas phase plasma chemistry.[10,11]

The last and fourth group includes discharge systems in which the *plasma forms an absorbing load at the end of the microwave transmission line*. Microwave plasma torches are typical examples.[12-15] These are coaxial structures opening into the surrounding atmosphere with the working gas flowing out and a torch, or "flame", forming at the end of the inner conductor. They have been in use for many years to provide excited particles for emission spectroscopy.

Unfortunately, despite the fact that the localized active zone discharges have been known and investigated for decades, no general theoretical description has yet been given. The geometrical simplicity of their field applicators is misleading; most such structures are in reality complex and difficult to analyze in terms of electromagnetic field theory. As we shall see, at least an approximate analysis is always possible as far as the maintenance processes in the discharge are concerned. However, major difficulties generally arise when attempting to construct an equivalent circuit in order to determine the electrodynamic characteristics of the plasma source. This concerns, in particular, the oldest and most commonly used cavity-based discharges.[16,17]

The next two sections describe various practical realizations of plasma sources excited within a microwave circuit. These are merely examples, as no attempt is made to present a full summary of the work done in the field. Instead, we concentrate on two types of plasma sources: those based on resonant cavities (Sec. 4.2) and those within a rectangular waveguide (Sec. 4.3). The former were chosen because they are often encountered both in scientific laboratories and in industry, the latter because they offer interesting possibilities in terms of equivalent circuit modeling. Section 4.4 describes our simplified modeling of low pressure HF discharges belonging to the localized active zone category. Finally, in Sec. 4.5, we apply this modeling to waveguide discharges, and compare it with experimental data.

4.2. Plasma sources based on resonant cavities

4.2.1. Equivalent circuits and impedance matching. We start by recalling a few facts about microwave cavities containing a plasma. The cavity is a hollow conducting enclosure with the electromagnetic energy confined on the inside,[18] and it is coupled to a feed line. When used as a field applicator to sustain a microwave plasma, it encloses the active zone of the discharge.

As far as the theoretical description of cavity discharges is concerned, there are only a few reports, e.g. refs. 16 and 17, where such analysis is even attempted. The equivalent circuit of such plasma devices is shown in Fig. 4.1a. Note that i) the circuit can only be used for a single mode cavity, and ii) its components correspond to distributed parameters of the microwave network and not to lumped elements, even though they are shown as such. Here G and $\mathcal{B}$ are conductances and susceptances respectively; G_R represents Joule losses in conductors and dielectrics within the resonator and $\mathcal{B}_R$ is the energy stored in the resonator electromagnetic field, while G_D and $\mathcal{B}_D$ account for the presence of the discharge (Sec. 2.3.3). The reactance X_c together with the ideal transformer (characterized by the "turns ratio" m_c) represent the coupling network between the cavity and the feed line. We neglect the losses in these coupling elements. The arrows on the circuit components express the fact that the absorbed power and the tuning affect their values. The considerations that are to follow will be facilitated if we further simplify the equivalent circuit by referring all the admittances to a common reference plane as in Fig. 4.1b. Then the characteristic

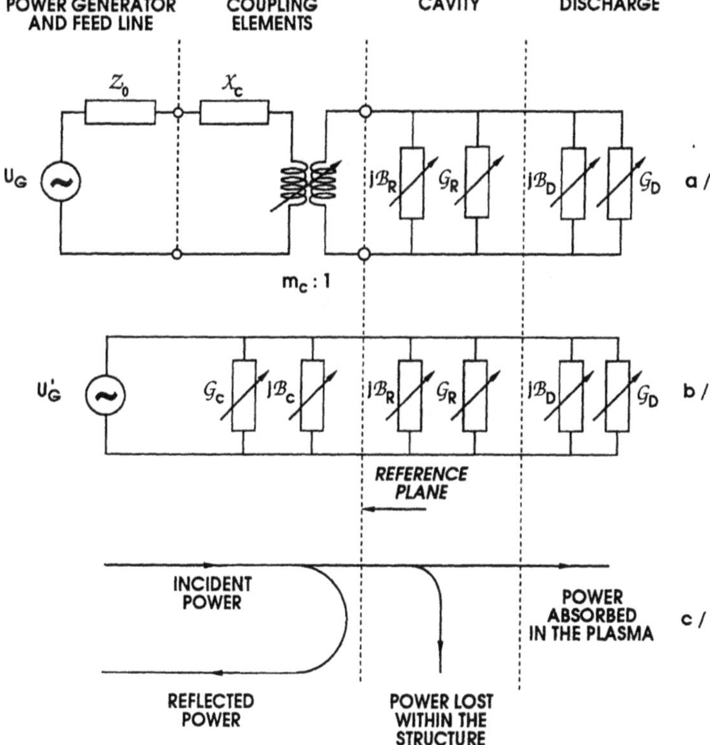

Fig. 4.1. Equivalent circuit of a single-mode resonant cavity containing a microwave discharge (a and b), and power flow (c). Here G and $\mathcal{B}$ are admittances (conductances and susceptances respectively) while Z_0 is the characteristic impedance of the feed line and X_c is the reactance of the coupling network.

impedance of the feed line, Z_0, and the parameters of the coupling network, X_c and m_c, are all represented by G_c and B_c.

Given this equivalent circuit, some intricacies of the cavity tuning, of impedance matching, and of power flow into and within the cavity can be discussed in a straight-forward way. Discharges sustained in cavities not equipped with intrinsic means of tuning[†] can be operated over only a narrow range of discharge conditions and input power, and most often with large reflections at the input.[17] Intrinsic tuning is obtained by varying the cavity dimensions, altering its shape or changing the coupling conditions with the feed line. As a result, it is possible to vary G_c, B_c, B_R or G_R and, therefore, the discharge system can be tuned as experimental conditions change. Recall that the circuit elements representing the discharge, B_D and G_D, under given discharge conditions, depend on the amount of power delivered to the plasma. This nonlinearity usually leads to hysteresis in the setting of the input power or in the impedance tuning procedure. This means that for a given tuning and a given power level, one can observe two different discharges depending on whether the power setting is reached by increasing or decreasing values.

The plasma loaded cavity is, by definition, resonant when the inward power flux through its input plane is real (Sec. 2.2.7). This obviously requires that the sum of all susceptances B in the circuit of Fig. 4.1b be equal to zero, and then the feed line sees a resistive load. Now, to achieve maximum power transfer from the feed line to the cavity input (reference plane in Fig. 4.1), the conductance G_c in Fig. 4.1 must be equal to $G_R + G_D$. When both these admittance conditions are satisfied, there is no reflected wave at the cavity input. One problem nevertheless remains: a careful observer may note a difference between the setting of tuning elements for zero reflected power at the cavity input and that, for example, for the maximum of light emission intensity that corresponds here to a maximum of absorbed power into the plasma. This is because the condition for maximum absorbed power in the plasma is $G_D = G_c + G_R$ and thus differs from the condition $G_c = G_R + G_D$ for maximum power transfer at the cavity input. It is only when $G_R \ll G_D$, i.e. when losses within the cavity structure are negligible, that the two conditions are satisfied for the same tuning. This phenomenon can be used for estimating the power loss in the structure. Clearly the objective in any practical realization should be to achieve $G_R \ll G_D$.

[†] Intrinsic tuning is opposed here to external tuning such as provided by putting stubs or tuning screws between the feed line and the plasma source. As a rule, losses are higher with external impedance matching means.

The requirement that G_R be much smaller than G_D is related to the quality factor[†] Q of cavities with and without a microwave discharge. Recall that resonators employed in classical microwave devices, for instance wavemeters, filters, always have very high Q factors, and this feature sets them apart from other microwave circuits. This is no longer true when they are used as field applicators to sustain microwave discharges: their Q value then drastically falls by a factor of the order of G_D/G_R, typically, by a factor of 100. To see this point, consider that in the presence of plasma, the energy stored in the cavity is comparable to that in the absence of plasma while losses due to the plasma are such that $G_D \gg G_R$. In fact, the distinction between cavity-induced discharges and other microwave configurations used to sustain plasmas is often somewhat artificial. Whenever matching conditions require that a standing wave be built up in "non-resonant" microwave discharge systems, there is a considerable amount of reactive energy stored. Thus the electrodynamic characteristics of the system closely resemble those of a discharge-loaded cavity: both act as low Q resonators and, in fact, their equivalent circuits are the same. For an excellent example of the resonant behavior of a "non-resonant" system, see Böhm,[19] who investigated experimentally a plasma column located in a waveguide. This setup served to study absorption of microwave power in DC plasma columns as well as to sustain overdense plasmas.[20]

4.2.2. Examples of practical cavity based-plasma sources. Resonant cavities have played an eminent role in microwave discharge studies and applications, mostly because they were previously well known and widely used devices. Field shaping could be achieved by choosing the right mode, and coupling and matching structures were available. However, the presence of plasma strongly affects the resonance characteristics and matching conditions. This effect is responsible for the two main shortcomings of the plasma sources employing resonant cavities, namely unstable operation resulting from even small changes in the discharge conditions or in the input power, and tedious tuning. Quite often, intrinsic tuning cannot ensure impedance matching, and external tuning at the cavity input must be used. This adds to the above shortcomings.

Many discharge cavity designs have been described in the literature. We have chosen a few typical examples that have become classics in the field. They usually operate at 2.45 GHz.

Rectangular cavity. These cavities are generally made from a section of standard rectangular waveguide, which is closed at both ends. The frequency of operation or the wide wall dimension of the waveguide is usually chosen such that the waveguide

---

[†] The quality factor of a resonant circuit is defined as the ratio of the reactive energy, inductive or capacitive, stored at resonance, to the energy loss per cycle. Thus, following Sec. 2.3.3, $Q = |\mathcal{B}|/\mathcal{G}$.

operates in the fundamental mode, yielding TE_{10p}[†] modes for the cavity. With such modes, the electric field is uniform along the narrow wall of the guide and there is a half-wavelength cosine variation along the wide wall. The empty cavity is at resonance when its length is $p\lambda_g/2$ where λ_g is the TE_{10} mode wavelength. As a rule, the cylindrical discharge tube is perpendicular to the wide wall, i.e. with its axis in the direction of the electric field, and at the position of the maximum field intensity.

Fehsenfeld et al.[21] devised and tested the TE_{103} mode discharge cavity shown in Fig. 4.2. Its design features ensure both efficiency and ease of operation. The discharge tube is inserted into a reduced-height section of the waveguide, perpendicular to its wide walls, at a $\lambda_g/4$ distance from its short-circuited end. The power is usually fed from a coaxial line to the standard guide portion by a coupling rod penetrating up to an optimum depth in the latter. The taper joining the reduced-height

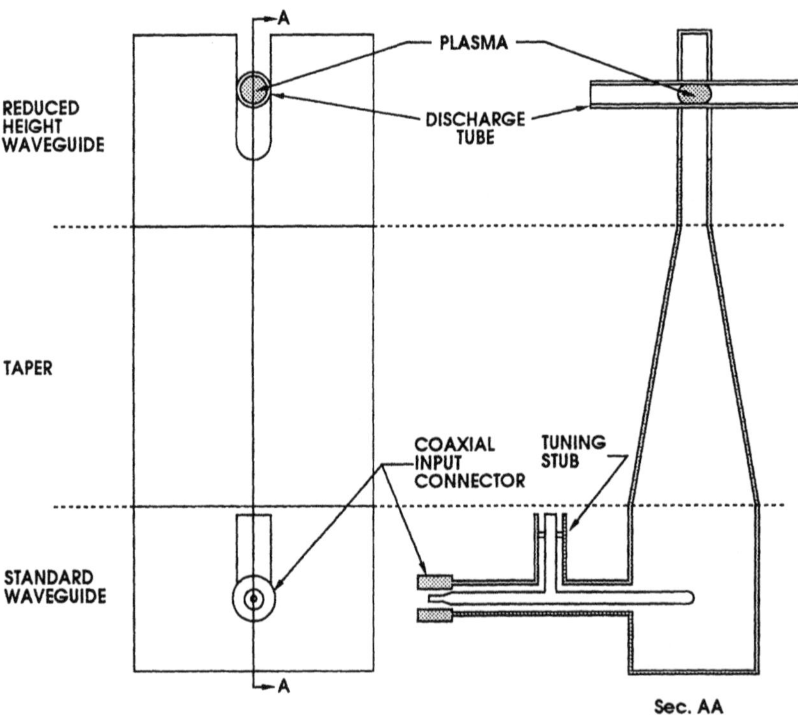

Fig. 4.2. Plasma source using a tapered rectangular cavity (after ref. 21).

[†] In waveguides, we have to consider TE (Transverse Electric) and TM (Transverse Magnetic) modes. The subscripts in the notation TE_{mnp} or TM_{mnp} for rectangular cavities refer to the number of half-wavelengths along the wide and narrow wall of the guide and along its axis respectively.

waveguide to the standard one serves as an impedance transformer to improve the matching;[†] it does not change the resonance length of the cavity, as λ_g depends only on its width for a TE_{10} mode. The slot-shaped receptacle for the plasma tube makes it possible to place the cavity in position without breaking vacuum. This discharge cavity performs best, for most gases, at pressures not higher than a few tens of torrs.

Foreshortened coaxial cavity. The resonator shown in Fig. 4.3 behaves like a section of coaxial line that is short-circuited at one end and loaded with a capacitor at the other. Openings made in the two walls that form the capacitive end allow a plasma tube to be inserted coaxially: the tube is thus surrounded by a ring-shaped gap through which the electric field extends from the resonator to sustain plasma.

The principle of the above field applicator is the basis of the foreshortened 3/4 wave coaxial cavity (Fig. 4.4) described in ref. 21. The resonator, formed by two concentric conducting tubes, is short-circuited at both ends. The discharge tube is placed inside the inner conductor. A gap in this inner conductor forms a plasma filled capacitor that is connected in series with the coaxial line. The power is supplied as shown in the figure. Tuning may be accomplished by varying both the gap length and the length of the tuning stub placed at the cavity input in parallel with the feed line. Various versions of this type of cavity are known collectively as Broida cavities. They are operated at gas pressures ranging from 1 torr to atmospheric pressure.

Circular cylindrical cavity. Consider a cavity formed by a section of a hollow cylindrical conductor, closed at both ends. In general, the resonant frequency and wave mode

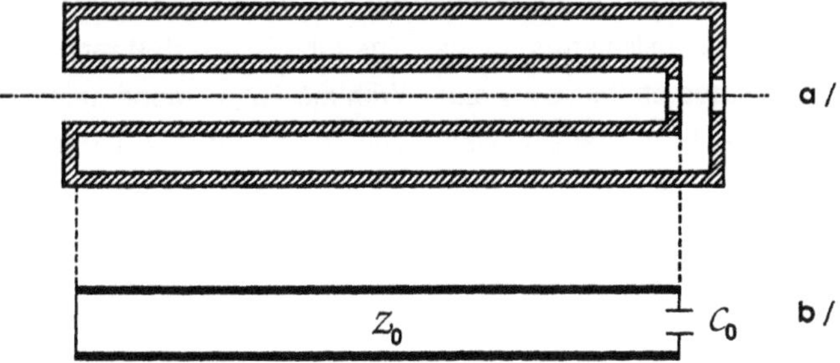

Fig. 4.3. Foreshortened coaxial cavity (a), and its equivalent circuit (b)where the parallel heavy lines represent a transmission line of characteristic impedance Z_0.

[†] Note that reducing the height of the cavity does not increase the field intensity within the plasma. As we pointed out in Sec. 2.2.9, the discharge maintenance field intensity is, to a first approximation, set by the particle loss mechanism. The manner in which the electromagnetic field is applied and the level of HF power absorbed are in that respect secondary.

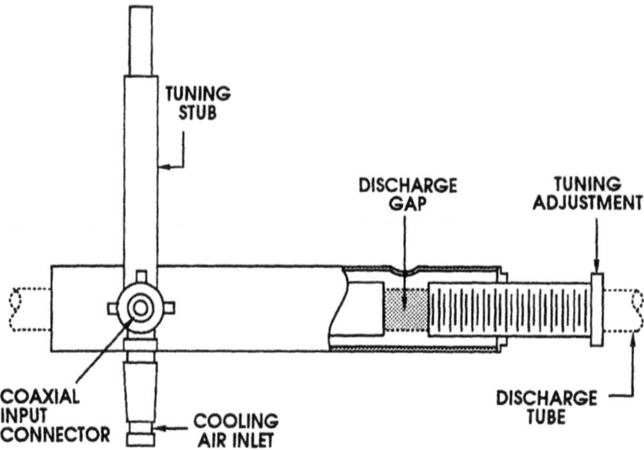

Fig. 4.4. Broida-type discharge cavity using the foreshortened coaxial cavity principle described in figure 4.3 (after ref. 21).

depend on all the dimensions of the cavity. The TM_{mnp} modes (Transverse Magnetic modes, where the subscripts denote in cylindrical coordinates the number of half-wavelengths in the azimuthal, radial, and axial directions respectively) with $p = 0$ are interesting: the resonance frequency of TM_{mn0} modes depends only on the cavity diameter and not on its length.

The short TM_{010} cavity used for plasma production shown in Fig. 4.5 has been popularized among analytical chemists by Beenakker.[22] It performs excellently when operated with atmospheric-pressure discharges in argon and in helium, sustained within small-diameter (few millimetres) tubes. The tube is inserted through holes cut in the flat walls and is located on the cavity axis. With the TM_{010} mode, this position is that of the maximum electric field intensity, which only varies radially and is parallel to the cavity axis. The power is supplied to the cavity from a coaxial feeder by means of a coupling loop. Two screws, shown in the figure, provide intrinsic tuning. A similar cavity was operated earlier by Fehsenfeld et al.[21] to sustain microwave discharges at reduced pressures.

4.3. Waveguide-based plasma sources

Various positions can be adopted for the discharge tube inside a rectangular waveguide. Depending on the orientation of the tube axis with respect to the direction of the microwave electric field, different discharge mechanisms may be involved. Recall that the electric field of the fundamental (TE_{10}) mode in a rectangular waveguide is perpendicular to the wide walls.

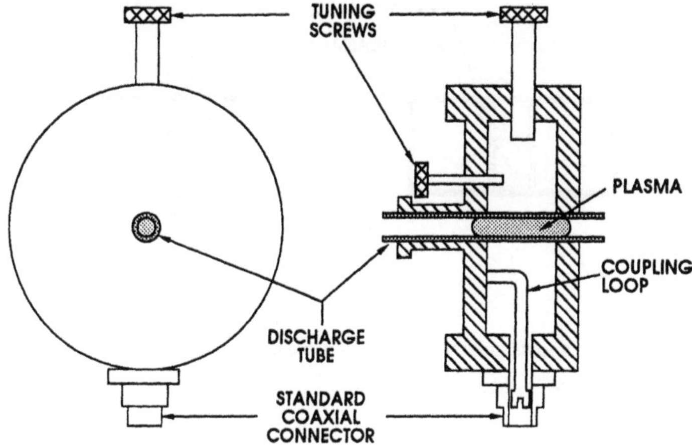

Fig. 4.5. Microwave discharge field applicator based on the TM_{010} mode cavity (after ref. 22).

The tube axis can be made perpendicular to the electric field in two interesting ways: i) the tube is placed along the length of the waveguide, which allows long, traveling-wave discharges;[23-25] ii) the tube is set across the waveguide and the discharge can then be maintained by a plasma resonance, as mentioned in Sec. 4.1.

When the tube axis is parallel to the electric field as in Figs. 4.6a and b, the discharge is easy to operate under a wide range of experimental conditions. This configuration is largely used,[10,11] both at reduced and atmospheric pressures. Note that the tube is either fully contained within the waveguide (Fig. 4.6a) or it runs across it through holes pierced in the wide walls (Fig. 4.6b). In Fig. 4.6b, we have not shown external metal sleeves or chokes that prevent the electromagnetic field from spreading into the room or from exciting surface-wave sustained plasmas. The extremity of the waveguide opposite to the power input is terminated either with a load of characteristic impedance Z_0 as in Fig. 4.7a, or with a movable short-circuiting plunger as in Fig. 4.7b.

A section of waveguide containing a plasma column can be analyzed as a two-port microwave network. The corresponding equivalent circuit can be readily inferred from solutions known in microwave circuit theory by considering the plasma column as a lossy dielectric post in a rectangular waveguide. Marcuvitz[26] presented a solution that is valid for a medium of arbitrary complex permittivity. In the case of a low pressure .

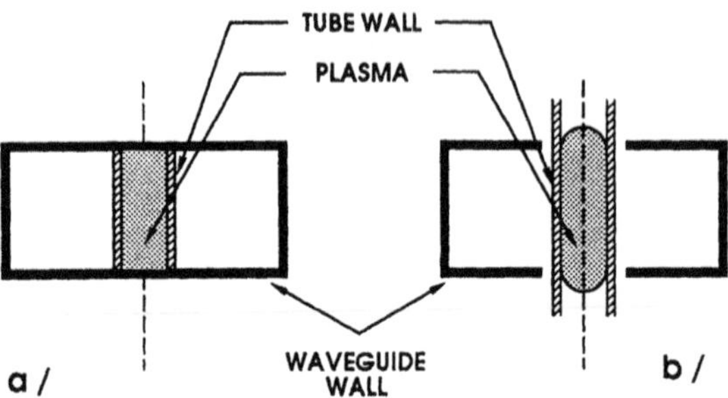

Fig. 4.6. Discharges inside a rectangular waveguide, with the tube parallel to the electric field: (a) *closed-cylinder* configuration, (b) *open-cylinder* configuration.

plasma of density not exceeding the critical value[†] by more than one or two orders of magnitude, and for 2R < 0.15 a_w (R - plasma radius, a_w - width of the waveguide), this equivalent circuit can be simplified as in Fig. 4.8, where a single impedance Z_p shunts the waveguide in the plane of the column axis. The dependence of this impedance on the plasma parameters is given approximately by[27]

$$Z_p/Z_0 = c_1 \frac{n_c}{\bar{n}} \frac{\nu}{\omega} + j\left(c_1 \frac{n_c}{\bar{n}} + c_2\right), \qquad (4.1)$$

where $\bar{n}$ is the average electron density in the plasma, ν is an average electron-heavy particle collision frequency for momentum transfer, and c_1, c_2 are shape coefficients

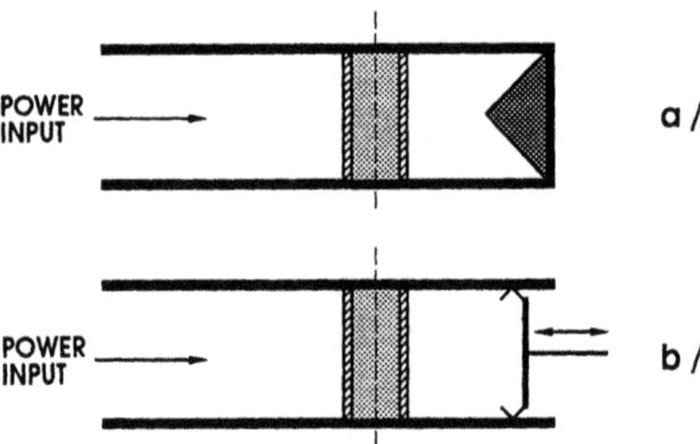

Fig. 4.7. Side view of the discharge configuration shown in Fig. 4.6a: (a) in a matched waveguide, (b) in a waveguide terminated with a movable short-circuit.

[†] The *critical electron density* n_c is defined as $n_c = \omega^2 \varepsilon_0 m_e / e^2$ where ω is the wave angular frequency, e and m_e are the charge and mass of electrons respectively, and ε_0 is the permittivity of free space. In fact, the value n_c is derived from the relation $\omega = \omega_{pe}$ where ω_{pe} is the electron plasma (angular) frequency (see Sec. 2.2.1). It yields $\omega^2/\omega^2_{pe} = n/n_c$ where n is the electron density.

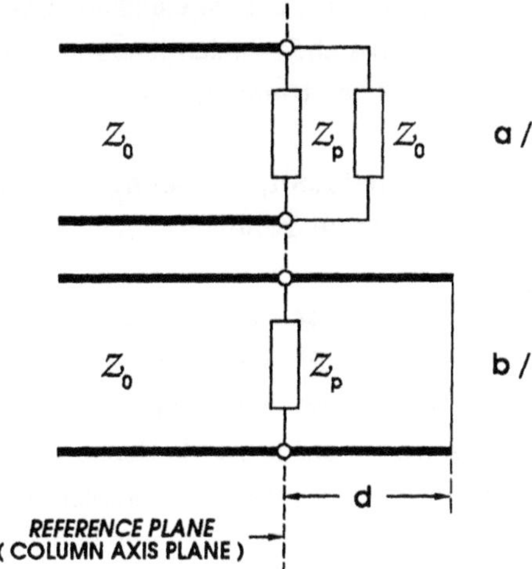

Fig. 4.8. Equivalent circuit of a plasma column located in a waveguide: (a) matched at the end, (b) terminated with a sliding short-circuit at a distance d from the reference plane. The parallel heavy lines represent a transmission line of characteristic impedance Z_0.

which, for a given waveguide and field frequency, only depend on the radius of the plasma column.

The above model has the virtue of simplicity, but its application is limited in terms of discharge conditions. More sophisticated equivalent circuits can be devised for these waveguide-based discharges that more closely reflect the various aspects of the experimental situations. For example, the equivalent circuit of Batenin et al.[11] includes, among others, elements representing the influence of the orifices, metal chokes as well as that part of the plasma column extending outside the waveguide.

Gerasimov et al.[28] and Dovzhenko et al.[29,30] have investigated experimentally reduced-pressure discharges with the configuration of Fig. 4.7a. The results obtained are interesting for the verification of discharge modeling and we shall present some of them in Sec. 4.5. Nevertheless, such a discharge is inefficient because part of the microwave power is either reflected back to the feed line or absorbed into the matching load that terminates the waveguide. Adding a tuner at the input of this system does not fully solve the problem because a significant part of the power is always absorbed in the matched load. This situation is improved by terminating the waveguide with a sliding short-circuit as in Fig. 4.7b. The system then has an intrinsic tuning element that ensures a reasonably good power transfer to the plasma over a wide range of discharge conditions and input power.[19,20,31] A further improvement consists in installing a tuner on the power input side of the discharge tube to optimize power transfer to the plasma. In such systems, a large standing wave may exist in the region between the tuner and the movable short, which leads to a resonant cavity situation and to the usual tuning and stability problems (Sec. 4.2).

4.4. Simplified (non self-consistent) modeling of reduced gas pressure discharges with a localized active zone

4.4.1. General procedure.

A full and exact discharge modeling necessitates a self-consistent formulation where the EM field and discharge equations are coupled. However, this usually requires complex numerical calculations, where the physical insight into the processes involved is obscured. In Sec. 2.4, we outlined the principles of a simplified discharge modeling that makes the problem more tractable and gives insight into the various factors influencing the discharge behavior. In contrast to a self-consistent treatment, this simplified modeling considers the discharge maintenance processes and the electromagnetic characteristics of the plasma source separately.

As discussed is detail in Sec. 2.2.9, the analysis of the discharge maintenance processes is based on two balance equations, namely that for charged particles (eqn. (2.19)) and that for electron energy (also known as the power balance per electron, eqn. (2.70)). For given discharge conditions, the first equation provides the relative spatial distribution, or profile, of the electron density, as well as the value of the maintenance field E_{rms} required to sustain the discharge (recall that the determination of E_{rms} from the charged particle balance equation necessitates that the EEDF be first determined by solving the homogeneous Boltzmann equation (2.32)). Then, the mean power absorbed or lost per electron, θ, is determined from the power balance per electron, using eqns. (2.70) and (2.71). The most complicated step in the above scheme is generally the determination of the EEDF since it requires sophisticated numerical codes to solve the Boltzmann equation, as well as a rather complete and accurate set of electron cross-section data which are often unavailable from the literature. For this reason, it may be useful, in some circumstances, to simplify the modeling procedure by assuming that the EEDF is for example a Maxwellian. In this case, for given discharge conditions, the charged particle balance equation determines the value of the electron temperature T_e instead of the maintenance field E_{rms}. The latter is then obtained through the power balance per electron: eqn. (2.71) yields θ_L as a function of T_e and eqn. (2.70) thus leads to E_{rms} as a function of T_e. Note that once θ is known, we have the total power lost by the electrons $P_L \equiv \bar{n} \mathbb{V}_D \theta$ expressed as a function of the electron density $\bar{n}$ averaged over the volume $\mathbb{V}_D$. Under steady-state conditions, the total power balance requires $P_L = P_A$, yielding $\bar{n}$. Then, the profile $n(r,z) / \bar{n}$ being previously known, we finally get the density distribution $n(r,z)$. The value of P_A, the total power absorbed in the plasma, is obtained from the second stage of our analysis.

The second part of the analysis concerns the plasma-loaded field applicator together with the matching network. It leads to $\eta(\bar{n})$, the efficiency of the microwave power transfer to plasma as a function of $\bar{n}$, which yields the power absorbed in the plasma since $P_A = \eta(\bar{n})P_I$, where P_I is the microwave power incident on the applicator. In many practical cases, $\eta(\bar{n})$ cannot be derived analytically, but is obtained directly from the measurement of the incident and reflected powers at the source input.

We apply in the following the simplified scheme just depicted to the specific case of discharges with a localized active zone. In particular, we shall determine the dependence of the plasma parameters on the discharge conditions and on the amount of microwave power delivered at the source input. This leads to the procedure shown schematically in Fig. 4.9. The initial data to this problem can be divided into three groups: discharge conditions, field applicator and matching network characteristics, and amount of power P_I incident at the source input. The end results sought are the electron temperature T_e, or average electron energy $<u> = (3/2) kT_e$ (assumed to be spatially constant throughout the plasma volume), the magnitude of the maintenance field E_{rms}, and the spatial distribution $n(r,z)$ of the electron density.

Now that the guidelines of our model have been presented, we apply it to the specific case of small diameter (i.e. negligible skin effect in the radial direction) discharges in microwave circuits; this leads us to discuss in more detail the charged particle balance in Sec. 4.4.2 and, in Sec. 4.4.3, the balance of the total microwave power and the stability of the discharge.

4.4.2. Charged particle balance. Consider in Fig. 4.10 the two idealized, possible configurations for discharge vessels in microwave circuits. The cylindrical vessel is either fully enclosed within the field applicator or it crosses it through orifices in the walls. In both cases, only the section delimited by the planes $z = \pm \ell/2$ is exposed to the microwave electric field. The latter is assumed i) to be parallel to the z axis and ii) to be uniform[†] within the whole vessel in the absence of plasma. These two idealized situations encompass most microwave discharges with a localized active zone. Examples of these are the discharges sustained in all the cavities described in Sec. 4.2.2 or in the rectangular waveguides of Sec. 4.3.

In modeling these discharges, for simplicity, we consider them to be sustained in the diffusion regime, recalling from Sec. 2.2.9.1 that the results obtained could nonetheless

---

[†] In most discharges sustained in microwave circuits, the tube diameter is much smaller than the cavity diameter or the waveguide wide wall. In fact, the tube diameter is typically less than 1/4 of these characteristic dimensions, hence the uniform field assumption in the absence of plasma. At 2.45 GHz, it generally corresponds to tube diameters of less than 15 mm.

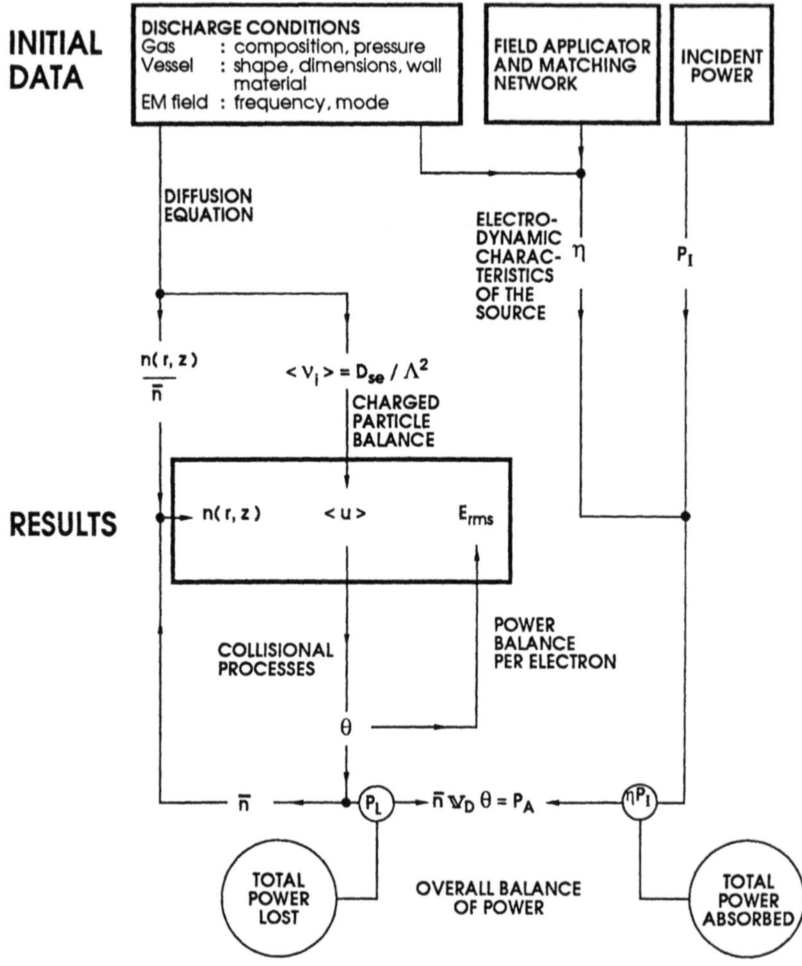

Fig. 4.9. Steps of the simplified model of microwave discharges with a localized active zone.

be easily extended to lower pressures (free-fall regime) in the form of a diffusion-type equation. The charged particle balance is then given by eqn. (3.3). In Chap. 2, we analyze the diffusion equation assuming a uniform field in the plasma and a long discharge tube. One objective of this section is to establish what happens to these results when the analysis of the diffusion equation is applied to a finite length discharge tube.

Charged particle balance: low electron density case. Let us first recall some facts about HF discharges under conditions where the electron density is so low that the presence of the plasma does not affect the imposed electric field distribution (assumed uniform here). This applies whenever the density is substantially smaller than the critical value (so-called underdense discharges).

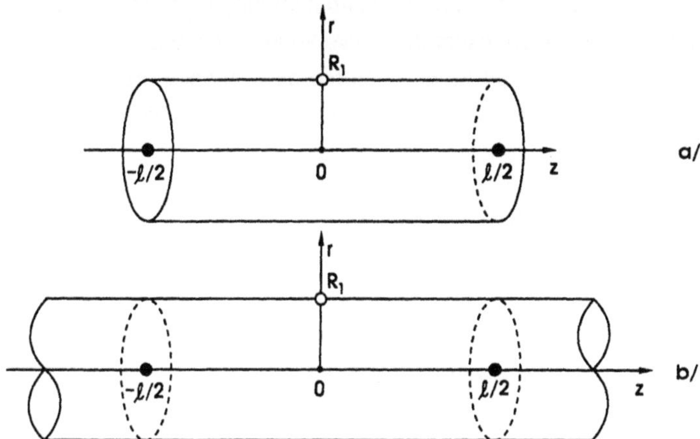

Fig. 4.10. Idealized configurations of discharge vessels in microwave circuits: (a) closed and (b) open-cylinder cases.

The solution of the diffusion equation for the *closed-cylinder* case is then the classical one that is found in plasma physics handbooks (see, for example, MacDaniel[32]). The electron density distribution is given by

$$n(r,z) = n_0 \cos(\pi z/\ell)\, J_0(x_0 r/R_1), \tag{4.2}$$

where n_0 is the electron density in the middle point of the vessel, and x_0 is the first zero of the Bessel function $J_0(x)$. In this case, the ionization frequency v_i is constant inside the plasma and satisfies the condition

$$<v_i>/D_a = 1/\Lambda^2 \equiv (x_0/R_1)^2 + (\pi/\ell)^2 = [1 + (\pi/2\kappa)^2]/\Lambda_\infty^2. \tag{4.3}$$

The diffusion length Λ depends only on the dimensions and shape of the vessel and for a long cylindrical tube, it is $\Lambda_\infty = R_1/x_0$. In eqn. (4.3), we have expressed Λ in terms of κ, a shape coefficient,[33] related to the extent of the active zone and defined as

$$\kappa \equiv x_0\ell/2R_1. \tag{4.4}$$

This shape coefficient tends to zero for very short cylinders and is large for long cylinders. Equation (4.2) shows that the electron density distribution in a closed cylinder, low density discharge is independent of κ.

In the *open-cylinder* case, the density distribution may be found by solving the time-independent diffusion equation inside the active zone and outside it. The radial density profile is the same as in eqn. (4.2) whereas the axial density profile now depends on the shape coefficient κ in contrast to the other cases to be considered below. The resulting $n(r,z)$ profile is[33]

$$n(r,z) = n_0 \cos (2\alpha_1 z/\ell) \, J_0(x_0 r/R_1), \quad \text{for } z \leq \ell/2 \tag{4.5}$$

and

$$n(r,z) = n_1 \exp \kappa \,(1 - |2z/\ell|) \, J_0(x_0 r/R_1), \quad \text{for } z > \ell/2, \tag{4.6}$$

with $\alpha_1 \tan \alpha_1 = \kappa$, $n_1 = n_0 \cos \alpha_1$. The electric field being uniform, the ionization frequency $\langle v_i \rangle$ is constant within the active zone and it is given by

$$\langle v_i \rangle / D_a = 1/\Lambda^2 \equiv (x_0/R_1)^2 + (2\alpha_1/\ell)^2 = \left[1 + (\alpha_1/\kappa)^2\right]/\Lambda_\infty^2. \tag{4.7}$$

The case of a short ($\kappa \ll 1$) open cylinder cannot be realized in practice because pronounced edge effects do not then permit one to assume, as required, that the field distribution in absence of plasma is uniform. For a slim open cylinder (large κ), $\alpha_1 \cong \pi/2$ and $n_1 \ll n_0$; therefore, the density distribution within the active zone closely resembles that in the closed-cylinder case and condition (4.3) applies.

The axial electron density profiles $\zeta(2z/\ell) \equiv n(r,z)/n_0 J_0(x_0 r/R_1)$ for both closed and open-cylinders are shown in Fig. 4.11a.

Charged particle balance: high electron density case. Now we turn to a more realistic situation where the electric field in the active zone of the microwave discharge is affected by the presence of the plasma (so-called overdense discharges). An exact analytical solution of the time-independent diffusion equation does not generally exist for this case. Such a problem was discussed by Allis et al.,[34] and later on by Bell,[35,36] in both cases for the so-called parallel-plate configuration where the discharge zone is enclosed between parallel metallic surfaces, either covered or not with a dielectric, with the transverse dimensions large compared to the mutual distance ($\kappa \ll 1$, closed cylinder case).

Recently, this high electron density case was considered[33] for the configurations of Fig. 4.10, under the following assumptions: i) the electron density within the active zone is substantially larger than the critical value, ii) the electric field is directed along the tube axis, iii) the plasma radius is small enough, as compared to the so-called skin depth, for the electric field to be radially uniform, provided the plasma-free field is

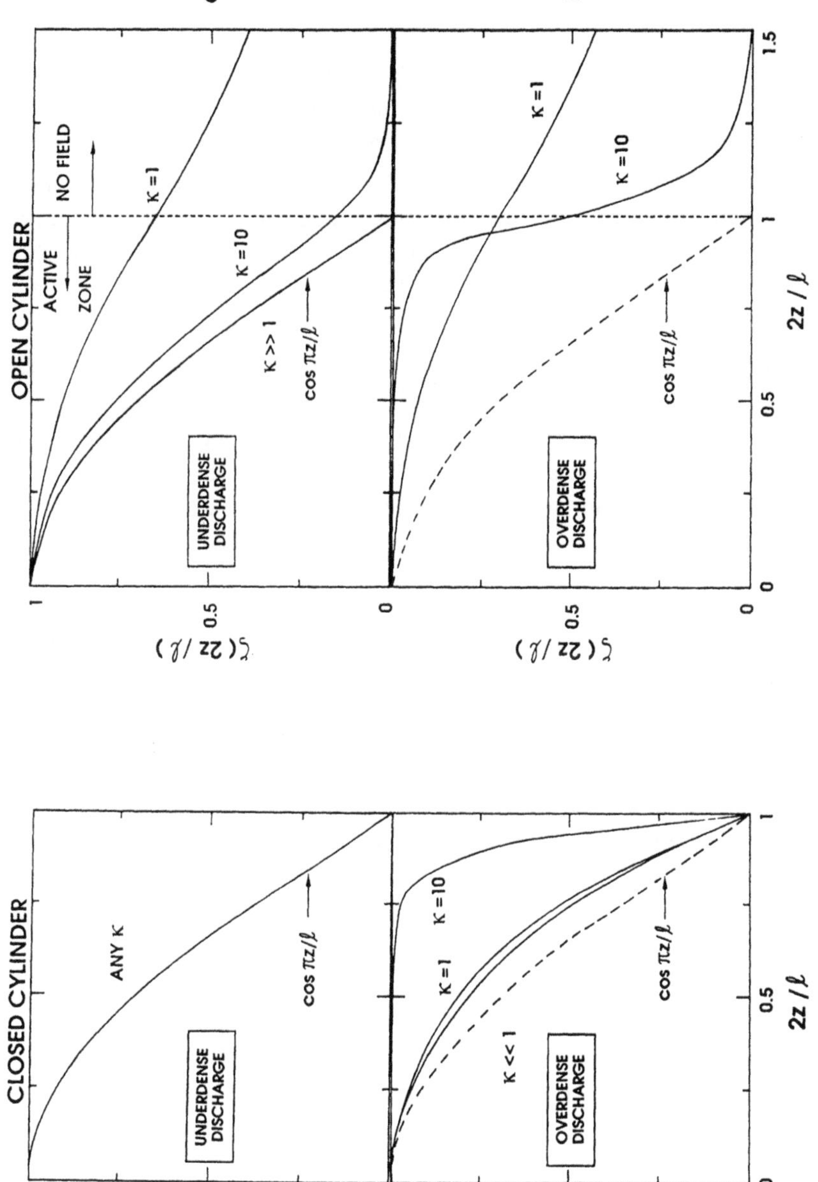

Fig. 4.11. Axial profile of the electron density in a cylindrical microwave discharge with a localized active zone, assuming the electric field to be uniform in absence of plasma: (a) undercritical density, or underdense, case; (b) overcritical density, or overdense, case, assuming negligible skin effect in the radial direction. See Fig. 4.10 for configurations and coordinate systems (from ref. 33).

uniform, iv) the ionization frequency depends upon the electric field intensity to the power $2\mathcal{T}$, where $\mathcal{T}$ is an adjustable parameter. Then, it can be shown that the electron density distribution is

$$n(r,z) = n_A(z)\, J_0(x_0 r/R_1), \tag{4.8}$$

where $n_A(z)$ is the density on the axis. The radial density profile $J_0(x_0 r/R_1)$ does not differ from that in the underdense case, whereas the axial profile $n_A(z)/n_0$, as we will see below, depends on the shape coefficient κ for both the closed and open-cylinder configuration. The influence of the electric field intensity, which is allowed to vary axially, is obtained by observing that the current must be continuous in the direction of the field, i.e. we have $\sigma_r(z)E(z) = \text{const.}$, where σ_r is the electric conductivity (eqn. (2.39)). It follows that the axial profile of the electric field obeys

$$E(z)/E_0 = n_0/n_A(z), \tag{4.9}$$

where $E_0 = E(0)$ and $n_0 = n(0, 0)$. Since the ionization frequency depends on the local value of the electric field, it also varies along the axis. The present formulation leads to consider the ionization frequency v_{i0} in the center point ($r = 0$, $z = 0$) of the steady-state discharge and we have

$$<v_{i0}> = D_a / \Lambda_D^2, \tag{4.10}$$

where Λ_D is an effective diffusion length. Figure 4.11b shows examples of axial density profiles while Fig. 4.12b shows the dependence of the ratio Λ_∞/Λ_D on the shape coefficient κ, both curves computed under the assumption that $<v_i>$ is proportional to E_{rms}^2 ($\mathcal{T} = 1$). This latter figure indicates[†] the extent to which the ionization frequency in the center of an overdense discharge deviates from that in a low density discharge sustained in a long vessel. Figures 4.11a and 4.12a compare the corresponding results for the low density case. We see that: i) the presence of an overdense plasma tends to flatten the axial electron density distribution, and consequently also that of the electric field and of the ionization frequency; this effect becomes more pronounced with increasing κ, so that discharges in long cylinders are approximately axially uniform; ii) there is no qualitative difference between the dependence of the characteristic diffusion length on κ in underdense and overdense discharges; iii) when κ increases, the effective diffusion length Λ_D in overdense discharges approaches the value Λ_∞ corresponding to the long-tube, low-plasma density case.

[†] Recall that provided D_a is independent of the electric field, then $\Lambda_D^2/\Lambda_\infty^2 = <v_i>_\infty / <v_{i0}>$, where Λ_∞ and $<v_i>_\infty$ are the diffusion length and the ionization frequency respectively in a long-tube, underdense discharge.

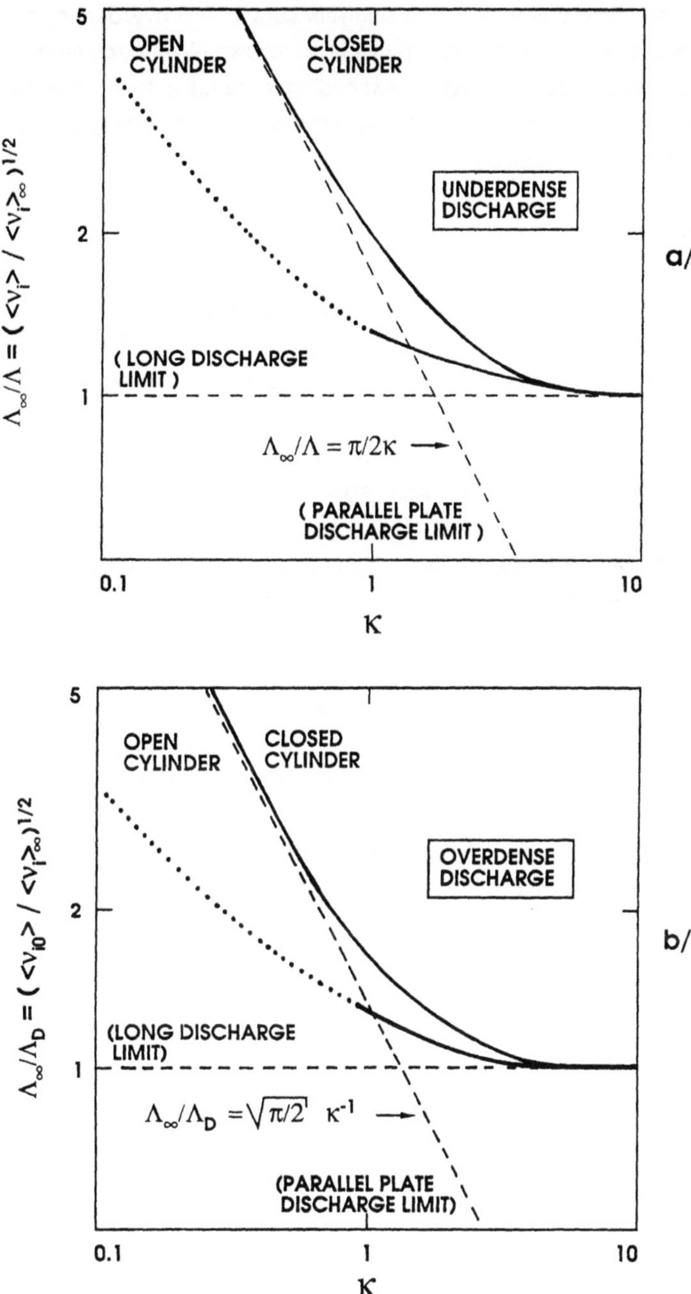

Fig. 4.12. Influence of the shape coefficient κ on the characteristic diffusion length Λ in an undercritical density discharge (a), and Λ_D in an overcritical density discharge (b). Note that Λ_∞ (and $<v_i>_\infty$) is the diffusion length in a long tube, low density discharge. The short, open cylinder situation, shown as dotted lines, cannot be realized in practice (from ref. 33).

The main practical outcome of the analysis concerning overdense, small diameter discharges sustained in microwave circuits is as follows. When the length of the active zone of such discharges increases, exceeding at least a few times their diameter ($\kappa \gtrsim 5$), then: i) unlike the underdense case, the electron density tends to be axially uniform and eqn. (4.8) reduces to

$$n(r,z) \simeq n_0 \, J_0(x_0 r / R_1);\tag{4.11}$$

ii) the electric field profile, which varies inversely as the density profile, also tends to be axially uniform; iii) the value $\Lambda_\infty / \Lambda_D$ tends toward unity, thus the field intensity, for given discharge conditions, then tends to be the same as in underdense discharges.

In summary, the results obtained in Sec. 2.2.9 under idealized conditions can be applied directly to overdense discharges sustained in microwave circuits, provided that the discharge tube, placed parallel to the electric field (axially uniform in the absence of plasma), is at least moderately long, namely $\kappa \gtrsim 5$, and that its diameter is small enough for the maintenance field to be radially uniform.

Average electron energy and the maintenance field intensity. Under the assumption (see Sec. 4.4.1) that the EEDF is a known function (for example a Maxwellian), then the average electron energy <u> in the steady-state discharge follows from the charged particle balance, which is an implicit equation in terms of <u>. Then, the average power θ loss or absorbed per electron comes out from eqn. (2.71). Finally, the maintenance field intensity E_{rms} is determined (from eqn. (2.70), using eqn. (2.41), from which it readily[†] follows that

$$E_{rms}^2 = \frac{m_e}{e^2 v}\,(\omega^2 + v^2)\,\theta\,.$$

$$\tag{4.12}$$

4.4.3. The balance of total microwave power and the stability of the discharge. The parameters that follow from the charged particle balance and the power balance per electron, namely <u> and E_{rms}, are fixed solely by the discharge conditions (see Fig. 4.9 for a definition of the latter); as discussed in Sec. 2.2.9, neither the field applicator design nor the power absorbed by the plasma affect, to a first approximation, these parameters. Finally, there remains to determine the average electron density $\bar{n}$, which depends on the power P_A absorbed in the plasma through the overall balance of the microwave power, as shown in Fig. 4.9.

---

† When using eqn. (2.71) either the shape of the electron energy distribution function must be known, or the approximation <vu> $\cong$ <v> <u> used.

In a steady-state discharge, the power ηP_I delivered to the plasma from the feed line must be equal to the power P_L lost by electrons in collisions of all kinds (Sec. 2.2.9). For a plasma sustained in a cylindrical vessel of length ℓ and internal radius R_1, we have

$$\eta(n)P_I = P_L \equiv \pi\, R_1^2\, \ell\, \theta\, \bar{n}, \tag{4.13}$$

where $\eta(n)$ is for a given plasma source. It depends on the discharge conditions, on the amount of power delivered to the plasma, and on the setting of the tuning elements in the plasma source. It follows from eqn. (4.13) that, whenever θ does not depend on the electron density, the average density

$$\bar{n} = \frac{\eta P_I}{\pi R_1^2 \ell \theta} \tag{4.14}$$

is proportional to $\eta\, P_I$, the power absorbed in the plasma. Figure 4.13 shows schematically how the power absorbed by the plasma from the feed line is finally lost to the environment.

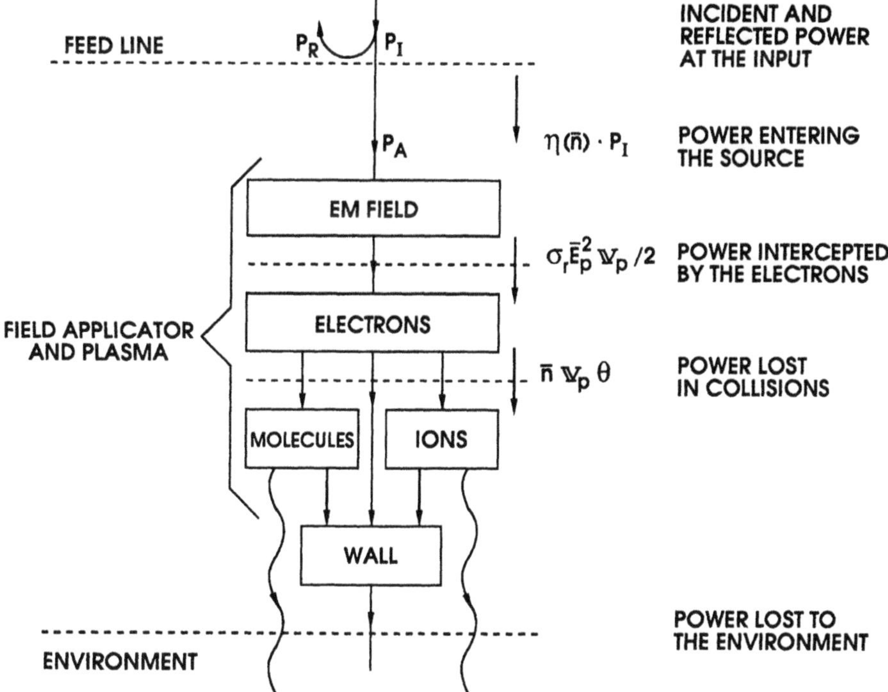

Fig. 4.13. Power flow in a microwave discharge.

The balance of power in a microwave discharge may turn out to be unstable, therefore, the energy balance condition (4.13) is necessary, but not sufficient, for a steady-state discharge to exist. Consider a microwave plasma source operated under given discharge conditions and with a particular setting of the matching elements. Then the efficiency of power transfer η can be expressed as a function of only the average electron density $\bar{n}$ (see further eqn. (4.21)). Figure 4.14 shows the power balance in a microwave discharge. The absorbed power curves $P_A = \eta(\bar{n})P_I$ correspond to a given P_I value and all have the same shape, with P_I as the multiplying constant. The loss curve P_L for given discharge conditions, and with θ independent of density, is a straight line. For each value of the power delivered to the plasma source, the average electron density in the plasma corresponds to the crossing point of the P_L and P_A curves. There is no discharge if the power delivered is too small (P_{I1}). With P_{I2} and P_{I3}, a steady-state discharge occurs at the points indicated by circles. There is no steady-state discharge at the points marked by squares. This behavior can be explained as follows.

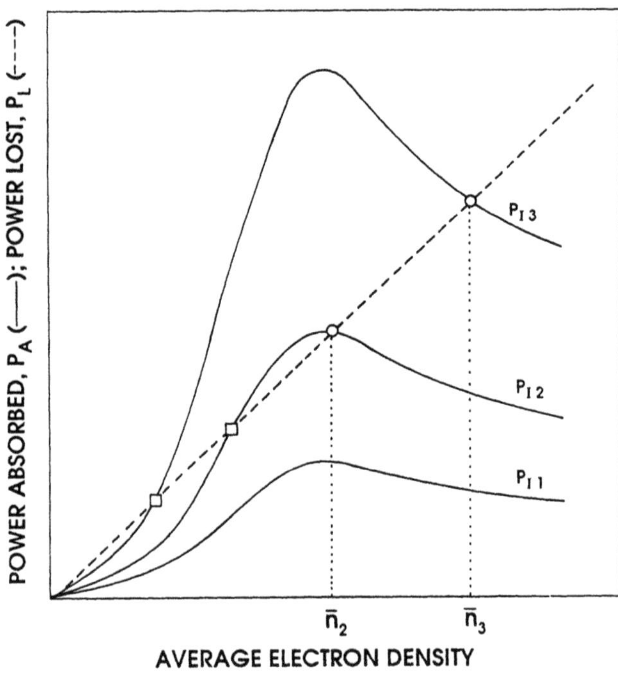

Fig. 4.14. Sketch illustrating the stable (o) and unstable (□) power balance in a steady-state HF discharge. Each absorbed power curve $P_A = \eta\,(\bar{n})\,P_I$ corresponds to a given incident microwave power P_I to the plasma source, where η is the efficiency of power transfer: these various curves are assumed here to have the same shape, P_I being the multipliying factor. Note that the power P_L lost by the electron in collisions of all kinds does not depend on P_A.

The discharge is stable if a random fluctuation of the electron density triggers a sequence of events that leads to the restoration of the original density. For example, a random decrease of the electron density causes the electrodynamic characteristics of the whole plasma-HF circuit to vary, i.e. $\eta(\bar{n})$ changes (θ may also change, if it is not density independent). If all this corresponds to a decrease of the power gained by the electrons relative to the power lost in collisions, a further decrease of the density occurs that ultimately leads to the extinction of the discharge. On the other hand, the original value of the electron density is restored when any decrease of the electron density causes the power absorbed from the electromagnetic field to prevail over the power loss due to electron collisions and, conversely, if the electron density was to increase from its stable point value. The following condition therefore must be fulfilled

$$dP_L/d\bar{n} > dP_A/d\bar{n}. \tag{4.15}$$

This condition was first formulated[37,8] on the basis of the stability of resonantly sustained plasmas. Within the present model, it becomes

$$\frac{d\eta(\bar{n})}{\eta(\bar{n})} - \frac{d\theta(\bar{n})}{\theta(\bar{n})} < \frac{d\bar{n}}{\bar{n}}. \tag{4.16}$$

The criterion (4.16) determines the range of electron density (and thus that of absorbed power) over which a steady-state discharge can be operated in a given plasma source and under given discharge conditions. In the particular case where θ is density independent, condition (4.16) reduces to

$$\frac{d\eta(\bar{n})}{d\bar{n}} < \frac{\eta(\bar{n})}{\bar{n}}, \tag{4.17}$$

where η and $\bar{n}$ are both positive quantities.

4.5. Modeling of waveguide-based discharges

4.5.1. General remarks. Discharges within waveguides seem to be the only ones among discharges with a localized active zone, that can be easily modeled to provide physical insight into the discharge phenomena. This is because their equivalent circuit, and therefore their electrodynamic characteristics, can readily be obtained for most discharge configurations. Section 4.5.2 presents an example of the modeling procedure. The electrodynamic characteristics of the discharge are derived analytically using a very crude equivalent circuit impedance for the plasma column, as given by eqn. (4.1). Although the characteristics thus obtained are rather qualitative, they suit our didactic purposes.

The group headed by Solntsev seems to have provided the most complete set of results concerning the modeling of waveguide-based discharges and its experimental verification.[28-30] Some of their results are presented in Sec. 4.5.3. All these discharges were sustained within a rectangular waveguide terminated with a matched load: this explains why we dwelt on this configuration in Sec. 4.3.

4.5.2. The analytical model. We now apply to waveguide-based discharges the simplified modeling procedure described in Sec. 4.4.

We have seen that the analysis of the charged particle balance and power balance per electron yields θ and E_{rms} respectively (Fig. 4.9), that these values are determined for given discharge conditions and are approximately independent of the power absorbed in the plasma and of the waveguide dimensions.

To obtain the electrodynamic characteristics of the plasma source, we shall use the simple equivalent circuit of Fig. 4.15. The waveguide is terminated with an impedance equal to its characteristic impedance Z_0. The discharge is represented by a single impedance Z_p, shunting the waveguide at the position of the tube axis, the value of Z_p being given by eqn. (4.1). Assuming that $c_1 \gg c_2 \, \bar{n}/n_c$, we have the normalized admittance of the plasma

$$y_p \equiv Z_0/Z_p \equiv g_p + jb_p \simeq \left(\frac{v}{\omega} - j\right)\bar{n} \left/ \left[n_c c_1 \left(1 + \frac{v^2}{\omega^2}\right)\right] \right. = \left(1 - j\frac{\omega}{v}\right)\bar{n}/n_p \,, \tag{4.18}$$

where

$$n_p \equiv n_c c_1 \left(1 + \frac{v^2}{\omega^2}\right)\frac{\omega}{v} \tag{4.19}$$

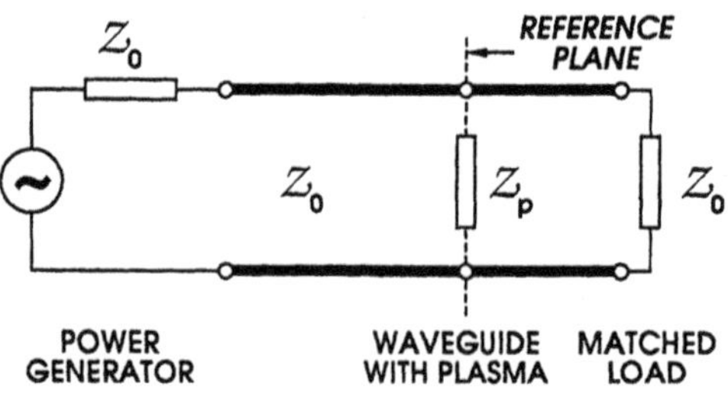

Fig. 4.15. Equivalent circuit of the waveguide-based plasma source terminated with a matched load. The parallel heavy lines represent a transmission line of characteristic impedance Z_0.

is the electron density at which the plasma conductance equals the characteristic admittance Z_0^{-1} of the waveguide.

For a given waveguide and a given frequency, the shape coefficient c_1 is a function of the ratio of the tube diameter to the waveguide width and is usually much larger than unity. It follows that n_p is always substantially larger than the critical density n_c. Let us further define P_p as the total power carried in the waveguide by a traveling wave whose electric field is assumed to have the same intensity as that of the maintenance field. Then,

$$P_p \equiv \frac{a_w\, b_w}{4Z_0}\, E_{rms}^2 , \tag{4.20}$$

where a_w and b_w ($a_w > b_w$) are the transverse dimensions of the waveguide. Both parameters n_p and P_p depend only on the discharge conditions, and not on the power delivered to the plasma.

The equivalent circuit of Fig. 4.15 together with eqn. (4.18) leads to an analytical expression for the efficiency of power transfer to the plasma, namely

$$\eta\,(\bar{n}) = P_A/P_I = 4\left\{g_p\left[(1 + 2/g_p)^2 + \omega^2/\nu^2\right]\right\}^{-1} , \tag{4.21}$$

where P_I is the power in the incident wave and P_A is the power absorbed in the plasma. The remaining wave power $(P_I - P_A)$ is either reflected by the discharge or passes through it and becomes absorbed in the matched load. Equation (4.21) expresses the fact that, under given discharge conditions, η depends only on the average electron density, which ultimately depends on the power absorbed in the plasma.

Recall that the absorbed microwave power P_A is intercepted only by the electrons of the discharge. In a steady state, P_A is equal to the power P_L lost by electrons in collisions of all kinds (eqn. (4.13)) and P_L is proportional to $\bar{n}$. Figure 4.16 shows P_A and P_L calculated as functions of $\bar{n}$, using the above analytical relations. The power and the density are normalized with respect to P_p and n_p respectively. This figure illustrates the overall balance of microwave power for the specific case of waveguide-based discharges. Figure 4.17 shows the efficiency of power transfer $\eta = P_A/P_I$ and the average electron density $\bar{n}$ as functions of the incident wave power P_I, also in a normalized form.

From these results, we draw the following important conclusions: i) the discharge in a waveguide exists only when the incident wave power exceeds a certain minimum value, which is equal to P_p; ii) the fraction of the incident power that is absorbed in the

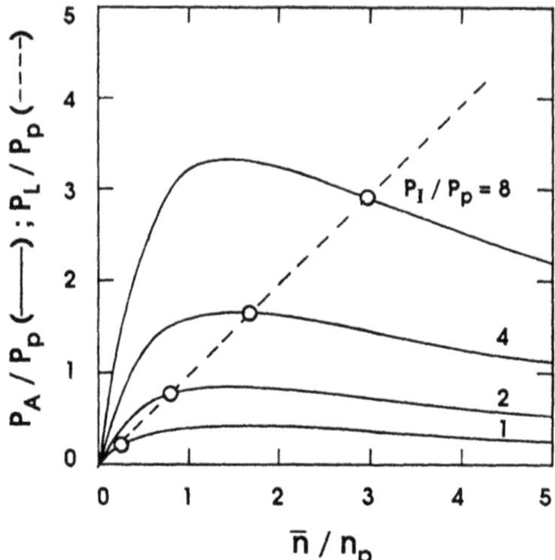

Fig. 4.16. The absorbed microwave power P_A and the power P_L lost by the electrons as functions of the average electron density $\overline{n}/n_p$ in the waveguide-based plasma source of Fig. 4.15, at various incident powers P_I. These curves are calculated analytically using the simple plasma admittance expression given by eqn. (4.18). The crossing points between curves P_A and P_L yield the electron density corresponding to a given incident power P_I. The quantities n_p and P_p are normalizing factors defined by eqns. (4.19) and (4.20) respectively and they do not depend on the power delivered to the plasma.

plasma never exceeds 1/2,[†] and it decreases with decreasing values of ν/ω; iii) the average electron density increases with increasing incident wave power; iv) any reasonable efficiency of power transfer to the plasma requires that the electron density significantly exceeds n_c (recall that always $n_p \gg n_c$). Also, as can be seen in Fig. 4.17, too small a value of $\overline{n}$ may lead to discharge stability problems, because of the steep slope of the η curve.

4.5.3. Experimental verification. All experiments[28-30] were performed with the following configuration. The discharge tube is a closed cylinder of 12 mm i.d., situated in a standard WR-284 waveguide ($a_w = 72$ mm, $b_w = 34$ mm).[††] The frequency was 3.3 GHz and the power could be varied between 3 and 100 W. The authors monitored the incident, reflected, and transmitted wave power. A triple Langmuir probe was used to measure plasma parameters, and the discharge was sustained in various noble gases at pressures from hundreds of mtorr up to tens of torr.

[†] Recall that the waveguide is terminated here by a matched load. As already mentioned, the efficiency improves when the load is replaced by a movable short.

[††] The corresponding shape coefficients are $\kappa \cong 7$ and $c_1 \cong 3.75$.

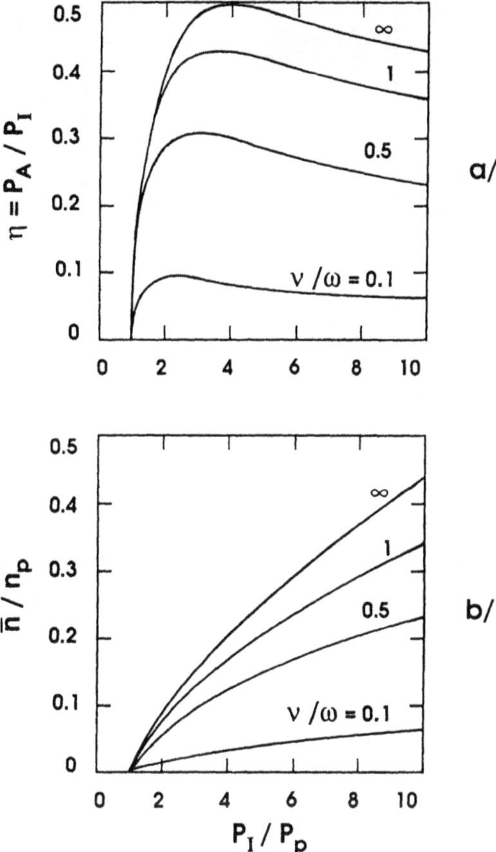

Fig. 4.17. The efficiency $\eta = P_A/P_I$ of the power transfer to the plasma (a) and the average electron density $\bar{n}$ (b) as functions of the incident wave power P_I, for the waveguide-based plasma source of Fig.4.15, at various ratios of the average collision frequency to the field angular frequency. These curves are calculated analytically using the simple plasma admittance expression given by eqn. (4.18).

The same authors[28-30] modeled the discharge, using a simplified (not self-consistent) procedure similar to that described in Sec. 4.4. However, in contrast to our calculations in Sec. 4.5.2, they used the complete Marcuvitz-type[26] equivalent circuit for the plasma column. They compared the results of the model with experimental data, as shown in Figs. 4.18 and 4.19.

Figure 4.18 shows the measurements of the electron density as a function of the incident wave power. One set of data was obtained from microwave power measurements which made possible to calculate the electron density and collision frequency in the plasma. The other set comes from probe measurements. The values obtained from microwave and probe diagnostics do not differ by more than 50%. The

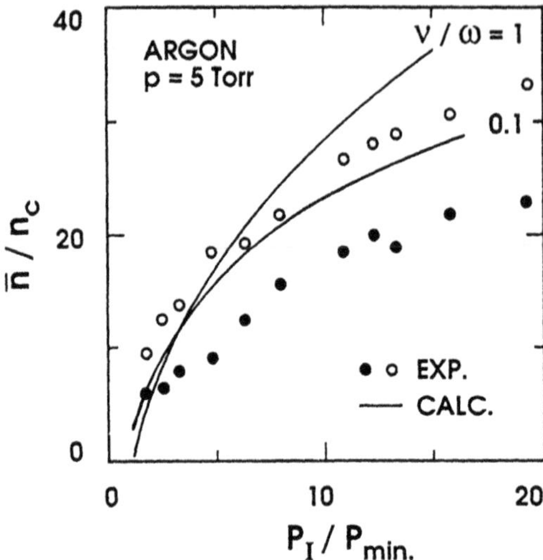

Fig. 4.18. Average electron density as a function of the incident wave power for the waveguide-based plasma source of Fig. 4.15: (o) determined from measurements of incident, reflected and transmitted wave power, (●) determined from probe measurements. The curves show the dependence calculated numerically from the complete Marcuvitz equivalent circuit for the plasma column, at two values of ν/ω.[28] P_{min} is the minimum power required to sustain the discharge and it corresponds to our P_p.

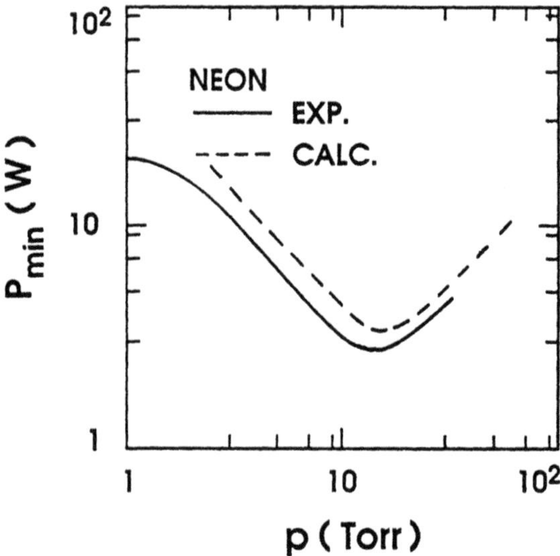

Fig. 4.19. Measured minimum power necessary to sustain a discharge in a waveguide-based plasma source (solid line) compared to values calculated from data obtained in a positive column under the same discharge conditions (dashed line).[28]

same figure also shows the results from their simplified modeling for two assumed values of v/ω. There is a good qualitative agreement between the calculated and measured data. Note that $\bar{n}$ is at least a few times n_c.

In comparing our analytical results in Fig. 4.17 with those computed numerically by Gerasimov et al.[28] in Fig. 4.18, one must take into account the difference in the normalization of the electron density, which is in terms of n_p and of n_c respectively. Then, the corresponding curves can be fitted to yield a common value of shape coefficient $c_1 \simeq 4.1$, which is in good agreement with the experimental value $c_1 \simeq 3.75$.

Figure 4.19 shows the minimum incident microwave power P_{min} necessary to sustain the discharge as a function of the gas pressure (to compare with the simple model of Sec. 4.5.2, note that our P_p is actually P_{min}). These values are compared with those observed in the positive column of a DC discharge. The close agreement between these two sets of results tends to support the idea developed in Sec. 2.2.7 that the DC electric field intensity and the effective value of the high frequency field are the same under the same discharge conditions (this is rigorously true only if v_c = const., see Sec. 3.2.2). In addition, these authors found that, in the present case,[†] the measured electron temperature in the microwave discharge was the same as in a DC positive column under the same discharge conditions. This supports the assumptions used in Sec. 4.4 that the radial field nonuniformity (assumed to be weak) resulting from the "skin" effect plays a secondary role in the present balance processes.

References

[1]   J. Marec, E. Bloyet, M. Chaker, P. Leprince and P. Nghiem, in *Electrical Breakdown and Discharges in Gases*, E.E. Kunhardt and L.H. Luessen eds. (Plenum, New York, 1982) p. 347 (part B).

[2]   A.T. Zander and G.M. Hieftje, Appl. Spectros. **35**, 357 (1981).

[3]   J.P. Matousek, B.J. Orr and M. Selby, Prog. Analyt. Atom. Spectrosc. **7**, 275 (1984).

[4]   S.R. Goode and K.H. Baughman, Appl. Spectrosc. **38**, 755 (1984).

[5]   P.E. Vandenplas, *Electron Waves and Resonances in Bounded Plasmas* (Wiley, London, 1968).

[6]   M. Moisan, Plasma Phys. **16**, 1 (1974).

[7]   A.M. Messiaen and P.E. Vandenplas, Appl. Phys. Lett. **15**, 30 (1969).

[8]   P. Leprince, G. Mathieussent and W.P. Allis, J. Appl. Phys. **42**, 412 (1971).

[9]   E. Bloyet, P. Leprince and H. Milleon, J. Physique. **34**, 1985 (1973).

---

† The degree of ionization in the microwave discharge and in the positive column was large enough to cause maxwellization of the EEDF as a result of electron-electron collisions, which thus excludes frequency effects on the EEDF.

[10]   L.S. Polak, A.A. Ovsiannikov, D.I. Slovietskii and F.B. Vurtsel, *Theoretical and Applied Plasma Chemistry* (Nauka, Moscow, 1975), in Russian.

[11]   V.M. Batenin, I.I. Klimovski, G.V. Lisov and V.N. Troiski, *Microwave Plasma Generators: Physics, Techniques, Applications* (Energy-Atomizdat, Moscow, 1988), in Russian.

[12]   J.D. Cobine and D.A. Wilbur, J. Appl. Phys. **22**, 835 (1951).

[13]   W. Schmidt, Electron. Rundschau **3**, 404 (1959).

[14]   J. Swift, Electron. Eng. **38**, 152 (1966).

[15]   G.S. Sobering, T.D. Bailey and T.C. Forrar, Appl. Spectrosc. **42**, 1023 (1988).

[16]   J. Asmussen, R. Mallavarpu, J.R. Hamann and H.C. Park, Proc. IEEE **62**, 109 (1984).

[17]   S. Whitehair, J. Asmussen and S. Nakanishi, J. Prop. and Power **3**, 136 (1987).

[18]   S. Ramo, J.R. Whinnery and T.V. Duzer, *Fields and Waves in Communication Electronics* (Wiley, New York, 1965).

[19]   G. Böhm, Z. Naturforsch. **3a**, 293 (1980).

[20]   H. Schlüter, University of Bochum, direct communication.

[21]   F.C. Fehsenfeld, K.M. Evenson and H.P. Broida, Rev. Sci. Instrum. **36**, 294 (1965).

[22]   C.I.M. Beenakker, Spectrochim. Acta **31B**, 483 (1979).

[23]   B. Kampmann, Z. Naturforsch. **34a**, 423 (1979).

[24]   V.A. Dovzhenko, D.O.M. Melnichenko and G.S. Solntsev, Sov. Phys.-Techn. Phys. **22**, 190 (1977).

[25]   Z. Zakrzewski, Czechosl. J. Phys. **B34**, 105 (1984).

[26]   N. Marcuvitz, *Waveguide Handbook* (McGraw-Hill, New York, 1951).

[27]   Z. Zakrzewski, Bull. Acad. Polonaises des Sci., Série Sci. et Techn. **14**, 939 (1966).

[28]   N.C. Gerasimov, V.A. Dovzhenko, T.P. Lebedeva and G.S. Solntsev, Proc. 10th Int. Conf. Phenom. in Ionized Gases, p. 1949 (London, 1971).

[29]   V.A. Dovzhenko and G.S. Solntsev, *Proc. 11th Int. Conf. Phenom. in Ionized Gases*, p. 158 (Prague, 1973).

[30]   V.A. Dovzhenko, G.S. Solntsev and V.I. Neshchadimenko, Radiotekhn. i Elektr. **9**, 1876 (1973), in Russian .

[31]   Z. Zakrzewski, M. Lubanski and T. Kopiczynski, J. Musil, Acta Phys. Slov. **32**, 201 (1982).

[32]   E.W. McDaniel, *Collision Phenomena in Ionized Gases* (Wiley, New York, 1964).

[33]   Z. Zakrzewski, M. Moisan, J. Margot and G. Sauvé, Plasma Sources Sci. Tech. **1**, 28 (1992).

[34]   W.P. Allis, S.C. Brown and E. Everhart, Phys. Rev. **84**, 519 (1951).

[35]   A.T. Bell, I and EC Fundamentals **9**, 160 (1970).

[36]   A.T. Bell, in *Techniques and Applications of Plasma Chemistry*, J.R. Hollohan and A.T. Bell eds. (Wiley, New York, 1974).

[37]   J. Taillet, Am. J. Phys. **37**, 423 (1969).

CHAPTER 5

SURFACE-WAVE PLASMA SOURCES [*]

5.1. Introduction

Electromagnetic surface waves can be used to sustain plasma columns. As a rule, the wave is excited at one extremity of the plasma column by a so-called *wave launcher*, and then travels along both the plasma column that it sustains and the dielectrics enclosing it, these media constituting the wave sole propagating structure.

Surface-wave produced plasmas are part of a larger class of high frequency (HF) discharges, termed traveling-wave discharges (TWD), which we briefly introduce in Sec. 5.2. The surface wave can propagate in different modes; in a discharge of cylindrical configuration, each mode is characterized by the variation of its electric field intensity upon the azimuthal angle. The most widely and easily achieved surface wave discharge calls on the azimuthally symmetric mode. Section 5.3 reviews according to theory the properties of such a wave when it propagates over a plasma column assumed for simplicity to be uniform, while Sec. 5.4 examines the characteristics of the discharge that it produces. The latter section also outlines the procedure for a simplified modeling of discharges sustained by traveling waves. The question of efficiently launching surface waves to sustain a plasma is developed in Sec. 5.5 and examples of practical realizations of surface-wave plasma sources are given in Sec. 5.6. In Sec. 5.7, experimental measurements are compared with some aspects of the modeling of surface-wave discharges. Finally, Sec. 5.8 summarizes the features of surface-wave plasma columns that are of interest for the user.

5.2. Surface-wave and other traveling-wave sustained discharges

The traveling-wave discharges constitute a category of HF discharges that started to be developed and investigated in the 1970s. Their common characteristic is that the energy delivered to the active plasma zone comes from the power flux carried by a traveling electromagnetic wave propagating along the plasma column. Within this category, one can distinguish two types of discharges: those that need a wave-guiding structure along the plasma column for the wave to propagate and those that do not, as is the case with surface-wave discharges. In both situations, there is no principle limitation to the plasma column length that can be achieved, but in practice limits are

---

[*] **Presented by M. Moisan and Z. Zakrzewski**

imposed by the power handling capability of the wave-guiding structure or of the wave launcher respectively. Such a feature contrasts with the generally limited plasma volume that can be obtained with the HF discharges presented in the previous chapter.

Various traveling-wave discharges have been described in the literature. For example, Bosisio et al.[1,2] studied long plasma columns sustained from the microwave field of an accompanying periodic, ladder-type slow-wave structure. Dovzhenko et al.[3] and Kampmann[4] obtained long plasma columns with discharge tubes placed axially within a rectangular and circular waveguide respectively. However, surface-wave discharges are by far the type of traveling-wave discharges that have attracted the most attention. As a matter of fact, surface-wave discharges have become the best documented, both experimentally and theoretically, of all kinds of microwave discharges, hence their considerable didactic potential. Also, their use as plasma sources is increasing in many fields. These facts led us to devote the present chapter to surface-wave discharges.

The fact that a non-ionizing electromagnetic surface wave can be guided along the interface between a plasma and its surrounding dielectric media has been known for many years.[5-8] It seems that Bulkin et al.[9] were the first to observe a discharge sustained by a surface wave when they reported the existence of a wave with a longitudinal electric field component along a pulsed discharge sustained in a waveguide. Tuma,[10] who employed a surface-wave discharge to excite a laser medium, was probably the first to clearly identify its nature. In fact, in the literature, one encounters reports on microwave produced plasmas that were obviously sustained by surface waves but not recognized as such (see, for example, refs. 11-13). Moisan et al.[14-16] developed a first series of simple, compact and highly efficient surface-wave launchers, based on coaxial and waveguide components, that were particularly suitable for the microwave generation of long plasma columns. The nature of the wave sustaining such plasma columns was definitively confirmed by Zakrzewski et al.[17] who measured its dispersion and attenuation characteristics and compare these with theory.

Though surface-wave discharges seem to be the best characterized and modeled of all types of microwave discharges, various aspects of their physics, techniques and applications are still the subject of interesting work and new results are continuously being reported. The progress made over the last decade can be found in a few review papers.[18-23]

5.3. Properties of the azimuthally symmetric surface wave propagating along uniform plasma columns (summary of theory)

5.3.1. Surface-wave modes.
A bounded plasma may support electromagnetic (EM) surface waves that are guided along the boundary surface, the energy flux of these waves being concentrated in the vicinity of this surface, hence their name. As a rule, the plasma is contained within a cylindrical dielectric tube, which is sometimes placed coaxially within a metal enclosure. However, the metal-free configuration is the one most often used.

As already mentioned, the possible modes of propagation of surface waves along a cylindrical medium are defined by their field dependence upon the azimuthal angle ϕ. Each of these propagation modes is determined by the integer m entering the amplitude factor $\exp(jm\phi)$ that affects the field intensity components.[†] The m = 0 mode, which exhibits an azimuthally symmetric field distribution, is the one mostly found in surface-wave plasma research and applications.

All the considerations and results presented in this section concern the wave propagation along plasma columns assumed for simplicity to be uniform. Such results may nevertheless be extended to describe the local properties of an axially nonuniform plasma column, providing the axial variation of the electron density is slow enough, namely when the condition

$$\frac{1}{\bar{n}(z)} \frac{d\bar{n}(z)}{dz} << \beta \tag{5.1}$$

is fulfilled, where β is the axial wavenumber or phase coefficient of the surface wave, z is the axial coordinate and $\bar{n}$, the cross-section average electron density. Note that $\beta = 2\pi/\lambda$ where λ is the surface wave wavelength. Equation (5.1) defines the so-called *local uniformity approximation*.[††]

5.3.2. Phase and attenuation characteristics of the m = 0 surface wave.
When considering traveling-wave discharges, the wave property of main interest, as we shall see in Sec. 5.4, is its attenuation characteristic, $\alpha(\bar{n})$. This dependence is easily

---

[†] We are assuming that the wave field intensity varies locally as $\exp[\,j\,(\omega t - \beta z + m\phi) - \alpha z\,]$, where $j = \sqrt{-1}$, β is the axial wavenumber, z is the axial coordinate, $\omega/2\pi$ is the wave frequency, t is time and α is the attenuation coefficient. We sometimes use $\gamma \equiv \alpha + j\beta$.

[††] Equation (5.1) states the condition of validity of the WKB approximation. It means that the wave field intensity can be considered to vary axially as $\exp[\,j\int_0^z \beta(z')\,dz'\,]$. For more detail, see ref. 24.

obtained experimentally.[17,25] Different methods may be used to calculate it,[25] and we present one which is straightforward and easy to perform with present computer techniques, in which the dispersion relation is solved using complex algebra. In the course of these calculations, one neglects the influence of the axial gradient of the electron density on the value of β and α (local uniformity approximation) and the effect of the radial density variation is taken into account by using $\bar{n}$, the average density over the plasma cross-section of radius R. This latter assumption is fully justified provided $\beta R < 1$,[6,20] as verified experimentally.[7,8,18]

In this method, the field components of the surface wave are complex algebra functions of complex arguments. Defining the complex wavenumber $\gamma = \alpha + j\beta$ and considering a cylindrical configuration, the axial component of the electric field intensity for the $m = 0$ surface-wave mode in the kth medium of relative permittivity ε_k is expressed as

$$E_{zk} = a_k J_0[(\beta_0^2 \varepsilon_k + \gamma^2)^{1/2} r] + b_k H_0^{(1)}[(\beta_0^2 \varepsilon_k + \gamma^2)^{1/2} r], \qquad (5.2)$$

where J_0 and $H_0^{(1)}$ are the zero-order Bessel and Hankel functions of the first kind respectively, and a_k and b_k are constants to be determined from boundary conditions; $\beta_0 = 2\pi/\lambda_0$ where λ_0 is the wavelength of the electromagnetic field in free space. The sign of the square roots is chosen so the imaginary part of the argument is positive. This restriction is necessary to obtain a zero field value with $H_0^{(1)}$ when r tends toward infinity. For similar physical reasons, in the case of a plasma tube surrounded by air (vacuum), $b_k = 0$ in the plasma and $a_k = 0$ outside it. The remaining field components E_{rk} and $H_{\phi k}$ are directly related to E_{zk} by the usual electromagnetic field relations. There are no other field components since the $m = 0$ surface wave is a TM wave. From the continuity of the field tangential components E_{zk} and $H_{\phi k}$ at the various medium interfaces, one obtains a set of linear, homogeneous equations which have a non-trivial solution only if their corresponding determinant $\mathcal{D}$ is zero. The condition $\mathcal{D}(\omega/\omega_{pe}, \nu, R_1; \alpha + j\beta) = 0$ for given ω (wave angular frequency), ν (average electron-neutral collision frequency for momentum transfer) and R_1 (discharge tube inner radius) as a function of ω_{pe},[†] yields two relations: ω/ω_{pe} - versus - β, called the *phase characteristic* (we use this term in the case of fixed ω, in contrast to the customary *dispersion characteristic* where ω varies and ω_{pe} is fixed), and ω/ω_{pe} - versus - α, the *attenuation characteristic*. Henceforth, we shall refer to the functional dependences $\beta(\bar{n})$ and $\alpha(\bar{n})$ as the phase characteristic and attenuation characteristic of the wave

---

[†] As discussed at the end of Sec. 2.2.1, the plasma frequency $\omega_{pe} = (ne^2/\varepsilon_0 m_e)^{1/2}$ is a fundamental quantity, which is determined only by the value of the electron density n. A complementary relation, namely $\omega = \omega_{pe}$, defines the critical electron density n_c as being given by $\omega = (n_c e^2/\varepsilon_0 m_e)^{1/2}$. Note that $\omega_{pe}^2/\omega^2 = n/n_c$.

respectively. This complex algebra method yields solutions, whatever the values of the plasma parameters, and in particular, whatever the value of α as compared to β.

Figure 5.1 shows an example of the phase characteristic of the azimuthally symmetric surface wave, where we have set $\nu = 0$ in the plasma permittivity (eqn. (2.14)). Under low pressure conditions, i.e. as long as $\nu^2 \ll \omega^2$, the phase characteristic is indeed practically unaffected by the presence of collisions.[5] Note that $\omega/\omega_{pe} = (1 + \varepsilon_g)^{-1/2}$ is the limiting value obtained when $\beta \to \infty$ (wave resonance) where ε_g is the tube relative permittivity.

Figure 5.2 shows the attenuation characteristic of the wave corresponding to the phase diagram in Fig. 5.1. For low gas pressures, a satisfactory analytical approximation for $\alpha(\overline{n})$ is[26]

$$\alpha(\overline{n}) = \frac{\mathcal{F}(\omega, R_1)\, \nu}{\overline{n} - n_D}, \qquad (5.3)$$

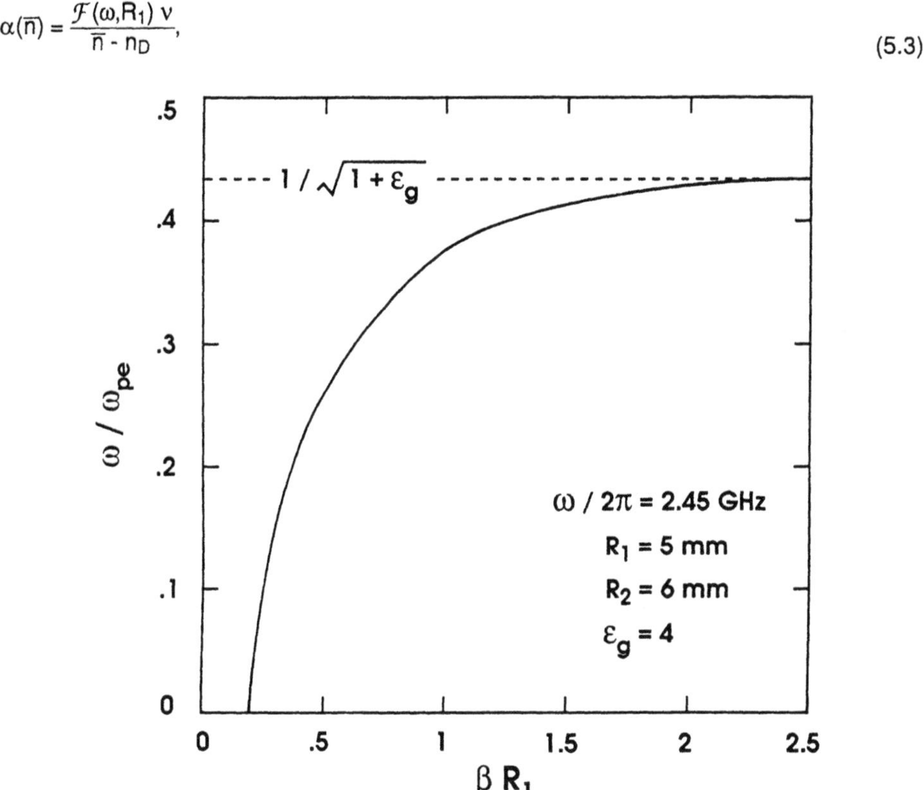

Fig. 5.1. Calculated phase characteristic of the azimuthally symmetric (m = 0 mode) surface wave propagating over a cylindrical plasma column, assuming the influence of collisions to be negligible. The dashed line corresponds to wave resonance ($\beta \to \infty$) conditions (from ref. 23).

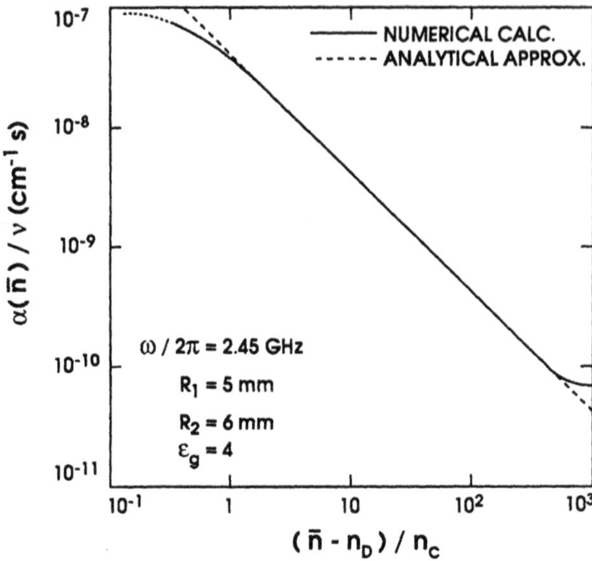

Fig. 5.2. Calculated attenuation characteristic of the azimuthally symmetric ($m = 0$ mode) surface wave as a function of the cross-section average electron density $\bar{n}$, corresponding to the phase diagram of Fig. 5.1. The plasma is assumed to be weakly collisional, namely $(\nu/\omega)^2 \ll 1$ where ν is the average collision frequency for momentum transfer; n_D and n_c are the resonance and critical densities respectively. The analytical approximation is that of eqn. (5.3) (from ref. 23).

where n_D is the electron density[†] when the surface wave is at resonance; $\mathcal{F}(\omega, R_1)$ is determined by fitting equation (5.3) to the attenuation characteristic computed for a given ω and tube inner radius R_1. The linear dependence of α on ν at low pressures permits us to plot the attenuation characteristic in a more general and useful way: instead of α vs. ω/ω_{pe} at a given ν value, we have α/ν vs. ω/ω_{pe}, which is valid for all ν values such that $(\nu/\omega)^2 \ll 1$. The value of ν depends on the nature of the gas and, to some extent, is proportional to its pressure. Note that we must have $\bar{n} > n_D$ and n_D thus defines the end of the plasma column. The above analytical formula displays explicitly the influence of the discharge conditions and plasma parameters on the attenuation coefficient. In particular, it shows that $\alpha(\bar{n})$ increases with decreasing electron density.

5.3.3. Characteristic impedance of the plasma column as seen by the $m = 0$ surface wave. As we shall show in Sec. 5.5, this impedance $\mathcal{R}_w$ is

---

[†] In the cold plasma approximation and for $\nu \ll \omega$, we have $n_D = (1 + \varepsilon_g)n_c$, where ε_g is the discharge tube relative electric permittivity and n_c is the critical density (see the preceding footnote). For values of the electron density too close to n_D, approximation (5.3) fails; it can however be shown that $\alpha(\bar{n})$ actually reaches its maximum value at approximately such a density value.

instrumental in the description of the energy transfer process from the wave launcher to the excited wave. It can be defined as

$$\mathcal{R}_w = U^2/2P \tag{5.4}$$

with

$$U = \int_0^\infty E_r(r)dr, \tag{5.5}$$

where $E_r(r)$ is the radial component of the wave electric field intensity, and $P(z)$ is the total power flux of the wave crossing the $z =$ constant plane. In the case of a weakly attenuated wave, the characteristic impedance can be considered a real quantity. Let us recall that the characteristic impedance for any wave other than the TEM mode cannot be uniquely defined. The definition (5.4-5.5) has been chosen because it yields the value of $\mathcal{R}_w$ which accounts for the power carried away by the excited wave in our equivalent circuit representation of the wave launchers. Figure 5.3 shows a typical

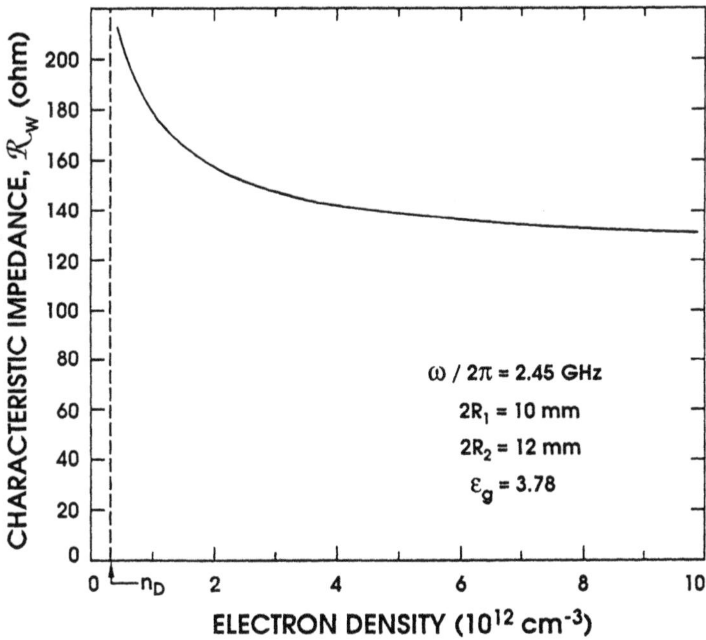

Fig. 5.3. Calculated characteristic impedance $\mathcal{R}_w$ of the plasma column as seen by a m = 0 mode surface wave, as a function of the cross-section average electron density. The plasma is assumed to be weakly collisional. The density n_D is that at the end of the plasma column (from ref. 23).

example of the dependence of $\mathcal{R}_w$ on $\bar{n}$. One notes that for electron density values exceeding a few times the resonance density n_D (the density at the end of the plasma column), the wave impedance does not vary much with $\bar{n}$. The stability of $\mathcal{R}_w$ under possibly varying discharge conditions ensures that the impedance matching of the plasma source with the generator does not vary, an important feature for most applications.

5.3.4. Radial distribution of the electric field intensity. The radial distribution of intensity for the two electric field components of the m = 0 surface wave is plotted in Fig. 5.4. As expected for a surface wave, each component grows larger close to the interface between the plasma and the surrounding dielectrics (glass and air). Within the plasma, assuming that it is homogeneous, the electric field intensity according to eqn. (5.2) is

$$E_z(r) = a I_0(\mathcal{H}_p r) , \tag{5.6}$$

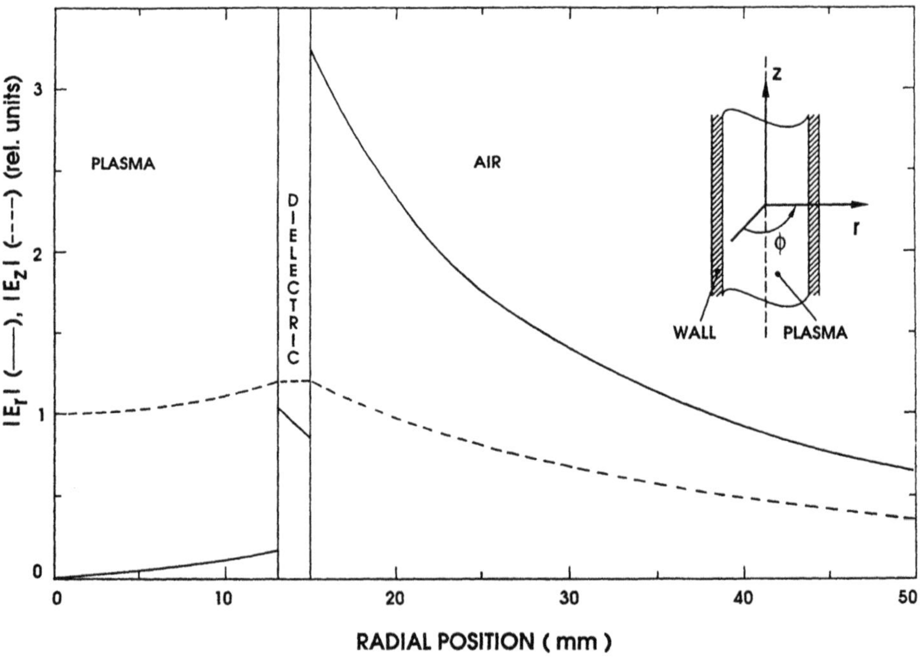

RADIAL POSITION (mm)

Fig. 5.4. Calculated intensity, as a function of radial position, of the radial and axial electric field components of the azimuthally symmetric surface wave propagating over a cylindrical, homogeneous plasma column enclosed in a glass tube surrounded by vacuum. The field frequency is 600 MHz, $\omega/\omega_{pe} = 0.2$ and $\varepsilon_g = 3.8$ (From ref. 8).

$$E_r(r) = b I_1(\mathcal{H}_p r),\qquad(5.7)$$

where a and b are coefficients set by boundary conditions, I_0 and I_1 are the modified Bessel functions of the first kind, of order zero and one respectively, and $\mathcal{H}_p^2 \equiv \beta^2 - \beta_0^2 \varepsilon_p$. Figure 5.5 shows the radial profile of E_z, which in the plasma is the largest of the wave's electric field components. We note that the steepness of this profile goes through a minimum as $\bar{n}$ increases. This behavior is due to $\mathcal{H}_p$ passing through a minimum as a function of $\bar{n}$ in eqn. (5.6). The existence of such a minimum depends upon wave frequency and upon tube nature and dimensions.

5.4. Maintenance processes and simplified modeling of the surface-wave discharge

5.4.1. General remarks. A simplified modeling procedure for HF discharges has been outlined in Sec. 2.4 and developed in Sec. 4.4 for those discharges with a localized active zone. Section 5.4 shows how it can be adapted to the case of traveling-wave discharges (TWD). In that respect, we note that the charged particle balance need not be discussed again since the analysis from Sec. 2.2.6 for a long, cylindrical plasma vessel fully applies here: all the quantities resulting from such an

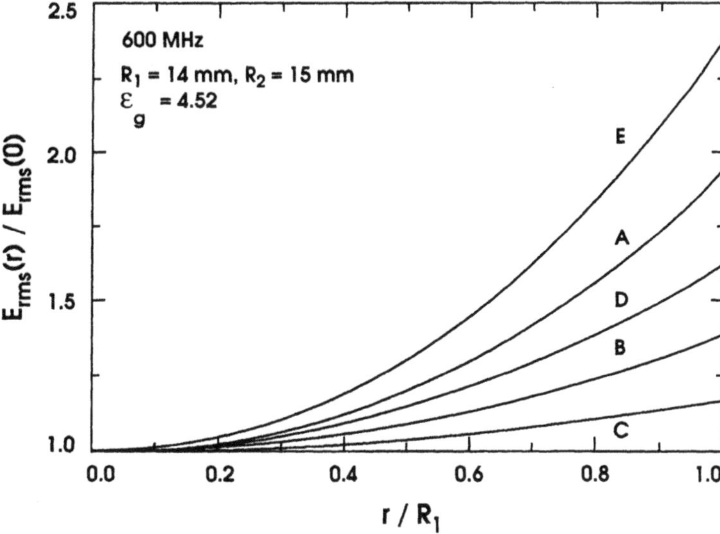

Fig. 5.5. Calculated radial profile of the electric field intensity in the plasma for the m = 0 mode surface wave, at the following values of electron density (10^{10} cm^{-3}): 44(A); 20(B); 4.9(C); 2.0(D); 1.7(E).

analysis can be assumed not to vary axially[†] in low pressure TWD (recall that these quantities are the average electron energy $<u>$, the power θ lost or absorbed per electron, the discharge maintenance field E_{rms} as well as the radial profile of the electron density). Thus, in Sec. 5.4.2, we pass on directly to the local power balance in steady-state TWD discharges whereas, in Sec. 4.4, we considered the overall (total) power balance. This stage of the discharge modeling requires special attention: the absolute value of electron density sought is therefore not simply the average value over the whole plasma volume but the axial distribution $\bar{n}(z)$ of the cross-section average density. Finally, in Sec. 5.4.3, we describe the full procedure to be followed when applying to TWD the approximate modeling outlined in Sec. 2.4.

Throughout Sec. 5.4, we discuss the modeling procedure with respect to TWD in general rather than specifically for surface-wave discharges. This is possible because, as we shall see shortly, the specific type of wave sustaining the discharge is only relevant through the wave attenuation dependence on $\bar{n}$.

5.4.2.   The local power balance in a steady-state traveling-wave discharge. The fact that in TWD, power is transferred to the plasma by a propagating wave really sets these discharges apart from conventional discharges sustained within HF circuits. We examine this matter, considering a cylindrical, dielectric discharge tube. The wave is launched at axial position $z = 0$ (Fig. 5.6) and propagates in the z direction. The total power flux of the wave decreases with growing distance z as the power is gradually expended in sustaining the discharge. The column ends at $z = \ell$ where the wave power drops below the level necessary to sustain the plasma. Because the power dissipated per unit length of the plasma column varies along the axis of the vessel, so do some of the parameters of the plasma.

The local power balance analysis of such discharges is facilitated by assuming, as a first approximation, that the influence of any axial gradient is negligible as far as the discharge processes and the wave propagation are concerned. This means that there is no net exchange of energy between consecutive elementary slabs of the plasma column, and the analysis can be performed separately for each slab. As a consequence, the analysis of an elementary plasma slab does not differ from the analysis of a uniform discharge.

Consider the power transfer from the wave EM field to plasma, which occurs continuously and gradually along the TWD. The amount of power transferred at any

---

† This is true as long as the electron energy distribution function (and hence θ) does not depend on the electron density.

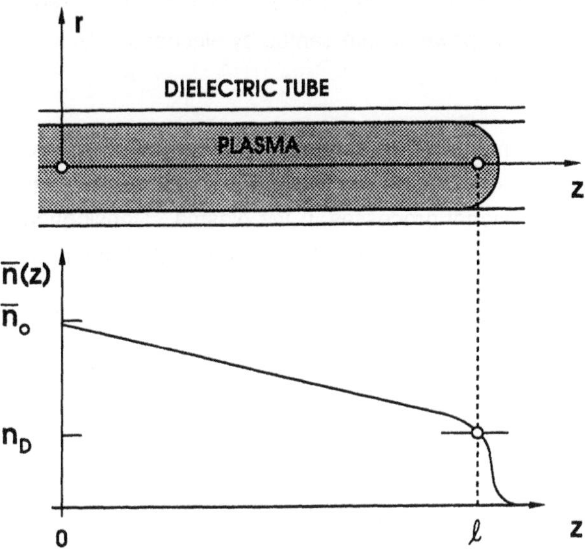

Fig. 5.6. Plasma column with its set of coordinates, and sketch of the axial distribution of the cross-section average electron density in a discharge of length ℓ sustained by a traveling wave. The wave ceases to propagate at $\bar{n} = n_D$.

position z is expressed by the wave attenuation coefficient α which, under given discharge conditions (provided the gas pressure is low, i.e. $\nu \ll \omega$), only depends on $\bar{n}(z)$, the local value of the cross-section average electron density (see Sec. 5.3.2). Representing the dissipated EM field energy as Joule losses, we have

$$\alpha \equiv -\frac{1}{2}\frac{1}{P(z)}\frac{dP(z)}{dz} = \frac{\pi \int_0^{R_1} \sigma(n)E_{rms}^2(r)r\,dr}{P(z)},\qquad (5.8)$$

where $\sigma(n)$ denotes the plasma electric conductivity and $E_{rms}(r)$ is the root-mean-squared value of the wave electric field intensity.

The power A absorbed from the field by the plasma, per unit length, at any position z along the column, is assumed to have been diverted from the wave power flux at the same position and we have

$$A(\bar{n}) = -dP(z)/dz = 2\alpha(\bar{n})P(z),\qquad (5.9)$$

where A and α depend on z through $\bar{n}(z)$. When the field frequency exceeds a few MHz, practically all this power is intercepted by electrons only.

While A is a parameter (to some extent externally set) that accounts for the power brought to the electrons, we know that these electrons are in turn losing this power through various collisional processes in the plasma. Considering that θ_L is the power lost per electron under collisions of all kinds (eqn. (2.71)), the power lost per unit length is then

$$L(\bar{n}) \equiv \pi\, R_1^2\, \bar{n}(z)\, \theta_L. \tag{5.10}$$

Ultimately, this power reaches the environment due, for example, to optical radiation and particle-wall interactions. When the electron energy distribution function is independent of the electron density, so is θ_L (see, for example, refs. 20-22, 27).

These processes of power transfer within a given region of space enclosed between the planes z and $z + \Delta z$, which delimit an elementary slab of the TWD, are illustrated in Fig. 5.7. The condition for the *local power balance in a steady-state* TWD is that the power extracted from the electromagnetic field equals that flowing out through the tube wall, namely $A(\bar{n}) = L(\bar{n})$, i.e.

$$2\alpha(\bar{n})P(z) = \pi R_1^2 \bar{n}(z)\, \theta_L, \tag{5.11}$$

at any axial position along the discharge. Finally, recalling that, in general in a steady-state discharge, the power θ_A absorbed per electron adjusts itself to compensate for the power θ_L loss per electron (eqn. (2.72)), we have $\theta_A = \theta_L = \theta$.

A direct consequence of eqn. (5.11) is the axial distribution of the electron density and that of the wave power flux. Taking the logarithmic derivative on both sides of this equation (with θ assumed density independent), after some rearrangements, one obtains

$$\frac{1}{\bar{n}(z)}\frac{d\bar{n}(z)}{dz} = -2\alpha(\bar{n})\left[1 - \frac{\bar{n}(z)}{\alpha(\bar{n})}\frac{d\alpha(\bar{n})}{d\bar{n}}\right]^{-1}. \tag{5.12}$$

The solution to this differential equation, provided $\alpha(\bar{n})$ is a known function, is the axial distribution $\bar{n}(z)$. The boundary condition $\bar{n}(0) = \bar{n}_0$ is imposed by the power $P_0 = P(0)$ delivered to the plasma column and is obtained from eqn. (5.11), which for $z = 0$ reads

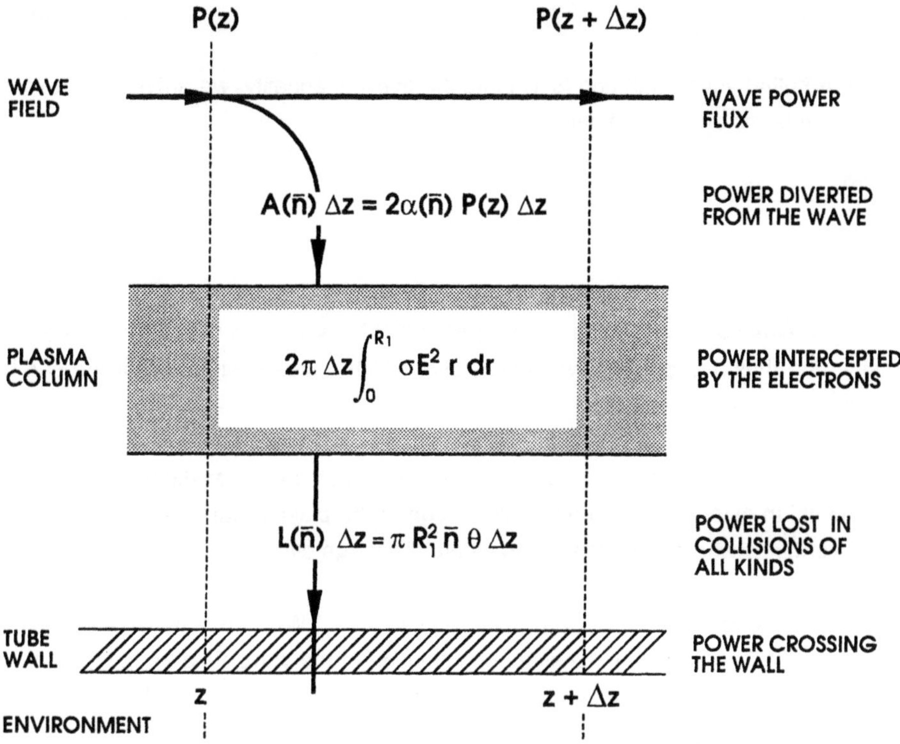

Fig. 5.7. Power flow in an elementary section of a traveling-wave-sustained plasma column (from ref. 23).

$$2\alpha(\bar{n}_0)P_0 = \pi R_1^2 \bar{n}_0 \theta. \tag{5.13}$$

As for the axial distribution of the wave power flux, it comes from eqns. (5.11) and (5.13) in the form

$$\frac{P(z)}{P_0} = \frac{\alpha_0}{\alpha(\bar{n})} \frac{\bar{n}(z)}{\bar{n}_0}, \tag{5.14}$$

where $\alpha_0 = \alpha(\bar{n}_0)$.

The local power balance condition (5.11) is necessary, but not sufficient, for a steady-state discharge to exist. Not only the power lost by electrons has to be compensated by the power transferred to the plasma from the electromagnetic field, but this power equilibrium has to be stable. This problem has been examined in Sec. 4.4.3 for discharges sustained within microwave circuits. The corresponding stability condition for steady-state TWD[23] is

$$\frac{dA(\bar{n})}{d\bar{n}} < \frac{dL(\bar{n})}{d\bar{n}}, \tag{5.15}$$

and, substituting eqns. (5.9) and (5.10) into the above inequality, we arrive at an explicit condition in terms of $\alpha(\bar{n})$, namely

$$\frac{d\alpha(\bar{n})}{\alpha(\bar{n})} < \frac{d\bar{n}}{\bar{n}}. \tag{5.16}$$

This criterion has first been formulated by Zakrzewski[28]: it determines the category of traveling waves that can be used to sustain a TWD, as well as the range of electron density (and thus that of the wave power) over which a steady-state discharge can be operated.

From the foregoing, it follows that the wave attenuation characteristic $\alpha(\bar{n})$ is not only instrumental in establishing the axial distribution of the plasma parameters in the TWD but it determines whether or not a steady-state discharge can exist.

5.4.3 Simplified modeling of reduced pressure TWD. A fully self-consistent model of a TWD requires the simultaneous solution of the wave field equations and of the equations describing the discharge maintenance processes. Such a procedure has been applied[20,27] to the particular case of discharges sustained by surface waves. Having the virtue of considerable accuracy, it also exhibits some disadvantages: i) the direct insight into the physical processes involved is often obscured; ii) the results are for discharges sustained in a given wave mode only. Thus, as in the case of discharges with a localized active zone, we shall turn to a simplified (non self-consistent) model, following the principles outlined in Sec. 2.4. This leads to the procedure presented schematically in Fig. 5.8, which we now discuss in detail.

As indicated in Fig. 5.8, the initial data of the problem are: discharge conditions, wave launcher characteristics and amount of power P_I incident at the launcher input. The end results sought are the same as for the localized active zone discharges described in Chap. 4: average electron energy <u>, maintenance field intensity E_{rms} (both these quantities being assumed to be constant within low pressure discharges), and spatial distribution $n(r,z)$ of the electron density. Figure 5.8 points out that the analysis of the discharge maintenance processes and that of the wave attenuation characteristics are performed separately, and only at the end merged through the local balance of power (5.11).

The analysis of the charged particle balance (continuity equation) and that of power balance per electron for TWD can be performed as in Sec. 2.2.7, provided that: i) the

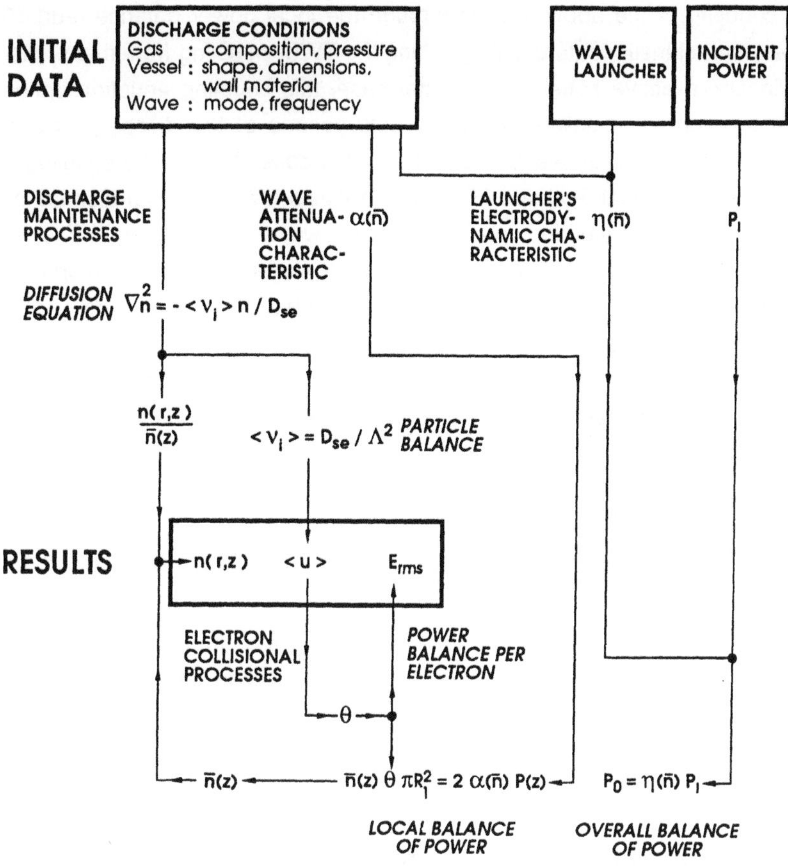

Fig. 5.8. Procedure followed in the simplified model of traveling-wave discharges.

discharge tube length is much larger than its diameter and ii) the influence of axial gradients of plasma parameters upon the maintenance processes can be neglected. It yields values of $<u>$, θ, E_{rms} and the radial profile of electron density

$$n(r,z)/n_A(z) = J_0(r/\Lambda_e),\tag{5.17}$$

where $n_A(z)$ is the electron density at the tube axis and position z along the column, and Λ_e is an effective diffusion length.[22]

The dispersion properties of the $m = 0$ surface wave are obtained as described in Sec. 5.3.2, yielding the required attenuation characteristic $\alpha(\bar{n})$.

The coupling of the above results through the local power balance (eqn. (5.11)) is the most characteristic feature distinguishing the TWD modeling from that of discharges with a localized active zone. In the latter case, this coupling equation concerns the balance of microwave power over the whole volume of the discharge active zone, yielding a volume average electron density. In the case of TWD, the coupling equation gives rise to differential equation (5.12) which yields $\bar{n}(z)$, the axial distribution of the cross-section average electron density. The total wave power $P_0 = \eta(\bar{n})\, P_I$ available at the beginning of the column is detemined by the electrodynamic characteristics $\eta(\bar{n})$ of the launcher and by the amount of incident power P_I in the feed line; through eqn. (5.13), P_0 imposes the boundary condition $\bar{n}_0 = \bar{n}(0)$ necessary to solve the differential equation for $\bar{n}(z)$. Once $\bar{n}(z)$ is known, the spatial distribution $n(r,z)$ in the plasma column follows and eqn. (5.17) can be expressed as

$$n(r,z) = \left[\frac{2}{R_1^2} \int_0^{R_1} J_0(r'/\Lambda_e)\, r'\, dr' \right]^{-1} \bar{n}(z)\, J_0(r/\Lambda_e). \tag{5.18}$$

The electrodynamic characteristics $\eta(\bar{n})$ of the wave launcher may be derived analytically in many cases. Otherwise, the value of η can be obtained directly from measurements of the incident and reflected wave power in the feed line. We shall return to this subject in Sec. 5.6, which deals with some practical realizations of wave launchers.

5.5. Launching and propagation of surface waves sustaining a plasma column

5.5.1. General remarks. There is an essential difference between the situation where the surface wave propagates along a plasma column created by some other means, as for example the positive column of a DC discharge, and the situation where the wave itself sustains the plasma. In the latter case, since the discharge and the wave field interact self-consistently, the plasma properties and those of the surface wave cannot be controlled independently. Another point worth mentioning is the necessity of using an efficient power transfer from the feed line to the plasma for surface-wave discharges, a factor that is generally not considered with non-ionizing surface waves.[7,8]

Surface-wave discharges (SWD) are this unique among HF plasmas: the wave builds up and sustains its waveguiding structure. The launching and propagation of these waves under such exceptional conditions is discussed below.

5.5.2. The initial stage of a surface-wave discharge. Although the present chapter is devoted to SWD in steady-state conditions, we nevertheless briefly examine the initial stage of such discharges.

A qualitative picture of the temporal evolution of a surface-wave discharge, starting with the initial breakdown until the establishement of a fully developed steady-state discharge, is as follows. The initial breakdown takes place in the discharge tube close to the launcher aperture. It occurs either spontaneously, when the intensity of the electric field extending from the launcher is large enough, or it must be triggered by an additional field coming from an auxiliary source, such as a Tesla coil. Due to existing field gradients, the electrons within the gap region of the exciter are expelled down the tube by the ponderomotive force.[29] Once the electron density $\bar{n}$ in the launching area exceeds n_D, a surface wave is excited and propagates until it is reflected back at the axial position where $\bar{n} \approx n_D$. At that point, the electrons are nevertheless ejected forward because of the large electric field gradient there, which gives rise to more plasma. This process allows the electromagnetic field to gradually extend away from the launcher, the ionization front moving along with the wave field. The discharge buildup goes on until the plasma column is fully developed. The total column length reached in this way is determined by the amount of power delivered to the launcher.

For a more detailed analysis of the initial stage of surface-wave sustained discharges, the interested readers are referred to articles specially devoted to this subject.[7,30-32]

5.5.3. The minimum input power required to sustain a discharge. The plasma column sustained by a surface wave ends when the wave power flux $P(z)$ drops below a given level, which is determined by the discharge conditions. Therefore, to sustain SWD, the power flux leaving the launcher, $P_0 = P(0)$, must exceed some threshold value, which we define as

$$\min\,[P_0] = \lim_{z \to z_D} P(z), \tag{5.19}$$

where z_D is the axial position of the end of the plasma column. The corresponding minimum electron density is a function of the wave dispersion and attenuation, which depend on discharge conditions. An approximate expression for this density is given in Sec. 5.3.2 for $\nu \ll \omega$.

To estimate the wave power providing the minimum density n_D, we return to eqn. (5.11) for the power balance in SWD. We then have[†]

$$\min [P_0] \cong \pi R_1^2 \, \theta \, n_D / 2\alpha(n_D). \tag{5.20}$$

In addition to the approximation of n_D as the density at the end of the column, a further limit on the accuracy of eqn. (5.20) must be considered. This is because eqn. (5.11) assumes that all the energy acquired by the electrons at any axial position is lost radially at that same axial position whereas, in the case of a very short plasma column, the axial loss of charged particles through the extremities of the column cannot be neglected, leading to a significant increase in the value of θ.

5.5.4. Setting a given mode of wave propagation. In plasma physics, the problem of launching surface waves was first encountered in experiments on the propagation of non-ionizing electromagnetic waves (see Moisan et al.[8] for a review). With such waves, the propagation mode of the launched wave appears to be fully determined by the field distribution imposed by the field applicator. However, this is no longer true with surface waves that sustain a discharge.

This problem was identified and studied in detail by Margot-Chaker et al.[33] for the $m = 0$ and $m = 1$ modes, and we summarize some of their findings: i) the $m = 1$ mode wave can be launched to sustain a plasma column only if the product of the wave frequency f times the discharge tube inner radius R_1 exceeds some critical value, namely 2 GHz-cm; ii) the $m = 0$ mode wave can propagate and sustain the plasma whatever the value of fR_1; iii) it is, however, difficult to launch the $m = 0$ wave when $fR_1 > 2$ GHz-cm. To bypass this obstacle, and sustain plasmas at microwave frequencies in large diameter tubes using the $m = 0$ wave, discharge tubes with a tapered transition section can be used (Sec. 5.6): the wave is launched over the small diameter section of the vessel and, provided enough power is delivered to the launcher, sustains a plasma also in the large diameter section.

5.5.5. The essential functions of a surface-wave plasma source system. So far, we have concentrated on various aspects of the physical processes that determine the ability of the surface wave to build its waveguiding structure. We now consider the various functions that must be performed by any practical configuration to efficiently generate a surface-wave plasma. Figure 5.9a shows the essential parts of

---

[†] Note that the exact functional dependence $\alpha(\bar{n})$, as calculated from Sec. 5.3.2, and not the approximate one from eqn. (5.3) that diverges at $n = n_D$, must be employed here.

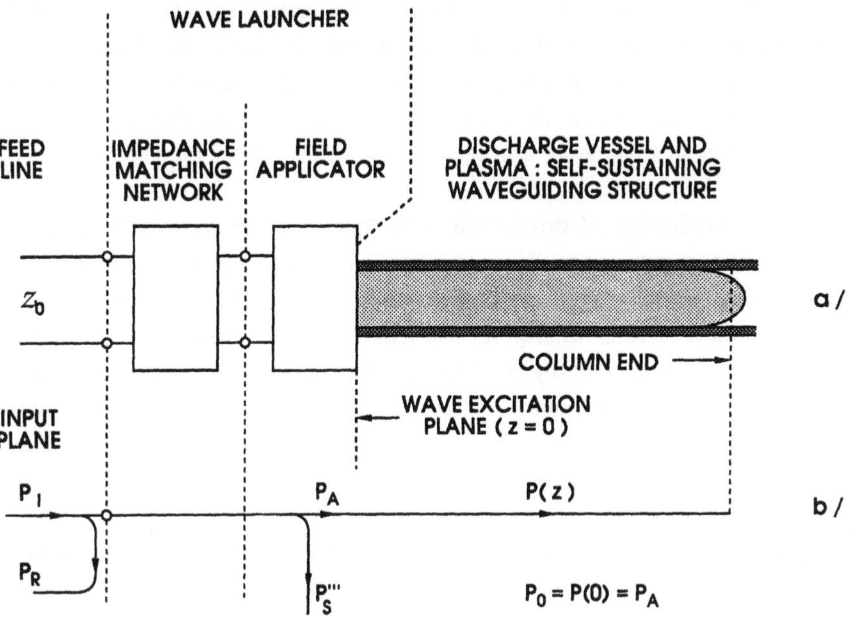

Fig. 5.9. Essential elements of a surface-wave plasma source (a) and power flow within it (b). P_I, P_R, P'''_S and P_A denote respectively the power incident in the feed line, that reflected at the launcher input, the radiated power, and the power absorbed in the plasma. $P(z)$ is the axial power flux of the surface wave, decreasing from P_0 at the launcher to almost zero at the end of the column (from ref. 23).

such a system. We briefly review their function before going into details in Sec. 5.5.6 and 5.5.7.

The role of the *field applicator* when used to launch a wave is to provide the electromagnetic field distribution favoring such a launching. As we shall see further on in this chapter, the field distribution required for launching an $m = 0$ wave can be achieved using a ring shaped gap formed by two conducting surfaces (see Fig. 5.11 in Sec. 5.6). The function of the *impedance matching network* is to optimize the power transfer to the plasma. The field applicator and the impedance matching network perform the fundamental functions associated with wave launching. Henceforth, they are jointly designated as the wave launcher; as we shall see, these two parts of a wave launcher cannot always be clearly distinguished.

The ideal launcher should accept the whole power coming from the feed line and fully convert it into a surface-wave power flux. Thus, in terms of HF circuit theory,[34] the launcher may be regarded as a "transducer", i.e. as a device for transferring power from a given mode in a waveguide to a given mode in another waveguide. Examined from the side of the feed line, the launcher together with the sustained plasma, constitute a lossy termination. In the following, we shall show that to optimize the power transfer from the feed line to the plasma, one must first of all make sure that very little or none of the power of the incident wave arriving at the launcher input is reflected back.

5.5.6 The efficiency of power transfer from the HF generator to plasma in surface-wave discharges. When surface waves are used to generate plasmas, in contrast to non-ionizing wave propagation experiments,[7,8] relatively high wave power levels may be involved. Therefore, the loss of power in a poorly designed plasma source not only becomes costly but it can result in damages to the personnel and to the equipment. The overall efficiency of the plasma source is then an important parameter. It is defined as the fraction of the power delivered by the HF generator at the source input that is absorbed into the plasma.

In a surface-wave plasma source, three loss mechanisms can divert the HF power from the discharge: losses in the dielectric and metal elements of the plasma source, reflections at the feed line-source interfaces, and radiation of non-guided waves into the surrounding space. By a careful design of the wave launcher, the losses of the first kind (i.e. those occurring within the field applicator and in the impedance matching network) can be kept small relative to the power absorbed into the plasma.

To quantatively analyze the power transfer process, we use the concepts of coupling and launching efficiencies, developed in connection with terrestrial surface waves.[35] Referring the reader to the power flow graph appearing in Fig. 5.9b, we note that P_I, P_R, P_A and P_S'' are respectively, the incident and reflected wave power flux in the feed line, the power absorbed into the plasma, and the power lost due to radiation (unguided wave).

The *coupling efficiency* is related exclusively to wave reflection at the feed line-launcher interface, and it is

$$\eta_C = (P_I - P_R) / P_I. \tag{5.21}$$

The *launching efficiency* is defined as the fraction of the total power leaving the launcher which is being converted into surface-wave power flux. The latter power is assumed to be fully[†] used to sustain the plasma column. Thus, the launching efficiency is

$$\eta_L = P_A / (P_A + P_S''') = P_A / (P_I - P_R). \tag{5.22}$$

The power that leaves the launcher as unguided radiation into space, P_S''', can most often be kept at a very low level, by a proper design of the launcher and by adequate operating conditions.

The *overall* efficiency of the surface-wave plasma source is then

$$\eta = \eta_C\, \eta_L = P_A / P_I \ . \tag{5.23}$$

5.5.7.   The equivalent circuit of a surface-wave plasma source.

Figure 5.10a shows that the wave launcher can be represented as a *two-port network* inserted between the feed line and the gap, the impedance seen at the latter being Z_g.

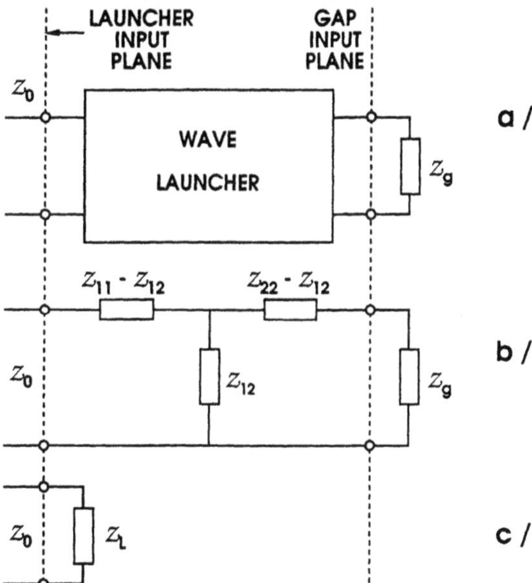

Fig. 5.10. The wave-launcher as a two-port network inserted between the feed line of characteristic impedance Z_0 and the gap impedance Z_g (a), its equivalent T-circuit (b), and the equivalent one-port termination Z_L representing the plasma-source impedance (c) (from ref. 23).

[†] This is generally true when using the m = 0 mode but not with higher modes. For example, with the m = 1 mode and when the product $f R_1$ is not much larger than 2 GHz - cm,[33] a significant part of the power flux remains past the column end.

Figure 5.10b shows one possible circuit representation of the launcher: the T-circuit impedance configuration. The customary notation for the corresponding impedances is chosen to yield Z_{11} at the input of the network with an open output, and Z_{22} in the reverse situation. Assuming the energy loss within the launcher to be negligible, the equivalent representation of such a lossless network has an important property: all impedances are imaginary ($Z_{11} = jX_{11}$, $Z_{22} = jX_{22}$, $Z_{12} = jX_{12}$). Note that Fig. 5.10b presents the wave launcher as an entity. When considering the field shaping structure and the impedance matching network as separate units as in Fig. 5.9, one ends with a representation in the form of tandem (cascade) networks. Clearly these can be transformed into a single network representation.

We now pass to the equivalent circuit representation for the impedance Z_g seen at the wave launching gap, in short, the *gap impedance*. Figure 5.11 is a sketch of the wave launching aperture most commonly used when sustaining a plasma column with an azimuthally symmetric (m = 0 mode) surface wave. This aperture is formed by a cylindrical metallic tube (generally surrounding the discharge tube) and a thin metallic plate placed perpendicular to the metallic tube axis and located at a few mm from it. The HF power is supplied to the launcher such that, in the gap region, a strong electric field exists that excites waves. When the launcher is properly designed, most of the power leaving the gap is carried away by two oppositely directed surface waves, and this power is gradually used up to sustain the discharge. In addition, some reactive energy is stored in the electromagnetic field at and in the vicinity of the gap, and a small amount of power is dissipated in the plasma located in the gap region. These various effects may be represented by different elements in the equivalent circuit, as shown in

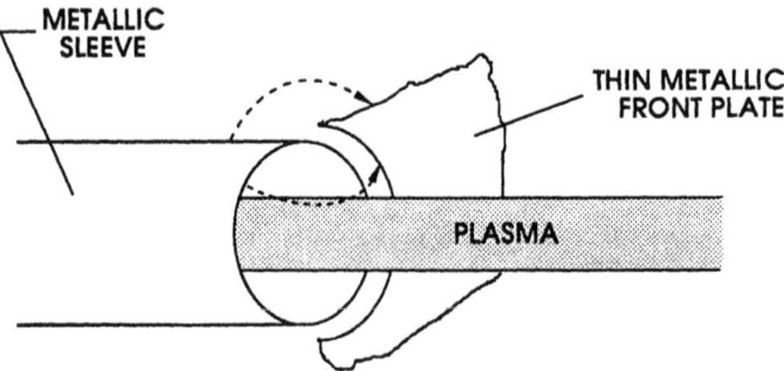

Fig. 5.11. Elements forming the so-called *launching gap* for exciting m = 0 mode surface waves. This field-shaping structure consists of a cylindrical metallic tube surrounding the discharge tube (or a dielectric-material link to the discharge tube) and, perpendicular to it, a thin metallic plate. With a tight-fitting metallic sleeve around the dielectric tube, the propagation of the wave excited at the gap towards the metallic sleeve is nearly choked; this usually gives rise to a longer plasma column on the other side of the gap, where the discharge tube is surrounded by air. The arrows show electric field lines (from ref. 23).

Fig. 5.12. $\mathcal{R}_w$ is the characteristic impedance of the plasma column (Sec. 5.3.3) and it represents the power carried away by the m = 0 surface wave, while C_w stands for the energy storage process associated with the wave launching; Z_p and Z_E represent both the dissipated and stored energy in the gap region, within the plasma and outside it respectively. The resulting gap impedance is defined with respect to a reference plane chosen in the coaxial or waveguide structure feeding the gap.

The power losses in the discharge tube walls and within the plasma in the gap region can be neglected in comparison with the power flux carried away by the wave.

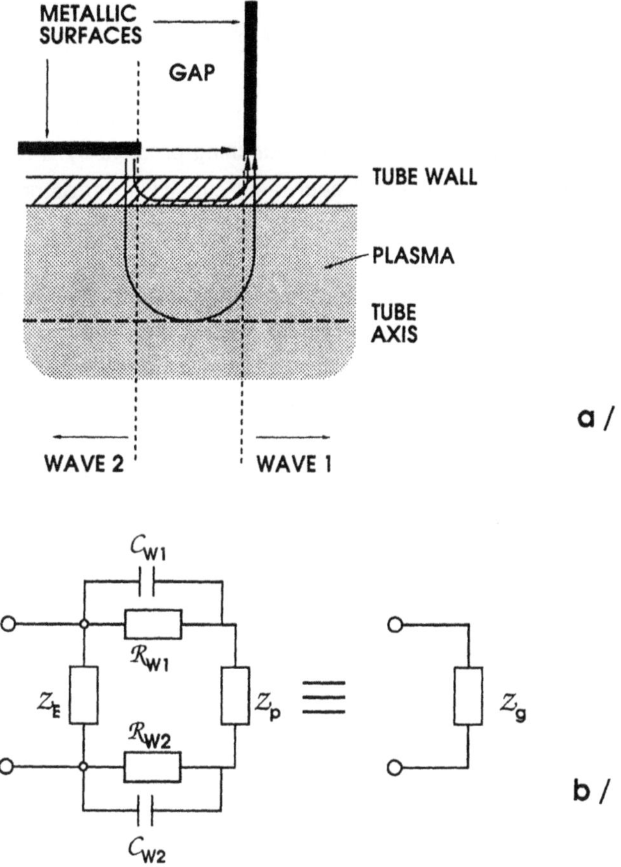

Fig. 5.12. Electric field distribution in the gap region where a surface wave is launched in both directions (a) and the corresponding equivalent circuit (b) representing the wave power transport ($\mathcal{R}_w$) and energy storage (C_w), the power dissipation ($\mathrm{Re}(Z_p)$ and $\mathrm{Re}(Z_E)$) and energy strorage ($\mathrm{Im}(Z_p)$ and $\mathrm{Im}(Z_E)$) processes in the gap region, within the plasma and outside it respectively. As usual, Re and Im denote the real and imaginary parts of a complex number respectively. The resulting impedance Z_g is called the gap impedance.

This implies that all the elements in the equivalent circuit of Fig. 5.12, excluding $\mathcal{R}_w$, can be considered imaginary ($Z_p = jX_p$, $Z_E = jX_E$). A somewhat simplified case results when the conditions of wave launching in both directions from the gap are approximately the same ($\mathcal{R}_{w1} \cong \mathcal{R}_{w2} = \mathcal{R}_w$ and $C_{w1} \cong C_{w2} = C_w$). The gap impedance is then given by

$$Z_g^{-1} = G_g + jB_g = \frac{1/\mathcal{R}}{1 + (X/\mathcal{R})^2} - j\left[\frac{1}{X_E} + \frac{X/\mathcal{R}^2}{1 + (X/\mathcal{R})^2}\right] \tag{5.24}$$

with

$$\mathcal{R} = 2\mathcal{R}_w/(1 + \omega^2 t_w^2) \quad \text{and} \quad X = X_p - 2\mathcal{R}_w \, \omega \, t_w/(1 + \omega^2 t_w^2), \tag{5.25}$$

where G_g and B_g are the gap conductance and susceptance respectively, and $t_w \equiv \mathcal{R}_w C_w$. The representation of the wave launching process by a resistance and a capacitance has been used earlier[35] in equivalent circuits of terrestrial radio surface-wave launchers.

The input impedance Z_L of the plasma source constitutes the load impedance for the feed line (Fig. 5.10c). Its value is fully determined by the gap impedance Z_g and by the parameters X_{11}, X_{12}, X_{22} characterizing the lossless launcher structure. The former depends on the discharge conditions and on the amount of power delivered to the plasma, the latter depend on the launcher design and on the tuning procedure. By altering at least two of these latter parameters, one can modify the value of the input impedance, so that $Z_L = Z_0$, thus complying with matching condition (2.87).

5.6. Practical realization of surface-wave plasma sources

5.6.1. General remarks. As with any microwave plasma source system, those using surface waves consist of the components described in Sec. 2.3.1 and shown schematically in Figs. 2.8 and 2.9, where the field applicator is specifically a wave launcher in this case. The launcher plays an essential part in the plasma source system: it determines the efficiency of the power transfer from the generator to the plasma and, to some extent, it imposes the wave mode but, otherwise, its design does not affect the plasma parameters. The wave launcher is really the centerpiece of any traveling-wave plasma source. The availability of efficient wave launchers which are easy to make and convenient to use was a key factor for the progress in SWD research and applications, rendering them competitive with other HF plasma sources. Section 5.6.2 presents a family of surface-wave launchers suitable for sustaining plasma at frequencies ranging from less that 1 MHz up to a few GHz.

The various possible designs of the plasma source vessel, as shown in Sec. 5.6.4, are a definite advantage of surface-wave discharges. The remaining parts of the plasma source are of a more or less conventional nature and may be dealt with by well established methods and procedures of RF and microwave engineering, which we shall not discuss.

5.6.2. A family of m = 0 mode surface-wave launchers.

Various devices, some of them quite ingenious, have been used for launching non-ionizing surface waves (for a review, see ref. 8). These devices are, in general, of limited use in plasma sources because they are intended for low power operation, and hence lack overall energy efficiency and power handling capability. To remedy this situation, various m = 0 surface-wave launchers had to be devised over the last few years.

The family of such launchers developed by the authors and their colleagues is shown in Fig. 5.13. Some of their design features, together with their operating frequency range, are listed in Table 5.1. We distinguish between two kinds of launchers: the modular design in which the field-shaping and intrinsic impedance-matching functions are performed by separate modules, and the integrated design where a single structure performs all the launcher functions.

In all these launchers, the field shaping function is achieved by a circular gap that provides an azimuthally symmetric field distribution, as required for a m = 0 mode wave. The power transfer is optimized by using tuning methods that are specific to each of these launchers. At the lowest frequencies, the matching networks consist of lumped elements in the form of coils and capacitors. With increasing frequency, distributed reactances[†] are used, first in the form of coaxial elements and then as waveguide elements.

The family of launchers presented ensures an efficient transfer of HF power from the generator to the plasma in complementary frequency domains, which all together cover approximately 1 MHz to 10 GHz. The possibility of sustaining surface-wave plasmas at much lower frequencies, namely at 200 kHz, was demonstrated,[36] but so far at lower efficiencies. From this reservoir of launchers, one can choose the appropriate one, according to the frequency of available HF generators or depending on the frequency required to optimize a given plasma process.[37]

[†] At frequencies of hundreds of MHz and higher, pure inductance (L) or capacitance (C) usually cannot be realized in the form of lumped elements since their dimensions become comparable with the wavelength.

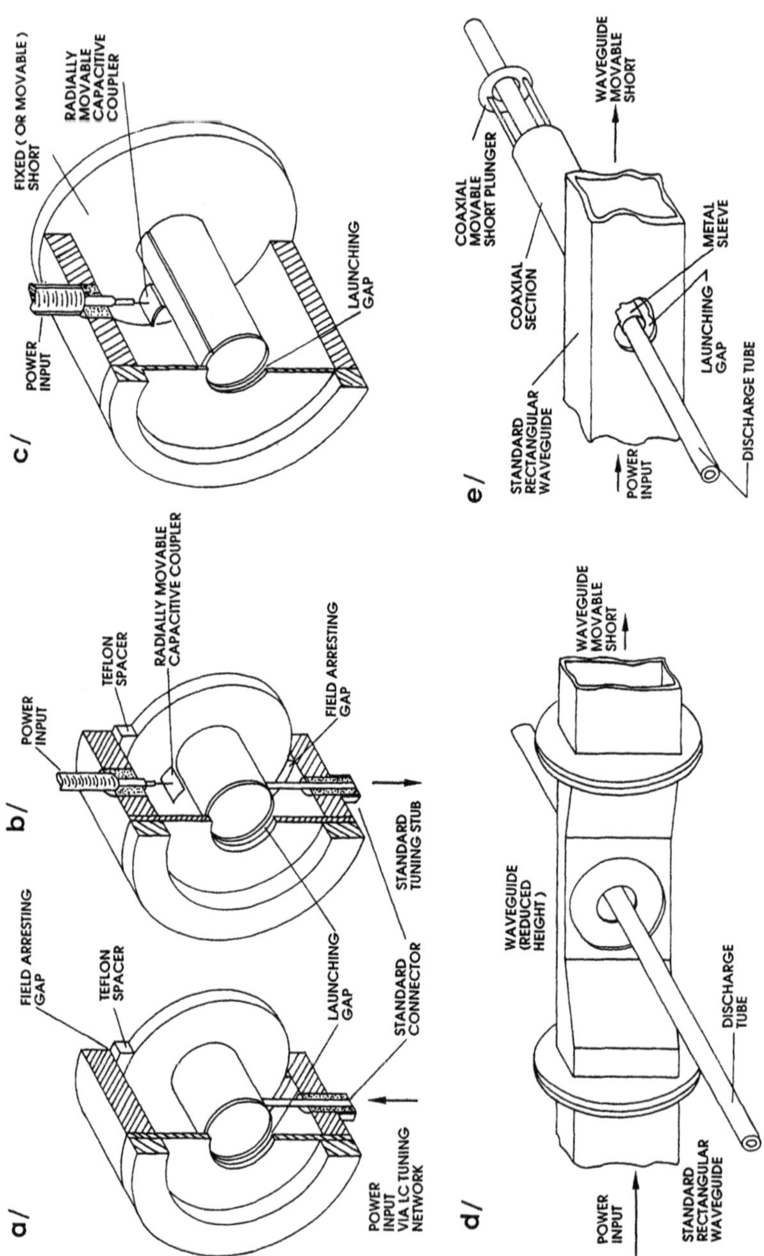

Fig. 5.13. A family of m = 0 mode surface-wave launchers: (a) the Ro-box field applicator, to be used with lumped elements impedance-matching network; (b) the Ro-box field applicator with a capacitive coupling to the feeder, to be used with a stub tuner; (c) the surfatron, (d) the surfaguide field applicator, to be used with a waveguide plunger (e) the waveguide-surfatron, to be used with a waveguide plunger (from ref. 23).

Table 5.1. Designation and distinguishing features of efficient surface-wave launchers for plasma generation (from ref. 23).

NAME	DESIGN	HF FEEDER	FREQUENCY RANGE (Hz)
RO - BOX (LC)	MODULAR*	COAXIAL	
RO - BOX (STUB)	MODULAR	COAXIAL	
SURFATRON	INTEGRATED**	COAXIAL	
WAVEGUIDE-SURFATRON	INTEGRATED	WAVEGUIDE	
SURFAGUIDE	MODULAR*	WAVEGUIDE	

* A MODULAR DEVICE IS MADE FROM SEPARATED, INTERCHANGEABLE FIELD SHAPING AND IMPEDANCE MATCHING UNITS

** AN INTEGRATED DEVICE PERFORMS BOTH FIELD SHAPING AND IMPEDANCE MATCHING FUNCTIONS

FREQUENCY RANGE CORRESPONDING TO ZERO REFLECTED POWER

The power handling capability depends strongly on the design of the launcher. With lumped element networks or with the coaxial structures that have been tested to date, it amounts to approximately 1 kW in the RF range ($\leq$ 300 MHz). With devices using coaxial cables, because dielectric losses increase with frequency, the power is limited to about 400 watts at 2450 MHz. The waveguide-based structures allow larger power levels; for example, an 8 GHz waveguide-surfatron has been operated at up to 5 kW.

5.6.3. Equivalent circuit description and analysis of the surfatron: an example. The members of the launcher family shown in Fig. 5.13 have been described in scientific journals and a detailed review and a unified discussion of these devices was recently published.[23] As an example of the analysis of such structures, we consider the surfatron. It was the first surface-wave launcher of the present family to be developed and it is still the most widely known.[14,15,38] The term "surfatron plasma" is even used by some authors to designate surface-wave sustained plasmas. This device found recognition as a source of long plasma columns for fundamental studies, but also in applications such as analytical chemistry, plasma processing and lasers.

Field shaping. The surfatron is an integrated wave launcher that performs both field shaping and impedance matching. Figure 5.14 shows its cross-section in an axial plane. The main body consists of two metal cylinders forming a section of coaxial line terminated by a short-circuit at one end and by a circular gap at the other. The gap has the required shape and symmetry so that the HF field coming out through it efficiently

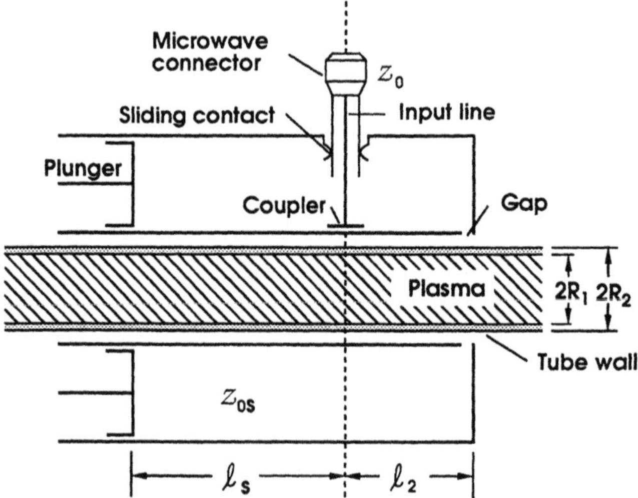

Fig. 5.14. Axial cross-section of a surfatron showing the field shaping structure (gap) and the intrinsic tuning means (coupler and plunger) (after ref. 38).

excites the azimuthally symmetric surface wave, sustaining plasma in the axially placed dielectric tube.

Impedance matching. The capacitive coupler is constructed from a section of semi-rigid coaxial cable, placed radially in the structure and extending outside it through a hole in the wall. There, it is connected to the input transmission line. Inside the structure, at the opposite end of the semi-rigid line section, the external conductor is partly removed and a plate is attached to the end of the inner conductor. A capacitance C_c therefore exists between this plate and the inner tube of the structure. The coupler is movable in the radial direction without rotation and a sliding contact ensures that it is electrically connected to the structure wall. Varying the insertion depth of the coupler affects the surfatron's input impedance (in fact, its imaginary part) and thus power coupling. Changing ℓ_s, the structure length, also affects the input impedance (mostly its real part). When operating it within its frequency range, in most practical situations, one can tune the surfatron such that the input power is totally absorbed by the plasma.

Equivalent circuit description. Having identified the various elements of this wave launcher, we arrive at the equivalent circuit shown in Fig. 5.15. The two heavy lines figure the equivalent transmission line of characteristic impedance Z_{0s} corresponding

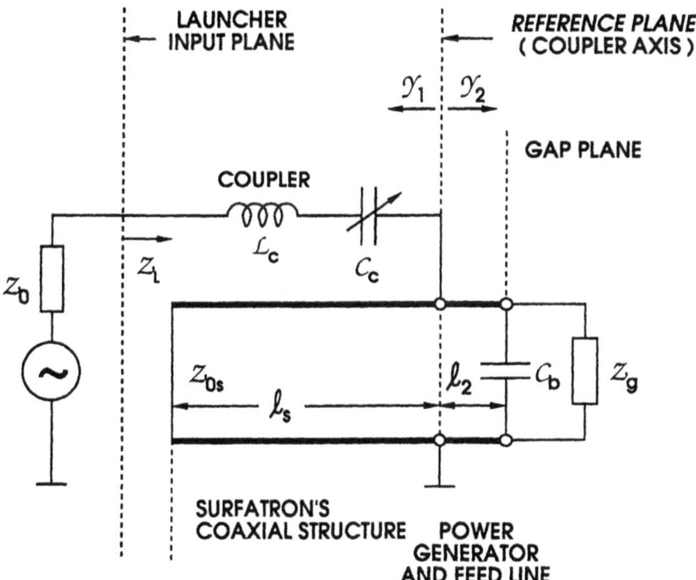

Fig. 5.15. Equivalent circuit of the surfatron plasma source. The two heavy lines figure an equivalent transmission line of characteristic impedance Z_{0s}, of lengths ℓ_s and ℓ_2 with respect to the reference plane (from ref. 23).

to the coaxial structure of the surfatron main body. For a short ℓ_2 ($\ell_2 \ll \lambda_0$) length, the input impedance of the launcher, normalized with respect to the feedline characteristic impedance Z_0, is

$$\frac{Z_L}{Z_0} \equiv \frac{R_L}{Z_0} + \frac{jX_L}{Z_0} = \frac{g_2}{g_2^2 + (b_2 - b_1)^2} + j\left(\frac{\omega L_c}{Z_0} - \frac{1}{\omega C_c Z_0} - \frac{b_2 - b_1}{g_2^2 + (b_2 - b_1)^2}\right), \tag{5.26}$$

where

$$b_1 = Z_0\left[Z_{0s} \tan\left(2\pi\,\ell_s/\lambda_0\right)\right]^{-1}, \quad g_2 = g_g = Z_0 G_g, \quad b_2 = Z_0(\omega C_b + B_g).$$

Here $G_g + jB_g \equiv Z_g^{-1}$ is the gap admittance (Sec. 5.5.7). As shown in Fig. 5.15, $y_1 = Z_0 Y_1 \equiv -jb_1$ and $y_2 = Z_0 Y_2 \equiv g_2 + jb_2$ are normalized admittances seen in each direction from the reference plane, which coincides with the coupler axis. The admittance Y_1 is imaginary for a lossless structure.

Matching conditions. Applying the general impedance-matching conditions (2.87) for the resistance and reactance at the input of a launcher to the surfatron equivalent circuit, one obtains

$$(b_2 - b_1)^2 = g_2 - g_2^2 \tag{5.27}$$

and

$$\omega L_c - \frac{1}{\omega C_c Z_0} = \pm\left(\frac{1}{g_2} - 1\right)^{1/2}. \tag{5.28}$$

Under given discharge conditions and for a given surfatron structure, the only two parameters affected by tuning are C_c and b_1. The value of C_c, which affects only the imaginary part of Z_L, varies with the radial position of the coupler while b_1 depends on length ℓ_s. As can readily be seen from the last two equations, it is only when $g_2 \leq 1$, i.e. $G_g \leq Z_0$, that matching conditions (5.27) and (5.28) can be met for full power transfer to plasma.

Tuning and frequency characteristics. Once the input impedance of the surfatron is known, the ratio P_R/P_I of the reflected to incident power in the input line can be readily obtained using eqns. (2.83) and then (2.85). For given R_L/Z_0, the reflected power is minimum when $X_L = 0$. This condition can always be met by adjusting the coupler insertion depth, and thus C_c, providing L_c is large enough. An important feature that can be seen from eqn. (5.26) is that, for given conditions, there is only one setting of the capacitive coupler yielding $X_L = 0$. This leads to simple and unambiguous tuning. In our experiments, the surfatron is always operated at that setting. With that important condition in mind, we shall refer to the dependence of P_R/P_I on the plunger position ℓ_s (at constant frequency) and on the frequency of operation (at constant ℓ_s) as the tuning characteristics and frequency characteristics of the surfatron respectively. Examples of such characteristics, measured and calculated according to the above formulas, are shown in Fig. 5.16 and Fig. 5.17.

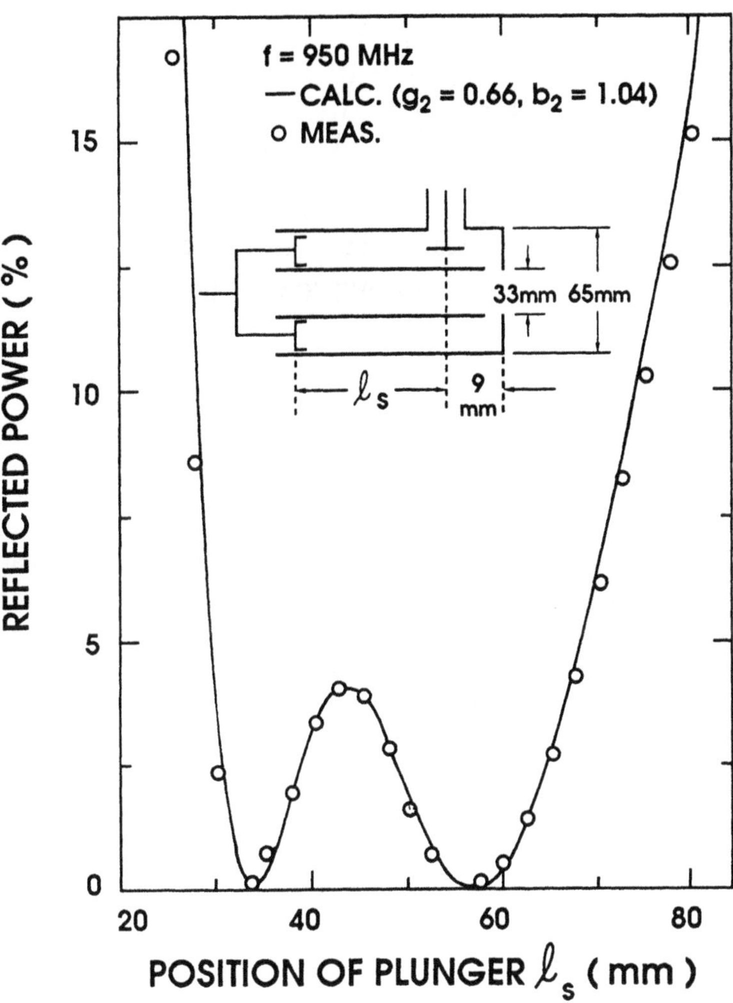

Fig. 5.16. Measured and calculated tuning characteristics of the surfatron when sustaining a plasma column. The theoretical curve is obtained from the equivalent circuit (Fig. 5.15) where the value of g_2 and b_2 have been calculated from the two ℓ_s positions at which there is zero reflected power. The insert shows the approximate dimensions of the launcher parts (after ref. 38).

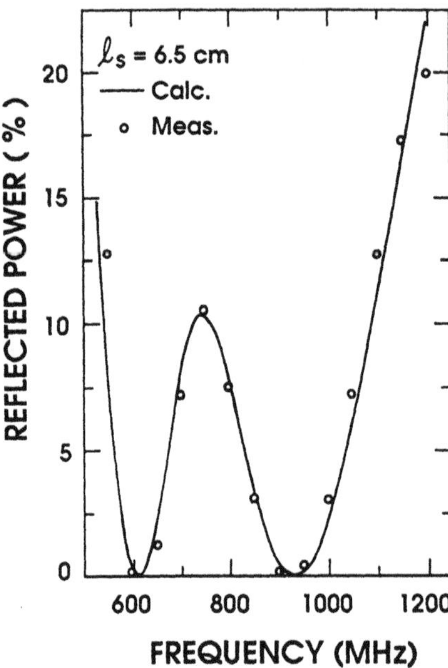

Fig. 5.17. Measured and calculated frequency characteristics of the surfatron when sustaining a plasma column. Launcher dimensions as in Fig. 5.16 (from ref. 38).

There is no lower limit to the operating frequency of a surfatron arising from its principle of operation. Rather, it results from practical considerations, namely the size and the weight of the structure, which make it combersome at lower frequencies. For example, the total length ($\ell_s + \ell_2$) of a surfatron working at 100 MHz is found to exceed 50 cm,[39] while, at 13.56 MHz, it is expected to reach approximately 4 to 5 m (This is the reason why we worked out the Ro-Box, a compact launcher for the frequency range < 100 MHz). The frequency upper limit of operation depends on the design of the gap-coupler part of the surfatron[38] and it decreases with increasing plasma tube radius. Above this frequency limit, the minimum reflected power rises with frequency, the reasons being the appearance of higher order coaxial line modes within the structure and the poor excitation of surface waves in $m \geq 1$ modes.[33] With tubes of diameters smaller than approximately 10 mm, surfatrons can be tuned for zero reflected power up to 2.5 GHz. In practice, 2.45 GHz seems to be the upper frequency for efficient operation of the surfatron.

The power handling capability of a surfatron is limited by the fact that HF energy is delivered via a capacitive coupler made from a coaxial feed line. The maximum input power is thus determined mainly by the quality of the cable and connector, and it decreases with increasing frequency.[40] From our experience, 500 W cw is really the power upper limit for a surfatron at 2450 MHz.

References to the detailed descriptions of the other surface-wave launchers shown in Fig. 5.13 are given in the quoted review paper.[23]

5.6.4. Plasma vessels. Using surface waves to generate plasmas provides more flexibility in terms of discharge vessel shape and volume than with any other type of discharge. This is because i) the wave follows the interface between the plasma and its surrounding dielectric medium, allowing smooth changes in vessel shape; ii) the plasma column length increases with increasing power delivered to the launcher.

To obtain a surface-wave sustained plasma, part of the discharge vessel must go into the launcher aperture (or be connected to it by a dielectric link). The tube walls must be made of a dielectric material (e.g. fused silica, glass, ceramic), preferably not too lossy to reduce overheating of the wall material. There is nonetheless always some wall heating that results from energy transfer from the plasma particles and from radiation. This loss is particularly noticeable within the field applicator whose metallic sleeve reduces the heat transport from the tube to the environment. Fortunately, for all the surface-wave launchers described in Sec. 5.6.2, the launching gap opening can be used for directing compressed air onto the discharge tube.

The specific design of the plasma vessel depends, of course, on its application. Figure 5.18 shows some vessel configurations that are suitable for surface-wave discharges. The simplest and most commonly used one is the straight cylindrical tube shown in Fig. 5.18a. Since we are dealing with guided waves, bending the tube does not prevent the generation of plasma.[15] However, abrupt changes in tube shape or size cause some wave reflection in the plasma and radiation into the surrounding space. The T-shaped vessel of Fig. 5.18b, with the wave launcher located at the base of the T, yields a plasma column that is symmetrical with respect to the junction: the electron density decreases away from it with the same axial gradient on both sides. By raising the input power, one increases the plasma length without losing this symmetry.[41]

To operate surface-wave discharges with diameters that are large relative to the wavelength and to the launcher dimensions, we recommend the configuration shown schematically in Fig. 5.18c.[42-43] The useful section of the discharge tube is tapered down at one end to a size that can be accommodated by the launcher. This design bypasses two limitations in the launching of surface waves over large diameter tubes: i) as a rule, for efficient operation, the launcher's aperture diameter should not exceed a given fraction of the free space wavelength; ii) the selective excitation of the $m = 0$

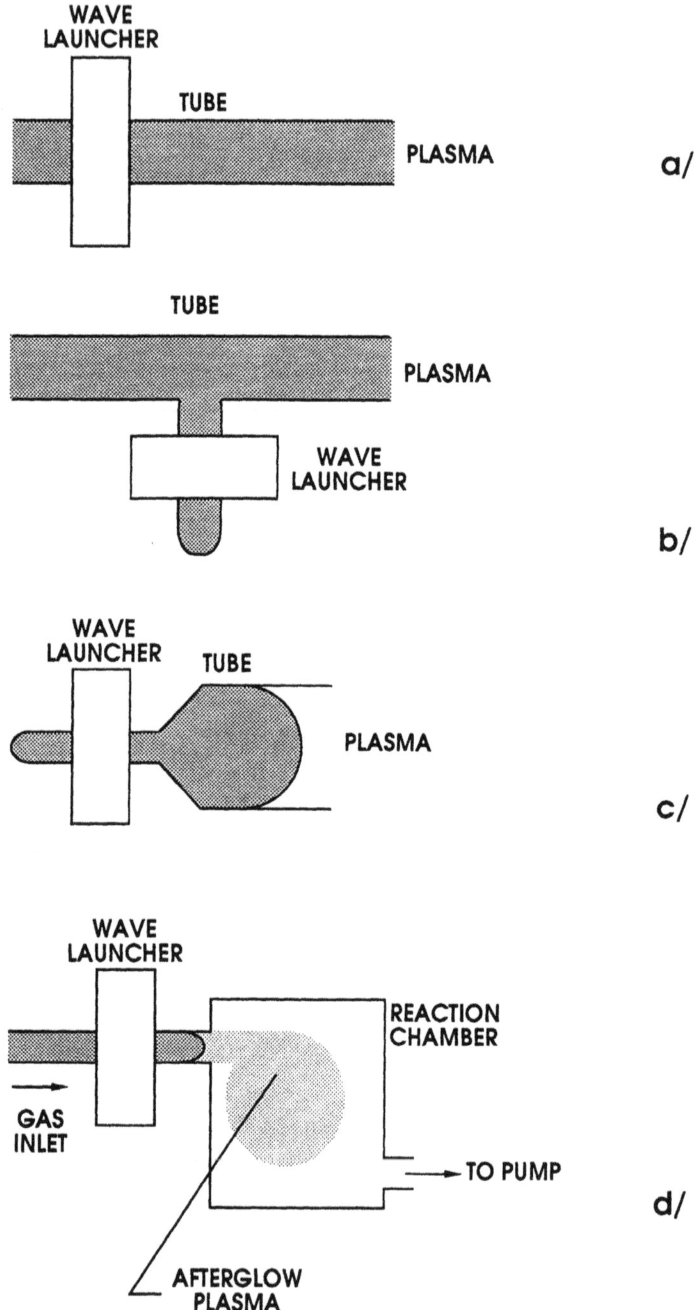

Fig. 5.18. Examples of vessels used in surface-wave plasma sources: cylindrical tube (a), symmetric T-tube (b), tapered tube (c) and reaction chamber for use with an afterglow plasma (d) (from ref. 23).

mode wave, as mentioned in Sec. 5.5.4, is no longer possible when $f R_1 \geq 2$ GHz-cm. In Fig. 5.18c, the excited wave reaches the large-diameter section of the tube via the conical transition. The length of the transition is the result of a compromise: if it is too short, the reflected wave power is excessive whereas if it is too long, too much power is expended in creating plasma outside the useful region.

Figure 5.18d shows a discharge vessel of particular interest for systems in which the afterglow (remote) plasma is used for surface processing. The chemically reactive particles are generated within the active zone and from there, they flow into the reaction chamber. Note that by increasing the HF input power to the launcher, the length of the plasma column increases and so does the residence time of particles in the discharge zone.

5.7. Experimental verification of the simplified discharge model

5.7.1. General remarks. In general, to check surface-wave discharge models experimentally, one needs to determine the dependence of plasma properties on both the discharge conditions and the power $L(\bar{n})$ dissipated per unit length of the plasma column. Whatever the type of TWD, only the discharge conditions and not $L(\bar{n})$ can be set by the operator: the dependence of power density L on $\bar{n}$ is independent of the input power P_0 but the extent of $\bar{n}$ values available along the plasma column increases with increasing P_0. This absence of control on L, as we shall show in the following, in fact makes the surface wave discharge an attractive subject for the verification of theoretical models, providing a deep insight into the physical processes governing HF discharges in general. To benefit from this feature of TWD, one has to measure the axial variation of electron density and wave power. This local "dissection" of the discharge permits one to study its properties at various levels of the power density $L(\bar{n})$ under the same discharge conditions. The sum of these results, i.e. the axial distribution of the plasma parameters, is also an important feature as it depends on processes governing the plasma column as an entity.

In Sec. 5.7.2, we describe a typical system and the procedures for the experimental investigation of surface-wave sustained plasmas. Our report is limited to surface waves propagating in the $m = 0$ mode. Sections 5.7.3 to 5.7.6 deal with the experimental verification of the simplified model presented in Sec. 5.4.

5.7.2. Experimental system and procedures. Figure 5.19 shows the diagram of a typical configuration for the experimental investigation of surface-wave discharges. Although the actual apparatus may differ, all the essential features of the systems used in previous experiments are represented in this figure. The TM_{010} resonant cavity is

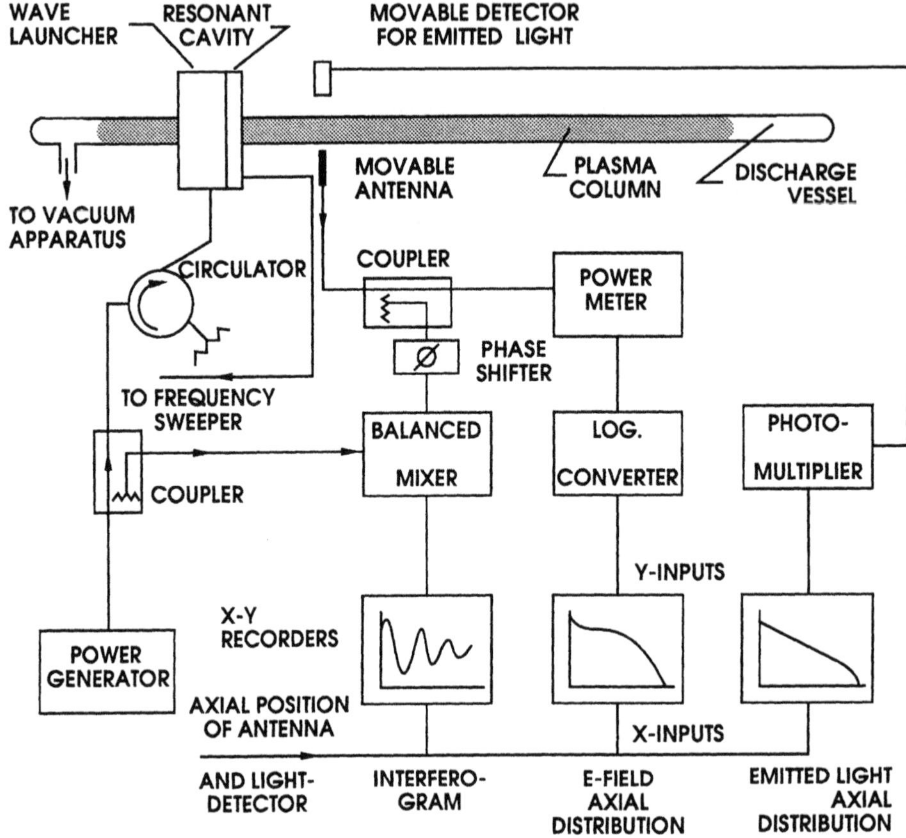

Fig. 5.19. Diagram of a typical experimental system for the investigation of surface-wave discharges.

employed for measuring electron density. This diagnostic method yields an average value over the plasma column contained within the cavity. In a first set of experiments, the cavity is moved axially along the plasma column to probe it;[14] however, once some general features of the plasma column have been established (see Sec. 5.7.4), it is sufficient to vary the input power to the launcher so that the plasma column moves through the cavity in a fixed position. To minimize the perturbation that it causes to the propagation of the wave, the passage holes through which the plasma column runs across the cavity are made 1.5 to 2 times larger than the plasma tube diameter. The resulting edge effects on the TM_{010} mode are taken into account by a calibration in which liquid dielectrics with known values fill the discharge tube. An alternate way to obtain the axial distribution of the electron density is to use the dispersive properties of the wave, namely the fact that its wavelength depends on the electron density. This is done by taking interferograms of the wave along the plasma column, using, for example, a double-balanced mixer and a reference signal from the generator output.[44] This yields $\beta(z) = 2\pi/\lambda(z)$ and, from the calculated phase characteristic, the value $\bar{n}(z)$. This latter method perturbs the plasma less and can be used at higher gas pressures and higher electron densities than the cavity method. It implies, however, that the validity of the calculated phase relation has been previously checked experimentally. The movable light detector system, when properly calibrated, provides alternate means of probing the axial distribution of electron density.[14]

Another diagnostic method consists of monitoring the radial component E_r of the electric field intensity outside the plasma (see the movable antenna in Fig. 5.19), as a function of either axial or radial position. In the first case, it yields the axial profile of the total power flux carried by the surface wave, since this power flux is approximately proportional to E_r^2 everywhere but near the end of the column. When this assumption can be made, this provides a convenient way of measuring $\alpha(z)$.[17] In the second case, $\beta(z)$ is derived from the radial dependence of E_r.

The above experimental apparatus and procedure were initially described by Zakrzewski et al.[17] and became a standard way of investigating surface-wave discharges (see, for example, refs. 25, 45-47).

5.7.3.   Identifying a surface-wave discharge. As we mentioned above, surface-wave sustained discharges had probably been obtained quite a while ago in the course of various experiments, without their physical mechanism being recognized. This is a conclusion that can be, a posteriori, drawn from the corresponding publications. A clear and simple way of identifying surface-wave discharges is to increase the amount of HF power delivered to the field applicator sustaining the plasma column under consideration and check i) that the plasma column extends beyond this applicator and ii) that the column length increases accordingly.

A complete experimental identification of a surface-wave discharge was first provided by Zakrzewski et al.[17] by showing that the phase characteristics of the wave sustaining the investigated discharge is, indeed, that of a $m = 0$ surface wave. Figure 5.20 shows a typical example of such observed characteristics.[18]

5.7.4.   The axial structure of the plasma column. The validity of our simplified model for surface-wave discharges is substantiated by comparing the predictions of this model with experimental data. To this end, we consider the axial distribution of electron density and wave power. Using the approximation expressed by formula (5.3) for the wave attenuation coefficient, and solving differential equation (5.12) with the boundary condition $n(\ell) = n_D$, we obtain $\bar{n}(z)$ and $P(z)$ in an analytic, although implicit, form as

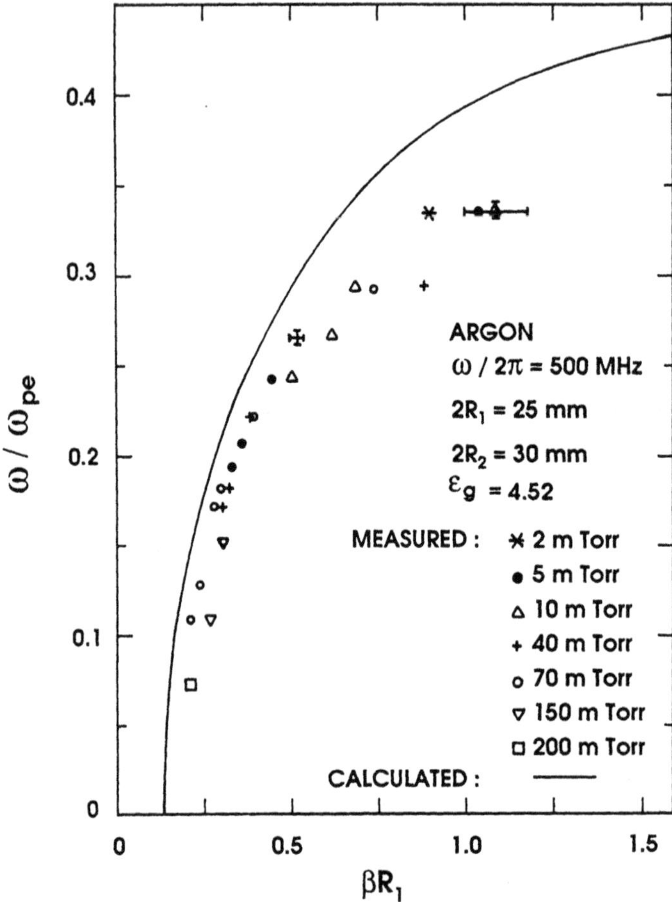

Fig. 5.20. Phase diagram obtained from measurements along an argon plasma column sustained at low pressures by a m = 0 mode surface wave. The full line is calculated from the full set of Maxwell's equations, assuming a homogeneous cold plasma (from ref. 18).

$$\alpha_D(\ell\text{-}z) = \frac{\overline{n}(z)}{n_D} - 1 - \frac{1}{2} \log_e\left(\frac{\overline{n}(z)}{n_D}\right),$$

$$(5.29)$$

and

$$\frac{P(z)}{P_D} = \frac{\overline{n}(z)}{\overline{n}_D}\left[\frac{\overline{n}(z)}{n_D} - 1\right].$$

$$(5.30)$$

Using $\mathcal{F}(\omega,R_1)$ introduced in eqn. (5.3), we have further added two new normalizing parameters $\alpha_D \equiv \mathcal{F}(\omega,R_1)\, v_m/n_D$ and $P_D \equiv \pi R_1^2\, \theta\, n_D/\alpha_D$, which correspond to the wave attenuation and power flux occurring at $n(z) = 2n_D$ respectively. The normalizing parameters n_D, α_D and P_D are constant under given discharge conditions, whatever the input power P_0. Thus, in the above equations, they represent the influence of the discharge conditions, while the input power determines the column length ℓ (see Fig. 5.22 and the text below in this section).

Figure 5.21 compares the measured distributions of electron density and wave power with those predicted by the simplified model.[26] The broken lines represent the best fit to the experimental data that is obtained from eqns. (5.29) and (5.30). There is only one fitting parameter, common to both curves, namely the effective collision frequency for momentum transfer ν_{eff} (Sec. 2.2.6).

Using the same formulas, one can readily derive the dependence, upon the imposed wave power P_0 at the origin of the column, of the corresponding electron density $\bar{n}_0$, and axial density gradient $[d\bar{n}(z)/dz]_0$, and column length ℓ (the power P_0 is assumed to be fully absorbed in the plasma column). The results of these calculations are shown in Fig. 5.22 together with experimental points.[26] The experimental conditions are listed in the figure. As already stated, the normalizing parameters n_D, α_D and P_D depend on the discharge conditions and are independent of P_0. The influence of P_0 can be seen directly from the figure. At low power levels, increasing P_0 serves mainly to increase

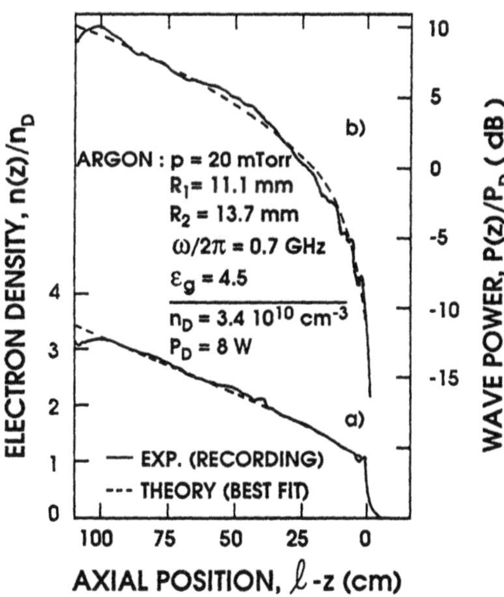

Fig. 5.21. Measured and calculated axial distributions of electron density (a) and wave power (b) along a plasma column sustained by a m = 0 mode surface-wave discharge. There is only one common fitting parameter for both curves (see text) (after ref. 26).

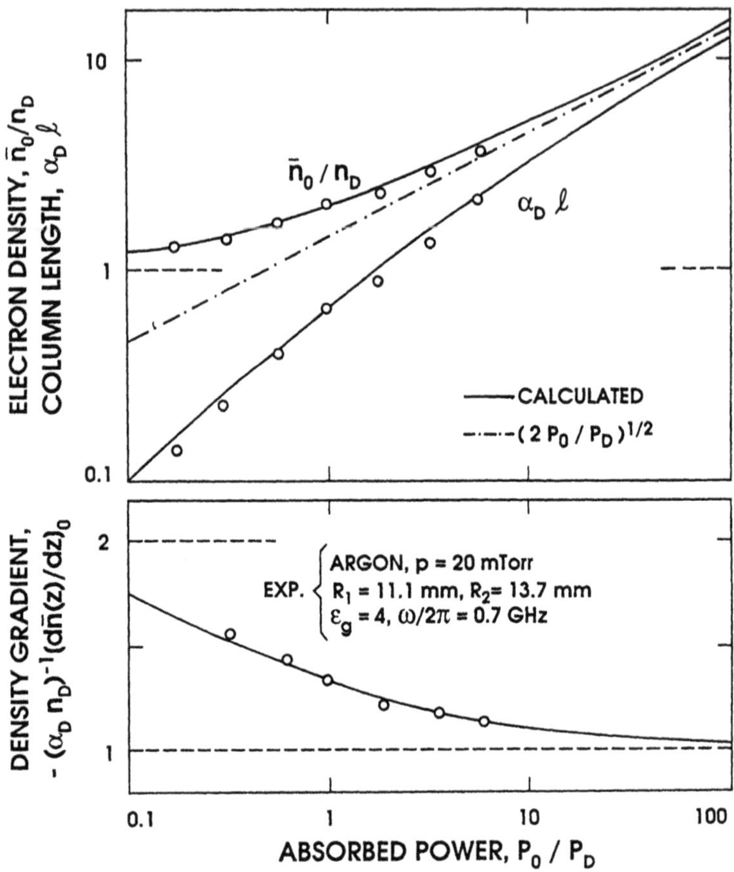

Fig. 5.22. Measured and calculated values of the plasma column length ℓ, of the electron density n_0 at the wave launcher position, and of the corresponding density gradient, as a function of the total power P_0 absorbed in the column when the plasma is sustained by a m = 0 surface wave (from ref. 26).

the column length, while at higher power, it contributes in approximately equal parts to increasing the column length ℓ and the electron density $\bar{n}_0$.

5.7.5. Electron collision frequency for momentum transfer. A further way of verifying the simplified model is to determine the electron collision frequency through its influence on the attenuation characteristic $\alpha(\bar{n})$ of the wave sustaining the discharge. This can be done either from the axial distribution of wave power $P(z)$[17] or from the axial distribution of electron density $\bar{n}(z)$.[25]

Figure 5.23 shows that the values of ν_{eff} determined from both methods are close to each other. The theoretical curve results from assuming that the electron energy

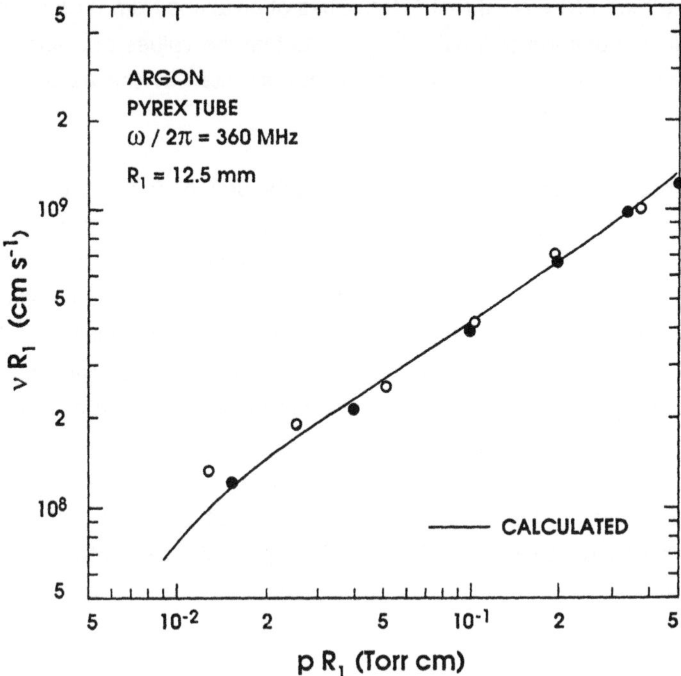

Fig. 5.23. Comparison of the calculated electron collision frequency for momentum transfer (Schottky theory assuming a Maxwellian EEDF) with the effective collision frequency for momentum transfer obtained from experimental data, from the electron density axial distribution (●) and from the wave power axial distribution (o) when the plasma column is sustained by a m = 0 surface wave (after refs. 17 and 25).

distribution function is Maxwellian and that its temperature is given by the Schottky theory for a diffusion controlled column.[48] The fact that the positive column collision frequency is close to that for SWD's supports our basic assumption that the charged particle loss is the determining factor of the discharge maintenance processes and thus that the way by which the discharge is sustained is secondary. Similar results were also reported by Chaker et al.[45]

5.7.6.   Miscellaneous experimental results. In the simplified model, the radial nonuniformity of the electron density is eliminated, as far as wave propagation characteristics are concerned, by the assumption that the plasma is radially uniform, the electron density being equal to the average value of the density over the cross-section. Darchicourt et al.[49] investigated the impact of this assumption on the values of the electron density and collision frequency inferred from measuring the phase and attenuation characteristics of the wave. This procedure requires calculating $\beta(\bar{n})$ and $\alpha(\bar{n},v)$ to yield $\bar{n}$ and v. These characteristics were computed for a radial distribution of density of the form $n(r) = n_0 J_0(ar/R_1)$, for values of a varying from 0 (uniform plasma) to 2.15 (strongly nonuniform plasma). They found that the values of n and θ obtained in this way from experiments may be significantly affected by the assumed value of a whereas the value of v is not.

Experimental studies conducted on the dependence of θ_A, the average power absorbed per electron, upon discharge conditions yielded some important results for the surface-wave discharge model (recall that under steady state, $\theta_A = \theta_L = \theta$). Moisan et al.[37] have shown that θ is constant along the plasma column, except close to its end. This result is illustrated in Fig. 5.24. It justifies the assumption of θ = constant in the derivation of eqn. (5.12). The validity, at a given wave frequency, of the similarity law θ/p vs. pR_1, predicted by the self-consistent theory of Ferreira and Loureiro,[50] has also been confirmed by the same authors. Figure 5.25 is a further example of their results where θ/p vs. pR_1 data, at different frequencies, are plotted. All these values fit between the theoretical θ/p vs. pR_1 curves for the case of a "DC" and a "Maxwellian" EEDF. As predicted by theory (Sec. 3.2.3), θ/p decreases with increasing wave frequency; at 2450 MHz, as a result of a higher electron density, thus of higher electron-electron collision frequency (Sec. 3.4), the measured values fit the theoretical curve obtained assuming a Maxwellian distribution.

Boisse-Laporte et al.[51] and Granier et al.[47] investigated surface-wave discharges in argon under conditions corresponding to v/ω varying from values much smaller than

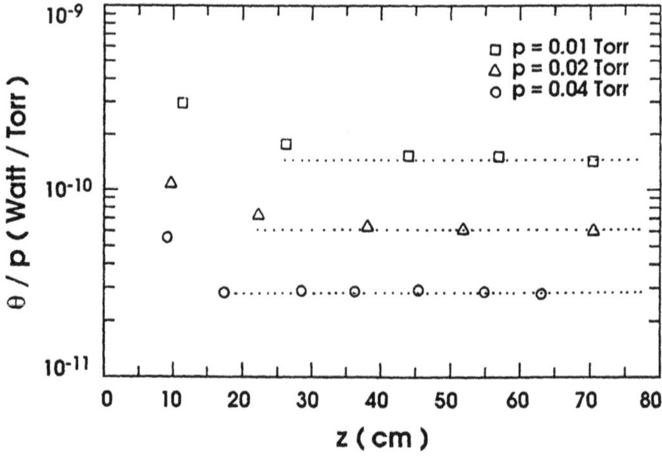

Fig. 5.24. Measured θ/p values along a surface-wave produced argon plasma column, as a function of axial position referred to the end of the column, at three different gas pressures (f = 200 MHz, R_1 = 13 mm, R_2 = 15 mm). The dotted line emphasizes the existence of an axially constant θ/p value, except close to the column end (from ref. 37).

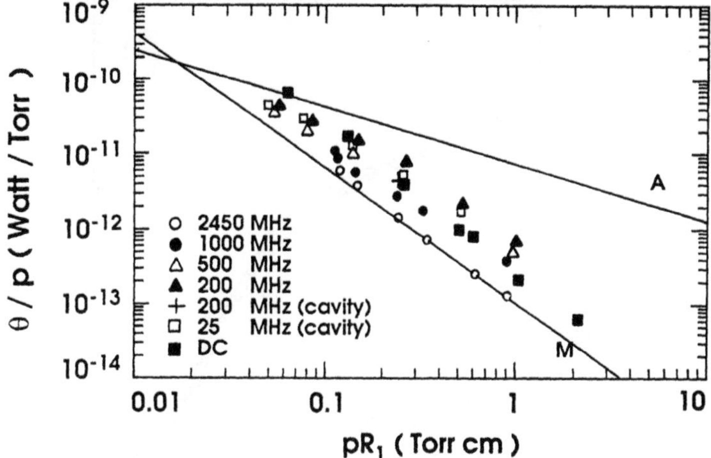

Fig. 5.25. Measured θ/p values obtained as a function of the pR_1 product, at different plasma stimulating frequencies in a surface-wave discharge and for a DC positive column. The two straight lines are theoretical, for two limiting situations, DC case (A) and Maxwellian EEDF (M). For surface-wave plasmas, the electron density has been determined by the axial phase variation of the surface wave or by a resonant cavity method (from ref. 37).

unity to values much larger than unity. They observed similarity relations involving θ, ν and the maintenance field intensity E_{rms}. In particular, they showed that the similarity law for θ/p vs. pR_1 predicted by Ferreira[27] is valid provided $pR_1 \leq 1$ torr cm.

Using second derivatives of Langmuir probe characteristics, Kortshagen et al.[52,53] measured the electron energy distribution function in argon plasmas sustained by surface waves. Their numerical solutions of the Boltzmann equation are in good agreement with experimental results. Figure 5.26 shows an example of such a comparison.

Solntsev et al.[54] observed effects resulting from the heating of the gas in surface-wave sustained plasma columns. The discharge tube was equipped with a set of thermocouples measuring the wall temperature at various points along the column as well as the local value of the heat flux per unit length. Care had been taken that these sensors did not disturb the wave propagation. They found that the axial gradient of gas temperature caused an axial variation of the gas density. Figure 5.27 shows an example of their results. Calculated and measured values of the axial density gradient at fixed positions along the discharge are compared. It follows that gas heating and the corresponding redistribution of gas density lead to a decrease in the axial density

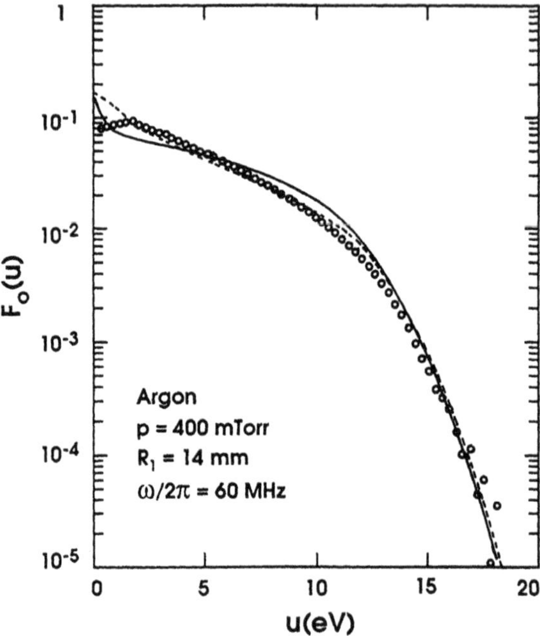

Fig. 5.26. Second derivatives of the Langmuir probe characteristics yielding the EEDF $F_0(u)$ in a surface-wave discharge: measured (circles) and calculated from homogeneous (full line) and inhomogeneous (broken line) Boltzmann equation (from ref. 53).

gradient. The conclusion is that in general gas temperature effects should be included in the modeling of surface-wave plasmas.

5.8. A practical guide to using surface-wave plasma columns

This section summarizes experimental results on surface-wave discharges to serve as a guide to potential users.

5.8.1. Range of discharge conditions. By discharge conditions, we refer throughout this book to those parameters set by the operator that comprise the wave frequency and mode, the inner and outer radius and permittivity of the tube, the gas nature and pressure. The HF power absorbed by the discharge is considered as a separate parameter.

Wave frequency and mode. As discussed in Sec. 5.5.4, in principle there is no frequency limit when using the m = 0 mode wave to sustain a discharge whereas a discharge with the m = 1 mode requires $fR_1 \geq 2$ GHz-cm (still higher values of the

fR$_1$ product are needed for m > 1 mode discharges[33]). The m = 0 mode surface-wave discharge has been obtained at frequencies as high as 10 GHz and as low as 200 kHz.[36] However, as mentioned above, achieving a m = 0 mode discharge in tubes with radius R$_1$ such that fR$_1$ ≥ 2 GHz-cm does not seem to be possible, the m = 1 mode discharge being preferentially excited. In that case, one needs to launch the m = 0 mode over a smaller diameter tube such that fR$_1$ < 2 GHz-cm, by using the tapered tube scheme described in Sec. 5.6.4. A priori, as compared to m ≥ 1 mode discharges, the m = 0 mode, because of its azimuthal symmetry, provides a more uniform plasma for applications. Note that once a discharge in a given mode is obtained, its characteristics are independent of the particular type of launcher used.

Tube diameter and gas pressure. Besides its role upon mode selection through the product fR$_1$, the diameter of the tube determines the gas pressure domain over which a stable surface-wave discharge can be obtained. This is shown in Table 5.2 for the

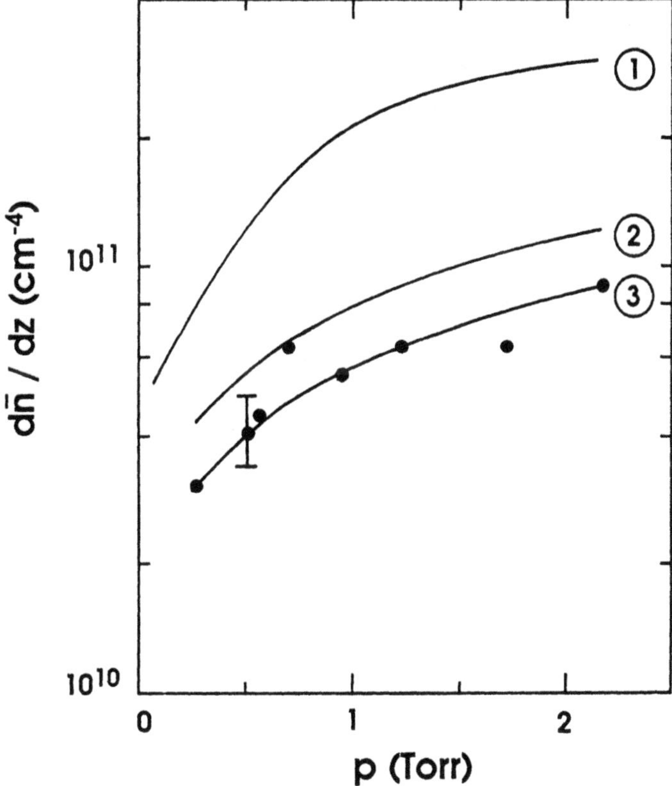

Fig. 5.27. Axial gradient of electron density in a surface-wave plasma: 1) as calculated for a gas filling pressure p ; 2) as calculated with allowance for the temperature gradient; 3) experimental results (tube radius R$_1$ = 0.24 cm, ω/2π = 3 GHz, incident wave power P$_I$ = 140 W) (from ref. 54).

case of argon gas. A stable surface-wave discharge can be achieved at pressures slightly lower than those attainable with DC discharges and, as a rule, the discharge tube heats less. As for the high pressure limit, stable argon plasmas can be obtained in tubes of internal diameters (i.d.) from 1 to 4 mm, at pressures up to ten times atmospheric pressure, this pressure limit having been set by us to avoid breaking the glass tube. As a rule, the plasmas produced at one atmosphere are stable provided that the tube i.d. is less than 5 to 7 mm; most generally, the plasma column is then constricted,[t] i.e. it does not fill the entire cross-section of the tube. It appears as a bright, straight filament, centered on the axis of the tube, with a diameter smaller than the tube bore. The fact that this filament is not in contact with the walls is commonly observed with other HF plasmas as well as with DC arcs when operated at atmospheric pressures.[55] Table 5.2 shows that a large range of plasma diameters has been investigated (0.5 to 150 mm). The upper diameter limit is the result of glass blowing constraints and costs. Column lengths up to 6 meters have been obtained in a 75 mm i.d. tube. A great variety of atomic and molecular gases have been used by various groups.

Table 5.2.   Pressure range for a stable surface-wave discharge in argon gas, according to the discharge tube diameter.

Tube inner diameter (mm)	Minimum pressure (torr)	Maximum pressure (torr)
0.5 to 7	$\sim 10^{-2}$	~ 700 (maximum tested value)
25	$\sim 10^{-3}$	~ 20
150	$\sim 10^{-4}$	$\sim 0.35^{*}$

* This value can increase by more than two orders-of-magnitude when sufficient gas flow is allowed.

5.8.2. Spatial distribution of electron density.

5.8.2.   Spatial distribution of electron density. The experimental results available on this matter concern almost exclusively the axial distribution of the cross-section average of electron density defined as

---

[t] Exceptions to this constriction effects are helium and hydrogen discharges when the HF power density in the column is not high enough.

$$\bar{n}(z) = 2\pi \int_0^{R_1} n(r,z) \, r \, dr / \pi R_1^2 . \tag{5.31}$$

Nevertheless, useful information on the radial distribution of n can be found examining the physics of the discharge. When diffusion to the wall is the main loss mechanism for charged particles, the radial profile of electron density is predominantly determined by the global movement of the particles diffusing toward the walls. As a consequence, this profile is mainly affected by the vessel shape and it is little influenced by radial variations of the ionization rate. In the specific case of surface-wave discharges where the wave electric field intensity increases with radial position, Ferreira's[27,56] modeling results show that in the ambipolar diffusion regime, the radial profile of electron density is, to a first approximation, only slightly flatter than that given by the Bessel function

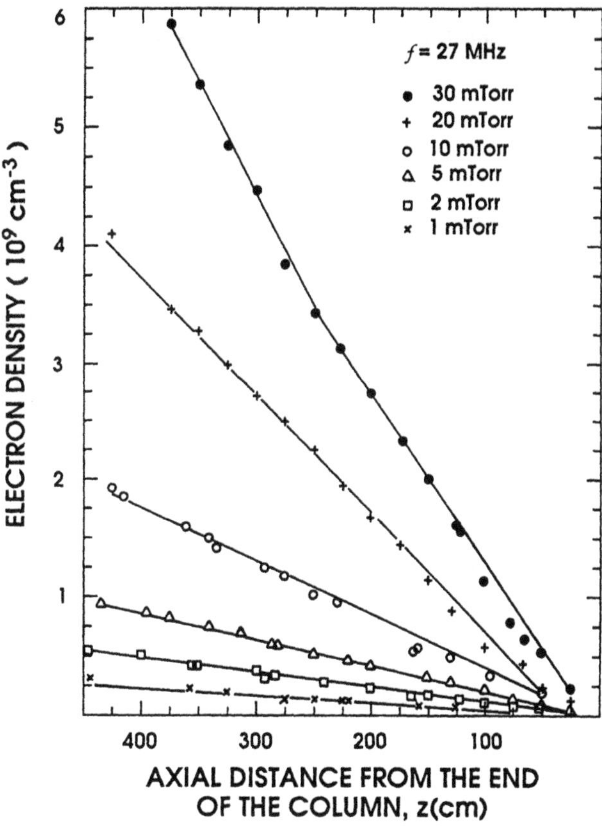

Fig. 5.28. Axial distribution of electron density $\bar{n}$ observed along a m = 0 surface-wave discharge, under traveling wave conditions (no reflection at the column end), at various pressures in argon (R_1 = 32 mm) (from ref. 39).

J_0 (2.4r/R) for the case of a discharge in a uniform electric field as, for example, in the positive column of DC discharges.

The axial distribution of electron density in surface-wave discharges is markedly different from that in a positive column plasma, where it is almost uniform. This is because traveling-wave discharges can only exist provided $\bar{n}$ always decreases with axial position away from the launcher.[†] In most cases, this density decrease is linear as illustrated in Fig. 5.28; the discharge ends at a density value close to n_D which for a cold, low collision plasma is

$$\bar{n}_D = 1.2 \times 10^4 \, (1 + \varepsilon_g) \, f^2 \, (\text{cm}^{-3}) \,, \tag{5.32}$$

where the wave frequency f is expressed in MHz (the relative permittivity ε_g of the tube wall is 4.52 and 3.78 for our Pyrex brand and for fused silica respectively). Figure 5.28 also shows that the axial gradient of density, $|d\bar{n}/dz|$, increases with increasing gas pressure. Figures 5.29 and 5.30 further indicate that the axial gradient grows with

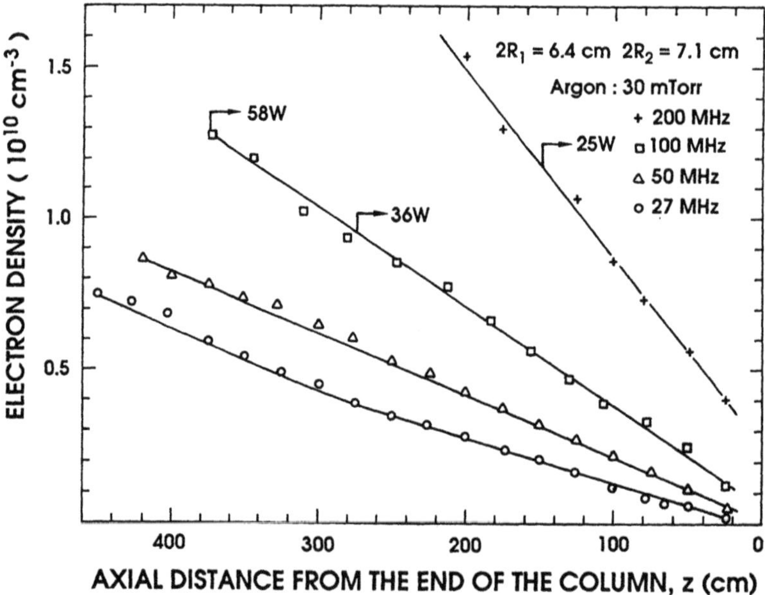

Fig. 5.29. Axial distribution of electron density $\bar{n}$ along a m = 0 surface-wave discharge, at various wave frequencies. The arrow shows the position of the wave launcher gap from the end of the plasma column at the indicated power P_0 applied to it (from ref. 57).

[†] This property comes from the discharge stability condition, eqn. (5.16), which requires the attenuation coefficient to increase with decreasing electron density, as is the case with surface waves.

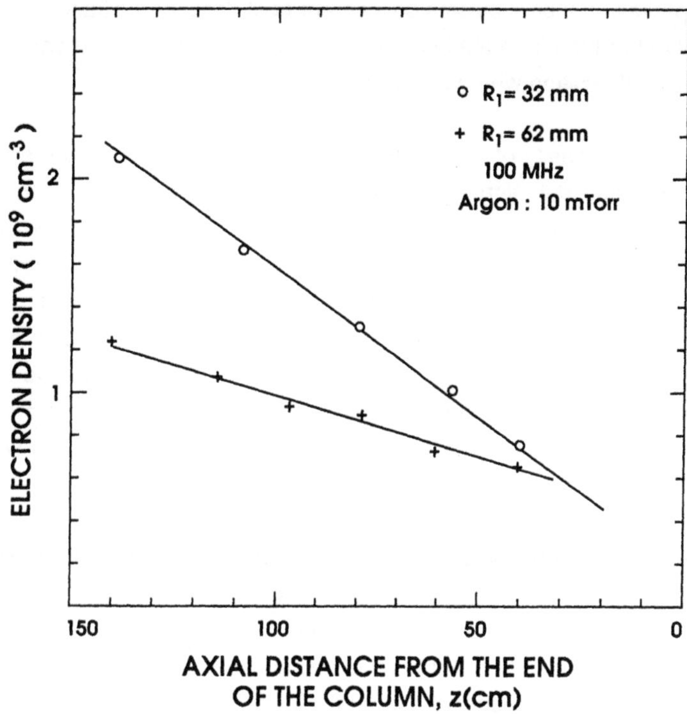

Fig. 5.30. Axial distribution of electron density $\bar{n}$ along a m = 0 surface-wave discharge, for two values of the tube diameter (from ref. 57).

increasing wave frequency while it decreases with increasing tube radius. The observed influence of gas pressure, wave frequency and tube radius upon the axial gradient of density can be expressed, in the electrostatic approximation (low phase velocity), in an explicit form as

$$d\bar{n}/dz = -0.73 \times 10^{-2} f v/R_1 \quad (cm^{-4}) \tag{5.33}$$

where f is in MHz and v in s^{-1}. This relation assumes a cold, low collision discharge under ambipolar diffusion.[58]

The effect of increasing the HF power transferred to the discharge is a peculiarity of surface-wave discharges: i) the plasma column length increases (provided the tube allows it), ii) the electron density axial gradient is barely modified. This can be seen from Fig. 5.29. Consider, for example, the 100 MHz curve and the increase of HF power from 36 to 58 W. The plasma column initially existing at 36 W remains unaffected by the increase of power: it becomes the tail end of the new column. This is why the properties of such a plasma column are best described axially by referencing it from the end rather than from the launcher. Note that in the 100 MHz case, $d\bar{n}/dz$ does not vary with axial position whereas it does slightly at 27 MHz.

The axial inhomogeneity of the surface-wave plasma columns can be a shortcoming in some applications but there exist possible remedies to this situation. One is to transfer enough HF power to the launcher so the surface wave is reflected at the end of the discharge tube. Then, on its way back toward the launcher, the reflected wave deposits most of its power where the electron density is the lowest, tending to level the axial inhomogeneity of the plasma produced when the wave is in a purely traveling mode. Another solution, probably more effective, is to make use of two launchers,[18] each located at one end of the plasma tube, and supplying them from two separate HF generators that are not phase related to avoid the formation of standing waves,[59,60,61] hence local inhomogeneities, along the column.

5.8.3. Spatial distributions of excited atom densities.

In contrast to the radial distribution of electron density, the radial distributions of excited atom density in surface-wave discharges can differ considerably from those observed in the positive column of DC discharges. In surface-wave discharges, a large variety of excited atom density profiles can be obtained that are of interest in applications such as, for example, gaseous lasers and lighting. We need to distinguish two types of excited atoms, namely those that are in a radiative state, i.e. those that once excited, emit light before suffering any collision, and those that are in a long-lived state and which, in the present case, lose their energy through collisions. The latter states correspond to atoms that are either in a metastable or in a resonant level of energy. One important difference between radiative and long-lived state atoms is that the radial distribution of the former is more closely dependent upon the radial distribution of the electric field intensity.

Radiative states. The experiments[62] were performed in a long, cylindrical Pyrex discharge tube (R_1 = 13 mm, R_2 = 15 mm) in argon gas. The pressure domain explored was 50 mtorr to 1 torr. The surface-wave plasmas were generated in the 200 to 900 MHz frequency range. The $f R_1$ product being well below 2 GHz-cm, only the azimuthally symmetric (m = 0) mode was excited (Sec. 5.5.4). A hot cathode DC discharge was also operated for comparison purposes. Both types of discharge were investigated by optical emission spectroscopy, using a tomography technique to obtain spatial resolution. It provided the radial profile of emitted light intensity from a given excited level, at a given axial position, and it is known that this emission intensity varies linearly with the excited atom density. Figure 5.31 shows the influence of the wave frequency (0, 200 and 600 MHz) on the radial profiles of emission intensity from a radiative state, at a constant cross-section average electron density $\bar{n}$: increasing f

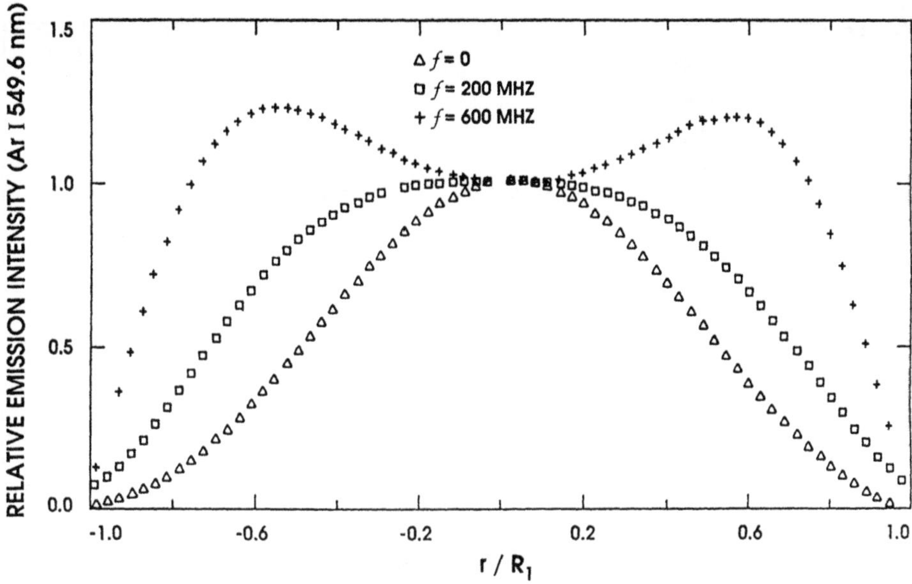

Fig. 5.31. Observed radial distribution of the emission intensity from an optically thin line, at a constant cross-section average electron density ($\bar{n}$ = 4.5 x 10^{10} cm^{-3}) and gas pressure (p = 150 mtorr), in a DC discharge (f = 0) and in a surface-wave discharge (f = 200 and 600 MHz) (from ref. 62).

generates wider profiles. The corresponding influence of the gas pressure, again at a constant $\bar{n}$ value, is shown for 600 and 900 MHz in Figs. 5.32a and 5.32b respectively. At 600 MHz, no significant change is observed as the pressure is raised from 50 to 500 mtorr whereas, at 1 torr, the plasma column shows a clear sign of constriction. At 900 MHz, increasing the gas pressure leads to wider radial profiles and to the appearance of off-axis maxima. Note that, in contrast to f = 600 MHz, the plasma at 1 torr is not constricted at 900 MHz.

Figures 5.33a to 5.33d show in a 3-dimensional manner the evolution of emission intensity of the same optically thin line as above, as a function of $\bar{n}$ and of radial position. We observe that, whatever the wave frequency, the relative intensity of the line emission grows up with $\bar{n}$. As for radial density profile in the positive column plasma (Fig. 5.33a), whatever the plasma density, it is maximum at the tube axis. As the frequency increases, the radial profiles get wider (Fig. 5.33b) and then flatten (5.33c). Occasionally, for certain values of the electron density and for high enough frequency, they even become concave, i.e. with a maximum off the plasma axis (Fig. 5.33d).

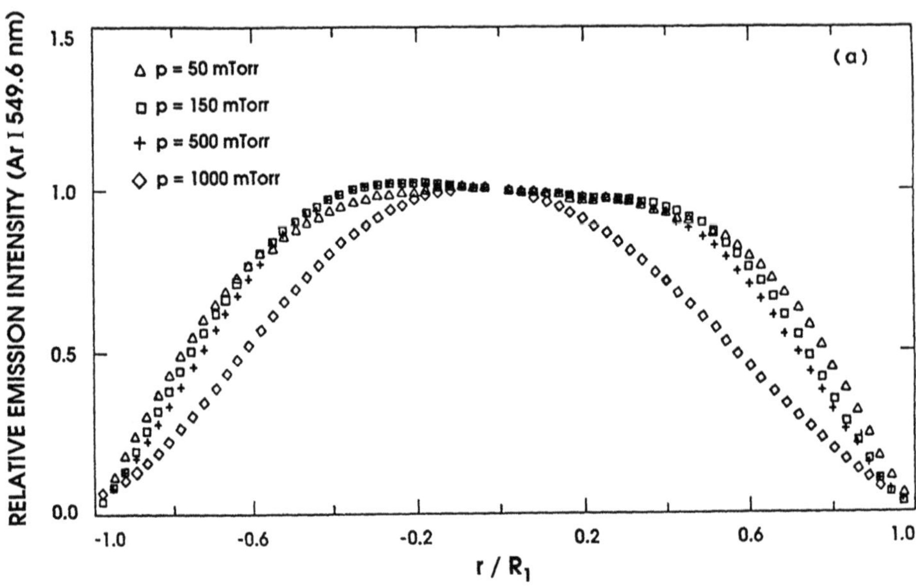

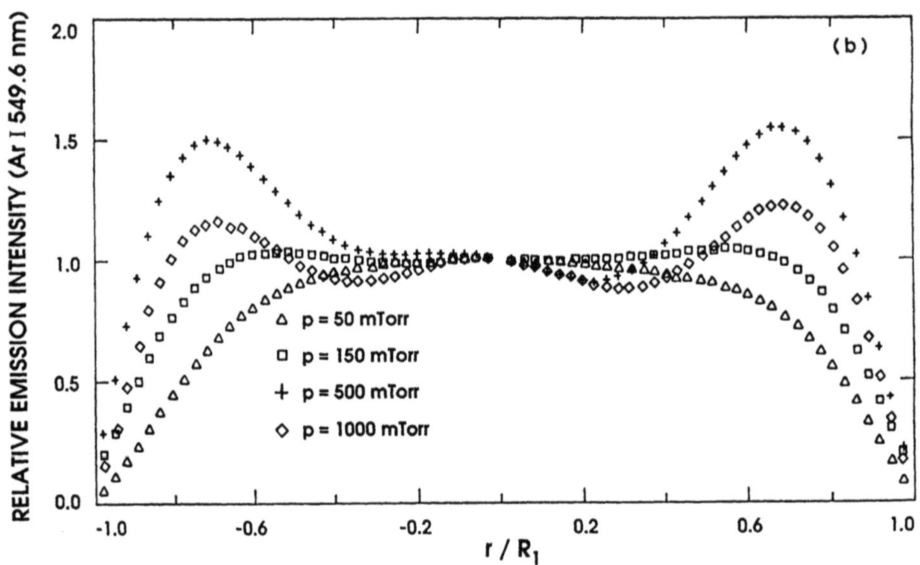

Fig. 5.32. Observed influence of the gas pressure on the radial distribution of the emission intensity from an optically thin line, at a constant cross-section average electron density ($\bar{n} = 1.1 \times 10^{11}$ cm^{-3}) and at a surface-wave frequency of (a) 600 MHz and (b) 900 MHz (from ref. 62).

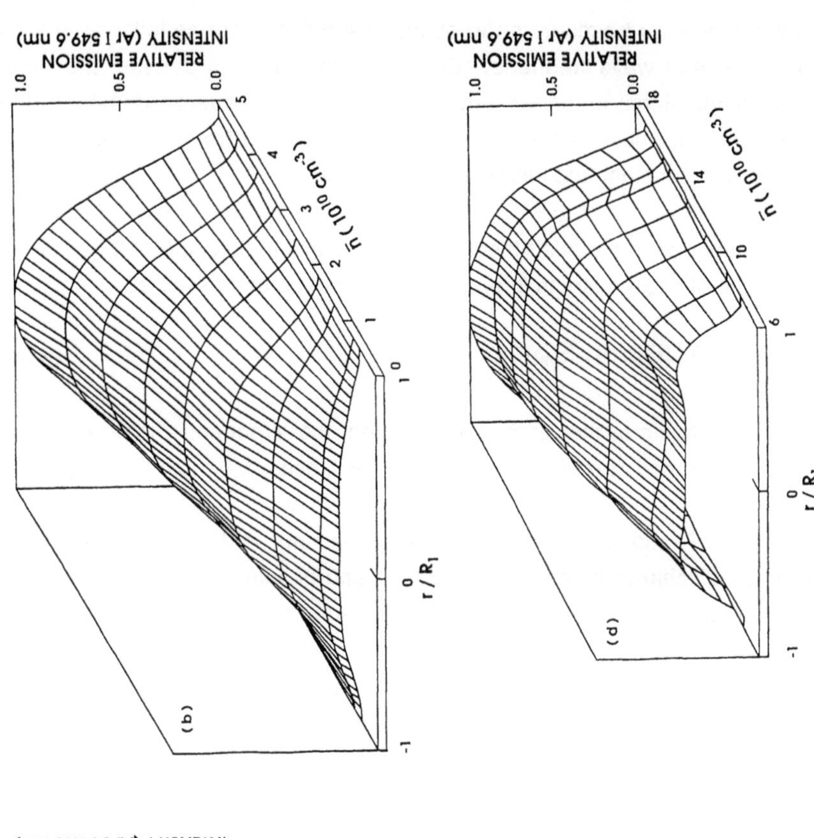
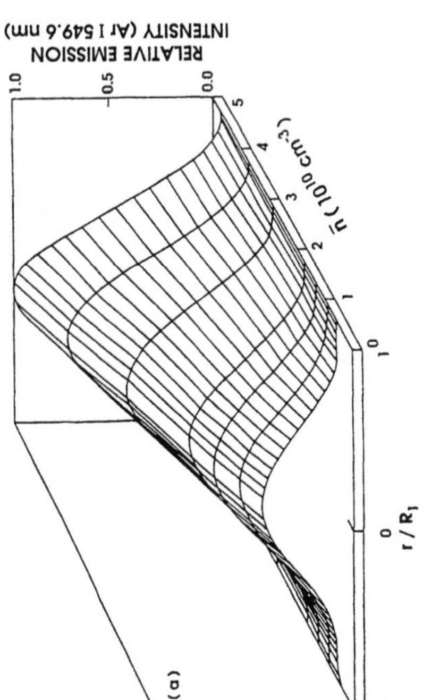
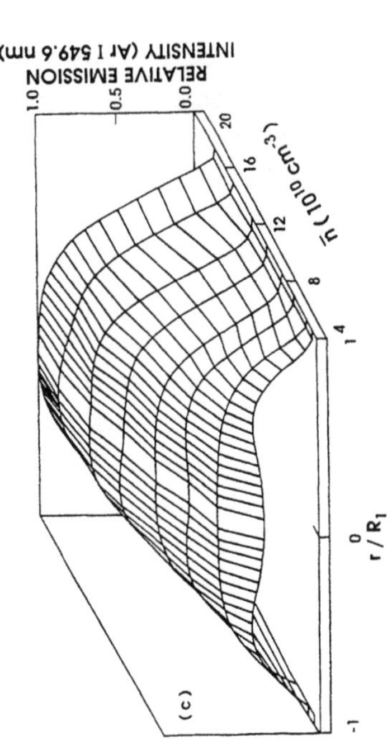

Fig. 5.33. Experimental 3-dimensional plots showing the emission intensity from an argon optically thin line, as a function of the radial position and cross-section average electron density. (a) Positive-column plasma; surface-wave-produced plasma at a frequency of (b) 200 MHz, (c) 600 MHz and (d) 900 MHz (from ref. 62).

In summary, the general trends from the above observations are as follows: i) for a given gas pressure and wave frequency, the higher the electron density, the more convex are the distributions; ii) for a given gas pressure and electron density $\bar{n}$, the higher the frequency, the less convex they are; iii) no general trend can be defined when $\bar{n}$ and f are kept constant and the pressure is varied.

These profiles have been modeled theoretically.[50] The final outcome of it is that the radial density profile of excited atoms in energy level k can be expressed as

$$n_k(r) / n_k(0) = [n(r) / n(0)] \, [E_{rms}(r) / E_{rms}(0)]^c. \tag{5.34}$$

For argon levels belonging to the $3p^56d$ configuration, the exponent c is equal to 2.5 when the EEDF is assumed Maxwellian. For more detail, see ref. 61 and those therein.

Long-lived states. These experiments[63] were carried out in 150 to 200 mm long $R_1 = 13$ mm Pyrex discharge tubes in argon, neon and helium gases. Compared to the radiative state data, the way of obtaining the present results differ in two respects: i) a spatial integration technique (so-called end-on measurements) was employed instead of the spatially resolved tomography technique; ii) line absorption measurements were performed, yielding absolute densities of excited atoms instead of relative densities as obtained from line emission recordings. The end-on measurement method requires both tube ends to be terminated by windows.[64] A collimator is placed in front of one of the windows, parallel to the tube axis and it is moved as a function of radial position. It collects the plasma light within a small diameter, axially oriented cylinder. Absorption measurements are realized by observing the attenuation through plasma of a given line emitted by a spectral lamp, located in front of the window opposite to that of the collimator. This method is particularly suited when the density of these excited atoms is high, i.e. when they do not lose energy by optical radiation (metastable state atoms) or when the photons that they emit do not fully escape from the plasma (resonant state atoms). Figure 5.34 shows the measured radial density distributions of the $^3P_2$ metastable and $^1P_1$ resonant state atoms in neon, in a surface-wave discharge at 900 MHz as compared with that in the positive column of a DC discharge. We observe that i) the radial distributions for the surface-wave plasma are flatter; ii) the maximum values attained radially do not differ markedly. It is slightly higher for the $^3P_2$ level in the positive-column plasma whereas it is the opposite for the $^1P_1$ level. A similar behavior has been observed in argon; iii) for the range of gas pressures investigated, the surface-wave plasma has a larger average cross-section density $\bar{n}_k$ ($\bar{n}_k$ is defined in a similar way as in eqn. (5.31)). Numerical integration of the distributions in Fig. 5.34 shows that $\bar{n}_k$ ($^3P_2$) and $\bar{n}_k$ ($^1P_1$) are 2.5 and 5.2 times larger respectively in the surface-wave plasma as compared to the positive-column plasma. A

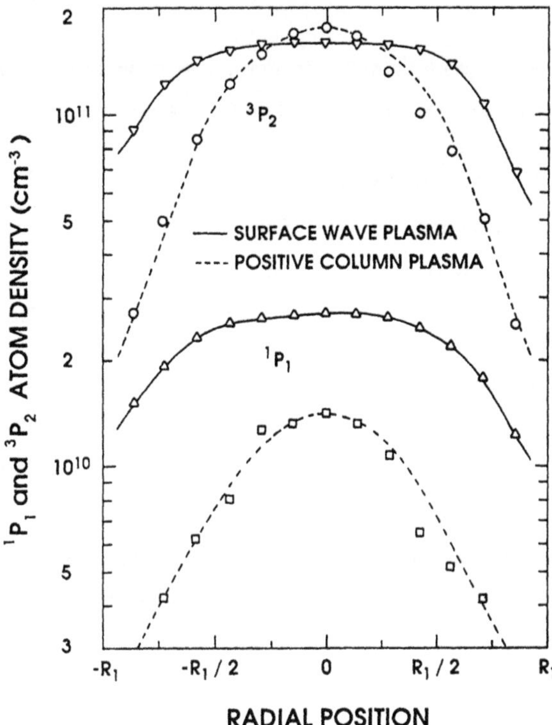

Fig. 5.34. Measured density of atoms in the $^3P_2$ metastable and $^1P_1$ resonant states in neon, as a function of radial position in a surface-wave-produced plasma (f = 900 MHz) and in a positive-column plasma, in the same discharge tube (R_1 = 13 mm), at the same gas pressure (p = 400 mtorr). The column length is 150 mm (from ref. 63).

similar advantage is also reported with the 2^3S and 2^1S metastable state atoms in helium.[63]

A larger yield of metastable atoms could, under certain circumstances, increase the reaction rate in plasma chemistry while the flatter radial distribution of resonant state atoms in a surface-wave plasma could lead to a more efficient UV lamp than with a positive column plasma. This increased UV efficiency would come from the fact that a larger number of photons, as compared to the case of the positive-column plasma, would leave the tube since their trapping in the volume of the plasma is less probable for those created close to the wall.[36] This problem is still under discussion.[65-68]

References

[1]   R.G. Bosisio, C.F. Weissfloch and M.R. Wertheimer, J. Microwave Power **7**, 325 (1972).

[2]   R.G. Bosisio, N. Nachmann and J. Spooner, IEEE Trans. Plasma Sci. **PS-2**, 273 (1974).

[3]   V.A. Dovzhenko, G.S. Solntsev and V.L. Neshchadimenko, *Contributed Papers from the XI Int. Conf. Phenom. Ionized Gases* (Prague, 1973) p. 158.

[4]   B. Kampmann, Z. Naturforsch. **34a**, 423 (1979).

[5]   A.W. Trivelpiece and R.W. Gould, J. Appl. Phys. **30**, 1784 (1959).

[6]   A.W. Trivelpiece, *Slow-Wave Propagation in Plasma Waveguides* (San Francisco University, San Francisco, 1967).

[7]   A. Shivarova and I. Zhelyazkov, J. Plasma Phys. **20**, 1049 (1978).

[8]   M. Moisan, A. Shivarova and A.W. Trivelpiece, Plasma Phys. **24**, 1331 (1982).

[9]   P.S. Bulkin, V.N. Ponomariev and G.S. Solntsev, Sov. Phys. - Tech. Phys. **8**, 911 (1964).

[10] D.T. Tuma, Rev. Sci. Instrum. **41**, 1519 (1970).

[11] F.C. Fehsenfeld, K.M. Evenson and H.P. Broida, Rev. Sci. Instrum. **36**, 294 (1965).

[12] J. Asmussen, R. Mallavarpu, J.R. Hamann and H.C. Park, Proc. IEEE **62**, 109 (1984).

[13] B. Vidal and C. Dupret, J. Phys. E: Sci. Instrum. **9**, 998 (1973).

[14] M. Moisan, C. Beaudry and P. Leprince, Phys. Lett. **50A**, 125 (1974).

[15] M. Moisan, C. Beaudry and P. Leprince, IEEE Trans. Plasma Sci. **PS-3**, 55 (1975).

[16] M. Moisan, C. Beaudry, L. Bertrand, E. Bloyet, J.M. Gagné, P. Leprince, J. Marec, G. Mitchel, A. Ricard and Z. Zakrzewski, *Gas Discharges*, IEE Conf. Publ. **143**, 382 (London, 1976).

[17] Z. Zakrzewski, M. Moisan, V.M.M. Glaude, C. Beaudry and P. Leprince, Plasma Phys. **19**, 77 (1977).

[18] M. Moisan, C.M. Ferreira, Y. Hajlaoui, D. Henry, J. Hubert, R. Pantel, A. Ricard and Z. Zakrzewski, Rev. Physique Appl. **17**, 707 (1982).

[19] J. Marec, E. Bloyet, M. Chaker, P. Leprince and P. Nghiem, in *Electrical Breakdown and Discharges in Gases*, E.E. Kunhardt and L.H. Luessen eds. (Plenum, New York, 1981) p. 347 (part B).

[20] C.M. Ferreira, in *Radiative Processes in Discharge Plasmas*, J.M. Proud and L.H. Luessen eds. (Plenum, New York, 1986) p. 431.

[21] M. Moisan and Z. Zakrzewski, in *Radiative Processes in Discharge Plasmas*, J.M. Proud and L.H. Luessen eds. (Plenum, New York, 1986) p. 381.

[22] C.M. Ferreira and M. Moisan, Physica Scripta **38**, 382 (1988).

[23] M. Moisan and Z. Zakrzewski, J. Phys. D: Appl. Phys. **24**, 1025 (1991).

[24] G.G. Lister and T.R. Robinson, J. Phys. D: Appl. Phys. **24**, 1993 (1991).

[25] V.M.M. Glaude, M. Moisan, R. Pantel, P. Leprince and J. Marec, J. Appl. Phys. **51**, 5693 (1980).

[26] Z. Zakrzewski and M. Moisan, in *Surface Waves in Plasmas and Solids*, S. Vukovic ed. (World Scientific, Singapore, 1986) p. 440.

[27] C.M. Ferreira, J. Phys. D: Appl. Phys **16**, 1673 (1983).

[28] Z. Zakrzewski, J. Phys. D: Appl. Phys **16**, 171 (1983).

[29]   E. Bloyet, P. Leprince, M. Llamas Blasco and J. Marec, J. Phys. Lett. **83A**, 391 (1981).

[30]   M. Llamas Blasco, V. Colomer and M. Rodriguez Vidal, J. Phys. D: Appl. Phys. **18**, 2169 (1985).

[31]   A. Gamero, A. Sola, J. Cotrino and V. Colomer, J. Appl. Phys. **65**, 2199 (1989).

[32]   A. Gamero, J. Cotrino, A. Sola and V. Colomer, J. Phys. D: Appl. Phys. **21**, 1275 (1988).

[33]   J. Margot-Chaker, M. Moisan, M. Chaker, V.M.M. Glaude, P. Lauque, J. Paraszcak and G. Sauvé, J. Appl. Phys. **66**, 4134 (1989).

[34]   D.M. Kerns and R.W. Beatty, *Basic Theory of Waveguide Junctions* (Pergamon, Oxford, 1967).

[35]   H.M. Barlow and J. Brown, *Radio Surface-Waves* (Clarendon, Oxford, 1962).

[36]   A. Ricard, C. Barbeau, A. Besner, J. Hubert, J. Margot-Chaker, M. Moisan and G. Sauvé, Can. J. Phys. **66**, 740 (1988).

[37]   M. Moisan, C. Barbeau, R. Claude, C.M. Ferreira, J. Margot, J. Paraszczak, A.B. Sá, G. Sauvé and M.R. Wertheimer, J. Vac. Sci. Technol. **B9**, 9 (1991).

[38]   M. Moisan, Z. Zakrzewski and R. Pantel, J. Phys. D: Appl. Phys. **12**, 219 (1979).

[39]   M. Chaker and M. Moisan, J. Appl. Phys. **57**, 91 (1985).

[40]   J. Hubert, M. Moisan and Z. Zakrzewski, Spectrochim. Acta **41B**, 205 (1986).

[41]   C. Moutoulas, M. Moisan, L. Bertrand, J. Hubert, J.L. Lachambre and A. Ricard, Appl. Phys. Lett. **46**, 323 (1985).

[42]   M. Moisan and Z. Zakrzewski, *Contributed Papers from the XVII Int. Conf. Phenom. Ionized Gases* (Budapest, 1985) p. 712.

[43]   M. Moisan and Z. Zakrzewski, U.S. Patent 4,810,933 (1989); U.S. Patent 4,906,898 (1990).

[44]   J. Margot-Chaker, M. Moisan, Z. Zakrzewski, V.M. Glaude and G. Sauvé, Radio Sci. **23**, 1120 (1988).

[45]   M. Chaker, P. Nghiem, E. Bloyet, P. Leprince and J. Marec, J. Phys. Lett. **43**, L-71 (1982).

[46]   P. Nghiem, M. Chaker, E. Bloyet, P. Leprince and J. Marec, J. Appl. Phys. **53**, 2920 (1982).

[47]   A. Granier, C. Boisse-Laporte, P. Leprince, J. Marec and P. Nghiem, J. Phys. D: Appl. Phys. **20**, 204 (1987).

[48]   S.C. Brown, *Basic Data of Plasma Physics* (M.I.T. Press, Cambridge, 1966).

[49]   R. Darchicourt, S. Pasquiers, C. Boisse-Laporte, P. Leprince and J. Marec, J. Phys. D: Appl. Phys. **21**, 293 (1988).

[50]   C.M. Ferreira and J. Loureiro, J. Phys. D: Appl. Phys. **16**, 2471 (1983).

[51]   C. Boisse-Laporte, A. Granier, E. Dervisevic, P. Leprince and J. Marec, J. Phys. D: Appl. Phys. **20**, 197 (1987).

[52]   U. Kortshagen, H. Schlüter and A. Shivarova, J. Phys. D: Appl. Phys. **24**, 1571 (1991).

[53]   U. Kortshagen and H. Schlüter, J. Phys. D: Appl. Phys. **24**, 1585 (1991).

[54]   G.S. Solntsev, P.S. Bulkin, M. Rakhman and L.I. Tsvetkova, Sov. J. Plasma Phys. **15**, 495 (1989).

[55]   M. Moisan, R. Pantel and J. Hubert, Contrib. Plasma Phys. **30**, 293 (1990).

[56]   A.B. Sá, C.M. Ferreira, S. Pasquiers, C. Boisse-Laporte, P. Leprince and J. Marec, J. Appl. Phys. **70**, 4147 (1991).

[57]   M. Chaker, M. Moisan and Z. Zakrzewski, Plasma Chem. Plasma Process. **6**, 79 (1986).

[58]   E. Mateev, I. Zhelyazkov and V. Atanassov, J. Appl. Phys. **54**, 3049 (1983).

[59]   J. Rogers and J. Asmussen, IEEE Trans. Plasma Sci. **PS-10**, 11 (1982).

[60]   J. Wolinska-Szatkowska, J. Phys. D: Appl. Phys. **21**, 937 (1988).

[61]   Z. Rakem, P. Leprince and J. Marec, Revue Phys. Appl. **25**, 125 (1990).

[62]   J. Margot, M. Moisan and A. Ricard, Appl. Spectrosc. **45**, 260 (1991).

[63]   A. Ricard, M. Moisan and J. Hubert, *Contributed papers from XVII Int. Conf. Phenom. Ionized Gases* (Budapest, 1985) p. 741.

[64]   R. Pantel, A. Ricard and M. Moisan, Beitr. Plasma Phys. **23**, 561 (1983).

[65]   A.T. Rowley, *43rd Annual Gaseous Electronics Conf.* (Urbana, Illinois, 1990) Abstract K-36.

[66]   D. Levy, *43rd Annual Gaseous Electronics Conf.* (Urbana, Illinois, 1990) Abstract GB-2.

[67]   C. Beneking and P. Anderer, *40th Annual Gaseous Electronics Conf.* (Atlanta, Georgia, 1987) Paper EB-7.

[68]   C. Beneking and P. Anderer, J. Phys. D: Appl. Phys. submitted (1992).

CHAPTER 6

PRINCIPLES OF MAGNETICALLY ASSISTED
MICROWAVE DISCHARGES [*]

6.1. Introduction

It is well known that relative to unmagnetized plasmas, plasmas submitted to a static magnetic field, the so-called *magnetoplasmas*, are characterized by a higher ionization efficiency. This useful property, resulting from reduced charged particle losses to the wall, is one of the many advantages of microwave-generated magnetoplasmas, others being, for example, an efficient microwave power transfer from the feed line to the discharge even at very low gas pressures, and, most often, the absence of electrodes in contact with the plasma. The combination of these advantages makes microwave magnetoplasmas a promising alternative to RF capacitive discharges for materials processing in the semiconductor industry.

Obtaining an efficient coupling of microwaves to the magnetoplasma and modeling such a discharge require more than simply considering particle losses. One must also take into account the properties of the wave sustaining the discharge, for example, its dispersion and field configuration, and determine how the wave energy is transferred to the discharge. Then one can obtain the spatial properties of the plasma along the wave propagation path.

Clearly a proper understanding of the physics of magnetized plasmas generated by an electromagnetic wave necessitates investigating the above aspects. In the following we therefore first examine the loss mechanisms of charged particles in a plasma submitted to a magnetic field. Then we review the characteristics of wave absorption by a magnetoplasma. Finally, as a working example, we study the characteristics of wave propagation in a magnetized, infinite plasma and then consider the influence of boundaries on their characteristics by looking at a waveguide partially or completely filled with plasma.

6.2. Charged particle losses in magnetoplasmas

In low-pressure bounded plasmas, volume recombination is negligible and the loss of charged particles mainly occurs through diffusion to the walls where recombination

---

[*] **Presented by J. Margot, T.W. Johnston and J. Musil**

occurs. In the absence of magnetic field, the way by which each kind of particle moves to the wall depends on the ratio of the charged particle-neutral atom (molecule) mean free path $\lambda_{\alpha n}$ (α = e for electrons and α = i for ions) to the vessel dimension $\mathbb{D}$. When $\lambda_{\alpha n}/\mathbb{D}$ is larger than unity, the plasma is said to be in the free-fall regime whereas, in the reverse condition, the movement of charged particles to the walls results from diffusion processes. A typical value of the mean free path of electrons in argon at 1 mtorr is approximately 1 meter, a distance larger than the usual plasma vessel dimensions. *In the presence of a significant magnetic field, $\lambda_{\alpha n}$ loses its meaning as a characteristic parameter since the electron trajectory is then, in fact, wrapped around field lines, and thus it is the Larmor radius which becomes the effective mean free path.* Under typical conditions, the Larmor radius for electrons is a few tenths of mm or less, which is much shorter than any vessel dimensions. This suggests that diffusion rather than free-fall conditions governs the physics of magnetoplasmas. For this reason, we examine in some detail below the various diffusion regimes.

6.2.1. Free diffusion

6.2.1.1. Diffusion coefficient and mobility. In the absence of a magnetic field, the diffusion coefficient D is a scalar quantity given by

$$D = \langle w^2 \rangle / 3\nu_{eff} \tag{6.1}$$

where w is the individual velocity of particles and $\langle \ \rangle$ denotes an average over the energy distribution function of the charged particle species under consideration, while ν_{eff} is an effective frequency (Sec. 2.2.6) of collisions between charged particles (ions or electrons) and other species.

We now assume that a static magnetic flux density $\mathbf{B_0}$, directed along the z-axis of a cartesian or cylindrical coordinate system, is superimposed on a plasma. The anisotropy generated by $\mathbf{B_0}$ makes the diffusion coefficient a tensorial quantity that in cartesian or cylindrical coordinates can be written as[1,2]

$$\underset{\sim}{D} = D \begin{pmatrix} \chi_1 & \chi_2 & 0 \\ -\chi_2 & \chi_1 & 0 \\ 0 & 0 & 1 \end{pmatrix} \equiv D\underset{\sim}{\chi} \tag{6.2}$$

where $\chi_1 = 1/(1 + \kappa_\alpha^2)$, $\chi_2 = \pm \kappa_\alpha \chi_1$ and $\kappa_\alpha = \omega_{c\alpha}/\nu_{eff}$, the quantity $\omega_{c\alpha} = |eB_0/m_\alpha|$ being the angular cyclotron frequency. The diffusion coefficients for electrons and ions are identified by setting α = e and α = i respectively. In the expression for χ_2, the + and - sign refers to positive and negative charges respectively. For low values of κ_α, i.e. at a low B_0 field strength or for a large collision rate, $\underset{\sim}{D}$ obviously tends toward the diagonalized tensor $D\underset{\sim}{I}$ (where $\underset{\sim}{I}$ is the identity tensor). In contrast, at high κ_α ratios, χ_1 and χ_2 tend toward $1/\kappa_\alpha^2$ and $1/\kappa_\alpha$ respectively. Since κ_α is inversely proportional to the mass of the particles, the magnetic field reduces more strongly the diffusion of electrons than that of ions. As a result, for a sufficiently strong magnetic field, the diffusion of ions in the direction perpendicular to $\mathbf{B}_0$ exceeds that of electrons in the same direction, a situation opposite to what is found when $B_0 = 0$ where $D_e > D_i$.

In the absence of any electric field, assuming a spatially uniform temperature, the particle flux $\boldsymbol{\Gamma} \equiv n\mathbf{v}$ due to diffusion, where n is the density of charged particles and $\mathbf{v}$ is the average diffusion velocity of a given species of particles, depends on the density gradient causing this diffusion through the relation

$$\boldsymbol{\Gamma} = - \underset{\sim}{D} \cdot \nabla n . \tag{6.3}$$

Note that $\underset{\sim}{D}$ being a tensor, relation (6.3) means that for a cylindrical plasma column, the presence of a density gradient in the radial direction generates a diffusion motion also in the azimuthal direction (analogous to the Hall current) in contrast with the unmagnetized case. Since the direction of diffusion depends on the charge's sign, the result is a diamagnetic current, whose effect on the magnetic field is nonetheless usually negligible.

When the plasma is in addition under the influence of an electric field $\mathbf{E}$, relation (6.3) becomes

$$\boldsymbol{\Gamma} = - \underset{\sim}{D} \cdot \nabla n \pm n\underset{\sim}{\mu} \cdot \mathbf{E} \tag{6.4}$$

where the + and - signs apply to ions and electrons respectively; $\underset{\sim}{\mu}$ is the mobility (Sec. 2.2.6) and is a tensor in an anisotropic plasma.[†] Note that any DC electric field (be it applied or resulting from space charge effects) can contribute to such a drift motion of the plasma bulk. For strong B_0 fields ($\kappa_\alpha^2 \gg 1$), the result is a common $\mathbf{E} \times \mathbf{B}_0/B_0^2$ drift velocity for ions and electrons, the so-called *electric field drift*.

6.2.1.2. Spatial distribution of density in free diffusion regime. The free diffusion regime applies when the charged particle density is so low (typically when $n\mathbb{D}^2 \leq 10^7$ cm^{-1} in unmagnetized plasmas) that electrons and ions can diffuse at their

---

[†] In a manner similar to eqn. (6.2), the mobility tensor $\underset{\sim}{\mu}$ can be expressed by $\mu\underset{\sim}{\chi}$ where the scalar quantity μ is the mobility in absence of B_0. This quantity is related to the scalar diffusion coefficient D (eqn. (6.1)) through the Einstein relation $D/\mu = u_k$ defining the characteristic energy in eV (Sec. 2.2.9).[2] The latter reduces to the particle's temperature (expressed in eV) for Maxwellian energy distributions.

own rate, which is substantially different. This means that no significant space charge electric field is created that would relate the movement of ions to that of electrons. In such a case, one defines a particle flux (eqn. (6.3)) for each species.

When free diffusion prevails, the time evolution of electron density is given by the *continuity equation*

$$\frac{\partial n}{\partial t} + \nabla \cdot \boldsymbol{\Gamma} = \langle v_i \rangle \, n, \tag{6.5}$$

where $\langle v_i \rangle$ is the average ionization frequency. Assuming that $\langle v_i \rangle$ is independent of spatial coordinates, using relation (6.3) and integrating (6.5) by separation of variables, it can be shown that in a cylindrical discharge tube (r, ϕ, z coordinates) where B_0 is directed along the tube axis z, the density $n(r, z, t)$ is a function of both the diffusion coefficient $D_{\parallel} \equiv D$ in the direction parallel to B_0 and the diffusion coefficient $D_{\perp}$ in the direction perpendicular to it, namely

$$n(r, z, t) = n_0 \, J_0(r/\Lambda_r) \sin(z/\Lambda_z) \exp \left(\langle v_i \rangle - \frac{D_{\parallel}}{\Lambda_z^2} - \frac{D_{\perp}}{\Lambda_r^2} \right) t, \tag{6.6}$$

where

$$D_{\perp} = \frac{D_{\parallel}}{1 + \mu^2 B_0^2} \tag{6.7}$$

and

$$\Lambda_r = R_1/x_m, \tag{6.8}$$

$$\Lambda_z = \ell/n\pi. \tag{6.9}$$

In these relations, x_m is the mth zero of the J_0 Bessel function of the first kind and order zero and n is an integer; R_1 and ℓ are the plasma vessel inner radius and length respectively. The values of Λ_r and Λ_z are imposed by the boundary conditions, taken here as $n(R_1, z, t) = 0$ and $n(r, \ell, t) = 0$, and these quantities are called the diffusion lengths in the radial and axial direction respectively. From eqn. (6.6), it follows that for a steady state

$$\langle v_i \rangle = \frac{D_{\parallel}}{\Lambda_z^2} + \frac{D_{\perp}}{\Lambda_r^2}. \tag{6.10}$$

Once the discharge is initiated, after a short transient time, there remains only the fundamental or dominant mode ($m = 0$, $n = 1$ with $x_0 = 2.405$), higher modes being damped due to smaller Λ_r and Λ_z values. Equations (6.8) and (6.9) provide the corresponding diffusion lengths. Equation (6.10), where $<v_i>$ and $D_\parallel$ depend on the electron energy distribution function (EEDF), determines the average energy (or the electron temperature) of the electron population.

In order to solve eqn. (6.5) analytically, we have considered $<v_i>$ to be spatially uniform. This assumption does not usually hold in magnetoplasmas sustained by electromagnetic waves, and eqn. (6.5) must then be solved numerically taking into account the spatial dependence of $<v_i>$. As a result, the spatial distribution of density can differ substantially from eqn. (6.6).

6.2.2. Diffusion of charged particles in magnetoplasmas under space-charge conditions.

The problem of non-free diffusion in magnetoplasmas has been a continuing subject of investigation over the last 40 years. The state of the question up to 1979 can be found in a review by Zhilinskii and Tsendin.[3]

One approach consists in assuming a priori classical ambipolar conditions, in particular the *congruence assumption*, namely $\boldsymbol{\Gamma_e} = \boldsymbol{\Gamma_i}$. Within such an approximation, the components of the ambipolar diffusion tensor are given by[1]

$$D_{a\parallel} = D_a \, , \tag{6.11}$$

$$D_{a\perp} = D_a / (1 + \mu_e \mu_i B_0^2) \, , \tag{6.12}$$

where D_a is the ambipolar diffusion coefficient in the absence of magnetic field (Sec. 2.2.6). One can show that for high enough B_0-field values, eqn. (6.12) implies that $D_{a\perp}$ is of the order of $D_{e\perp}$.

The congruence approximation used above originates from a 1-dimensional plasma where $\nabla \cdot (\boldsymbol{\Gamma_e} + \boldsymbol{\Gamma_i}) = 0$ strictly leads to $\boldsymbol{\Gamma_e} = \boldsymbol{\Gamma_i}$. When modeling two or three-dimensional plasmas, it is customary to use this approximation; however, when $B_0 \neq 0$, the anisotropy induced by B_0 differs for each species and the scalar condition $\nabla \cdot (\boldsymbol{\Gamma_e} + \boldsymbol{\Gamma_i}) = 0$ on the flux vectors does not require $\boldsymbol{\Gamma_e}$ and $\boldsymbol{\Gamma_i}$ to be parallel and thus equal.

In a short note published in 1955, Simon[4] suggested that in finite size plasmas, the diffusion across a magnetic field may not be ambipolar but would occur at a rate much faster than predicted by eqn. (6.12), due to what he called the "short-circuit" effect. This model postulates that charge separation between radially spaced plasma zones could be neutralized through electron currents flowing first parallel to the B_0-field and then through the end walls enclosing the discharge. This effect would result from a much higher electron conductivity in the parallel than in the perpendicular direction and, hence, of much larger currents flowing in this direction. The transverse diffusion coefficient is then predicted to be essentially that of ions

$$D_{S\perp} \simeq D_{i\perp},   \tag{6.13}$$

where s is for Simon's value. Since $D_{i\perp} \gg D_{e\perp}$, Simon's theory yields a diffusion coefficient which is much larger than that predicted by the classical ambipolar theory, leading to a better agreement with most experimental observations (see further).

Monroe[5] has developed a formal 3-dimensional theory where a set of 14 macroscopic plasma equations (continuity and momentum equations for both electrons and ions plus Maxwell's equations) is transformed into an equivalent set of 4 integral equations. According to Monroe, plasma currents perpendicular to B_0 modify the diffusion process in such a way that it is described by the usual ambipolar coefficient in the direction parallel to the magnetic field and, in the direction perpendicular to it, by a coefficient $D_{M\perp}$, where M stands for Monroe,

$$D_{M\perp} = (\frac{\mu_i D_e}{b_e} + \frac{\mu_e D_i}{b_i}) \, / \, (\mu_i + \mu_e) \, ,   \tag{6.14}$$

where $b_e \equiv 1 + \mu_e^2 B_0^2$ and $b_i \equiv 1 + \mu_i^2 B_0^2$. We note that $D_{M\perp}$ reduces to D_a for B_0 tending toward zero and to $D_{S\perp}$ for large B_0-field intensities since then $D_{M\perp} \simeq D_{i\perp}$. Because $\mu_e \gg \mu_i$, we further see that $D_{M\perp}$ can be rewritten as

$$D_{M\perp} \simeq D_{e\perp} \frac{\mu_i}{\mu_e} + D_{i\perp},   \tag{6.15}$$

showing that $D_{M\perp}$ is in general larger than $D_{S\perp}$.

Experimentally, it is observed that the perpendicular diffusion coefficient is larger, often by one or two orders of magnitude, than the ambipolar value $D_{a\perp}$ of relation (6.12), indicating that the congruence assumption may not be verified. A series of experiments conducted by Geissler[6,7] has led to the conclusion that the actual value of the perpendicular diffusion coefficient is intermediate between that predicted by the classical ambipolar theory and that from Simon's model. Since we have shown that $D_{a\perp} < D_{S\perp} < D_{M\perp}$, this rules out Monroe's theory. It was further noted by Geissler that

$D_\perp$ could be written as the sum of a classical part and of a non-classical part, the magnitude of the latter depending on the conductivity of the plasma vessel. For plasmas contained in a conducting vessel, the non-classical part can be one order of magnitude larger than for a glass vessel of the same dimensions. This underlines the importance of the nature of the container's walls on the operation of magnetoplasmas. Recall that Simon's theory considers currents which need to flow from one plasma zone to the other through electrical circuits that include the walls, implying that these walls must be conducting.

Our short review of diffusion phenomena in magnetoplasmas has shown that no satisfactory theory of diffusion exists up to now except in the limit of 1-dimensional plasmas, which require $R_1^2/D_\perp \ll \ell^2/D_\parallel$.[†] Because of the reduction of diffusion in the direction perpendicular to the B_0-field, this condition is however much more difficult to satisfy than in absence of magnetic field, and it may be violated in a number of experimental situations. Finally, one must bear in mind that we have considered the diffusion coefficient and the ionization frequency independent of the spatial coordinates, an assumption that must be reevaluated when the plasma is sustained by an HF electric field.

6.3. Radial density profiles in magnetoplasmas from diffusion-controlled to free-fall regime

The modeling of long cylindrical plasma columns in a static magnetic field has been mainly developed in the 1960s and 1970s. These models neglect the loss of particles in the direction parallel to B_0 and thus apply to plasma columns where the ratio ℓ / R_1 is substantially higher than ω_{ce}/ν_{eff}. In practice, such theories do allow one to describe the radial structure of the plasma column far enough from its extremities, to the extent that the situation can be actually reduced to a 1-dimension case. Obviously, the condition of applicability of these theories is less stringent as pressure increases since then ν_{eff} increases.

In absence of magnetic field, there exist models that are particularly useful in describing long cylindrical plasma columns in two limiting cases: i) the Schottky ambipolar diffusion theory[8] applies to the case where ion and electron-neutral mean free paths are small in comparison with the plasma vessel radius. For a cylindrical plasma column submitted to a radially uniform electric field, it yields a radial density profile of the J_0 Bessel function-type; ii) in the opposite case, where mean free paths are large, the free-fall theory[9] is appropriate. A theory covering these two limiting cases

---

[†] This inequality can be written equivalently as $\ell/R_1 \gg \omega_{ce}/\nu_{eff}$.

and the intermediate range of pressures has been developed by Self and Ewald[10] and further improved by Ingold.[11]

The influence of an axial static magnetic field on a DC discharge has been studied by Persson[12] for the high pressure situation, considering ambipolar diffusion assumptions. The range from high to low pressures was subsequently covered by Forrest and Franklin[13], Ewald et al.[14] and Ilic.[15] All these theories are based on fluid equations. Forrest and Franklin[13], and Ewald et al.[14] use continuity and momentum transfer equations for both electrons and ions while Ilic[15] additionally considers energy transfer equations. The latter model allows one to take into account self-consistently the radial inhomogeneity of the ionization frequency resulting from spatially varying electron and ion temperatures. Nonetheless, both models yield similar results for the radial density and space-charge field distributions. For example, the numerical results obtained by Ilic[15] in helium show that the increase of the magnetic field intensity leads to a contraction of the radial density profile (see Fig. 6.1), which is approximately the same as that produced by a gas pressure increase. However, while in absence of a B_0 field the plasma potential increases negatively with radius up to the plasma-sheath boundary, in contrast, for large enough values of B_0, this potential increases positively with radius, reaching a maximum near the plasma-sheath boundary where it drops suddenly to negative values. This inversion of potential results from the fact that electron diffusion in the direction perpendicular to the magnetic field decreases much more rapidly with increasing magnetic field than ion diffusion. Thus a space-charge field builds up in such a way that ions, and not electrons, are slowed down, in contrast to the unmagnetized plasma situation. This potential inversion also modifies the radial

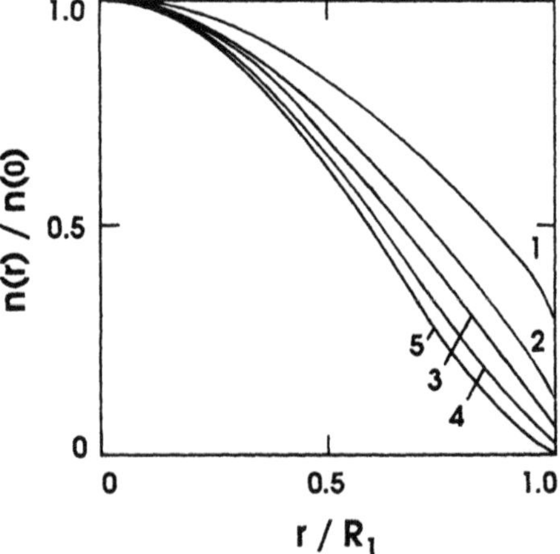

Fig. 6.1. Calculated radial profile of plasma density, $n(r)/n(0)$, at $\omega_{ce}/\nu_{eff} = 0$, 50, 100, 300, 1000 for curves 1, 2, 3, 4, 5 respectively (from ref. 15).

profile of the ion temperature found in the $B_0 = 0$ case: in a magnetoplasma, since the electric field changes sign at some radial position (where the potential goes through an extremum), the ion motion reverses direction at that position. Thus, the ions not only cease to be attracted to the wall but lose energy as they move toward it against the space-charge electric field.

These results for DC discharges need to be adapted to the case of HF sustained plasmas. The problem is that in the latter case, the ionization frequency is usually radially non-uniform, the degree of non-uniformity varying with the magnetic field intensity, the plasma conditions and the way the electromagnetic energy is fed to the plasma (see ref. 16 and Sec. 8.2). A second point that needs to be clarified is the influence of the limited axial extension of the plasma, which requires one to solve a 2-dimensional rather than a 1-dimensional problem. Only a few attempts to model 2-dimensional HF magnetoplasmas have been pursued to date[17,18] due to the complex geometries and conditions typical of the systems used for industrial applications (see Chap. 7). For example, Porteous and Graves[18] have solved the two-dimensional problem for a realistic cylindrical geometry ($\ell = 4R_1$) with conducting walls, by treating electrons as a fluid and ions as particles. The wave power P is assumed to be transferred to electrons uniformly as a function of axial position, i.e. dP/dz = constant. To determine the influence of the power deposition radial profile on plasma density distribution, the authors have assumed two distinct radial profiles of power deposition, as shown in Fig. 6.2. The important conclusion of this study is that the charged particle fluxes are not given by the classical ambipolar model since electrons are lost exclusively axially. These results can be interpreted in terms of the short-circuit effect for conducting walls, as suggested initially by Simon.[4]

To improve our understanding of HF magnetoplasmas, more refined models are required that take into account the exact power deposition profile, i.e. models in which Maxwell's and plasma equations are coupled. We also need to consider in a more realistic way the actual magnetic field configuration and the plasma boundary. Work in this direction is in progress in a number of laboratories and should yield significant results in the near future.

6.4. Wave-to-plasma power transfer mechanisms

Electromagnetic power can transfer to plasma in several ways. In the following sections, we provide elementary considerations on two such important mechanisms: collisional and collisionless absorption. In both cases, the average power density gained by electrons from the wave is given by

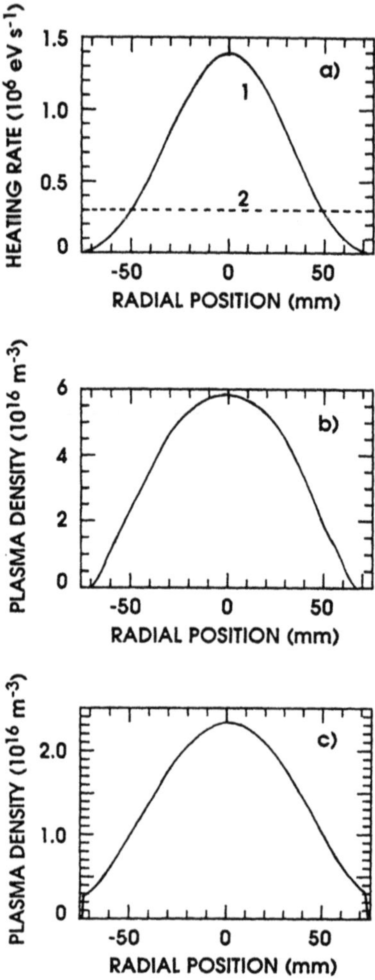

Fig. 6.2. a) Assumed radial profiles 1 and 2 of power deposition used to calculate the radial profile of plasma density (b) and (c) respectively (from ref. 18).

$$\mathbb{P} = \frac{1}{4} [\mathbf{E} \cdot \mathbf{J}^* + \mathbf{E}^* \cdot \mathbf{J}] \, , \tag{6.16}$$

where $\mathbf{E}$ is the complex amplitude of the wave electric field and $\mathbf{J}$ is the current density related to $\mathbf{E}$ through the conductivity tensor, namely

$$\mathbf{J} = \underline{\sigma} \cdot \mathbf{E} \, . \tag{6.17}$$

Recall that the symbol * denotes the complex conjugate value and the dot · the scalar product.

The absorption of energy by the charged particles causes the wave attenuation, and the absorbed energy is redistributed into thermal motion of the plasma particles.

6.4.1. Collisional absorption. When the electrons of a plasma are submitted to a HF electric field, they oscillate at the same frequency as the field (Sec. 2.2.1). In absence of collisions with other particles, the average energy acquired over a field period by the electrons is zero. However, each time an electron suffers a collision, its periodic motion is temporarily interrupted and it accumulates a small amount of power from the wave (Sec. 2.2.2). After several collisions, some of the electrons will have gained a large enough amount of energy to produce ionizing collisions and thus contribute to the maintenance process of the plasma.

To further discuss the collisional absorption process, the following assumptions will be made: i) the wave frequency $\omega/2\pi$ is well above the ion plasma frequency. Thus, ions can be considered immobile in the wave electric field; ii) the wave phase velocity is much larger than the electron thermal velocity. This assumption allows one to describe the plasma by means of hydrodynamic equations without retaining the kinetic pressure gradient term depending on the thermal electron motion. It is called the *cold plasma approximation*; iii) the wave electric field strength is weak enough so as to ignore nonlinear effects in the plasma equations; iv) the microscopic electron collision frequency for momentum transfer ν_c is weakly dependent on the energy u of electrons. In such a case, the effective collision frequency for momentum transfer, ν_{eff}, can be regarded as a scalar.

Conductivity tensor. Under the above assumptions, in a plasma submitted to a magnetic induction vector $\mathbf{B}_0$ directed along the z axis of a cylindrical coordinate system, the conductivity tensor is

$$\underline{\underline{\sigma}} = \begin{pmatrix} \sigma_1 & -\sigma_2 & 0 \\ \sigma_2 & \sigma_1 & 0 \\ 0 & 0 & \sigma_3 \end{pmatrix}. \tag{6.18}$$

When considering an electromagnetic field varying as $\exp(j\omega t)$, under the above assumptions, the elements σ_1, σ_2 and σ_3 are

$$\sigma_1 = \varepsilon_0 \omega \xi^2 \{1 / [j(1+\tau) + \delta] + 1 / [j(1-\tau) + \delta]\} / 2, \tag{6.19}$$

$$\sigma_2 = j\varepsilon_0\omega\xi^2 \{1 / [j\,(1+\tau) + \delta] - 1 / [j\,(1-\tau) + \delta]\} / 2, \qquad (6.20)$$

$$\sigma_3 = \varepsilon_0\omega\xi^2 / (j + \delta), \qquad (6.21)$$

with $\xi \equiv \omega_{pe}/\omega$, $\delta \equiv \nu_{eff}/\omega$ and $\tau \equiv \omega_{ce}/\omega$; ε_0 is the free-space permittivity, and ω_{pe} is the electron plasma angular frequency.

Wave field polarization and power density in the plasma. Using (6.16) and (6.17), the power density $\mathbb{P}$ can be expressed as

$$\mathbb{P} = \frac{1}{2}\Big[\mathrm{Re}(\sigma_1)\,|E_\perp|^2 + \mathrm{Re}(\sigma_3)\,|E_z|^2 - 2\,\mathrm{Im}(\sigma_2)\,|E_r E_\phi|\,\sin\Theta\Big], \qquad (6.22)$$

where $\mathrm{Re}(\sigma_k)$ and $\mathrm{Im}(\sigma_k)$ $(k = 1,\,2,\,3)$ designate the real and imaginary part of σ_k respectively; $|E_\perp|^2 \equiv |E_r|^2 + |E_\phi|^2$ is the square of the electric field perpendicular component. The angle Θ is termed the polarization angle of the electric field in the direction perpendicular to $\mathbf{B}_0$ (Sec. 8.2.1). It is given by

$$\tan\Theta = \mathrm{Im}\,(E_r/E_\phi) / \mathrm{Re}\,(E_r/E_\phi). \qquad (6.23)$$

For $\Theta = 0$, $E_\perp$ is said to be linearly polarized (constant orientation in space) while for $\Theta = \pm\,\pi/2$ and $|E_r| = |E_\phi|$, it is right-hand (+) and left-hand (-) circularly polarized. A further point that comes out from eqn. (6.22) is related to $\mathrm{Im}(\sigma_2)$ being negative as can be seen from eqn. (6.20). Then for given $|E_r|$ and $|E_\phi|$, the power density gained by electrons is maximum for $\Theta = (2n + 1/2)\,\pi$ $(n = 0,\,1,...)$, i.e. when the wave is right-hand polarized. It is minimum for $\Theta = (2n - 1/2)\,\pi$, i.e. for a left-hand polarized wave. In the case of a linear polarization $(\Theta = n\pi)$, the contribution of the third term of the right-hand side of eqn. (6.22) vanishes. Equation (6.22) is a generalization of previous expressions for $\mathbb{P}$ given, for example, by Shkarofsky et al.[2] or by Asmussen,[19] which corresponded specifically to a linearly polarized wave.

There are other interesting features to be deduced from (6.22). When the wave electric field's only component is E_z, $\mathbb{P}$ does not depend on the magnetic field. The absorption of power is then similar to that in an unmagnetized plasma. Clearly the influence of the B_0-field on plasma shows up only through the components of the wave electric field that are perpendicular to the magnetic field.

Wave-to-plasma power transfer. The power dP/dz transferred from the wave to the plasma per unit length dz, can be derived from eqn. (6.22) by integrating this expression over the plasma cross-section $\mathbb{S}$. To simplify the discussion, we assume that the electron density and each component of the wave electric field are spatially uniform. Then, dP/dz is simply

$$\frac{dP}{dz} = \text{Re}(\sigma_{eff}) \left| E_{rms}^2 \right| \mathcal{S}, \tag{6.24}$$

where $\left| E_p \right| \equiv \left[\left| E_r \right|^2 + \left| E_\phi \right|^2 + \left| E_z \right|^2 \right]^{1/2} = \sqrt{2}\, E_{rms}$ is the wave total electric field strength and

$$\text{Re}(\sigma_{eff}) = \left\{ \text{Re}(\sigma_1) \left| E_\perp \right|^2 + \text{Re}(\sigma_3) \left| E_z \right|^2 - 2\,\text{Im}\,(\sigma_2) \left| E_r E_\phi \right| \sin\Theta \right\} / 2E_{rms}^2, \tag{6.25}$$

σ_{eff} defining an *effective value of the plasma conductivity*. To take into account realistic nonuniform plasmas and fields, one could readily introduce a cross-section averaged conductivity.[16]

We note from relations (6.19) to (6.21) that for $\nu_{eff} = 0$, σ_1 and σ_3 are purely imaginary while σ_2 is real. According to (6.25), in such a case, except at $\omega = \omega_{ce}$, dP/dz = 0: no power transfer occurs. The case where $\omega = \omega_{ce}$ requires a specific treatment that we discuss below.

Particular case: $\omega = \omega_{ce}$. Because the $\text{Re}(\sigma_1)$ and $\text{Im}(\sigma_2)$ terms in eqn. (6.25) reach a maximum at $\omega = \omega_{ce}$, σ_{eff} goes through a maximum at or near electron cyclotron frequency. This results in a plasma conductivity showing a resonant behavior as a function of ω_{ce}/ω. The corresponding physical mechanism is the following. When submitted to a B_0-field, electrons having a velocity component perpendicular to $\mathbf{B_0}$ rotate circularly around its field lines at the angular frequency ω_{ce}, undergoing the so-called *cyclotron motion*. Let us now assume that a wave of angular frequency $\omega = \omega_{ce}$ propagates into the plasma. Provided this wave has an electric field component in the direction perpendicular to $\mathbf{B_0}$ and if this component rotates as a function of time in the same direction as electrons in their circular motion, then in their reference system, the electrons "see" a DC electric field by which they are continuously accelerated. This suggests that electrons could be accelerated indefinitely. However, in reality, the energy gain is limited through collisions of electrons with other particles. This resonant absorption mechanism will be treated in Sec. 6.4.3. As shown in Sec. 8.4.4, the fact that σ_{eff} is maximum at ECR leads to a minimum of the electric field in the discharge, making it easier to initiate and sustain a plasma.

6.4.2. Collision frequency of electrons with ions compared to that of electrons with neutral particles.

The collision frequency entering the plasma conductivity is some effective collision frequency. It should account for collisions of electrons with neutral particles, ions and electrons, the contribution of which varies with the degree of ionization. The problem of taking into account all the kinds of collisions is far from being simple but has been treated extensively by Shkarofsky et al.[2] They showed that the electric conductivity is a function of a *total effective collision frequency* $v_{eff}^t \equiv v_{eff} + v_{ei}$, where v_{eff} is the effective electron-neutral collision frequency for momentum transfer (Sec. 2.2.6) calculated in the limit $v_{eff} << |\omega \pm \omega_{ce}|$ and v_{ei} is the electron-ion collision frequency (Coulombian interaction) given by

$$v_{ei} = \sqrt{\frac{2}{\pi}} \, \omega_{pe} \, \frac{\log_e \Lambda_c}{\Lambda_c} , \qquad (6.26)$$

where Λ_c, the *screening factor*, can be expressed as

$$\Lambda_c = \frac{3}{2} \frac{(4\pi \, \varepsilon_0 \, k_B \, T_e)^{3/2}}{Z \, e^3 \, (\pi \, n_e)^{1/2}} , \qquad (6.27)$$

where Z is the charge number of the ion and k_B is the Boltzmann constant.

In practice, electron-ion collision effects become important whenever $v_{eff}/v_{ei} \leq 10$, that is, under typical conditions, above ionization degree percentages lying in the range 0.01 and 1 %. In the following, the designation v_{eff} will be used to qualify the total effective collision frequency.

6.4.3. Collisionless absorption. Collisionless absorption results from a resonant wave-particle interaction which causes damping of the wave. To quantify this mechanism, one must use a kinetic model of the plasma (Vlasov theory). A well-known example of collisionless absorption is the Landau damping.[20] The underlying physical mechanism is that the electrons of the EEDF moving at velocities close to the wave phase velocity strongly interact with the wave. Electrons whose velocity is slightly slower than the wave phase velocity gain energy while those that are slightly faster lose energy. In general, the EEDF contains more slow electrons than fast ones ($\partial F_0/\partial w < 0$), and the result is a net gain of energy for electrons. This absorption process leads to a modification of the EEDF in the velocity space. Nonlinear models must be used to predict how this energy is subsequently distributed into the plasma.

For $B_0 \neq 0$, two types of collisionless absorption mechanisms can be identified:[21] i) the case of electrons with a microscopic velocity component w_z along $\mathbf{B_0}$, and which interact resonantly with the wave electric field component along that same direction. The corresponding absorption is termed *Cerenkov absorption* and requires the condition $\omega \pm \beta w_z = 0$, where β is the component of the wave vector along $\mathbf{B_0}$. For a wave propagating strictly along $\mathbf{B_0}$, it reduces to Landau absorption; ii) the case of electrons rotating at an angular frequency ω_{ce} around B_0-field lines and traveling with a microscopic velocity w_z along $\mathbf{B}_0$. When the wave has a right-hand circularly polarized electric field component in the plane perpendicular to B_0, these electrons "see" this component at a Doppler-shifted frequency $\omega' = \omega \pm \beta w_z$. For $\omega' = \omega_{ce}$, the electrons are submitted to a constant electric field and gain energy from the wave, as discussed in Sec. 6.4.1. The resonant electrons thus obey the condition $\omega - \omega_{ce} \pm \beta w_z = 0$ and the corresponding absorption process is called *cyclotron absorption*. It is accompanied by an increase of the electron velocity component in the direction perpendicular to $\mathbf{B}_0$. This energy increase is subsequently shared within the whole plasma volume through collisions or nonlinear processes. When the wave propagates nearly perpendicular with respect to $\mathbf{B}_0$, collisionless absorption is also possible at the harmonics of ω_{ce} and the resonant electrons are those for which $\omega - n\omega_{ce} \pm \beta w_z = 0$ ($n = 1, 2, ...$).[21]

The cyclotron absorption mechanism that we just discussed is exactly the same as the resonant absorption mechanism at $\omega = \omega_{ce}$ in a cold ($w_z = 0$) collisionless ($\delta = 0$) plasma, which we described in Sec. 6.4.1.

It is customary in the recent literature concerning HF magnetoplasmas to use the term electron cyclotron *resonance* (ECR), formally implying $\omega = \omega_{ce}$, to qualify the collisionless wave-plasma interaction. In fact the absorption mechanism does not necessarily occur at exactly $\omega = \omega_{ce}$, as we discussed above, but close to it. This terminology can be confusing also because electrons can interact resonantly with a wave that is not necessarily at resonance ($\beta \to \infty$). This is the case with the cyclotron absorption of a wave propagating along $\mathbf{B}_0$. The resonant electron velocity is $w_z = \pm (\omega - \omega_{ce})/\beta$ and, even for finite β values, i.e. not at a wave resonance, electron populations exist that can be in resonance with the wave. The cyclotron absorption is however an exception in that respect, since, in the case of Landau or Cerenkov absorption, resonant electrons are those for which the wave phase velocity is near electron thermal velocity: this requires large β values, a condition fulfilled close to wave resonances only.

6.5. Using an electromagnetic plane wave to sustain a magnetoplasma: a working example

To fully characterize the properties of microwave sustained magnetoplasmas, it is essential to know the type of waves that propagate into the plasma, how these waves can be excited, what is their E-field polarization and in what regions they are absorbed. The most simple situation to model is that dealing with plane waves. Clearly, such a model is only valid provided the presence of boundaries does not reflect on the wave propagation.

6.5.1. Wave dispersion and field polarization in a collisionless, homogeneous cold magnetoplasma. For the purpose of discussion, the $\mathbf{B_0}$ field is taken parallel to the z axis while the propagation vector $\boldsymbol{\beta}$ is assumed to be in general in the yz plane and inclined at an angle Ψ with respect to $\mathbf{B_0}$. We also assume that the plasma is cold and collisionless, the influence of thermal motion being examined in Secs. 6.5.4 and 6.5.5. Under these conditions, the wave dispersion relation can be written as[22]

$$\tan^2\Psi = -\frac{\varepsilon_3\left(\mathcal{N}^2 - \varepsilon_R\right)\left(\mathcal{N}^2 - \varepsilon_L\right)}{\left(\mathcal{N}^2 - \varepsilon_3\right)\left(\varepsilon_1\mathcal{N}^2 - \varepsilon_R\varepsilon_L\right)}, \tag{6.28}$$

where we define the *plasma refraction index* as $\mathcal{N} \equiv \beta c/\omega$ (c is the speed of light in free space) and where the permittivity terms are

$$\varepsilon_R = 1 - \frac{\xi^2}{1 - \tau}, \tag{6.29}$$

$$\varepsilon_L = 1 - \frac{\xi^2}{1 + \tau}, \tag{6.30}$$

$$\varepsilon_1 = \frac{\varepsilon_R + \varepsilon_L}{2}, \tag{6.31}$$

$$\varepsilon_3 = 1 - \xi^2. \tag{6.32}$$

Relation (6.28) can be equivalently represented by the Appleton-Lassen equation

$$a\mathcal{N}^4 - b\mathcal{N}^2 + c = 0, \tag{6.33}$$

where

$$a = \varepsilon_1 \sin^2 \Psi + \varepsilon_3 \cos^2 \Psi, \tag{6.34}$$

$$b = \varepsilon_R \varepsilon_L \sin^2 \Psi + \varepsilon_3 \varepsilon_1 (1 + \cos^2 \Psi), \tag{6.35}$$

$$c = \varepsilon_R \varepsilon_L \varepsilon_3. \tag{6.36}$$

Solutions to dispersion relation (6.28) can be found for two situations: i) infinite plasma. There are no boundary conditions; ii) bounded plasma. Boundary conditions need to be satisfied by the field components. In this case, it is better to use the

Appleton-Lassen equation, as discussed in Sec. 6.6. In what follows, we first examine the infinite plasma situation, i.e. we consider plane waves.

Plane wave solutions of (6.28) are simple to discuss for the so-called *principal waves*, i.e. when $\Psi = 0$ (propagation is parallel to $\mathbf{B}_0$) and $\Psi = \pi/2$ (propagation is perpendicular to $\mathbf{B}_0$). For other values of Ψ, the situation is more complicated but also less useful when describing the most common experimental situations.

6.5.1.1. Propagation parallel to $\mathbf{B}_0$. This situation refers to the case where the numerator of eqn. (6.28) is zero. There are then three obvious solutions: $\varepsilon_3 = 0$, $\mathcal{N}^2 = \varepsilon_R$ and $\mathcal{N}^2 = \varepsilon_L$, which can be more conveniently written respectively as

$$\omega = \omega_{pe}, \tag{6.37}$$

$$\beta^2 = \beta_0^2 \, \varepsilon_R, \tag{6.38}$$

$$\beta^2 = \beta_0^2 \, \varepsilon_L, \tag{6.39}$$

where $\beta_0 = \omega/c$ is the free-space wavenumber. The case $\omega = \omega_{pe}$ does not correspond to a wave since the group velocity $\partial\omega/\partial\beta = 0$, but rather to a plasma oscillation with its electric field parallel to $\mathbf{B}_0$. The second and third solutions are both transverse waves, polarized in the plane perpendicular to $\mathbf{B}_0$ ($E_z = 0$), yielding the right-hand and left-hand waves respectively. In absence of magnetic field, these two waves reduce to a unique solution: it is the presence of B_0 that induces the plasma anisotropy causing this mode to split.

Analysis of the field equations shows that both transverse waves are circularly polarized.[21,22] The right-hand wave has an electric field which rotates in the same direction as the electrons in the B_0-field while the left-hand wave rotates in the same direction as the ions.

To propagate in a collisionless plasma, i.e. no imaginary contributions being allowed in the permittivities, we need $\beta^2 > 0$. Then the right-hand wave requires $\omega_{pe}^2 < \omega(\omega - \omega_{ce})$ for $\omega_{ce} < \omega$, but there is no restriction on plasma density for $\omega_{ce} > \omega$. In contrast, whatever ω_{ce}, the propagation of the left-hand wave is only possible for $\omega_{pe}^2 < \omega(\omega + \omega_{ce})$.

6.5.1.2. Propagation perpendicular to $\mathbf{B}_0$. In such a case, the denominator of eqn. (6.28) must be equal to zero, leading to two solutions,

$$\beta^2 = \beta_0^2 \, \varepsilon_3 \tag{6.40}$$

and

$$\beta^2 = \beta_0^2 \frac{\varepsilon_R \, \varepsilon_L}{\varepsilon_1} \,. \tag{6.41}$$

It can be shown[22] that the first solution corresponds to a wave linearly polarized along B_0. In such a case, the electrons that oscillate in the wave E-field are not affected by B_0. This explains why the dispersion equation of this wave is the same as that of a transverse wave in absence of magnetic field. For this reason, it is called the *ordinary wave*. From eqn. (6.40), it can be seen that it only propagates in an underdense plasma ($\omega_{pe} < \omega$).

The second wave is termed the *extraordinary wave*. It is neither a pure transverse nor a pure longitudinal wave. The polarization of its electric field in the plane perpendicular to B_0 is generally elliptic. The propagation of this wave requires $\varepsilon_R \varepsilon_L / \varepsilon_1 > 0$.

6.5.2. Cutoff and resonance frequencies for propagation along B_0.

The term *cutoff* is used for the condition in which the phase velocity of the wave is infinite ($\beta \to 0$) and the term *resonance* for the condition in which it is zero ($\beta \to \infty$). The cutoff frequencies ω_R and ω_L of the right-hand (eqn. (6.38)) and left-hand (eqn. (6.39)) waves are thus determined by the conditions $\varepsilon_R = 0$ and $\varepsilon_L = 0$ respectively. This yields

$$\omega_R = \frac{\omega_{ce} + \sqrt{\omega_{ce}^2 + 4\,\omega_{pe}^2}}{2} \tag{6.42}$$

and

$$\omega_L = \frac{-\,\omega_{ce} + \sqrt{\omega_{ce}^2 + 4\,\omega_{pe}^2}}{2} \,. \tag{6.43}$$

The resonance frequencies are set by the conditions ε_R and $\varepsilon_L \to \infty$. For the right-hand wave, it is $\omega = \omega_{ce}$: the electron cyclotron resonance (ECR). There is no resonance for the left-hand wave since ε_L always remains finite.

An example of the right-hand and left-hand wave dispersion curves[†] is shown in Fig. 6.3. The important point to be noted is that the right-hand wave splits into two branches, a high frequency one ($\omega \geq \omega_R$) and a low frequency one ($\omega \leq \omega_{ce}$), the latter also being called the whistler wave.[††] Note that for the high frequency branch, as ω is decreasing, the wave is cutoff before the wave resonance ($\beta \rightarrow \infty$) can be reached simply because ω_R is always larger than ω_{ce}.

Since the right-hand circularly polarized (RHCP) wave has the adequate polarization for a resonant power transfer to plasma and because it corresponds to the common case of wave propagation along $\mathbf{B}_0$, we discuss in more detail the properties of this wave below.

6.5.3. Phase diagram of the RHCP wave. In HF sustained plasmas, ω is constant and it is ω_{pe} that varies. It is thus more useful to examine phase rather than dispersion diagrams. Figure 6.4 shows such a phase diagram for the RHCP wave, in

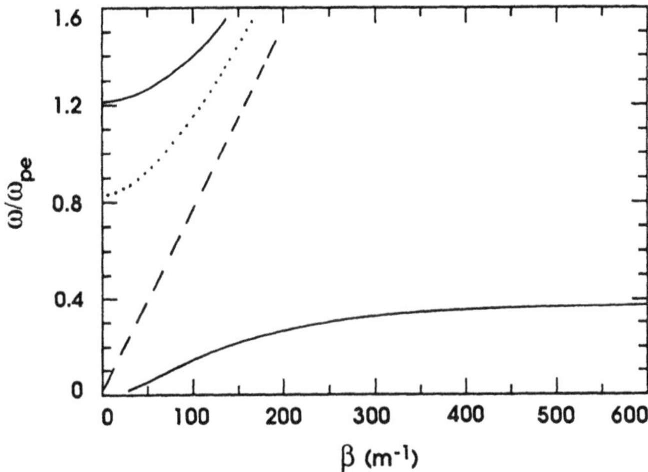

Fig. 6.3. Calculated dispersion diagram of the right-hand (—) and left-hand (···) circularly polarized waves for $n = 5 \times 10^{17} \text{m}^{-3}$ and $B_0 = 875$ G. The dashed line represents the dispersion of an electromagnetic wave propagating in free-space.

[†] Recall that a dispersion curve implies that ω_{pe} is fixed and ω varies, in contrast to the reverse situation in a phase diagram.

[††] Strictly speaking, the whistler wave corresponds to the right-hand wave within the frequency range $\omega_{ci} \ll \omega \ll \omega_{ce}$. However, it now commonly designates the right-hand wave in the range $\omega_{ci} \ll \omega \leq \omega_{ce}$.

the case of ω_{ce}/ω well below the ECR condition (Fig. 6.4a) and when it is well above it (Fig. 6.4b). For $\omega_{ce}/\omega < 1$, the wave is fast ($\beta < \beta_0$) and propagation is restricted to underdense plasmas ($\omega_{pe} < \omega$). For $\omega_{ce}/\omega > 1$, the wave is slow and its phase velocity increases with decreasing plasma densities: in the case of a wave propagating in a plasma of slowly decreasing density, this means that the wavelength progressively

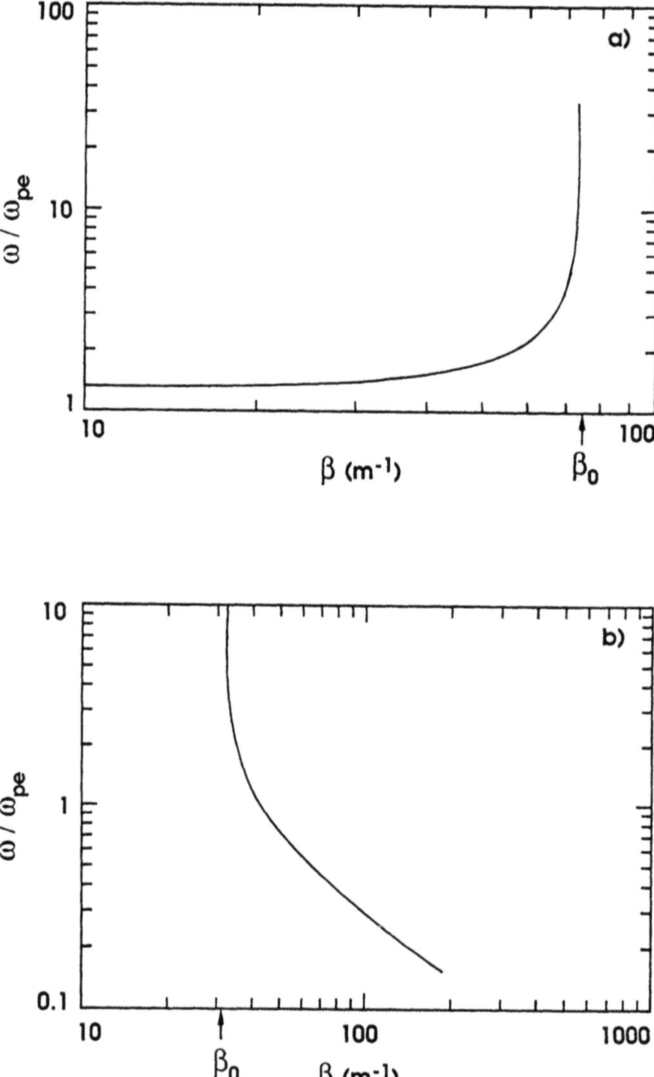

Fig. 6.4. Calculated phase diagram of the right-hand circularly polarized plane wave for $B_0 = 875$ G ($\omega_{ce}/2\pi = 2.45$ GHz), (a) at $\omega/2\pi = 3.5$ GHz and (b) $\omega/2\pi = 1.5$ GHz, where the free-space wavenumber $\beta_0 = \omega/c$.

increases. Comparison of Fig. 6.4 with Fig. 6.3 shows that phase and dispersion diagrams are really different, and thus that *dispersion curves should not be used to make experimental predictions on HF sustained magnetoplasmas.*

6.5.4. Collisional absorption of the RHCP wave. To take collisions into account, the permittivity element ε_R related to the right-hand wave must be written as

$$\varepsilon_R = 1 - \frac{\xi^2}{1 - \tau - j\delta} \tag{6.44}$$

instead of (6.29). Dispersion relation (6.38) then requires that the wavenumber be complex and we thus introduce the complex wavenumber $h = \beta - j\alpha$, where β is the (real) wavenumber and α is the wave attenuation coefficient. Separating eqn. (6.38) in terms of β and α yields for α

$$\left(\frac{\alpha}{\beta_0}\right)^2 = \frac{1}{2}\left\{-\left[1 - \frac{\xi^2\,(1-\tau)}{(1-\tau)^2 + \delta^2}\right] + \sqrt{\left[1 - \frac{\xi^2\,(1-\tau)}{(1-\tau)^2 + \delta^2}\right]^2 + \left[\frac{\delta\xi^2}{(1-\tau)^2 + \delta^2}\right]^2}\right\}. \tag{6.45}$$

Solution far from ECR. We first note that for $\delta = 0$ ($\nu_{\rm eff} = 0$), α is zero except at $\omega_{ce} = \omega$ (see below). For a low collision rate, namely $\delta \ll |1 - \tau|$, eqn. (6.45) reduces to

$$\frac{\alpha}{\beta_0} \approx \sqrt{\frac{\delta}{2}} \, \frac{\xi}{|1 - \tau|}. \tag{6.46}$$

Clearly, the wave attenuation coefficient increases with plasma density, a situation, according to the discharge stability criterion discussed in Sec. 5.4.2, that does not allow this wave to sustain a plasma. Note that relation (6.46) indicates that α increases when the wave frequency or the B_0-field intensity is varied in such away that τ tends toward unity.

Solution at ECR. In the case where $\tau = 1$, eqn. (6.45) becomes

$$\left(\frac{\alpha}{\beta_0}\right)^2 = \frac{1}{2}\left(-1 + \sqrt{1 + \frac{\xi^4}{\delta^2}}\right). \tag{6.47}$$

For large plasma densities and low collision rates ($\xi^2 \gg \delta$), this relation further reduces to

$$\frac{\alpha}{\beta_0} \simeq \frac{\xi}{\sqrt{2\delta}} \,. \tag{6.48}$$

Relation (6.48) shows that in absence of collisions, α tends toward infinity at $\omega = \omega_{ce}$ (recall that in this case, β would also be infinite (Sec. 6.5.2)). These singularities of α and β at $v_{eff} = 0$, which do not correspond to a realistic situation, are removed as soon as a finite, even small, value of collision frequency is considered. Solving eqn. (6.38) in the presence of collisions also smooths the sudden splitting of the RHCP wave in two branches.

Examples of the variation of β/β_0 and α/β_0 as a function of ω/ω_{ce} are presented in Figs. 6.5a and 6.5b respectively, at constant ω and for two values of ω_{pe}. We see that both β and α reach a maximum at $\omega_{ce} = \omega$ and we could show that whatever the value of v_{eff}/ω, we have $\beta = \alpha$. This means that the wave is then strongly attenuated over a distance of the order of one wavelength.

6.5.5. Collisionless cyclotron absorption of the right-hand wave. The cyclotron attenuation of the right-hand wave through collisionless mechanisms can be calculated in the context of the Vlasov theory, i.e. using a kinetic rather than a fluid description of the plasma. When the condition $\omega_{ce}/\beta v_{th} \gg 1$ is met,[†] where $v_{th} = \sqrt{2k_B T_e/m_e}$ is the so-called *electron thermal velocity* (most probable velocity), then the refraction index of the right-hand wave is[21]

$$\mathcal{N}^2 \simeq \xi^2 \, \frac{\omega}{\beta v_{th}} \, \mathbb{Z}\!\left(\frac{\omega - \omega_{ce}}{\beta v_{th}}\right), \tag{6.49}$$

where $\mathbb{Z}$ is termed the (complex) *plasma dispersion function* (sometimes known as Fried and Conte's function) and is[23] (with the dx integral along the Landau contour)

$$\mathbb{Z}(z) = \frac{1}{\sqrt{\pi}} \int_{-\infty}^{+\infty} \frac{\exp(-x^2)}{x - z} \, dx. \tag{6.50}$$

In general, eqn. (6.49) must be solved numerically. It yields the complex wavenumber $h = \beta - j\alpha$. Figure 6.6 shows the variation of β and α as a function of $|\omega - \omega_{ce}|/\beta_{th}v_{th}$, for $v_{th}\,\xi/c\tau \ll 1$, a condition fulfilled in practice. We used the normalizing parameter $\beta_{th} = \beta_0\,(\xi^2\tau\,c/v_{th})^{1/3}$. We note the following: i) both β and α

[†] This assumption allows one in particular to neglect the contribution of ions in the expression of the refraction index.

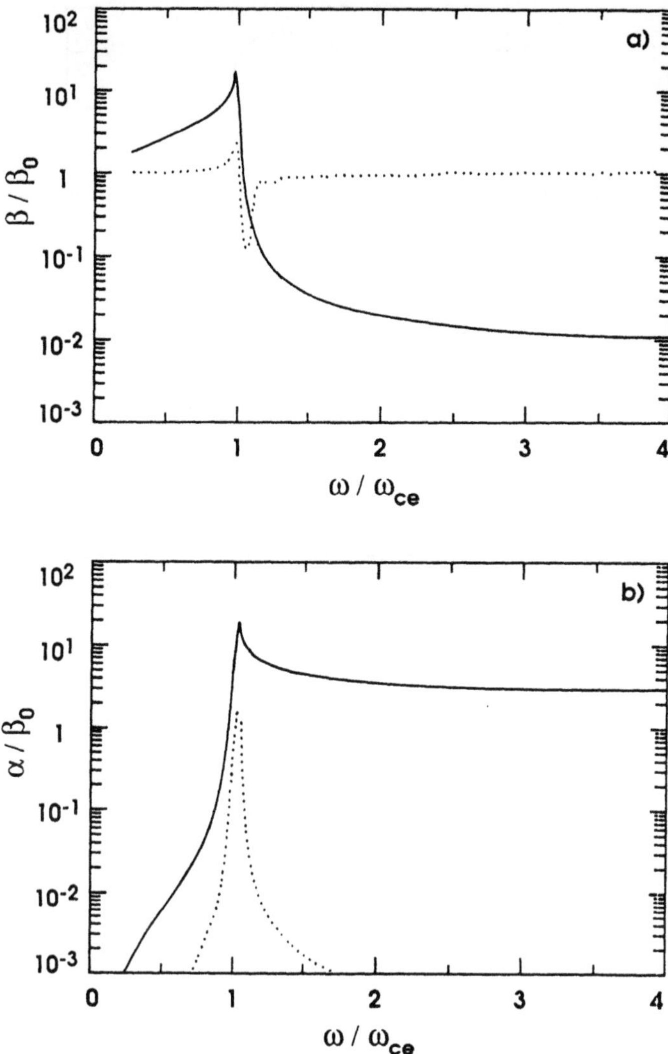

Fig. 6.5. Calculated values of the wavenumber β (a) and attenuation coefficient α (b) for the right-hand polarized plane wave as a function of ω/ω_{ce}, for $\omega/2\pi = 2.45$ GHz and $\nu_{eff}/\omega = 5 \times 10^{-3}$ (collisional case), at two different values of plasma density: $n = 5 \times 10^{17}\mathrm{m}^{-3}$ (—) and $n = 5 \times 10^{15}$ (···). Both β and α are normalized to the free-space wavenumber β_0.

take finite values whatever the frequency, i.e. the singularities of the collisionless fluid model (Fig. 6.3) have been eliminated; ii) the maximum value of the wavenumber is approximately equal to β_{th} and the attenuation factor reaches a maximum when

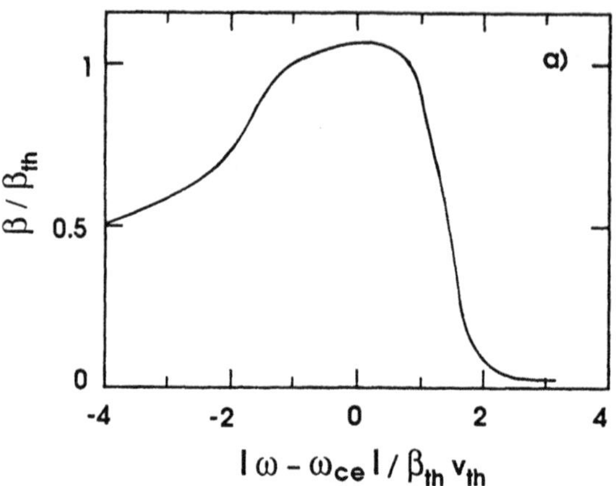

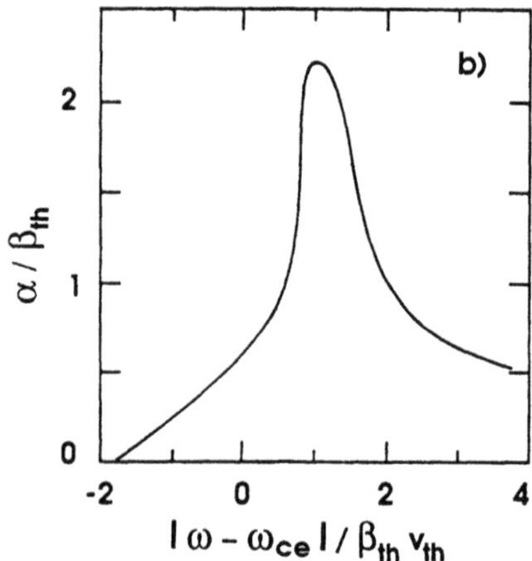

Fig. 6.6. Calculated values of the axial wavenumber β (a) and attenuation coefficient α (b) of the right-hand polarized wave as a function of $|\omega - \omega_{ce}| / \beta_{th}v_{th}$ (see text) in the collisionless case. Both β and α are normalized to the parameter β_{th} (from ref. 21, p. 290).

$\omega - \omega_{ce} \simeq \pm \beta v_{th}$. This shift of the absorption curve with respect to $\omega = \omega_{ce}$ occurs because the largest number of resonant electrons is found at velocity v_{th}; iii) the attenuation curve exhibits Doppler broadening: the larger v_{th}, the broader the attenuation curve.

It is instructive to compare collisionless cyclotron absorption and collisional absorption at $\omega \simeq \omega_{ce}$ by considering the attenuation coefficients from Figs. 6.5b and 6.6b. Assuming a wave frequency of 2.45 GHz, an electron temperature of 2 eV and a plasma density of 5×10^{11} cm^{-3}, it is found that the collisional absorption is still comparable to the cyclotron absorption even at ν_{eff}/ω ratios as low as 5×10^{-3}. It results that in most practical situations encountered in plasma processing, collisions always contribute significantly to the transfer of energy from the wave to the plasma. Under such conditions, a model taking into account both electron collisions and thermal motion should thus be used.

6.6. Wave propagation in waveguides completely or partially filled with plasma

The finite extension of laboratory plasmas makes the plane-wave (or 1-dimensional) approximation for wave propagation inappropriate in most practical cases. The presence of boundaries affects the field distribution within the plasma and influences the wave propagation characteristics as well. It is only for plasma dimensions much larger than a free-space wavelength that the electromagnetic field is only weakly disturbed by the walls, except in their immediate vicinity, and that the propagation properties approach those of the plane wave model. Nonetheless, it is customary to use the plane wave approximation to describe wave propagation in a cylindrical waveguide filled with plasma, which thus requires assuming its radial dimension to be a few free-space wavelengths. At 2.45 GHz, where the free-space wavelength is about 12 cm, this implies the radial extension of the plasma to be at least a few tens of centimeters, a value which is well above the dimensions generally feasible with present ECR plasma sources. This situation worsens at lower frequencies.

In general, solving Maxwell's equations for a bounded plasma, say, a cylindrical waveguide completely or partially filled with a uniform, axially (z-axis) magnetized plasma is a complicated affair that exceeds the scope of this book. One can nonetheless obtain analytically some useful information about the existing wave modes by examining the Appleton-Lassen dispersion equation (6.33) parametrized by the (generally complex) *transverse wavenumber* $h_\perp$ defined by $h_\perp c/\omega = \mathcal{N} \sin \Psi$.[†] This wavenumber represents the component of the wave vector in the direction perpendicular to B_0 and reflects the boundary conditions. Equation (6.33) then comes out as

$$h_\perp^4 - a' h_\perp^2 + b' = 0, \tag{6.51}$$

where

[†] This concept originated in the ray-tracing for the ionosphere where the relevant equation is called the Booker quartic.[24]

$$a' = \beta_0^2 \left(\varepsilon_3 + \frac{\varepsilon_R \varepsilon_L}{\varepsilon_1} \right) - h^2 \left(1 + \frac{\varepsilon_3}{\varepsilon_1} \right), \tag{6.52}$$

$$b' = \frac{\varepsilon_3}{\varepsilon_1} \left(\beta_0^2 \varepsilon_R - h^2 \right) \left(\beta_0^2 \varepsilon_L - h^2 \right), \tag{6.53}$$

h being the axial component of the wave vector. In the following, to simplify the discussion, we assume the plasma to be cold and collisionless. Let us first consider the case where $h_\perp = 0$, i.e. the wave vector is parallel to $\mathbf{B}_0$. From eqn. (6.51), $b' = 0$ and we have the three plane wave modes discussed in Sec. 6.5.1.1. Similarly, when the wave vector is perpendicular to $\mathbf{B}_0$ (h = 0), we recover the ordinary and extraordinary waves. These situations are easy to study because either h or $h_\perp$ is zero. In the more complicated case where the propagation occurs at an angle with respect to $\mathbf{B}_0$, neither h nor $h_\perp$ is zero but it is always possible to find a coordinate system where the transverse (i.e. perpendicular to the direction of propagation) wavenumber is zero.

Influence of boundaries. In contrast to plane wave propagation, for guided waves propagating along $\mathbf{B}_0$, the value of $h_\perp$ cannot be set to zero in eqn. (6.51) as it depends on bounday conditions. The relation between $h_\perp$ and h is obtained from the dispersion equation (6.51) and eqns. (6.52) and (6.53). Because the dispersion equation is of fourth order in $h_\perp$, in general there are two values of $h_\perp^2$ that satisfy it, except for $B_0 = 0$ where it reduces to one solution. These pairs of values usually form a discrete infinite series.

In the case of a cylindrical radially uniform plasma, it can be shown that the axial electric field component E_z in the plasma is[22,25-29]

$$E_z = \left[a_1 J_m (h_{\perp 1} r) + a_2 J_m (h_{\perp 2} r) \right] \exp \left[j (\omega t - hz + m\phi) \right], \tag{6.54}$$

where a_1 and a_2 are constant coefficients, h is the axial wavenumber, the integer m is the azimuthal wavenumber, $h_{\perp 1}$ and $h_{\perp 2}$ are the two transverse wavenumbers, and J_m is the m-th order Bessel function of the first kind. A similar expression with different coefficients is found for the wave axial magnetic field component H_z. The remaining field components can be obtained from Maxwell's equations as a combination of E_z and H_z fields and their derivatives.

By requiring that boundary conditions be satisfied by the relevant field components, a set of linear, homogeneous equations is obtained. Their solution is non-trivial provided the determinant of the field components' coefficients is zero. This determinantal equation must be solved self-consistently with eqn. (6.51) to yield h, $h_{\perp 1}$, $h_{\perp 2}$ as functions of ω, ω_{ce}, ω_{pe}, plasma dimensions and nature of the boundaries (conductor or dielectric). Although the two perpendicular wavenumber values are usually quite different, one proves to involve the dominant wave function while the other really helps in satisfying the boundary requirements.

6.6.1. Cylindrical waveguide completely filled with an axially magnetized plasma.

For a given mode of azimuthal configuration, m, the boundary conditions imposed by the wall of the waveguide of radius R_1, assumed to be a perfect conductor, are $E_z = E_\phi = 0$ at $r = R_1$ for all values of ϕ. This leads to the determinantal equation

$$\mathcal{g}_1 h_{\perp 1} J'_m (h_{\perp 1} R_1) J_m (h_{\perp 2} R_1) - \mathcal{g}_2 h_{\perp 2} J'_m (h_{\perp 2} R_1) J_m (h_{\perp 1} R_1) =$$

$$m \beta_0^2 \frac{(\varepsilon_R - \varepsilon_L)}{2R_1} \left(h_{\perp 2}^2 - h_{\perp 1}^2 \right) J_m (h_{\perp 1} R_1) J_m (h_{\perp 2} R_1), \tag{6.55}$$

where the prime indicates the derivative of the Bessel function J_m with respect to its argument, and

$$\mathcal{g}_{1,2} \equiv \left(\beta_0^2 \varepsilon_1 - h^2 \right) \left(\beta_0^2 \varepsilon_1 - h^2 - h_{\perp 1,2}^2 \right) + \beta_0^4 \varepsilon_2^2. \tag{6.56}$$

Since $E_z(r = R_1) = 0$, we note from expression (6.54) that the coefficients a_1 and a_2 are related through

$$\frac{a_2}{a_1} = - \frac{J_m (k_{\perp 1} R_1)}{J_m (k_{\perp 2} R_1)}, \tag{6.57}$$

their absolute value remaining undetermined.

For given conditions (ω, ω_{pe}, ω_{ce}, R_1), each set of solutions h, $h_{\perp 1}$, $h_{\perp 2}$ corresponds to a wave mode characterized by a specific spatial configuration of the field components. This configuration depends strongly on the nature of $h_{\perp 1}$ and $h_{\perp 2}$ which can both be real, complex or imaginary, or one real and the other imaginary. Figure 6.7 delimits in the plane (ω/ω_{pe}, β) where ω_{pe} is fixed and ω varies, the four different zones that result from the possible combinations of the complex nature (real, imaginary and complex) of $h_{\perp 1}$ and $h_{\perp 2}$. Each zone corresponds to a particular type of solution of the dispersion equation, yielding three families of waves that are briefly reviewed below. All these waves are hybrid, i.e. they are a linear combination of TM(E) and TE(H) waves,[†] except in the vicinity of cutoffs and resonances.

---

† A TM or E wave has all its components of magnetic field in the plane perpendicular to the direction of propagation whereas a TE or H wave has all its electric field components in that same perpendicular plane.

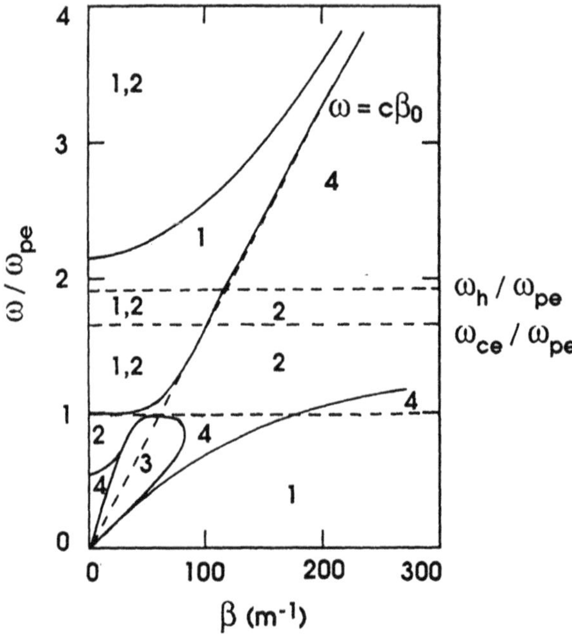

Fig. 6.7. Delimitation of the various types of wave dispersion solutions, according to the complex nature (real, imaginary, complex) of the transverse wavenumbers $h_{\perp 1}$ and $h_{\perp 2}$, for ω_{pe} and ω_{ce} constant ($\omega_{pe} = 0.6\,\omega_{ce}$). Zone 1: $h_{\perp 1}$ is real; zone 2: $h_{\perp 2}$ is real; zone 3: $h_{\perp 1}$ and $h_{\perp 2}$ are complex; zone 4: $h_{\perp 1}$ and $h_{\perp 2}$ are imaginary. The horizontal dashed lines, starting from below, correspond to $\omega = \omega_{pe}$, $\omega = \omega_{ce}$ and $\omega = \omega_h$ where $\omega_h \equiv \sqrt{\omega_{pe}^2 + \omega_{ce}^2}$ is the upper hybrid angular frequency (from ref. 26).

i) For $\omega < \min(\omega_{pe}, \omega_{ce})$, the waves are called *plasma modes* because their existence fully depends on the presence of plasma. Following Ivanov and Alexov,[26] we designate these modes by the symbol EH^P_{mn} where the integer m, the azimuthal wavenumber, and n, the radial wavenumber, characterize the azimuthal and radial configuration of the electromagnetic field respectively. The designation EH rather than HE indicates that these modes turn into TM(E) waves near resonances. For the EH^P_{mn} plasma modes with $n > 1$, the transverse wavenumbers are such that the dispersion curve entirely lies inside region 1 in Fig. 6.7, i.e. at least $h_{\perp 1}$ is real ($h^2_{\perp 1} > 0$). For $n = 1$ and increasing frequency, the corresponding dispersion curve crosses successively region 3 where both $h_{\perp 1}$ and $h_{\perp 2}$ are complex (and conjugate), region 4 where both wavenumbers are imaginary and region 1 where at least $h_{\perp 1}$ is real. Note that the fact that there exist modes for which $h_{\perp 1}$ and $h_{\perp 2}$ are complex (region 3) was ignored for a long time until powerful computers and routines for complex Bessel functions became available. Because they can propagate in (and consequently sustain) an overdense plasma, the study of these modes is of major interest for understanding microwave excited magnetoplasmas.

ii) For min $(\omega_{pe}, \omega_{ce}) < \omega < \omega_h$, where $\omega_h \equiv \sqrt{\omega_{pe}^2 + \omega_{ce}^2}$ is the *upper hybrid angular frequency*, we find the *cyclotron modes* HE^C_{mn}. These waves are limited to a narrow band of frequencies and reduce to Langmuir oscillations for $\omega_{ce} \to 0$. Such modes are characterized by at least one of their transverse wavenumber being real.

iii) Finally, for $\omega > \omega_{pe}$ and at least $h_{\perp 1}$ real, we have the *waveguide modes*. These modes are represented by the symbols EH_{mn} and HE_{mn}. In absence of plasma, EH_{mn} waves are TM_{mn} modes while HE_{mn} waves are TE_{mn} modes: these modes are thus those of an empty circular waveguide, which are modified by the presence of the magnetoplasma. The EH_{mn} waves propagate for $\omega > \omega_{pe}$ and the HE_{mn} waves for $\omega > \omega_h$.

An example of the radial distribution of E_z for the EH^P_{01} plasma mode is shown in Fig. 6.8 at different frequencies. Curve 1 is obtained in that portion of the dispersion curve where $h_{\perp 1}$ and $h_{\perp 2}$ are complex (zone 3): the E_z component is then concentrated in a layer near the waveguide boundary. Curve 2 corresponds to a somewhat higher frequency at which both $h_{\perp 1}$ and $h_{\perp 2}$ are imaginary (zone 4). In this case, E_z is still maximum near the boundary but it penetrates more deeply in the plasma volume than in the case of curve 1. Finally, curve 3 which corresponds to $h_{\perp 1}$ being real (zone 1), yields a maximum of E_z at the plasma axis. Figure 6.8 thus illustrates how a change in wave frequency influences the electric field distribution in the plasma.

The dispersion curves obtained for the first three plasma modes of azimuthal symmetry EH^P_{0n} ($n = 1, 2, 3$) are shown in Fig. 6.9 where they are compared with the dispersion curve of the low frequency branch of the right-hand plane wave, the whistler wave. We see that the dispersion of the infinite plasma RHCP plane wave is close to

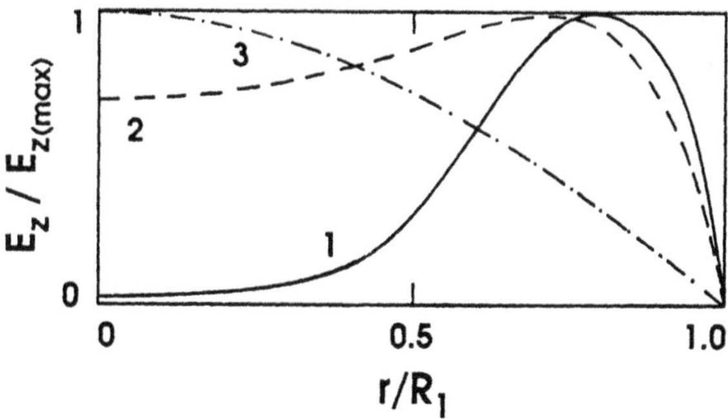

Fig. 6.8. Calculated radial distribution of the electric field axial component of the EH^P_{01} plasma mode for $n = 3.68 \times 10^{19} m^{-3}$, $B_0 = 21.4$ kG and $R_1 = 6.5$ mm; 1): $\omega/2\pi = 10$ GHz, 2): $\omega/2\pi = 33$ GHz, 3): $\omega/2\pi = 40$ GHz (from ref. 26).

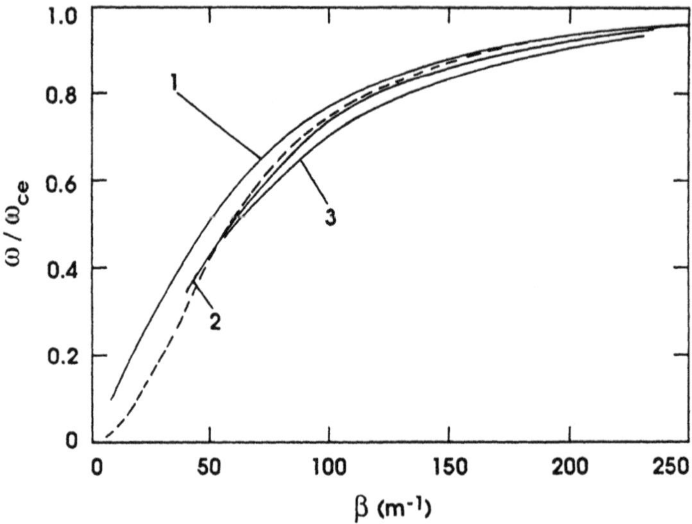

Fig. 6.9. Dispersion diagram of the first ($n = 1,2,3$) azimuthally symmetric plasma modes EH^P_{0n} propagating in a cylindrical waveguide of radius 8.5 cm completely filled with plasma, at $B_0 = 200$ gauss and $n = 10^{17} m^{-3}$. 1): $n = 1$; 2): $n = 2$; 3): $n = 3$. The dashed line represents the dispersion curve of the lower branch of the right-hand polarized plane wave, or whistler wave (from ref. 28).

that of the m = 0 plasma modes. Calculations of the electric field components show that near $\omega = \omega_{ce}$, the transverse component of these plasma modes is circularly polarized over the major part of the waveguide radius (except near the wall where the radial component E_r increases rapidly, leading to a radial polarization).[28] However, the radial distribution of their electric field intensity is far from being uniform as in the plane wave case, and depends significantly on the radial wavenumber n.

Note that in the quasi-static approximation ($\nabla \times \mathbf{E} = 0$), i.e. assuming that the electric field derives from a potential, the EH^P_{mn} plasma modes are often called Trivelpiece-Gould modes.[30]

6.6.2. Plasma partially filling a circular waveguide. When the plasma does not completely fill the waveguide (e.g. due to a static magnetic field confinement), or when the waveguide is coated with some dielectric material, the analysis presented in this section remains valid, but the determinantal equation differs from that given by eqn. (6.55) because of boundary conditions that now require the continuity of the wave tangential field components at the plasma-dielectric interface.[31] Nonetheless, the results are qualitatively similar to those presented in Sec. 6.6.1. In particular, we still have the plasma modes. These modes, which disappear in completely filled waveguides when $B_0 = 0$, are then present as surface waves. The influence of plasma anisotropy on surface waves is discussed in ref. 16.

The presence of a conducting enclosure around a plasma partially filling it also allows waveguide modes. Removing this enclosure obviously prevents these waves from propagating, but the plasma modes are only qualitatively affected.

Note finally that the so-called *helicon waves*,[32] which belong to the category of plasma modes (for a completely or partially filled waveguide), are obtained in the limit $\omega_{pe} \gg \omega$ (and $\omega < \omega_{ce}$). Therefore these waves constitute the low-frequency or high-density limit of the EH^P_{mn} modes. In partially-filled waveguides or in absence of metallic waveguide, they turn into pure surface waves when $B_0 = 0$.

The above presentation has shown that the presence of boundaries enclosing the discharge substantially alters the oversimplified description obtained when assuming plane waves to propagate in HF magnetoplasmas. The variety of modes which can be found in bounded systems certainly complicates the interpretation of results but only such a realistic approach can lead to a better understanding of magnetically assisted HF plasmas.

References

[1]   S.C. Brown, *Introduction to Electrical Discharges in Gases* (John Wiley, N.Y., 1966).

[2]   I.P. Shkarofsky, T.W. Johnston and M.P. Bachynski, *The Particle Kinetics of Plasmas* (Addison-Wesley, Reading, Mass., 1966).

[3]   A.P. Zhilinskii and L.D. Tsendin, Sov. Phys. Usp. **23**, 331 (1980) and references therein.

[4]   A. Simon, Phys. Rev. **98**, 317 (1955).

[5]   R.L. Monroe, Can. J. Phys. **51**, 564 (1973).

[6]   K.H. Geissler, Plasma Phys. **10**, 127 (1968).

[7]   K.H. Geissler, Phys. Fluids **13**, 935 (1970).

[8]   W. Schottky, Phys. Z. **25**, 635 (1924).

[9]   L. Tonks and I. Langmuir, Phys. Rev. **34**, 876 (1929).

[10] S.A. Self and H.N. Ewald, Phys. Fluids **9**, 2486 (1966).

[11] J. Ingold, Phys. Fluids **15**, 75 (1972).

[12] K.B. Persson, Phys. Fluids **5**, 1625 (1962).

[13] J.R. Forrest and R.N. Franklin, Br. J. Appl. Phys. **17**, 1061 (1966).

[14] H.N. Ewald, F.W. Crawford and S.A. Self, J. Appl. Phys. **38**, 2753 (1967).

[15] D.B. Ilic, J. Appl. Phys. **44**, 3993 (1973).

[16] J. Margot and M. Moisan, J. Phys. D: Appl. Phys. **24**, 1765 (1991).

[17] M. Hussein and G.A. Emmert, J. Vac. Sci. Technol. **A8**, 2913 (1990).

[18] R.K. Porteous and D.B. Graves, IEEE Trans. Plasma. Sci. **19**, 204 (1991).

[19] J. Asmussen, J. Vac. Sci. Technol. **A7**, 883 (1989).

[20] N.A. Krall and A.W. Trivelpiece, *Principles of Plasma Physics* (McGraw-Hill, New York, 1973).

[21] D. Quémada, *Ondes dans les plasmas* (Hermann, Paris, 1968).

[22] W.P. Allis, S.J. Buchsbaum and A. Bers, *Waves in Anisotropic Plasmas* (MIT press, Cambridge, Mass., 1963).

[23] B.D. Fried and S.D. Conte, *The Plasma Dispersion Function* (Academic, New York, 1961).

[24] K.G. Budden, *The Propagation of Radio Waves* (Cambridge University, 1985).

[25] L.A. Ferrari, Nuovo Cimento **9**, 212 (1987).

[26] S.T. Ivanov and E.G. Alexov, J. Plasma Phys. **43**, 51 (1990).

[27] S.T Ivanov, E.G. Alexov and P.N. Malinov, Plasma Phys. Controlled Fusion **31**, 941 (1989).

[28] S. Pasquiers, J. Appl. Phys. **67**, 7246 (1990).

[29] V. Bevc and T.E. Everhard, J. Electron. Control **13**, 185 (1962).

[30] A.W. Trivelpiece and R.W. Gould, J. Appl. Phys. **30**, 1784 (1959).

[31] H.S. Uhm, K.T. Nguyen, R.F. Schneider and J.R. Smith, J. Appl. Phys. **64**, 1108 (1988).

[32] F.F. Chen, Plasma Phys. Controlled Fusion **33**, 339 (1991).

CHAPTER 7

OPERATION AND PROPERTIES OF MAGNETICALLY ASSISTED HIGH FREQUENCY DISCHARGES INTENDED FOR APPLICATIONS [*]

7.1. Introduction

It is now well recognized that magnetically assisted high frequency plasmas exhibit a number of characteristics which are of interest for the reproduction of narrow patterning during the fabrication of electronic components. The Japanese were the first to propose such systems in the seventies as a substitute for reactive ion etching (RIE).[1-3] Since then, interesting performance has been obtained in etching,[4-13] thin film deposition[14-20] and ion implantation.[21-24] Such plasma reactors are in use in Japanese factories but there are open questions concerning pattern profile, selectivity and rate of processes, uniformity, rate and damage control of the surfaces treated. Therefore, additional research work is required before the long term viability of such machines can be definitely proven.

For use in plasma deposition and etching, magnetically assisted HF plasmas have to fulfill several requirements, in particular the control of ion and electron energies. In that respect, the ion velocity distribution function (IVDF) is probably the most important parameter.[25] Additional requirements are the production of a high density plasma to ensure high process rates, a uniform flux to provide uniform treatment, and finally an easy to maintain, low cost plasma device.

In this context, the interest in so-called electron cyclotron resonance (ECR) plasmas essentially comes from their low operating pressure. This helps to minimize loading of reactive species and linewidth loss by ensuring that ion transport through the sheath is both collisionless and nearly anisotropic. In spite of their low pressure, these plasmas can be used to generate substantial concentrations of charged particles by achieving high degrees of ionization.

In this chapter, we examine in detail the means to maintain efficiently magnetized HF discharges (HF magnetoplasmas) as well as the characteristics of such plasmas. We will consider several kinds of HF magnetoplasmas including ECR plasmas. Their applications to microcircuit fabrication are described in Chap. 15.

---

[*] **Presented by J. Margot and R.A. Gottscho**

7.2. Static magnetic field configuration in HF discharges

As already discussed in Chap. 6, the two basic functions of the magnetic field are i) to confine the plasma, i.e. to maintain charged particles as much as possible within the plasma bulk and ii) to allow an efficient transfer of the electromagnetic energy to the plasma. Because charged particle transport occurs mostly along magnetic field lines, to some extent the geometry and extension of the region filled with plasma can be controlled by a convenient choice of the spatial configuration of B_0, the static magnetic field. In addition, the plasma can be accelerated or decelerated along decreasing or increasing B_0 respectively. In cases where an ECR effect is desired, varying the magnetic field configuration intensity makes it possible to move the position of the resonance point along and across the plasma source and thereby influence other plasma parameters such as charge density uniformity.[26] Finally, an auxiliary magnetic field can be applied to achieve better control of the downstream plasma density distribution.

HF magnetoplasma discharges can be generated under various vessel arrangements and static magnetic field configurations. There are two basic confinement systems: i) the magnetic field is dipolar and directed essentially parallel to the plasma axis. At some point along the vessel, the field lines are either allowed to diverge or forced to converge. In the first situation, a plasma stream will be induced owing to the B_0 gradient. In the convergent field case, a magnetic mirror is created to maintain high charge density within the plasma vessel. Such configurations are generally obtained by using a set of electromagnet coils around the plasma vessel, in a plane parallel to the tube axis. The required B_0 configuration is controlled by the spacing between coils and/or by the current (direction and intensity) flowing through them. Examples of magnetoplasmas using such kinds of magnetic fields will be described in the following sections; ii) multicusp or multipolar magnetic fields are used. This kind of plasma which includes those sustained through the so-called distributed ECR (DECR) is extensively treated in Chaps. 10 to 14 and will not be discussed further here.

7.3. Electromagnetic wave coupling to a magnetoplasma

To sustain efficiently a magnetoplasma with electromagnetic energy one has first to determine the mechanisms of wave absorption. As discussed in Chap. 6, even though collisions are essential to the cascade process sustaining the plasma, at very low gas pressures they only weakly contribute to the electromagnetic energy absorption. In such a case, one must then operate under conditions where collisionless absorption is possible.

Besides the wave-to-plasma absorption mechanism, the dispersion and field configurations of the waves play an essential role in sustaining a magnetoplasma. For example, in the case of ECR absorption, we need a wave with an adequate polarization, namely a right-hand wave. In practice, however, such a wave cannot be selectively launched for several reasons: i) as discussed in Chap. 6, the presence of boundaries does not allow the propagation of pure transverse electromagnetic waves, thus the wave electric field always possesses an axial component which does not contribute to ECR absorption; ii) right-hand circular polarization cannot be achieved over the entire plasma diameter. Nonetheless recall that the E-field of the wave in the plane perpendicular to $\mathbf{B}_0$ can always be decomposed into right-hand and left-hand circularly polarized components. Only the right-hand wave component is absorbed through resonant absorption and the remaining fraction (left-hand polarized and axial component) of the wave power is either absorbed through collision processes or flow through the discharge without being absorbed. In the latter case, the wave is reflected somewhere in the device, penetrates the plasma again with a modified polarization and so on. After a few such passes through the plasma, most of the wave energy will be finally absorbed. Thus, to produce ECR plasmas, one does not really need to carefully select the wave mode (though it can help) but concentrate rather on the design of an efficient coupling structure and matching system. For example, the system should be optimized to minimize incident power reflection at the field applicator and spurious radiation at the location where the wave-to-plasma energy transfer occurs.

7.4. Devices to achieve HF magnetoplasma discharges

Several setups have been proposed in the literature to obtain HF magnetoplasma discharges, each having both advantages and drawbacks. These devices differ both in the way that electromagnetic energy transfers to the plasma and in the configuration of the static magnetic field. Because of the availability of low cost "oven" magnetrons and the belief that microwave frequencies are more efficient than radio frequencies, most of these devices use the ISM frequency of 2.45 GHz. Besides "helicon" sources (see Sec. 7.4.4), there are only a few studies of discharges operated at lower frequencies.[27] At 2.45 GHz, the magnetic field intensity for ECR in the cold plasma approximation is 875 G (see Sec. 6.4.3).

The design of a HF magnetoplasma source suitable for applications in the field of surface treatment must satisfy at least two requirements: i) as mentioned above, the energy of ions must be conveniently controlled to ensure adequate linewidth; ii) microwave radiation should generally be excluded from the reactor zone as it can heat and damage some materials. One exception to this is diamond thin film coating that needs high substrate temperatures (Sec. 15.5.2). As for the influence of the static magnetic field on surface processing, it is not well documented. In what follows, we describe the various setups proposed to meet such requirements and discuss their characteristics.

7.4.1. ECR plasmas generated within waveguides.

The most common microwave based discharges use rectangular or circular waveguides, operated preferentially in the fundamental mode (TE_{10} mode for rectangular waveguides and TE_{11} for circular waveguides).[28-50] Most experimental results on ECR-type plasmas have been obtained from such systems.

The ECR-type plasmas sustained by an incident wave propagating in a rectangular waveguide were primarily studied during the sixties as plasma sources for space propulsion.[51] In more recent years, the number of publications on this kind of device has drastically increased as shown in Fig. 7.1. At the same time, a strong emphasis has been placed on the development of commercial reactors operating on this principle. Before considering in more detail the basic configurations of such systems, let us stress that their main advantage over other ECR setups at 2.45 GHz stems from the fact that electromagnetic energy is being transported in waveguides instead of coaxial lines. This allows use of much higher power levels. Powers of several kW can be carried at 2.45 GHz without any significant heating of the waveguide.

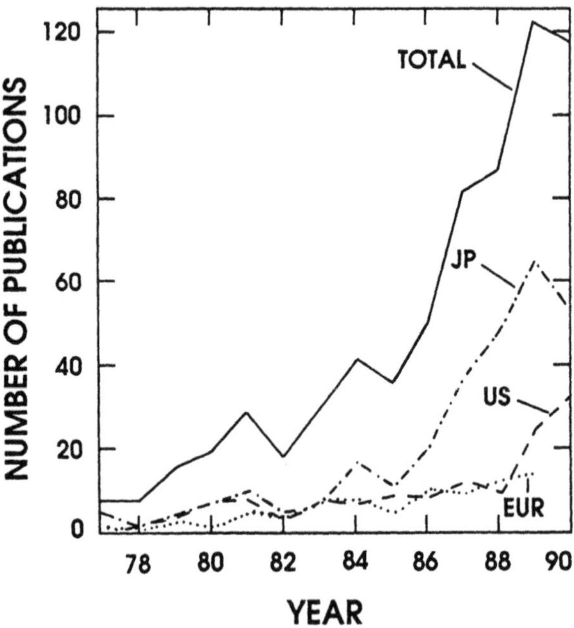

Fig. 7.1. Number of publications on ECR plasmas in Europe, United States and Japan (from ref. 52).

7.4.1.1. The experimental device. A typical setup is shown in Fig. 7.2. As a rule, except for the commercial Hitachi source, the waveguide is separated from the plasma chamber, referred to as the source or sometimes but often erroneously as the cavity,[†] by a dielectric window (fused silica, sapphire, ceramics...). The source, usually of cylindrical geometry, is surrounded by magnetic coils that in most cases yield an axially nonuniform static magnetic field. The current in the coils is set such that the resonance point is located at a few centimeters from the window. The resulting plasma generally diffuses into a second chamber of larger dimensions, the *plasma reactor* (process chamber), designed for surface treatment. Ideally, as mentioned above, microwaves should not leak into the reactor from the plasma source.

In the simplest coil configuration, the magnetic field lines in the reactor are divergent so that the plasma expands in the radial direction after leaking from the source. An additional coil can be located at the reactor end and its current varied independently to obtain better control of the plasma shape; permanent magnets generating multicusp fields have also been proposed for that same purpose.

7.4.1.2. Properties of waveguide based ECR plasmas. Due to their wide use these systems are probably the best documented to date. While the resulting set of

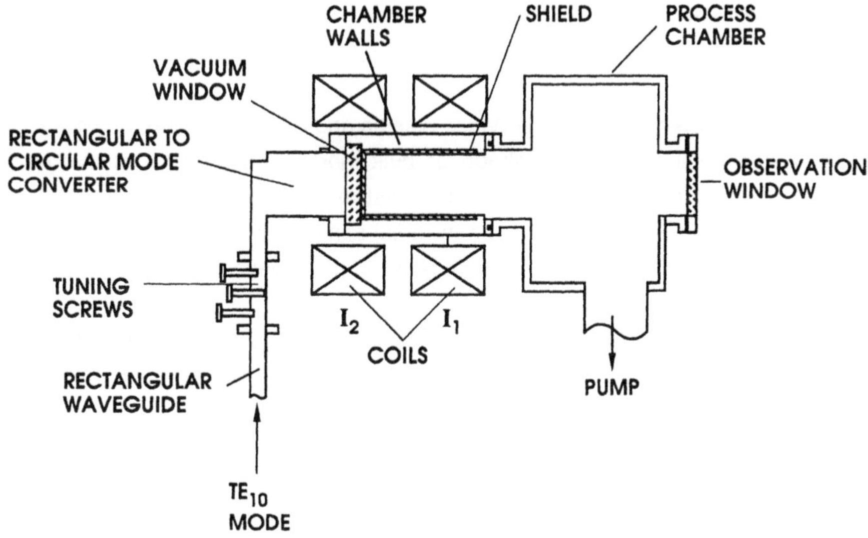

Fig. 7.2. Schematic of a waveguide ECR source (adapted from ref. 40).

[†] Strictly speaking, an electromagnetic cavity consists of a piece of waveguide, most often of circular cross-section, closed at both extremities by conducting flanges.

data are somehow confusing because of wide differences in magnetic field configurations, incident wave power level, gas pressure and composition, and dimensions of the system, there are some basic features which appear to be common to all these plasmas.

Most authors agree on the fact that densities of 10^{11} - 10^{12} electrons/cm^3 are easily obtained in the source under typical operating conditions of approximatly 1 kW absorbed power at a few mtorr argon pressure in a 2-4 liters volume.[34,35,37-40] A lower density regime is also reported in which the plasma is underdense, i.e. the density is much lower than the critical density. In the latter case, wave absorption apparently takes place near the dielectric window, presumably indicating that the wave is unable to propagate in the plasma and is thus evanescent at the dielectric-plasma boundary.[37] This means that the electromagnetic energy is only partially transferred to the plasma and this transfer occurs over a thin layer corresponding to the wave penetration depth (skin depth). A second important characteristic of ECR plasmas is the volume the glow occupies. It can adopt several shapes depending on the external parameters, mainly B_0, the source geometry and the wave power. One can observe donut shape (or ring mode) or column shape (also denoted bulk mode).[34] In the first case, one actually sees a plasma ring centered around the discharge axis, with its maximum intensity off this axis. Measurements show that this maximum of emission intensity approximately corresponds to a maximum plasma density. In the column shape situation, the maximum emission comes from a region around, and containing, this axis. Combinations of these two extremes have also been observed.[37] The ring mode seems to be associated with an underdense plasma and a larger amount of reflected or radiated power while the column shape or bulk mode is related to an overdense discharge. Such changes in the radial distribution of electron density are representative of corresponding changes in radial configuration of the plasma electric field and are not consistent with modeling in terms of planewaves.

Similar ECR plasma characteristics were observed when exciting the discharge with other waveguide modes such as, for example, the circular TE_{11} and the rectangular TE_{10} modes. Production of plasmas with the TM_{01} mode in a circular waveguide was also demonstrated. All these results tend to show that the way in which the wave energy is absorbed into the plasma chamber does not critically depend on how this energy is injected. These observations support the idea that wave channelling through the discharge and thus field distributions are essentially governed by the presence of the plasma so that the initial feeding structure only plays a secondary role. As an example, we mention a paper by Popov[40] in which it is shown that the radial profile of the plasma density remains the same whether the plasma is initially supplied by a TE_{10} or a TM_{01} circular waveguide mode. Consider, for example, the generation of plasma by an electromagnetic wave propagating in a circular waveguide excited in the fundamental TE_{11} mode. The fact that it is a TE mode means that it has no axial component of electric field. When coupling a wave to the discharge chamber through a dielectric window, all field components tangential to the window must be continuous across it. In the present TE_{11} mode case, a transverse electric field should thus persist once the plasma is present in the chamber. However, from the theory of wave propagation in waveguides presented in Chap. 6, we know that the electromagnetic modes guided by circular waveguides filled with a magnetoplasma are all characterized by the existence of an axial electric field component. Consequently, a rearrangment of the E-field necessarily occurs at the plasma-dielectric window interface, generating an axial field component on the plasma side.

As mentioned in Chap. 6, among the three types of wave solutions obtained, the waveguide modes cannot propagate in an overdense plasma. We thus conclude that such modes cannot be responsible for generation of high density plasma in the overdense regime. However, we surmise that such waves are at the origin of the underdense regime. The transition from the underdense to the overdense regime would occur at some threshold density or, equivalently, at some wave power level, leading to a conversion from a waveguide mode to a plasma mode, that would explain the sudden change in plasma conditions observed when wave power is increased above some level (Fig. 7.3).[49]

Recent experimental results obtained by Stevens et al.[26] show that absorption occurs at higher B_0-fields than the value given by the $\omega = \omega_{ce}$ condition. This frequency shift may be interpreted in terms of a Doppler broadening of the collisionless cyclotron absorption due to a finite electron temperature. Nonetheless, at pressures of a few mtorr, collisions can also contribute to the wave absorption. Thus a definite quantitative answer necessitates to compare the relative contribution of both collisional and collisionless absorption for the relevant wave propagating in the plasma. Achieving this implies an efficient control of the propagation mode and the ability to calculate the wave attenuation in the warm collisional plasma model.

7.4.2. ECR plasmas generated using non waveguide-based excitation structures. A modification to the waveguide based ECR plasma is provided by Sakudo et al.[21] These authors use a waveguide to transport electromagnetic energy but the wave coupling to the discharge chamber is realized through a coaxial antenna rather than through a dielectric window, thus mode configuration is not preserved.

A resonant cavity can also be used as a wave coupling device, as reported for example by Bardet et al.[53] In this system, a cylindrical cavity excited in the TE_{111} mode

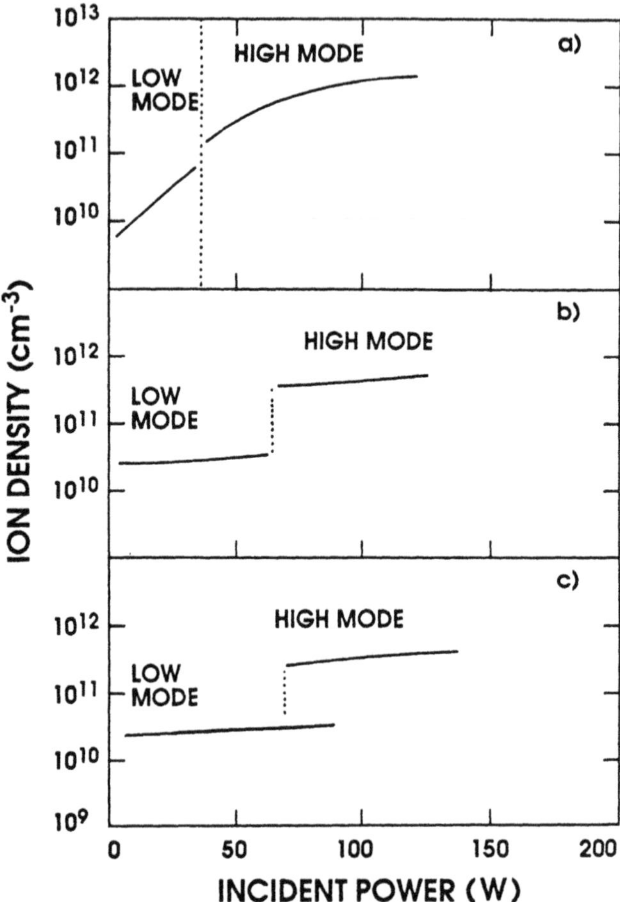

Fig. 7.3. Variation of ion density in a waveguide ECR plasma as a function of wave power, at different coil currents and different positions from the plasma-vacuum transition; (a) 150 A, $z = 6.3$ cm, (b) 125 A, $z = 10.2$ cm, (c) 110 A, $z = 10.2$ cm (adapted from ref. 49).

is located in a decreasing magnetic field such that the B_0 field intensity in the center of the cavity yields a cyclotron frequency equal to the resonance frequency of the cavity.

A slotted Lisitano[54] coil is also used to generate ECR plasma, as described in references 55-57. The originality of this system is that the excitation device is directly located within the plasma. The drawback is that the coil could contribute to the introduction of impurities in the system especially when the system is used with reactive gases such as Cl_2.

Microwave horns can also be utilized as coupling devices to excite ECR discharges.[58,59] As compared to those above, such devices are not well documented. Their advantage relies essentially on the fact that the cross-section of the horn can in principle be increased indefinitely, thus allowing generation of large diameter plasmas.

7.4.3. Magnetoplasmas sustained by modified surface waves. From Chap. 5 we know that in the absence of a static magnetic field B_0, surface waves can efficiently sustain long plasma columns. When applying an axial B_0 field to these columns, as discussed in Chap. 8, two important effects occur: i) the wave degeneracy related to the azimuthal wavenumber m is removed: waves with either positive or negative m values can propagate; ii) the wave field configuration is no longer necessarily that of a surface wave, i.e. its total electric field intensity can reach a maximum elsewhere along the radius than at the tube wall. The common feature of these "generalized" surface waves is that they are guided plasma waves. Based on the fact that these guided waves can be efficiently excited with the same wave launchers as those for unmagnetized surface-wave plasmas and because the advanced modeling achieved with these plasmas can be extended to magnetoplasmas, such waves have been recently proposed as an alternative way to generate ECR plasmas and as a means to better characterize them.[60]

The experimental device is shown schematically in Fig. 7.4. The plasma source is generated in a fused silica tube through the propagation of a modified electromagnetic surface wave. The latter is launched by a gap-type exciter (a waveguide-surfatron[61] for excitation at 2.45 GHz) of high coupling efficiency. The plasma tube penetrates into a larger diameter metallic chamber. The magnetic field is a divergent one but its gradient is fairly smooth (about 200 G/m) except near the last magnetic coil.

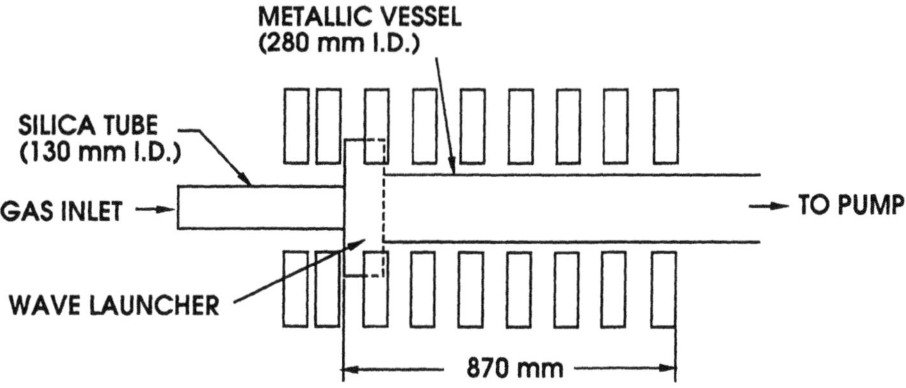

Fig. 7.4. Schematic for producing magnetoplasmas using modified surface waves.

Experiments show that to achieve an efficient coupling between the wave and the plasma, the static magnetic field must be adjusted so that its intensity at the launcher gap is at least 17% higher than the cold plasma ECR value. At 2.45 GHz, the emitted light intensity collected perpendicularly to the B_0-field lines at 40 cm from the field applicator is maximum when a 1 kG magnetic field intensity is applied at the field applicator. In contrast to a number of experimental results reported for ECR discharges, no local enhancement either of plasma density or of electron temperature is observed in the discharge.

Preliminary measurements at 600 MHz show that an efficient power transfer to the plasma can be achieved using the same magnetic field configuration and intensity as at 2.45 GHz (1 kG at the launcher gap) and yields a similar total number of electrons in the plasma for the same absorbed power. The latter observation indicates that the power balance per electron (parameter θ: see Sec. 2.2.9.2) is approximately the same at those two frequencies. This is not a surprise since, as discussed in Chap. 3, the power balance of the discharge does not depend much on the frequency when the EEDF is a Maxwellian, as expected at such high electron densities.

7.4.4. Plasma sources using helicon waves. Dense plasma production can be achieved by using helicon waves.[7,62,63] These waves belong to the category of whistler waves which in free-space are right-hand electromagnetic waves. However, in bounded plasmas, these helicon waves, as all bounded plasma waves, are not pure electromagnetic waves since they have an axial E-field component (Sec. 6.6.2).

These plasmas exhibit high degrees of ionization in the vicinity of the axis. We stress that, for similar absorbed power per unit volume and for a given magnetic field intensity, the total number of electrons in helicon wave high density plasmas is close to that in high density microwave sustained magnetoplasmas, meaning that the overall ionization efficiency is approximately the same at 13.56 and 2450 MHz. This also agrees with observations made in the previous section for plasmas sustained by generalized surface waves. In this sense, helicon plasmas are not superior to other kinds of HF magnetoplasmas, but they have the advantage of being produced in a wider range of magnetic fields than that allowed for waveguide ECR plasmas.

7.5. Ion transport in HF magnetoplasma discharges

As already stressed, ion energy control in HF magnetoplasma discharges is a requirement for materials processing applications such as etching, deposition, cleaning, and passivation. When performing anisotropic etching, highly directional ions are needed. In the absence of collisions in the sheath, the ions strike the surface at a "pitch" angle Ψ_p given by $\tan \Psi_p = v_\perp / v_\parallel$ where $v_\perp$ and $v_\parallel$ denote the ion velocity in the direction perpendicular and parallel to the surface respectively. When Ψ_p departs too much from 90°, a significant loss in dimensional control of patterns to be replicated can occur.

To meet this anisotropy requirement, two ion characteristics must be considered: their temperature and their directed velocity. The temperature relates to the thermal motion of ions, which is nearly isotropic, while their directed velocity accounts for a bulk motion. In order that the ions strike the surface vertically, we need to reduce their transverse motion as much as possible, i.e. minimize both the temperature and the possible transverse directed motion induced by magnetic field or density radial gradients. One should also minimize ion energy to avoid damaging the substrates to be treated. If the ions are created with both a large directed velocity and a large spread in energy, the ability to optimize etching characteristics is seriously compromised. Finally, the neutral gas temperature is also of concern since the presence of hot neutrals can contribute to isotropic etching in thermally activated processes. Thus the possibility that in ECR plasmas ions and neutrals become too hot, as reported by McKillop et al.,[64] has been a concern to those interested in applying these sources to ultra large scale integration (ULSI).

To understand how ions are created in the plasma source and then transported, several experiments have been realized in a variety of gases. As a rule, the ion velocity distribution function (IVDF) is obtained using two techniques: ion energy analyzers[65] or laser induced fluorescence (LIF). The main drawback of ion analyzers is their limited angular and spatial resolutions, which make accurate measurements of both $v_\perp$ and $v_\parallel$ difficult. These devices are not costly and relatively easy to implement in the discharge. In contrast, LIF techniques are relatively expensive and not universal, and they require equipment and know-how that are not commonly available. However, they offer a non-intrusive in-situ method yielding high spatial and angular resolutions. A suitable choice of the laser lines allows 3-dimensional measurements of ion and neutral velocity distribution functions averaged over the velocity component perpendicular to the laser propagation direction. One can find reviews on LIF techniques in refs. 66 and 67. LIF often limits one to making measurements on minor species. This can be both an advantage and a disadvantage. For example, metastable ions usually have a lower collision rate than ground-state ions and are thus more convenient for studying ion generation and transport. On the other hand, their IVDF is not necessarily representative of the total ion population and can be misleading if not properly interpreted.

Recent measurements of the ion and neutral velocity distribution functions in waveguide-based ECR discharges in Ar,[68] Ar-He,[69] Cl_2[70] and N_2[71-73] have shown that the situation is not as desperate as reported earlier since neutral temperatures seem to range between 300 and 800 K, i.e. close to room temperature. As for ions, their perpendicular temperature $T_{i\perp}$ appears to vary between 0.1 and 0.6 eV, depending on discharge conditions and position within the plasma.

It is instructive to know how the IVDF evolves spatially. Sadeghi et al.[74] report spatially resolved velocity distributions of Ar_m^+ metastables in Ar-He gas mixtures. Their main observation is that along the magnetic field axis, the ion distribution in the reactor has two distinct components, a fast one and a slow one (Fig. 7.5). The fast component presumably originates from the source and represents ions accelerated toward the end of the reactor, while the slow component could be related to ions directly created where the source expands into the reactor.

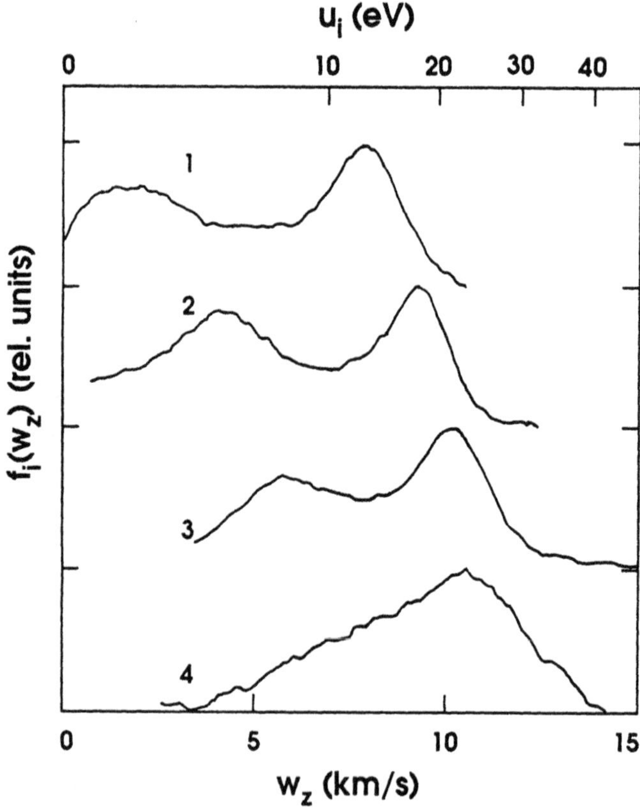

Fig. 7.5. Axial ion energy distribution function, at different axial positions from the plasma-vacuum transition, in an argon/helium waveguide ECR discharge; (1) 48.2 cm, (2) 55.2 cm, (3) 56 cm, (4) 56.1 cm (from ref. 74).

From the above, we see that the potential drop between the plasma source and the substrate should be minimized: strongly accelerating the ions emanating from the source toward the substrate is neither required nor desirable since these ions could then arrive at the substrate sheath edge with several tens of eV. In addition, the transverse motion of ions should be reduced by avoiding diverging magnetic field lines. This leads us to conclude that the present design of ECR reactors is poorly suited for etching applications. The next generation of such systems should thus process surfaces in a region where the plasma is constrained to follow magnetic field lines that are perpendicular to the substrate. This contrasts with the present use of divergent field lines, which allows the plasma to expand; this is in fact a trick to increase the plasma transverse dimension, making it compatible with the substrate dimension required by manufacturers. However, as we have seen, this scheme is not suitable for achieving adequate plasma characteristics. Note further that the stronger this radial expansion of the plasma at the source exit, the more substantial is the decrease of plasma density and the generation of strong density gradient in the axial direction. Avoiding this expansion reduces the space-charge electric field due to accelerated electrons and consequently the downstream velocity of ions. Obeying these rules might explain the success of the production-line ECR system sold by Hitachi, where the wafer is placed directly in the source.

7.6. Conclusion

One common feature of magnetically assisted HF discharges is certainly the possibility of generating high density plasmas at low gas pressures. Whether systems other than the Hitachi reactor will emerge with characteristics suitable for plasma process applications is still an open question. We believe that the characteristics of the ion velocity distribution will be decisive in achieving the best reactor. Hence, not only the nature of the source but also the complete design of the reactor constitute determinant parameters.

From the above discussion based on experimental results obtained in HF magnetoplasmas, it appears that new proposals for ECR reactor designs should consider the following guidelines: i) maintain the magnetic field lines as parallel as possible to the axis of the whole system, i.e. from the plasma source to the wafer location (the latter being placed perpendicularly to the axis), to avoid any gain in the perpendicular component of ion velocity; ii) achieve from the start plasma source diameters as large as the substrates; iii) reduce the build-up of a space-charge electric field oriented perpendicularly to the axis by achieving a flat electron density profile in that direction. This can be done by depositing preferentially electromagnetic energy close to the source's wall rather than in the volume, so as to "fight" against diffusion (see Chap. 9). Obtaining a flat electron density profile perpendicular to the discharge vessel axis is thus of interest not only to obtain process homogeneity, but also to reduce the transverse component of the ion velocity.

One of the key points in the development of magnetized HF discharges is a full understanding of the characteristics of the wave propagating in the plasma, since it determines where and how the electromagnetic energy is deposited. In our opinion this is a major weakness in the present state of ECR plasmas and we believe that a strong effort is needed to overcome it.

Clearly adding a magnetic field to HF plasmas allows one to considerably lower the operating pressure. This is certainly an essential feature of magnetized plasmas. As far as the ionization efficiency is concerned, no particular setup seems to have a definite advantage over others, at least for the time being. In that respect, systems with a good flexibility in terms of plasma stimulating frequencies could bring new perspectives to the future of reactors.

References

[1]   Y. Okamoto and H. Tamagawa, Rev. Sci. Instrum. **43**, 1193 (1972).

[2]   K. Suzuki, S. Okudaira, N. Sakudo and I. Kanomota, Jap. J. Appl. Phys. **16**, 1979 (1977).

[3]   K. Suzuki, S. Okudaira, I. Kanomota, J. Electrochem. Soc. **126**, 1024 (1979).

[4]   M. Miyamura, O. Tsukakashi and S. Komiya, J. Vac. Sci. Technol. **20**, 986 (1982).

[5]   T. Ono, M. Oda, C. Takahashi and S. Matsuo, J. Vac. Sci. Technol. **B4**, 696 (1986).

[6]   A.S. Yaspir, G. Fortuño-Wiltshire, J.P. Gambino, R.H. Kastl and C.C. Parks, J. Vac. Sci. Technol. **A8**, 2939 (1990).

[7]   A.J. Perry, D. Vender and R.W. Boswell, J. Vac. Sci. Technol. **B9**, 310 (1991).

[8]   S. Samukawa, T. Toyosato and E. Wani, Appl. Phys. Lett. **58**, 896 (1991).

[9]   E. Ghanbari, T. Nguyen, S. Bui and E. Ostan, J. Vac. Sci. Technol. **A8**, 2945 (1990).

[10]   G. Fortuño-Wiltshire, J. Vac. Sci. Tecnol. **A9**, 2356 (1991).

[11]   J.M. Cook, D.E. Ibbotson, P.D. Foo and D.L. Flamm, J. Vac. Sci. Technol. **A8**, 1820 (1990).

[12]   S. Salimian, C.B. Cooper III and A. Ellingboe, Appl. Phys. Let. **56**, 1311 (1990).

[13]   R.K.F. Teng and H.G. Yang, J. Vac. Sci. Technol. **B9**, 380 (1991).

[14]   S. Matsuo and M. Kiuchi, Jap. J. Appl. Phys. **22**, L210 (1983).

[15]   Z.H. Zhou, F. Yu and R. Reif, J. Vac. Sci. Technol. **B9**, 374 (1991).

[16]   G.T. Salbert, D.K. Reinhard and J. Asmussen, J. Vac. Sci. Technol. **A8**, 2919 (1990).

[17]   D.A. Carl, D.W. Hess and M.A. Lieberman, J. Vac. Sci. Technol. **A8**, 2924 (1990).

[18]   M. Matsuoka and K. Ono, Jap. J. Appl. Phys. **28**, L503 (1989).

[19] M. Matsuoka and K. Ono, J. Vac. Sci. Technol. **A9**, 691 (1991).

[20] K. Wakita and S. Matsuo, Jap. J. Appl. Phys. **23**, L556 (1984).

[21] N. Sakudo, K. Tokiguchi, H. Koike and I. Kanomata, Rev. Sci. Instrum. **48**, 762 (1977).

[22] N. Sakudo, K. Tokiguchi, H. Koike and I. Kanomata, Rev. Sci. Instrum. **49**, 940 (1978).

[23] Y. Torii, M. Shimada, I. Watanabe, J. Hipple, C. Hayden and G. Dionne, Rev. Sci. Instrum. **61**, 253 (1990).

[24] J. Hipple, C. Hayden, G. Dionne, Y. Torii, M. Shimada and I. Watanabe, Rev. Sci. Instrum. **61**, 294 (1990).

[25] D.M. Manos and D.M. Flamm, in *Plasma Etching, An Introduction*, D.M. Manos and D.M. Flamm eds. (Academic, San Diago, CA, 1989).

[26] J.E. Stevens, Y.C. Huang, R.L. Jarecki and J.L. Cecchi, J. Vac. Sci. Technol. **A10**, 1270 (1992).

[27] S. Oda, J. Noda and M. Matsumara, Jap. J. Appl. Phys. **28**, L1860 (1989).

[28] S.R. Mejia, R.D. McLeod, K.C. Kao and H.C. Card, Rev. Sci. Instrum. **57**, 493 (1986).

[29] K. Shirai, T. Iizuka and S. Gonda, Jap. J. Appl. Phys. **28**, 897 (1989).

[30] J. Forster and W. Holber, J. Vac. Sci. Technol. **A7**, 899 (1989).

[31] O.A. Popov and H. Waldron, J. Vac. Sci. Technol. **A7**, 914 (1989).

[32] Y.H. Lee, J. E. Heidenreich III and G. Fortuno, J. Vac. Sci. Technol. **A7**, 903 (1989).

[33] J.S. McKillop, J.C. Forster and W.M. Holber, J. Vac. Sci. Technol. **A7**, 908 (1989).

[34] O.A. Popov, J. Vac. Sci. Technol. **A7**, 894 (1989).

[35] E. Ghanbari, I. Trigor and T. Nguyen, J. Vac. Sci. Technol. **A7**, 918 (1989).

[36] H. Nihei, J. Morikawa and N. Inoue, Jap. J. Appl. Phys. **29**, L822 (1990).

[37] S.M. Gorbatkin, L.A. Berry and J.B. Roberto, J. Vac. Sci. Technol. **A8**, 2893 (1990).

[38] S.M. Rossnagel, S.J. Whitehair, C.R. Guarnieri and J.J. Cuomo, J. Vac. Sci. Technol. **A8**, 3113 (1990).

[39] O.A. Popov and A.O. Westner, Rev. Sci. Instrum. **61**, 303 (1990).

[40] O.A. Popov, J. Vac. Sci. Technol. **A8**, 2909 (1990).

[41] S. Miyake, W. Chen and T. Ariyasu, Jap. J. Appl. Phys. **27**, 2491 (1990).

[42] T.W. Jolly and P. Blackborrow, Rev. Sci. Instrum. **61**, 297 (1990).

[43] S.M. Rossnagel, K. Schatz, S.J. Whitehair, R.C. Guarnieri, D.N. Ruzic and J.J. Cuomo, J. Vac. Sci. Technol. **A9**, 702 (1991).

[44] J.L. Cecchi, J.E. Stevens, R.L. Jarecki Jr. and Y.C. Huang, J. Vac. Sci. Technol. **B9**, 318 (1991).

[45] J.E. Stevens, J.L. Cecchi, Y.C. Yuang and R.L. Jarecki Jr., J. Vac. Sci. Technol. **A9**, 696 (1991).

[46] M. Shimada, I. Watanabe and Y. Torii, J. Vac. Sci. Technol. **A9**, 707 (1991).

[47] T. Oomori, M. Tuda, H. Ootera and K. Ono, J. Vac. Sci. Technol. **A9**, 722 (1991).

[48] C.A. Outten, J.C. Barbour and W.R. Wampler, J. Vac. Sci. Technol. **A9**, 717 (1991).

[49] D.A. Carl, M.C. Williamson, M.A. Lieberman and A.J. Lichtenberg, J. Vac. Sci. Technol. **B9**, 339 (1991).

[50] O.A. Popov, J. Vac. Sci. Technol. **A9**, 711 (1991).

[51] D.B. Miller and E.F. Gibbons, AIAA J. **2**, 35 (1964).

[52] *Plasma Processing of Materials: Scientific Opportunities and Technological Challenges*, National Research Council (National Academy Press, Washington, DC, 1991).

[53] R. Bardet, T. Consoli and R. Geller, Nuclear Fusion **4**, 48 (1964).

[54] G. Lisitano, R.A. Ellis Jr., W.M. Hooke and T.H. Stix, Rev. Sci. Instrum. **39**, 295 (1968).

[55] A. Yonesu, Y. Takeuchi, A. Komori and Y. Kawai, Jap. J. Appl. Phys. **27**, L1746 (1988).

[56] Y. Kawai, A. Komori, H. Ikeda, K. Kishimoto, M. Murata and S. Uchida, Jap. J. Appl. Phys. **29**, 2487 (1990).

[57] A. Komori, Y. Takada, A. Yonesu and Y. Kawai, J. Appl. Phys. **69**, 1974 (1991).

[58] M. Geisler, J. Kieser, E. Räuchle and R. Wilhelm, J. Vac. Sci. Technol. **A8**, 908 (1990).

[59] G. Neumann and K.H. Kretschmer, J. Vac. Sci. Technol. **B9**, 334 (1991).

[60] J. Margot and M. Moisan, J. Phys. D: Appl. Phys. **24**, 1765 (1991).

[61] M. Moisan, M. Chaker, Z. Zakrzewski and J. Paraszczak, J. Phys. E: Sci. Instrum. **20**, 1356 (1987).

[62] R.W. Boswell and R.K. Porteous, Appl. Phys. Lett. **50**, 1130 (1987).

[63] R.W. Boswell, A. Perry and M. Emami, J. Vac. Sci. Technol. **A7**, 3345 (1989).

[64] J.S. McKillop, J.G. Forster and W.H. Holber, Appl. Phys. Lett. **55**, 30 (1989).

[65] S. Samukawa, Y. Nakagawa and K. Ikeda, Jap. J. Appl. Phys. **29**, L2319 (1990).

[66] R.A. Gottscho and T.A. Miller, Pure & Appl. Chem. **56**, 189 (1984).

[67] R.W. Dreyfus, J. M. Jasinski, R.E. Walkup and G.S. Selwyn, Pure & Appl. Chem. **57**, 1265 (1985).

[68] D.J. Trevor, N. Sadeghi, T. Nakano, J. Derouard, R.A. Gottscho, P.D. Foo and J.M. Cook, Appl. Phys. Lett. **57**, 1188 (1990).

[69] T. Nakano, N. Sadeghi and R.A. Gottscho, Appl. Phys. Lett. **58**, 458 (1991).

[70] T. Nakano, N. Sadeghi, D.J. Trevor, R.A. Gottscho and R.W. Boswell, J. Appl. Phys., submitted (1992).

[71] E.A. Den Hartog, H. Persing and R. Claude Woods, Appl. Phys. Lett. **57**, 661 (1990).

[72] E.A. Den Hartog, H. Persing, J.S. Hamers and R. Claude Woods, *10th International Symposium on Plasma Chemistry*, Bochum, Germany (1991) p. 2.1-3 p.1.

[73] R. Claude Woods, R.L. McClain, L.J. Mahoney, E.A. Den Hartog, H. Persing and J.S. Hamers, *SPIE, Process Module Metrology, Control, and Clustering* **1594**, 366 (1991).

[74] N. Sadeghi, T. Nakano, D.J. Trevor and R.A. Gottscho, J. Appl. Phys. **70**, 2552 (1991).

CHAPTER 8

**SURFACE-WAVE-SUSTAINED PLASMAS IN STATIC MAGNETIC FIELDS
FOR THE STUDY OF ECR DISCHARGE MECHANISMS** [*]

8.1. Introduction

Researchers in the field of microelectronics and materials who wish to improve plasma processes cannot completely ignore the work being done on electron cyclotron resonance (ECR) discharges. Though this type of plasma is certainly not the solution to all problems as we already know, interesting developments can still come out from further experimental and theoretical investigations. This eventuality is based on the observation that the understanding of ECR discharges is still far from being satisfactory and that novel ECR configurations are continuously being disclosed.

The interest in ECR discharges stems from a combination of features: i) it is the only type of high frequency (HF) discharges that can be operated efficiently at such low gas pressures (typically 1 mtorr). This means that there are few collisions in the substrate sheath, a fact that favors the directionality of the particles accelerated within the sheath; ii) the ECR mechanism leads to a strong attenuation of the feeding wave, i.e. a strong local absorption of HF power. This yields high plasma densities in a limited volume from which large ion current can then be extracted for application purposes; iii) most ECR systems use an HF field applicator that is not in contact with the plasma. Then, any isolated surface immersed in plasma is at floating potential, i.e. the energy of ions striking it is relatively low (a few tens of eV). However, when needed, the ion bombardment energy can be increased, independently of the plasma parameters, by applying a biasing (DC or HF) voltage to the substrate, providing in this way a better flexibility in setting it than with RF capacitive plate discharges.

Despite the technical and experimental efforts deployed, the modeling of ECR discharges is still elementary: crude approximations are often used (e.g. assuming plane wave propagation in a spatially bounded medium) or important physical phenomena are neglected (e.g. the existence of multimode wave propagation in the discharge chamber). The difficulty in achieving such a modeling results from the complexity of ECR reactors with respect to boundary conditions, geometry, static magnetic field configuration and wave feeding, and it explains that quite generally these systems are optimized only empirically.

---

[*] **Presented by J. Margot and M. Moisan**

The problems affecting the analysis and operation of ECR discharges can be greatly reduced by extending to these discharges the theoretical and experimental expertise developed with surface-wave plasmas, which has led to a better understanding of HF discharges in general. For similar reasons, by investigating surface-wave plasmas when they are submitted to a static magnetic field B_0, we should be able to improve our understanding of HF magnetoplasmas in general and ECR plasmas in particular.

The present chapter reviews the first step made in that direction. It focuses on the wave-to-discharge power transfer and it takes into account the role of the wave field frequency and spatial distribution. At this stage, for simplicity and also for clarity of exposition, we use a hydrodynamic description of the plasma, though only a kinetic approach could provide a full answer to ECR discharge mechanisms, as discussed in Sec. 6.4.3. Nonetheless, as will be shown, the results obtained in this way yield a fair insight in the basic power transfer mechanisms of HF magnetoplasmas.

8.2. Wave-to-plasma power transfer: basic mechanisms and parameters

8.2.1. Field configuration and field polarization of guided waves propagating parallel to B_0.

At microwave frequencies, because of the very rapid change of the field intensity with time, the ions with their large mass can be considered at rest, only the electrons responding to the wave field. These electrons gain energy from the oscillating electric field through collisional or collisionless absorption mechanisms, as described in Sec. 6.4.

Consider a guided wave[†] propagating in the same direction as B_0 (z-axis) within a cylindrical coordinate system where r and ϕ, the radial position and the azimuthal angle respectively, are in a plane perpendicular to B_0. Since these waves are generally hybrid, we need to use all the electric field components, namely E_z, E_r and E_ϕ. In the plane perpendicular to B_0, these are

$$E_r = |E_r| \exp [j(\omega t - \beta z + m\phi + \psi_r)] \exp(-\alpha z), \qquad (8.1)$$

$$E_\phi = |E_\phi| \exp [j(\omega t - \beta z + m\phi + \psi_\phi)] \exp(-\alpha z), \qquad (8.2)$$

[†] The term *guided wave* designates in this chapter an electromagnetic wave propagating along a plasma column bounded by a dielectric material, surrounded or not by a metallic enclosure. In the presence of a magnetic field B_0, we limit this name to such waves for which the attenuation coefficient α goes to zero as collisions vanish. In absence of metallic enclosure and in the case where $B_0 \to 0$, the only guided waves are surface waves. For more on guided plasma waves, see Sec. 6.6 and ref. 1.

where β is the axial wavenumber or phase coefficient ($\beta = 2\pi/\lambda$, λ is the guided wave wavelength) and α is the attenuation coefficient, m is an integer called the azimuthal wavenumber, and the angles ψ_r and ψ_ϕ are phase angles; the value $\Theta \equiv \psi_r - \psi_\phi$ defines the *polarization angle of the electric field in the plane perpendicular* to $\mathbf{B}_0$. The components E_r and E_ϕ generate the so-called *perpendicular electric field vector* $\mathbf{E}_\perp$.

Concentrating on the movement of the $\mathbf{E}_\perp$ vector tip with time, we have the following possible situations: i) Θ is equal to zero or some integral multiple of π: the vector $\mathbf{E}_\perp$ keeps a fixed orientation in space, this orientation depending on the ratio $|E_r|/|E_\phi|$. The wave is linearly polarized; ii) when Θ takes some other value, the $\mathbf{E}_\perp$ field rotates with time and its tip describes an ellipse. For $0 < \Theta < \pi$, the wave is said to have a right-hand polarization while for $\pi < \Theta < 2\pi$, it is left-handed. iii) Θ is equal to $\pi/2$ or $3\pi/2$ and $|E_r| = |E_\phi|$. In this very special case, the tip of $\mathbf{E}_\perp$ generates a circle and the wave is circularly polarized, right-handed for $\Theta = \pi/2$ and left-handed for $\Theta = 3\pi/2$.

As shown in Sec. 6.4, a wave having a right-hand circularly polarized electric field component in the plane perpendicular to $\mathbf{B}_0$, can be strongly damped in the vicinity of $\omega = \omega_{ce}$ through both collisional and collisionless mechanisms.

8.2.2. Wave-discharge power balance. Consider a cylindrical plasma column enclosed in a dielectric tube and sustained by a traveling electromagnetic wave guided by this column. The tube is assumed to be surrounded by air but the same type of calculations could be made when it is enclosed within a cylindrical conductor. The wave power flux P(z) decreases with increasing z as the wave moves away from the launcher since it loses power to sustain the discharge. The corresponding wave attenuation is characterized by the *attenuation coefficient* $\alpha(z)$ which depends on both the plasma and wave properties, and on the wave absorption mechanism (collisional or collisionless). The power lost by the wave over the axial distance z, z + dz is by definition

$$\left|\frac{dP(z)}{dz}\right| \equiv 2\alpha(z)\,P(z). \tag{8.3}$$

We assume that this power, taken from the wave by the electrons, is fully used to sustain the discharge over the same axial interval z, z + dz. This yields axially a *local balance of power* given by

$$2\alpha(z)\,P(z) = \bar{n}\,(z)\,\theta_A\,\mathcal{S}, \tag{8.4}$$

where $\bar{n}(z)$ is the cross-section average electron density at z, S is the plasma cross-section, and θ_A is the microwave power absorbed on the average per electron. Under steady-state conditions, θ_A must exactly compensate θ_L, the average power loss per electron in collisions of all kinds,[†] and we thus have $\theta_A = \theta_L = \theta$. The value of θ_A can be obtained, for example, from eqn. (2.18) while θ_L is given by eqn. (2.70) or (2.71). Finally, the local power balance comes to

$$2\alpha(z)\,P(z) = \bar{n}(z)\,\theta\,S, \tag{8.5}$$

as in Sec. 5.4.2. The important point that we want to make in the remaining part of this section is that θ does not depend on the wave properties.

Under ambipolar diffusion conditions and for surface-wave-sustained plasmas when $B_0 = 0$, it was shown from theory that θ does not depend on n, and thus, not on z (except at the very end of the plasma column), provided that two conditions are met:[2] i) the excitation and ionization of atoms (molecules) results predominantly from a single electron collision on their ground level, i.e. step-wise processes are negligible; ii) the EEDF does not vary with n. The latter condition implies that either electron-electron or electron-neutral atom collisions are prevailing. It excludes any transition between these two collisional regimes, as such a transition is controlled by the degree of ionization, hence by n for given discharge conditions. As discussed in Sec. 2.2.9, determining the value of θ requires calculating the EEDF. In a steady-state plasma, this is achieved by combining the Boltzmann equation for electrons and the continuity equation, the latter under ambipolar diffusion conditions leading to the diffusion equation

$$<v_i> = D_a/\Lambda^2, \tag{8.6}$$

where D_a is the ambipolar diffusion coefficient (a function of the EEDF and of the gas nature and pressure) and Λ is the diffusion length (in a long cylindrical tube, for example, $\Lambda \simeq R_1/2.405$, where R_1 is the tube inner radius). Note that eqn. (8.6) means that ionization (left-hand side) must compensate for losses (right-hand side) of charged particles by diffusion to the wall where they recombine; when $<v_i>$ depends on radial position, Λ in eqn. (8.6) must be replaced by some effective diffusion length. The main point coming out of this analysis[3] is that the EEDF and, hence, θ are essentially determined by the gas nature and pressure, and by the vessel dimensions. In other

---

[†] One can look at θ_L as the power required to sustain an electron-ion pair in the discharge. It relates to the ionization efficiency: the smaller θ_L, the larger the ionization efficiency.

words, the value of θ is set by the charged particle losses and *not by the wave parameters.* This model has been well verified experimentally for $B_0 = 0$.[3]

A theoretical analysis similar to that presented above for θ remains to be made for surface-wave-sustained plasma columns when they are submitted to an axial B_0 field. The physical reasoning is the same but a more complex situation arises because of the anisotropy of the charged particle diffusion. It requires considering the diffusion coefficient as a second order tensor, not simply as a scalar quantity (Sec. 6.2). At this stage of our presentation, we turn to experimental results to go on characterizing θ. We consider the case of a plasma column sustained by a guided wave of azimuthal symmetry ($m = 0$), in the dominant mode ($n = 1$, see Sec. 8.3.2) at $\omega/2\pi = 600$ MHz, in a tube with $R_1 = 13$ mm, in argon at pressures p from 5×10^{-3} to 1 torr.[4] The experimental plot of θ/p as a function of ω_{ce}/ω in Fig. 8.1 shows that i) the lower the gas pressure, the steeper the decrease of θ with increasing ω_{ce}/ω; ii) the value of θ does not exhibit any extremum at $\omega_{ce}/\omega = 1$ down to the lowest gas pressure examined, namely 5 mtorr. The decrease of θ with increasing B_0 simply means that the average electron energy decreases as a result of an overall reduction of the

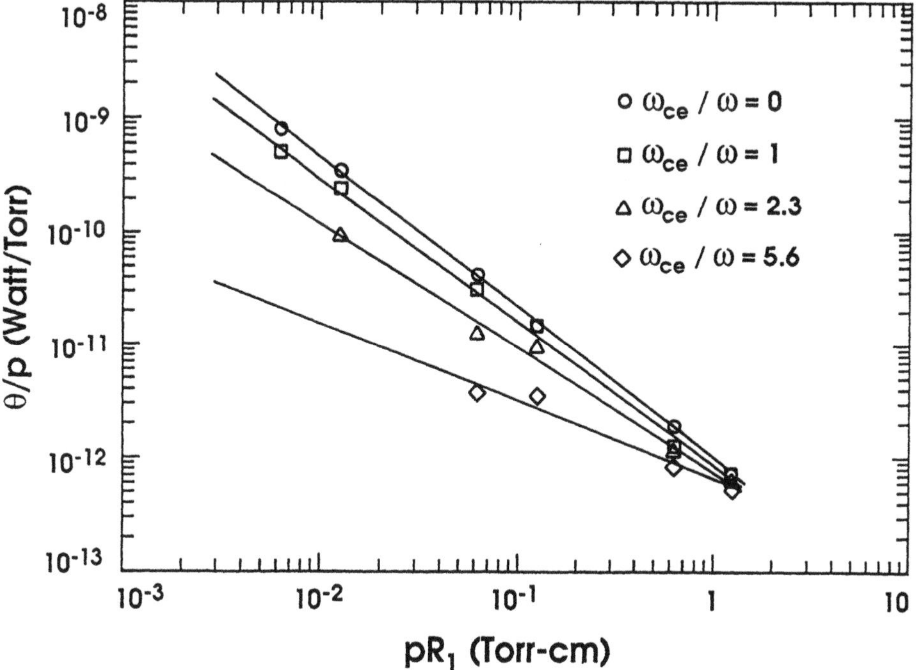

Fig. 8.1. Measured average power loss per electron over gas pressure as a function of gas pressure times discharge tube inner radius ($\omega/2\pi = 600$ MHz, $R_1 = 13$ mm, argon gas, HE_{01} fundamental mode).

charged particle losses to the wall. The fact that θ does not show a maximum at $\omega_{ce} \simeq \omega$ is an indication, based on eqn. (2.71), that the shape of the EEDF is not drastically modified as ω_{ce} reaches such a value; a major modification of the EEDF would yield a different sharing of the input energy between the various collisional processes (e.g. ionization and excitation), resulting in a different value of θ. The influence of a varying EEDF on θ can be observed, for example, when the plasma excitation frequency is varied (Sec. 3.2).

To further characterize the wave-to-plasma power transfer, we need the wave dispersion and attenuation properties as well as its electric field spatial distribution. These are obtained below within the hydrodynamic cold plasma description.

8.3. Surface-wave dispersion and attenuation properties along a plasma column when submitted to a static, axial magnetic field B_0

In a plasma, a true *electromagnetic surface wave* is a wave guided by the interface between the plasma and a containing dielectric vessel, and obeying two criteria: i) it is a slow wave, i.e. its phase velocity is smaller than the speed of light in free space; ii) in a cold collisionless plasma, its transverse wavenumber (Sec. 6.6) is strictly imaginary in the plasma and in the far most outer medium. In the absence of B_0, the guided waves sustaining a plasma column bounded by a dielectric without a metallic enclosure are all surface waves. In what follows, we examine the transformation of these surface waves as a function of the applied static magnetic field B_0 and we restrict the term guided waves to these resulting "generalized surface waves". The presentation is limited to a summary of the properties of these waves, the reader being directed to ref. 1 for details.

8.3.1. Hydrodynamic description of a plasma submitted to B_0.
We consider the plasma as a fluid of cold electrons. As mentioned in the introduction, such an approach cannot provide all the physics of ECR discharges but it nevertheless leads to a clearer view of the mechanisms driving HF sustained magnetoplasmas. In the present context,[†] the plasma is described as a dielectric material characterized by a permittivity tensor $\underline{\varepsilon}_p$ that takes into account the anisotropy introduced into the discharge by B_0. In a cylindrical plasma column submitted to an axial static magnetic field directed along the z-axis, this tensor has the form

---

[†] When discussing collisional absorption in Sec. 6.4.1, we have used the "charges in a free space" description, which leads to the conductivity tensor $\underline{\sigma}$ instead of $\underline{\varepsilon}_p$ used here. Since $\underline{\sigma}$ and $\underline{\varepsilon}_p$ are related by eqn. (2.5), clearly the structure of the matrices representing $\underline{\sigma}$ in eqn. (6.18) and $\underline{\varepsilon}_p$ in eqn. (8.7) is the same.

$$\underline{\underline{\varepsilon}}_p = \begin{pmatrix} \varepsilon_1 & -\varepsilon_2 & 0 \\ \varepsilon_2 & \varepsilon_1 & 0 \\ 0 & 0 & \varepsilon_3 \end{pmatrix}, \tag{8.7}$$

where the elements are given by

$$\varepsilon_1 = 1 - \xi^2 (1-j\delta) / [(1-j\delta)^2 - \tau^2], \quad \varepsilon_2 = -j\tau\xi^2 / [(j+\delta)^2 + \tau^2], \quad \varepsilon_3 = 1 - \xi^2 / (1-j\delta), \tag{8.8}$$

with $\xi^2 = \omega_{pe}^2/\omega^2$, $\delta = v_{eff}/\omega$ and $\tau = \omega_{ce}/\omega$, ε_0 being the free space permittivity and v_{eff}, the effective collision frequency for momentum transfer; ω_{pe} is the electron plasma angular frequency and $\omega_{pe}^2 = ne^2/m_e\varepsilon_0$. At $\omega_{ce} = \omega$ ($\tau = 1$) and for $v_{eff}/\omega \to 0$ ($\delta \to 0$), the elements ε_1 and ε_2 are discontinuous, tending to infinity. We thus have to check what happens to β and α under these circumstances. In what follows, we consider a cylindrical plasma column contained in a dielectric vessel surrounded by air.

8.3.2. Dispersion equation.

Determining and solving the *determinantal equation* $\mathcal{D}(\omega,h) = 0$ where $h = \beta - j\alpha$, lead to β and α as a function of ω/ω_{pe}. The procedure is as follows: i) Maxwell's equations provide in each medium (plasma, glass, air) the various E and H components of the electromagnetic field. For example, when the plasma is assumed radially uniform, E_z is given by

$$E_z = [a_1 J_m (h_{\perp 1} r) + a_2 J_m (h_{\perp 2} r)] \exp[j(\omega t + m\phi) - hz], \tag{8.9}$$

where J_m is the Bessel function of the first kind and of order m, and $h_{\perp 1}$ and $h_{\perp 2}$ are the transverse wavenumbers (see Sec. 6.6); ii) then one requires continuity across the various medium interfaces, of the E and H components that are directed tangentially to the interface, namely E_r, E_ϕ, H_r and H_ϕ; iii) this yields a set of continuity equations (serving to determine the a_k coefficients), which is linear and homogeneous in terms of the E and H components. This system of equations is indeterminate and thus has a non-trivial solution (E_s, $H_s \neq 0$ where $s = r,\phi,z$) only if the corresponding determinant is zero. This determinantal equation provides β and α as a function of ω/ω_{pe}; here ω is kept fixed and ω_{pe} is varied, yielding phase diagrams rather than dispersion diagrams as obtained when ω is varied and ω_{pe} fixed, the former situation corresponding to the experimental situation.

The determinantal equation has at least one solution for each value of the integer m. More specifically, it can be shown[1] that: i) in the case $\omega_{ce}/\omega < 1$, there exists only one solution for any given m value. These are called *fundamental hybrid modes* and are designated by the symbol HE_{m1}. Each such m mode tends continuously toward the surface wave of the same azimuthal symmetry as $\omega_{ce} \to 0$ (hence the name fundamental) and they are neither pure TM nor pure TE modes (hence the term hybrid mode). The most commonly observed ones are the $m = 0$ and $|m| = 1$ modes; ii) when $\omega_{ce}/\omega \geq 1$, to a given value of m corresponds an infinity of solutions, each denoted by the positive integer n. These n-sub modes are ordered by increasing values of β as obtained from the dispersion equation for a given ω/ω_{pe} ratio and they are characterized, in particular, by the radial distribution of their electric field intensity. They are called *magnetic hybrid* modes because they do not exist for $B_0 = 0$ and are symbolized by HE^*_{mn}, where the star serves to distinguish the magnetic modes from the fundamental modes.

8.3.3. Influence of ω_{ce}/ω on the axial wavenumber β of the guided waves.

Figure 8.2 shows the predicted influence of ω_{ce}/ω on the phase diagram for the HE_{01} mode. There is no singularity for β at $\omega_{ce}/\omega = 1$. Another important point is that the maximum value ω/ω_{pe} at which the wave can propagate increases with ω_{ce}/ω and, thus, the minimum plasma density n_{min} at which the wave can sustain the plasma decreases as B_0 increases.

Figure 8.3 shows β as a function of ω_{ce}/ω, at a given ω/ω_{pe} value, for the first ($m = 0$, ± 1) fundamental modes and the corresponding first magnetic modes with $n = 1$. We

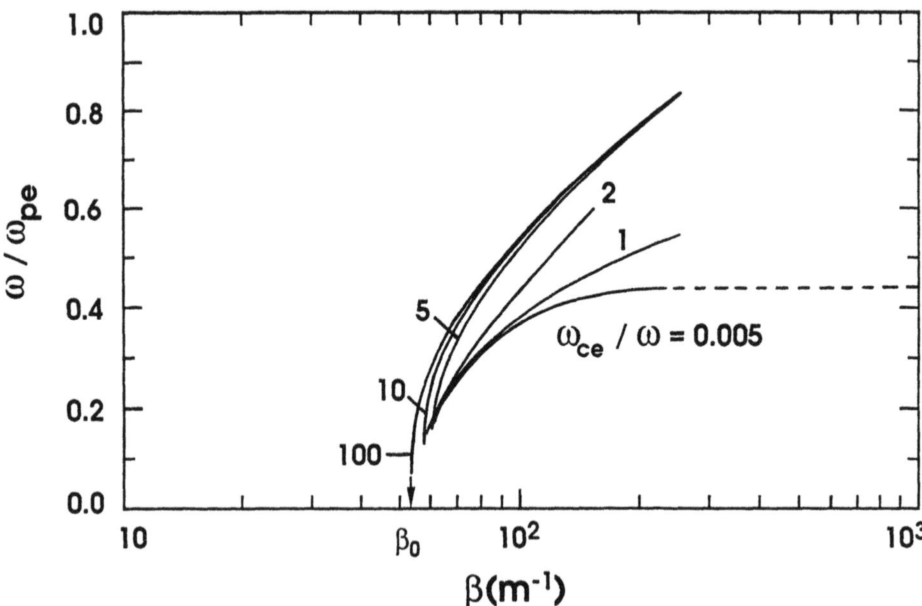

Fig. 8.2. Influence of ω_{ce}/ω on the phase diagram of the HE_{01} fundamental mode. Parameters of the calculation: $R_1 = 13$ mm, $R_2 = 15$ mm, discharge tube relative permittivity $\varepsilon_g = 4.52$, $\omega/2\pi = 2.45$ GHz, $\nu_{eff}/\omega = 10^{-3}$ (from ref. 1).

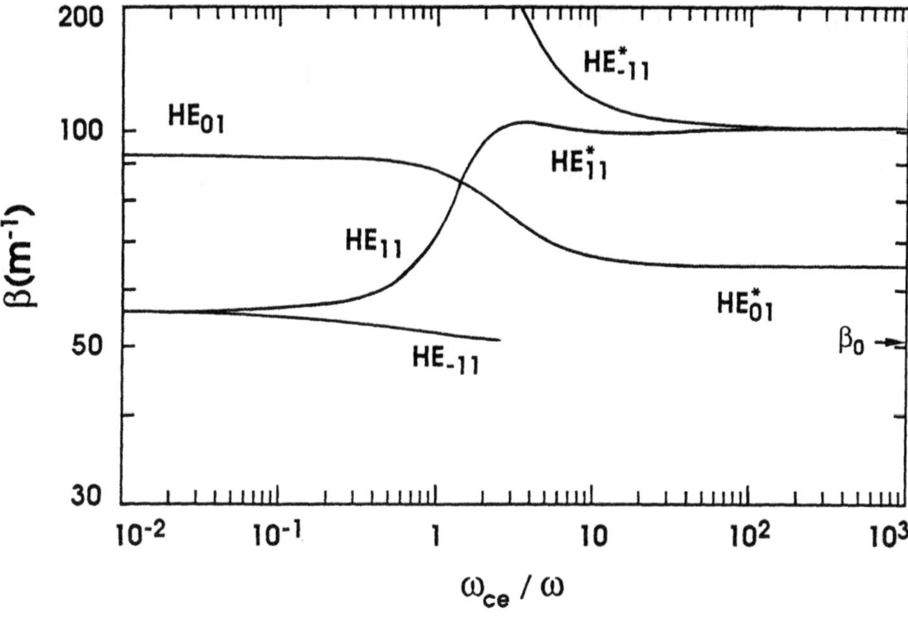

Fig. 8.3. Axial wavenumber β as a function of ω_{ce}/ω for the first ($m = 0, \pm 1$) HE$_{ml}$ fundamental modes and HE$^*_{ml}$ magnetic modes, at $\omega/\omega_{pe} = 0.35$ ($\omega/2\pi = 2.45$ GHz). Other conditions of the calculation as in Fig. 8.2 (from ref. 1).

see that the $m = 0$ and $m = 1$ azimuthal modes with a $n = 1$ radial configuration yield a β value that goes continuously across $\omega_{ce} = \omega$ as the intensity of B_0 is varied. As for the HE$_{-11}$ mode, with increasing B_0, it tends continuously toward $\beta_0 \equiv \omega/c$ (c is the speed of light in free space) for $\omega_{ce}/\omega \gtrsim 1$. Finally, the HE$^*_{11}$ mode exhibits a β value that strongly increases as ω_{ce}/ω decreases toward unity. Recall that the fact that β tends to infinity is characteristic of a wave resonance but this resonance is limited here by the inclusion of collisions in the permittivity ε_p. Figure 8.3 also demonstrates that although ε_1 and ε_2 tend to infinity at $\omega_{ce} = \omega$ ($\delta \to 0$), the waves then do not necessarily pass through resonance ($\beta \to \infty$).

8.3.4. Influence of ω_{ce}/ω on the attenuation coefficient α.

We remind that the diagrams that we are going to consider have been calculated assuming collisional absorption. Figure 8.4 shows no singularity in α for the fundamental modes at $\omega_{ce}/\omega = 1$ and we note that the HE$_{11}$ mode achieves the largest attenuation for $\omega_{ce}/\omega \geq 1$. As for the HE$_{-11}$ mode, its attenuation coefficient keeps on decreasing with increasing ω_{ce}/ω, attaining values at least one order of magnitude smaller than for the HE$_{01}$ and HE$_{11}$ modes. Recall that eqn. (8.5), for a given P(z) power (say at the wave launcher exit), requires the value of α to be large enough to yield a density $n \geq n_{min}$,

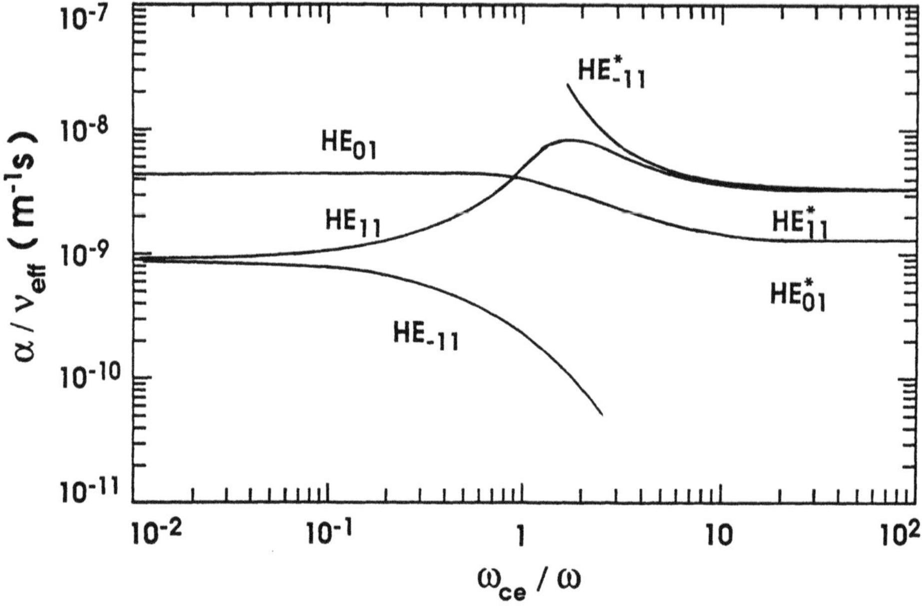

Fig. 8.4. Attenuation coefficient α normalized to the effective collision frequency ν_{eff} as a function of ω_{ce}/ω ($\omega/2\pi$ = 2.45 GHz) for the first ($m = 0, \pm 1$) HE_{m1} fundamental modes and magnetic HE^*_{m1} modes, at ω/ω_{pe} = 0.35. Other conditions of the calculation as in Fig. 8.2 (from ref. 1).

otherwise the plasma column cannot be sustained: this may occur with the HE_{-11} mode at large enough ω_{ce}/ω. On the other hand, the larger the local value $\alpha(z)$, the larger the plasma density $n(z)$ for a given $P(z)$. Finally, examining the HE^*_{11} mode, we see that α (as well as β) strongly increases as ω_{ce}/ω decreases toward unity, indicating that this mode would resonate ($\beta \to \infty$) in absence of collisions.

8.4. Characteristics of the wave electric field intensity in the plasma

As we just saw, varying B_0 can strongly affect β and α. It also influences the radial distribution of the total electric field intensity in the plasma, $E_p = \sqrt{|E_r|^2 + |E_\phi|^2 + |E_z|^2} = \sqrt{2}\, E_{rms}$. In what follows, we only need to consider the relative radial variation, or radial profile, of the maintenance field intensity.

8.4.1. Radial profile of E_{rms}. We limit our considerations to the $m = 0, \pm 1$ fundamental modes. *HE_{01} mode.* - Figure 8.5 shows the radial profile of E_{rms}, as normalized to its value at the wall. Whatever the ω_{ce}/ω value, the radial profile is akin to that of a surface wave with $B_0 = 0$: E_{rms} grows from the axis toward the wall. *HE_{-11} mode.* - The E_{rms} radial profile (not shown) is then similar to that of the HE_{01} mode.

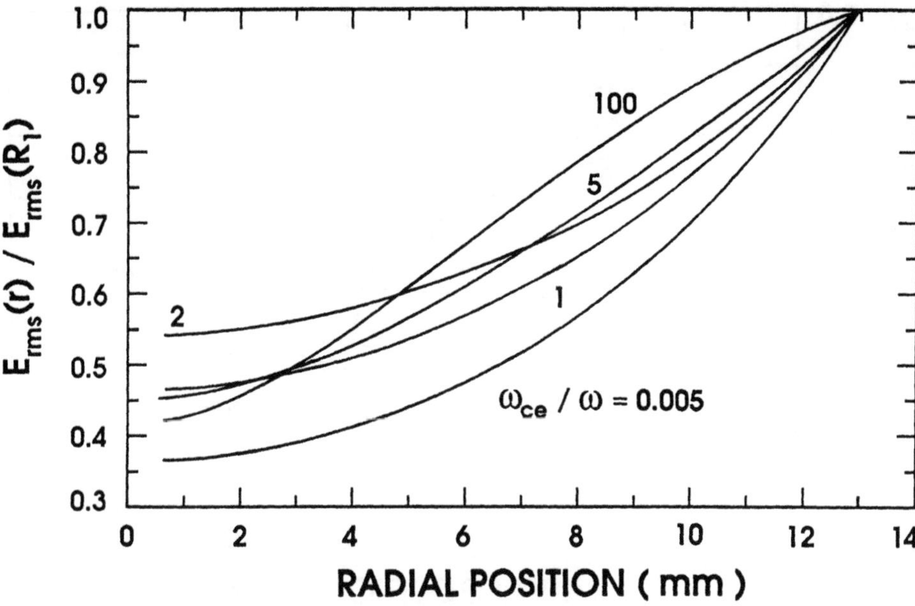

Fig. 8.5. Influence of ω_{ce}/ω ($\omega/2\pi$ = 2.45 GHz) on the radial profile of the plasma maintenance field intensity with the HE_{01} mode at ω/ω_{pe} = 0.35. Other conditions for the calculation as in Fig. 8.2 (ref. 1).

HE$_{11}$ mode. - Figure 8.6 shows that for $\omega_{ce}/\omega \geq 2$, under the present discharge conditions,[†] the field profile is no longer similar to that of a surface wave since E_{rms} is then maximum at the axis.

8.4.2. Influence of the electric field intensity radial profile on the radial profile of density of excited atoms and radicals.

We first recall the case of surface-wave-sustained plasmas with $B_0 = 0$. Under gas pressure conditions such that the particle mean free path between collisions is much smaller than the tube radius (for the field to be in local equilibrium with the plasma), it was then shown from theory,[5] assuming atom excitation by electron impact directly from the ground state, that the radial profile of excited atom density is well described by the empirical relation

$$n_k(r) / n_k(0) = [n(r) / n(0)] \, [E_{rms}(r) / E_{rms}(0)]^c, \qquad (8.10)$$

[†] The *discharge conditions* include: gas nature and pressure, tube radii R_1 and R_2, wave frequency $\omega/2\pi$ and mode (m, n), and B_0 intensity; the wave power is considered as a separate parameter.

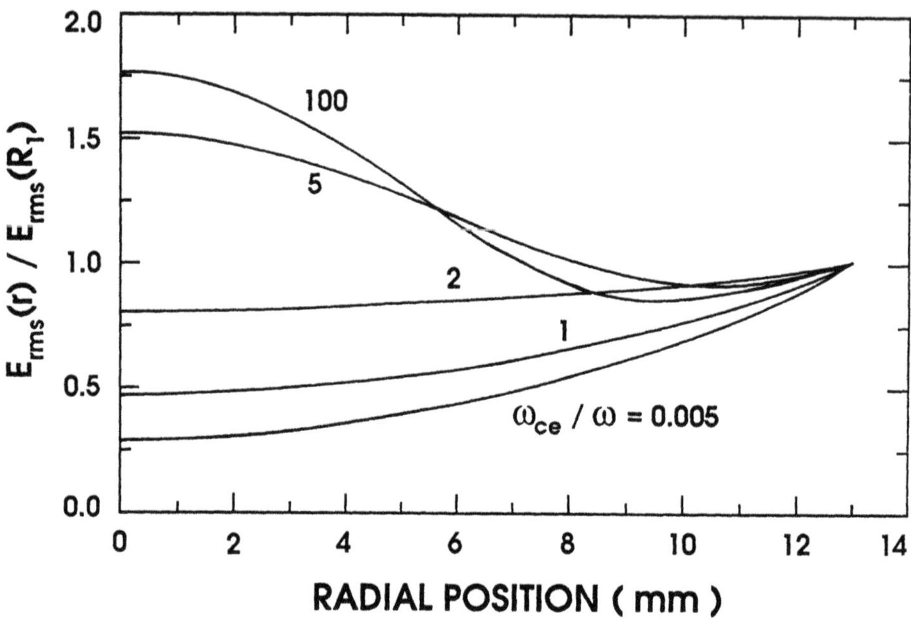

Fig. 8.6. Influence of ω_{ce}/ω ($\omega/2\pi = 2.45$ GHz) on the radial profile of the plasma maintenance field intensity with the HE_{11} mode at $\omega/\omega_{pe} = 0.35$. Other conditions for the calculation as in Fig. 8.2.

where k relates to the k - th energy level of the excited atom and c is a constant, fitted from numerical results of the Boltzmann equation, which increases with the energy of the level considered (e.g. $c = 2.9$ for the $3p^56d$ argon levels).[5] Since n is a decreasing function of radius while, in contrast, the electric field is an increasing one, eqn. (8.10) suggests (and this is supported by experiments[6]) that discharge conditions can be optimized to yield an almost constant radial profile of excited atom density.

When submitting surface-wave-sustained plasmas to a B_0 field, we may assume that eqn. (8.10) remains valid. The problem is to find the exact value of c since the corresponding theory is not fully developed yet. However, we have experimental, results that show what happens as a function of ω_{ce}/ω, as illustrated in Fig. 8.7. These results have been obtained with the HE_{01} mode at 200 mtorr of argon; curve no. 3 clearly shows that at $\omega_{ce}/\omega \simeq 1.1$, under the present discharge conditions, one can achieve an almost constant radial density profile of excited argon atoms in a given energy state. This method of obtaining radial uniformity with uncharged particles thus appears as an alternative to that which requires particle mean free paths between collisions to be much larger than the tube radius. The latter conditions are met under true ECR conditions where the gas pressure is of the order of a few mtorr or less. The alternate method presented here can be of importance in applications where the

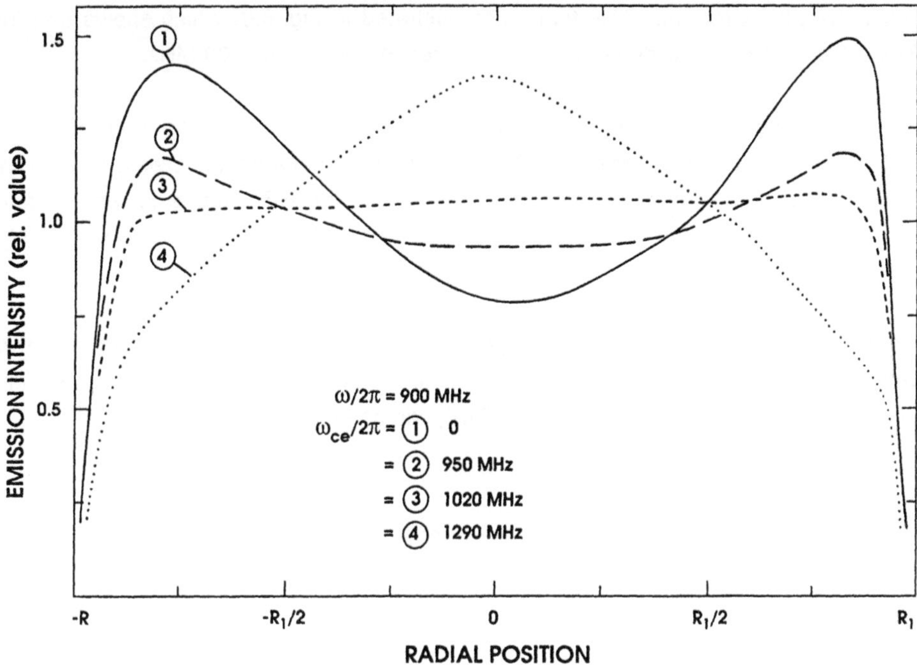

Fig. 8.7. Observed relative emission intensity of the Ar I 549.6 nm thin line as a function of radial position, in an argon (200 mtorr) plasma column (R_1 = 13 mm) sustained in the HE_{01} mode in an axial, static magnetic field of cyclotron frequency $\omega_{ce}/2\pi$ (from ref. 6).

required active (uncharged) species are more efficiently produced at 200 mtorr than at 1 mtorr. It would be interesting to carry similar experiments with the HE_{11} mode since, according to Fig. 8.6, its electric field intensity goes through a richer set of radial profiles than the HE_{01} mode (Fig. 8.5) as ω_{ce}/ω is varied.

8.4.3. Polarization of the wave electric field in the plane perpendicular to B_0.

We have seen that the polarization angle Θ, defined from the ratio of the complex values of E_r and E_ϕ (eqn. (6.23)), must be equal to $\pi/2$ and $|E_r|/|E_\phi|$ equal to unity (so-called right-hand circular polarization) for electrons to gain energy in their cyclotron motion. Summarizing the results of our numerical calculations at $\omega_{ce} = \omega$, for the fundamental modes (the magnetic modes do not exist then), we have that: i) whatever the azimuthal wavenumber m, these modes all yield $|E_r|/|E_\phi| \simeq 1$; ii) whatever the electron density and the radial position, for the various discharge conditions examined, we find the HE_{11} mode to be systematically right-hand circularly polarized ($\Theta = \pi/2$); iii) the polarization angle of the HE_{-11} mode varies from $-\pi/2$ to $\pi/2$ along the plasma radius; iv) the HE_{01} mode is right-hand circularly polarized provided

$(\omega/2\pi)\, R_1 \geq 0.6$ GHz-cm. This feature is illustrated in Fig. 8.8, which shows that for $R_1 = 1.3$ cm, Θ starts to depart from $\pi/2$ for frequencies below 500 MHz.

In conclusion, as far as achieving circular polarization is concerned, both the HE_{11} and HE_{01} modes can be employed but with certain reservations. Owing to stability conditions of $|m| = 1$ mode discharges,[7] the use of the HE_{11} mode is restricted to $(\omega/2\pi)\, R_1 \gtrsim 2$ GHz-cm whereas the HE_{01} mode has no such stability limit but requires $(\omega/2\pi)\, R_1 \gtrsim 0.6$ GHz-cm for Θ to be equal to $\pi/2$.

8.4.4. Breakdown and maintenance electric field intensities in the discharge.

The power absorbed by the electrons from the microwave field, per unit length of the discharge (eqns. (8.3) and (8.4)), can be expressed in the form

$$\frac{dP}{dz} = \bar{n}(z)\, \theta_A\, \mathcal{S} \equiv \frac{1}{2} \int_{\mathcal{S}} \mathrm{Re}(\mathbf{E}^* \cdot \mathbf{J})\, d\mathcal{S}, \tag{8.11}$$

where the asterisk denotes a complex conjugate and $\mathbf{J} = \underline{\sigma} \cdot \mathbf{E}$ is the current density. The electric conductivity tensor $\underline{\sigma}$ for a collisional plasma can be found in Sec. 6.4.1.

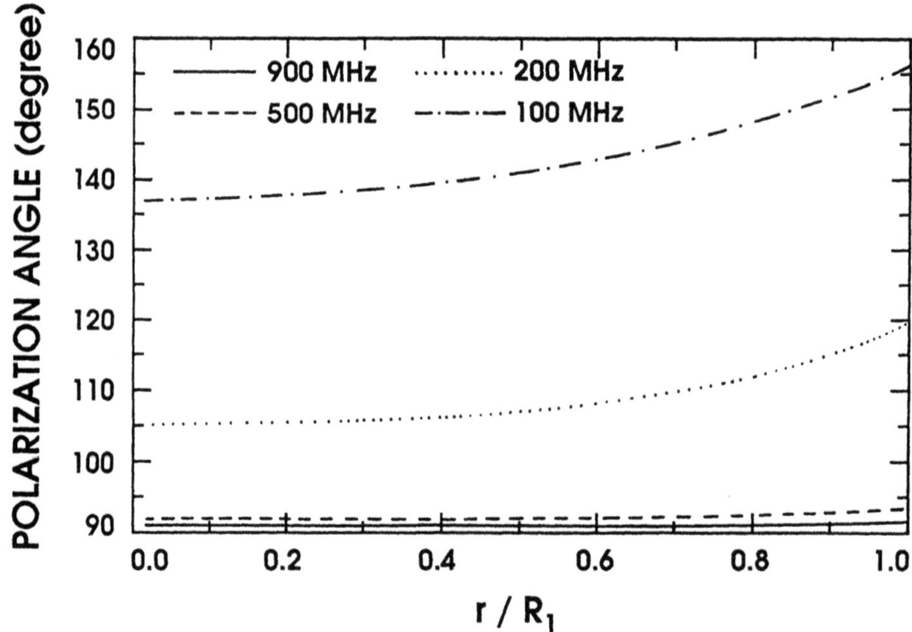

Fig. 8.8. Radial variation inside the plasma of the wave polarization angle Θ in the plane perpendicular to B_0, for the HE_{01} fundamental mode, for $\omega_{ce}/\omega = 1$ and $\omega/\omega_{pe} = 0.3$, at four wave frequencies. Other conditions of the calculation as in Fig. 8.2 (from ref. 1).

Alternately, $\underline{\sigma}$ being related to $\underline{\varepsilon}_p$ through $\underline{\varepsilon}_p = 1 - (j\underline{\sigma}/\omega\varepsilon_0)$, its elements can also be obtained from eqns. (8.7) and (8.8). In any case, substituting $\mathbf{J} = \underline{\sigma} \cdot \mathbf{E}$ in eqn. (8.11), we get

$$\left|\frac{dP}{dz}\right| = \frac{1}{2}\int_{\mathcal{S}} \{\mathrm{Re}(\sigma_1)\,[\,|E_r|^2 + |E_\phi|^2] + \mathrm{Re}(\sigma_3)|E_z|^2 - 2\,\mathrm{Im}(\sigma_2)\,|E_r E_\phi|\,\sin\Theta\}\,d\mathcal{S}, \qquad (8.12)$$

where Re and Im denote the real and imaginary part of a quantity respectively. Two of the three terms in eqn. (8.12) are related to components of the transverse electric field. From the expression of σ_2 in eqn. (6.20), one can check that $\mathrm{Im}(\sigma_2)$ is always negative. Thus, the contribution to the power transfer from the third term in eqn. (8.12) is maximum when $\sin\Theta = 1$, i.e. when the wave is right-hand (circularly or elliptically) polarized, whereas it is zero under linearized polarization ($\Theta = k\pi$). Finally, in the particular case where the electric field would only have an E_z component ($E_r = E_\phi = 0$), one obviously recovers the power transfer relation for an isotropic plasma.

To go on with our analysis of the power transfer, we note that eqn. (8.12) can be put in the form

$$\left|\frac{dP}{dz}\right| = \mathrm{Re}(\sigma_{eff})\,E_{rms}^2\,\mathcal{S}, \qquad (8.13)$$

introducing the novel concept of *effective electric conductivity* $\mathrm{Re}(\sigma_{eff})$. The use of this effective conductivity yields some insight into the power transfer without having to know the absolute value of E_{rms}, in contrast with the effective electric field representation. From eqns. (8.3), (8.5) and (8.13), we then arrive at

$$\mathrm{Re}(\sigma_{eff})\,E_{rms}^2 = \theta\,\bar{n}. \qquad (8.14)$$

Figure 8.9 shows $\mathrm{Re}(\sigma_{eff})$ as a function of ω_{ce}/ω for the HE_{01}, HE_{11} and HE_{-11} fundamental modes, as calculated for $\omega/2\pi = 2.45\ \mathrm{GHz}$, $R_1 = 1.3\ \mathrm{cm}$ and $\nu_{eff}/\omega = 10^{-3}$. One observes that the effective conductivity goes through a marked maximum at $\omega_{ce} = \omega$ with both the HE_{11} and HE_{01} modes whereas there is no trace of a maximum with the HE_{-11} mode. As a matter of fact, it could be shown that the closer the mode to a perfectly right-hand circularly polarized wave, the more intense is the effective conductivity peak. Figure 8.10 illustrates the influence of the wave frequency on $\mathrm{Re}(\sigma_{eff})$ in the case of the HE_{01} mode: too low a frequency with this mode and the conductivity resonance peak is lost.

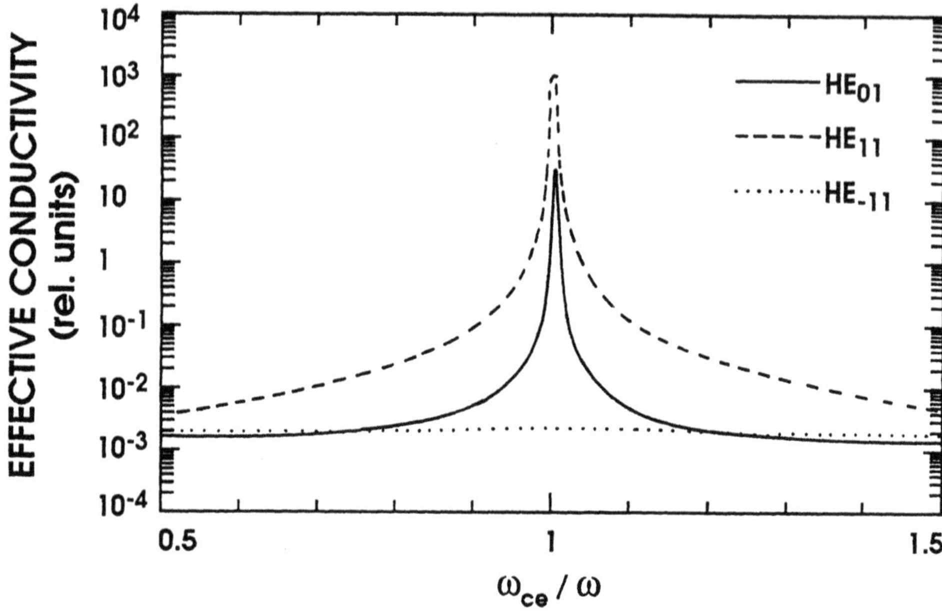

Fig. 8.9. Real part of the effective electric conductivity as a function of ω_{ce}/ω ($\omega/2\pi = 2.45$ GHz) for the HE_{01}, HE_{11} and HE_{-11} fundamental modes, at $\omega/\omega_{pe} = 0.3$. Other conditions of the calculation as in Fig 8.2.

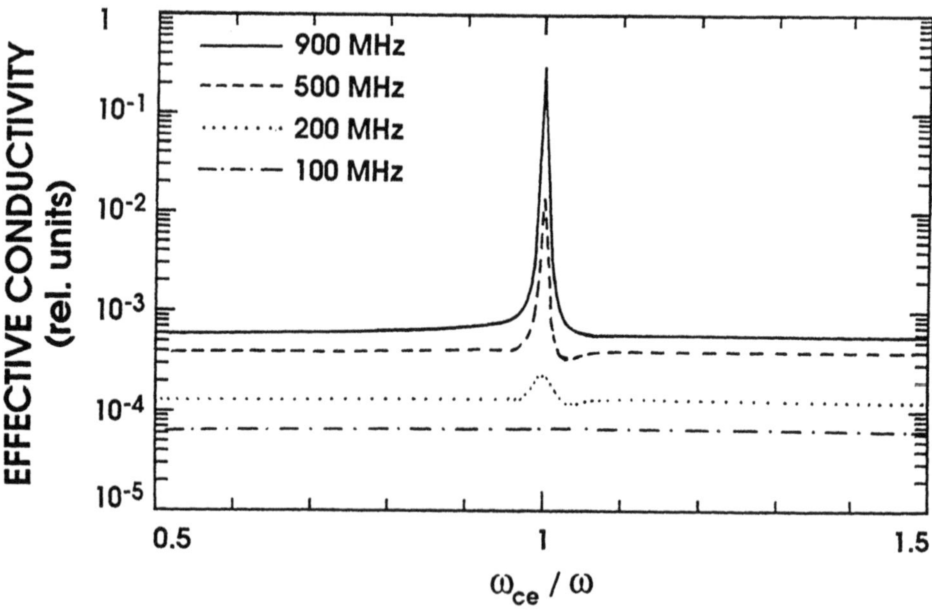

Fig. 8.10. Real part of the effective electric conductivity as a function of ω_{ce}/ω for the HE_{01} fundamental mode, at four wave frequencies $\omega/2\pi$, for $\omega/\omega_{pe} = 0.3$. Other conditions of the calculation as in Fig.8.2

The conductivity peak obtained here at $\omega_{ce}/\omega = 1$ is not associated with a larger power absorption in the discharge: this absorption, as shown by (eqn. 8.3), is governed by the attenuation coefficient, which does not go through a maximum at $\omega_{ce}/\omega = 1$ for these modes (see Fig. 8.4). It rather means that the electric field intensity required for the breakdown of the discharge or its maintenance is lower at $\omega_{ce} = \omega$, even in a collisional absorption regime[8]. To see this point, recall (Sec. 8.2.2) that θ is a monotonically decreasing function of increasing ω_{ce}/ω, with no extremum at $\omega_{ce}/\omega = 1$; the same behavior thus applies to the left-hand side of eqn. (8.14), for a given electron density. Thus, if $\mathrm{Re}(\sigma_{eff})$ in the left-hand side of eqn. (8.14) passes through a maximum at $\omega_{ce} = \omega$, then E_{rms} must go through a corresponding minimum. This shows that provided ν_{eff}/ω is not too large, the condition $\omega_{ce} = \omega$ leads to a reduction in the electric field intensity required to achieve breakdown or to sustain the discharge.

8.4.5.   Dependence of the maintenance electric field intensity E_{rms} on the electron temperature T_e.

For exposition purposes, we assume the EEDF to be Maxwellian, a distribution which is then fully characterized by its temperature T_e. From eqns. (8.11) and (8.14), we can write the power balance equation in the form

$$\theta_A \equiv \frac{1}{2\bar{n}S} \int_S \mathrm{Re}(\mathbf{E}^* \cdot \mathbf{J})\, dS = \frac{1}{\bar{n}S}\, \mathrm{Re}\,(\sigma_{eff})\, E_{rms}^2 = \theta_L. \tag{8.15}$$

In the case $B_0 = 0$, it can be shown that $\mathrm{Re}(\sigma_{eff})/\bar{n} \equiv \mathrm{Re}(\sigma)/\bar{n}$ is ultimately a function of discharge conditions only. As for θ_L, the power loss per electron, it is also a function of discharge conditions only,[9] these conditions determining T_e. Under steady-state conditions, for $B_0 = 0$, we thus have the functional relation

$$\theta_A\,(\text{disch.},\ E_{rms}^2) = \theta_L\,(\text{disch.},\ T_e), \tag{8.16}$$

where disch. stands for discharge conditions. This relation implies a biunivocal (one-to-one) relation between T_e and E_{rms}^2, the former determining the latter.

In the case $B_0 \neq 0$, $\mathrm{Re}(\sigma_{eff})/\bar{n}$ is a function of discharge conditions (comprising B_0) but eqn. (8.12) shows that it also depends on Θ and on E_s/E_t (s, $t = r, \phi, z$). Thus, under steady-state conditions, the power balance functional relation in the presence of B_0 is

$$\theta_A\,(\text{disch.},\ E_{rms}^2,\ \Theta,\ E_s/E_t) = \theta_L\,(\text{disch.},\ T_e), \tag{8.17}$$

which shows that E_{rms}^2 is not a biunivocal function of T_e: for example, for a given T_e value, the maintenance field depends on Θ and it is the lowest for $\Theta = \pi/2$.

When the EEDF is not a Maxwellian, the same conclusions are arrived at in terms of the average electron energy rather than T_e.

8.5. Summary and conclusion

By extending the wave and plasma analysis developed with surface-wave discharges to the case where these cylindrical columns are submitted to an axial, static magnetic field B_0, we have put into relief certain properties that apply to all HF sustained magnetoplasmas and we have identified the parameters that are of importance in modeling these discharges, as summarized below. Regarding our conclusions on ECR discharge mechanisms, we remind that they were obtained assuming collisional absorption and thus that they need to be completed by a kinetic model of the plasma to describe the collisionless absorption.

8.5.1. Identification and role of the main parameters in HF magnetoplasmas.

Power absorbed per electron, θ_A. Under steady-state conditions, this parameter is determined by the charged particle losses. It can also be considered as the power required to maintain an ion-electron pair in the discharge. It decreases with increasing B_0 due to the reduced diffusion losses to the wall. It does not depend on the wave properties, and it shows no extremum at $\omega_{ce} = \omega$.

Wave attenuation coefficient, α. The larger the coefficient α, the larger the power absorbed per unit length of the discharge. The local power balance shows that, given θ and the power flux $P(z)$, the plasma density increases with α. The value of α depends strongly on the wave field configuration and it varies with ω_{ce}/ω .

To emphasize the meaning of the above results on α, we consider the conceptual case where a given amount of power is dissipated into plasma but in several conditions, according to the value of the wave attenuation coefficient. Since θ is independent of the wave properties and provided θ does not vary with n, clearly the total number of electrons created in the various columns is the same. However, the larger the value of α, the smaller the volume of plasma within which these electrons are obtained.

Minimum electron density to sustain a discharge. When using guided plasma waves to sustain a discharge bounded by a dielectric material (surrounded or not by a metallic enclosure), the electron density must exceed a minimum value n_{min} for the wave to be able to propagate. This n_{min} value decreases with increasing ω_{ce}/ω but it generally does not go below n_c, the critical density. Taking this into account and the fact that θ decreases with increasing B_0, it means that the minimum power required to sustain a discharge decreases with increasing B_0.

Total electric field intensity for discharge breakdown and maintenance. To examine this matter as a function of ω_{ce}/ω, we have introduced the novel concept of effective electric conductivity. It permits one to determine what occurs at $\omega_{ce} = \omega$ without knowing the absolute value of the total electric field intensity, which is required though when using the effective electric field representation. From thus, we have shown that in the collisional absorption regime, provided $\nu_{eff}/\omega \ll 1$, the breakdown and maintenance fields go through a minimum at $\omega_{ce} = \omega$. Such a result agrees with the experimental fact that it is easier to initiate a discharge at $\omega_{ce} \approx \omega$. The sharpness of this minimum depends on the wave polarization. The important point to be made is that, in the collisional absorption regime, this phenomenon is not related to extrema in α and θ at $\omega = \omega_{ce}$.

Wave polarization angle Θ in the plane perpendicular to B_0. A given wave can always be decomposed into a right-hand and a left-hand circularly polarized wave, each having then a certain amplitude which depends on the nature of the wave considered. We have shown that the larger the contribution from the right-hand circularly polarized wave, the sharper is the maximum in the effective electric conductivity at $\omega_{ce} = \omega$.

Radial density profile of excited atoms and molecules, and of radicals. When operating specifically under collisional absorption conditions, such a density profile strongly depends on the radial profile of the total electric field intensity. The latter profile can be adjusted by selecting an adequate wave mode and varying ω_{ce}/ω, providing excited atom density distributions that are radially uniform. This method of achieving radially uniform profiles of uncharged active species at gas pressures much higher (typically 100 times larger) than that of usual ECR discharges, can be an interesting alternative solution to the latter discharges for uniform isotropic plasma processing when a higher yield of active neutral species is obtained at high pressures.

8.5.2. Conclusion. We have analyzed the plasma columns sustained in dielectric vessels by guided plasma waves in the presence of a static, axial magnetic field to determine the wave-to-plasma power transfer and to identify the main parameters of such discharges. We have also clarified certain aspects of ECR discharges. The overall results obtained show the potential, both theoretical and experimental, of such an approach.

References

[1]   J. Margot and M. Moisan, J. Phys. D: Appl. Phys. **24**, 1765 (1991).

[2]   C.M. Ferreira, J. Phys. D: Appl. Phys. **22**, 705 (1989).

[3]   M. Moisan, C. Barbeau, R. Claude, C.M. Ferreira, J. Margot, J. Paraszczak, A.B. Sá, G. Sauvé and M.R. Wertheimer, J. Vac. Sci. Technol. **B9**, 8 (1991).

[4]   J. Margot and M. Moisan, 8e Colloque sur les procédés plasmas (Antibes, France). Comptes rendus des travaux du CIP 91, pp. 47-58 (Société Française du Vide, Paris, 1991).

[5]   C.M. Ferreira and J. Loureiro, J. Phys. D: Appl. Phys. **16**, 2471 (1983).

[6]   M. Moisan, R. Pantel and A. Ricard, Can. J. Phys. **60**, 379 (1982).

[7]   J. Margot-Chaker, M. Moisan, M. Chaker, V.M.M. Glaude, P. Lauque, J. Paraszczak and G. Sauvé, J. Appl. Phys. **66**, 4134 (1989).

[8]   B. Lax, W.P. Allis and S.C. Brown, J. Appl. Phys. **21**, 1297 (1950).

[9]   C.M. Ferreira and M. Moisan, Phys. Scr. **38**, 382 (1988).

CHAPTER 9

INTEREST OF PLASMA CONFINEMENT AND ITS LIMITS *

9.1. Introduction

The previous chapters have been devoted to the production of plasmas from microwave sustained discharges that are generally not homogeneous in volume and most often small in size. However, plasma processing at the industrial level usually requires large volumes or large areas of homogeneous plasmas. The aim of this chapter is to analyze, based on the most simple assumptions, the various possibilities of confining charged particles to obtain homogeneous plasmas and, further, to identify the conditions required to achieve a uniform interaction between plasma and surfaces immersed into it.[†]

We begin the chapter by elementary theoretical considerations on the problem of homogeneity in collisional and non-collisional plasmas. Then, we examine the cases of infinite, semi-infinite and finite plasmas with their specific boundary conditions. In particular, we investigate the influence of the spatial distribution of the ionization sources within the discharge vessel. We also consider the effect of walls and substrates upon the plasma homogeneity and upon the uniformity of plasma interaction with surfaces. Next, we show that by essence multipolar magnetic field confinement leads to boundary conditions that improve plasma homogeneity. Finally, we examine the role of elastic and inelastic collisions and the limitations they introduce, as far as obtaining homogeneous dense plasmas or achieving a uniform and intense plasma interaction with surfaces is concerned.

9.2. Background

9.2.1. Collisionless plasmas. We consider a two-fluid (electrons and ions) description of the plasma from the most basic, simple physical assumptions: ionization and volume recombination as well as negative ions are ignored; the electron and ion temperatures are supposed to be spatially constant and the plasma to be unmagnetized. The plasma obeys the continuity and momentum equations for

---

[†] In the following, the word *homogeneous* will be used to characterize plasma parameters that are constant in the volume of the plasma. The word *uniform* will be employed for plasma parameters that are constant on a surface immersed in the plasma.

* Presented by J. Pelletier, Y. Arnal and M. Moisan

electrons and ions (where there are no collision related terms) while the electrostatic potential is governed by Poisson's equation. In a steady state, the complete set of equations is

$$\nabla \cdot n_e \mathbf{v}_e = \nabla \cdot n_i \mathbf{v}_i = 0, \tag{9.1}$$

$$n_e m_e \,(\mathbf{v}_e \cdot \nabla)\, \mathbf{v}_e + k_B T_e \,\nabla n_e + n_e \, e\mathbf{E} = 0, \tag{9.2}$$

$$n_i m_i \,(\mathbf{v}_i \cdot \nabla)\, \mathbf{v}_i + k_B T_i \,\nabla n_i - n_i \, e\mathbf{E} = 0, \tag{9.3}$$

$$\nabla \cdot \mathbf{E} = \frac{e}{\varepsilon_0} \,(n_i - n_e), \tag{9.4}$$

$$\mathbf{E} = -\nabla V, \tag{9.5}$$

where m_i and m_e are the ion and electron masses, n_i and n_e are the ion and electron densities, T_i and T_e are the ion and electron temperatures, $\mathbf{v}_i$ and $\mathbf{v}_e$ are the ion and electron *directed velocities* due to ∇n and $\mathbf{E}$, $\mathbf{E}$ being the electric field in the plasma, and V the corresponding electric potential.

Neutrality of the plasma implying $n_i = n_e$, its homogeneity then requires

$$n_i = n_e = n_h, \tag{9.6}$$

where n_h is a given density. Clearly, one possible solution to eqns. (9.1) to (9.4) is

$$\mathbf{v}_e = \mathbf{v}_i = 0 \tag{9.7}$$

and

$$\mathbf{E} = 0, \tag{9.8}$$

i.e. the plasma is isotropic and no space-charge field exists.

To show that the above solution $\mathbf{v}_e = \mathbf{v}_i = 0$ is the only solution, we consider the transition region between two plasmas of different densities. Since the electron mobility is larger than that of ions, a charge separation occurs. The electric field created by the space charge slows down the electrons and produces a decrease in the plasma potential while ions are accelerated towards the low density region. Expressions for plasma density and velocity in terms of selected plasma parameters can be obtained by integrating the continuity and momentum equations. A usual assumption, called the *congruence approximation* (Sec. 6.2.2), consists in equating charged particle fluxes, namely

$$n_e v_e = n_i v_i \, , \tag{9.9}$$

where $n_e v_e$ (and $n_i v_i$) is in general a function of spatial coordinates $\mathbf{r}$. When quasineutrality of the plasma is assumed everywhere including the transition zone ($n_e \approx n_i = n(\mathbf{r})$), eqn. (9.9) implies that the directed electron and ion velocities also have quasi-identical values ($v_i \approx v_e = v(\mathbf{r})$). Assuming that the directed velocity v_e is negligible with respect to the electron thermal velocity, the first term of eqn. (9.2) can be neglected with respect to all the other terms in eqn. (9.2) and to those of eqn. (9.3) since $m_e \ll m_i$. Hence, eqns. (9.1) to (9.3) can then be rewritten as

$$\nabla \cdot n\mathbf{v} = 0 \, , \tag{9.10}$$

$$k_B T_e \, \nabla n = en \, \nabla V \tag{9.11}$$

and

$$n \, m_i \, (\mathbf{v} \cdot \nabla) \, \mathbf{v} + k_B T_i \, \nabla n + en \, \nabla V = 0 \, . \tag{9.12}$$

Integration of eqn. (9.11) shows that electrons can be considered to form a neutralizing gas obeying the Boltzmann relation

$$n = n_h \exp \left[\frac{e \, (V - V_h)}{k_B T_e} \right] . \tag{9.13}$$

The correlative variation of the plasma velocity as a function of plasma potential or of plasma density is then obtained by integration of eqn. (9.12), yielding respectively

$$v^2 - v_h^2 = \frac{2e}{m_i} \, (V_h - V) \left(1 + \frac{T_i}{T_e} \right) , \tag{9.14}$$

and

$$n = n_h \exp \left(\frac{v_h^2 - v^2}{2 c_i^2} \right) , \tag{9.15}$$

where n_h, v_h and V_h in eqns. (9.13) to (9.15) are the density, velocity and potential of the plasma at some given point, and c_i is the ion-acoustic velocity or Bohm speed

$$c_i = \left(\frac{k_B T_e + k_B T_i}{m_i} \right)^{1/2} .$$

(9.16)

Equation (9.15) shows that the velocity gained during plasma expansion through the transition region is of the order of, or a few times the ion-acoustic velocity. However, when the plasma is assumed to be fully homogeneous ($n(r) = n_h$), then $v(r) = v_h = 0$ is the only solution of eqn. (9.15) compatible with eqn. (9.10), which governs the dependence of nv on spatial coordinates.[†] Hence, $v = 0$ is a necessary condition for plasma homogeneity in a collisionless plasma.

9.2.2. Collisional plasmas: the hypothesis of ambipolar diffusion. The starting hypotheses are the same as for collisionless plasmas: isothermal approximation throughout the plasma volume, no ionization, no ion recombination and a plasma free of negative ions. Therefore, the continuity equation (9.1) remains valid. Collisions are present through plasma diffusion and drift motion, and the diffusion equations for electrons and ions are then (see Sec. 2.2.6)

$$n_e v_e = - D_e \nabla n_e - n_e \mu_e E,$$

(9.17)

$$n_i v_i = - D_i \nabla n_i + n_i \mu_i E,$$

(9.18)

where D_e and D_i are the scalar diffusion coefficients for electrons and ions, and μ_e and μ_i are the associated electron and ion mobilities.

As with collisionless plasmas, we could show that the only solution compatible with plasma homogeneity ($n_e = n_i = n_h$) that satisfies eqns. (9.1), (9.4), (9.17) and (9.18) is also given by eqns. (9.7) and (9.8). In the case of a non homogeneous plasma such as at the transition between two plasmas of different densities, the hypothesis of quasi-neutrality ($n_e \approx n_i = n(r)$) together with that of congruence ($n_e v_e = n_i v_i$) leads to the equality of ion and electron velocities (recall that plasma neutrality and joint diffusion of electrons and ions are the basic hypotheses of perfect ambipolar diffusion). Thus, eqns. (9.17) and (9.18) can be rewritten in the classical form

$$nv = - D_e \nabla n - n\mu_e E,$$

(9.19)

---

† We exclude the solution $v = v_h$ in a one-dimensional configuration as being trivial.

$$n\mathbf{v} = -D_i \nabla n + n\mu_i \mathbf{E} \tag{9.20}$$

and the electric field is then given by

$$\mathbf{E} = -\frac{D_e - D_i}{\mu_e + \mu_i} \frac{\nabla n}{n}. \tag{9.21}$$

Since $D_e \gg D_i$ and $\mu_e \gg \mu_i$, the electric field reduces to the expression

$$\mathbf{E} = -\nabla V = -\frac{k_B T_e}{e} \frac{\nabla n}{n}, \tag{9.22}$$

which, by integration, again leads to the Boltzmann density distribution relation (eqn. (9.13)). Note that expression (9.22) does not depend on the electron flux, to the extent that the transport coefficients (D_i, μ_i) for ions are negligible with respect to those for electrons.

Substituting back eqn. (9.21) in eqn. (9.20) gives either

$$n\mathbf{v} = -D_a \nabla n, \tag{9.23}$$

where D_a is the ambipolar diffusion coefficient

$$D_a = \frac{\mu_e D_i + \mu_i D_e}{\mu_e + \mu_i} \approx D_i \left(1 + \frac{T_e}{T_i}\right) \tag{9.24}$$

or

$$\mathbf{v} = \mu_a \mathbf{E}, \tag{9.25}$$

where μ_a is the ambipolar mobility

$$\mu_a = \frac{\mu_e D_i + \mu_i D_e}{D_e - D_i} \approx \mu_i \left(1 + \frac{T_i}{T_e}\right). \tag{9.26}$$

The fact that the ambipolar diffusion coefficient (eqn. (9.24)) and the ambipolar mobility (eqn. (9.26)) ultimately depend only on ion coefficients clearly shows that it is the slowest species that limits the plasma flux. As we have already noted from eqn. (9.22), the dependence of $n(\mathbf{r})$ on plasma parameters is the same as in a collisionless plasma (eqn.(9.13)), the essential difference being that the plasma directed velocity in collisional plasmas is proportional to the electric field (eqn. (9.25)): thus, in the field-free region past the transition, the plasma directed velocity tends again to zero, illustrating the role of collisions in the slowing down of the electron and ion directed motion.

9.2.3. Boundary conditions: electrostatic sheaths.

Sheaths on passive electrodes. Electrostatic sheaths develop wherever a significant space charge ($n_e - n_i \neq 0$) appears in a plasma, as is the case at a plasma boundary. Walls, plasma sources, substrates, electrodes, probes, all introduce local electrostatic perturbations to the plasma. An important property of sheaths is their efficient screening, over a few Debye lengths, of any electric perturbation brought to the plasma. Thus, excluding a very limited region of space, this screening ensures that plasma neutrality is preserved. Nonetheless, there remains to find out to what extent given boundary conditions, and more generally, the introduction of an object in a plasma, can modify plasma characteristics, such as, for example, plasma potential or plasma density (see also Sec. 10.6.2).

For the sake of simplicity, in the following we only consider those surfaces in contact with plasma that have a curvature radius much larger than the Debye length, which allows them to be assimilated to plane surfaces. In the case where the hypotheses generally adopted in the theory of plane probes are valid, the results from such a model can therefore be applied to these surfaces. These hypotheses are as follows: i) the plasma is illimited, ii) the plasma is isotropic and the distribution of electron velocities is Maxwellian, iii) the sheath is collisionless (no collision, ionization or ion recombination within the sheath), iv) all charges impinging on the surface are collected, v) the charges impinging on the surface do not produce emission of secondary electrons. This set of assumptions defines a *passive electrode* (a Langmuir probe or a reactor wall, for example), as opposed to an excitation electrode where different boundary conditions prevail.

The current-voltage characteristics of an ideal plane probe are well known. They are summarized below in the case of a plasma free of negative ions. When the surface is biased positively with respect to the plasma potential V_p, the ions, assumed to be cold ($T_i \ll T_e$), are repelled and an electron sheath appears, a few Debye length thick. The electrons that cross the sheath edge are accelerated onto the surface ($V > V_p$) and collected. The corresponding saturation current density J_{es}, independent of the acceleration voltage, is proportional to the random thermal flux of electrons, yielding

$$J_{es} = - e n_e \frac{v_{T_e}}{4} = - e n_e \left(\frac{k_B T_e}{2\pi m_e}\right)^{1/2}, \tag{9.27}$$

where v_{T_e} is the average thermal velocity of electrons. When the surface is biased negatively ($V < V_p$), part of the electrons are repelled and an ion sheath appears; only those electrons that are energetic enough to cross the potential barrier $V - V_p$ are collected. The collected electron current density then reduces to

$$J_e = - en_e \left(\frac{k_B T_e}{2\pi\, m_e}\right)^{1/2} \exp\left[\frac{e(\,V - V_p\,)}{k_B T_e}\right]. \tag{9.28}$$

For positive ions, when $V < V_p$, the potential is attractive. The corresponding ion saturation current density J_{is} is obtained by application of the *Bohm criterion*.[1] This criterion states that the charge separation at the sheath edge corresponds to a potential difference (of the order of a fraction of $k_B T_e/e$) for which the relative velocity between electrons and ions is of the order of the ion-acoustic velocity c_i (eqns. (9.14) and (9.16)). For the sake of simplicity, we adopt†

$$J_{is} = en_i \left(\frac{k_B T_e}{m_i}\right)^{1/2} \tag{9.29}$$

as the ion saturation current density

Boundary conditions. Consider a plasma in contact with a surface. A likely boundary condition is to assume, as for probes, the neutralization of the charged species collected on the surface. In that respect, two positions (see Sec. 10.6.3) can be adopted: i) the first ignores the existence of a plasma sheath at the plasma-wall interface and assumes zero plasma density at the surface (Fig. 9.1a). This approach generally yields rather good qualitative results for plasma profiles, but may introduce problems such as a diverging electric field at the interface; ii) the second approach takes into account the plasma sheath at the interface between plasma and the surface (Fig. 9.1b). The problem of plasma homogeneity in the presence of a plasma sheath can be fully solved from general equations of electrons and ions (momentum and Poisson equations). Two cases can then be envisaged: i) the plasma is quasi-homogeneous and the current density values are those given by eqns. (9.27) to (9.29), ii) the plasma develops a density gradient towards the surface and thus the boundary conditions must include flux conservation.

Another important related question concerns the extent of the perturbation caused to the plasma as a whole by the introduction of an object at a given point. As discussed

---

† The Bohm criterion does not result from a rigorous calculation, and various approximations have been used in the literature for the derivation of the ion saturation current density. Expression (9.29), which does not fundamentally differ from others usually adopted, must clearly be considered as an approximation.

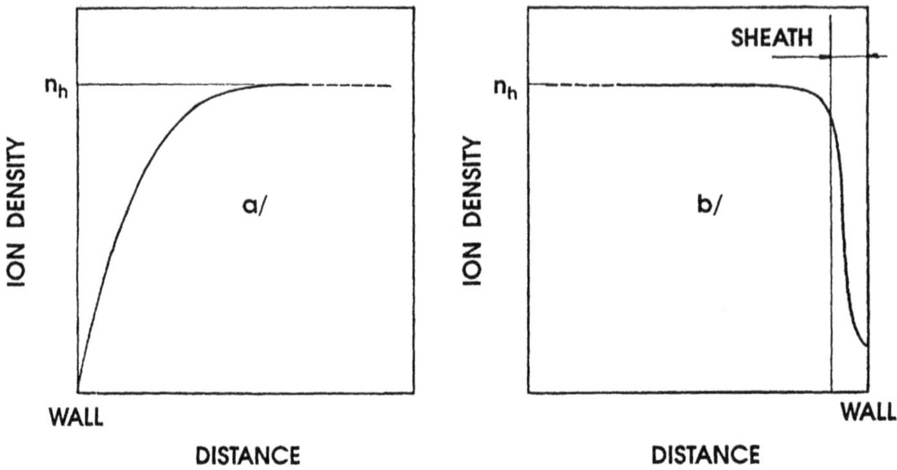

Fig. 9.1. Sketch of the ion density profile in the vicinity of a surface in contact with plasma: a) when ignoring the plasma sheath; b) when including it in the model.

later in this chapter and in the following chapters, we consider that a real perturbation results only when the object immersed in the plasma significantly modifies the balance between the loss and creation terms of the plasma. Expressed differently, perturbation to the plasma as a whole is generally not the result of local electrostatic perturbations, but is rather produced by the modification of the discharge balance.

9.3. Uniformity of plasma-surface interaction

9.3.1. Generalities. Now that we have presented and discussed the basic equations that determine the behavior of a plasma and its interaction with surfaces, the objective is to look for different means to achieve a homogeneous plasma or a uniform plasma-surface interaction. A uniform interaction with a plane probe (see Sec. 9.2.3) or with a surface that can be assimilated to a plane probe (curvature radius >> Debye length) can be obtained: i) when the plasma is homogeneous, under any conditions; ii) when the plasma is inhomogeneous, provided that the fluxes of charged species at the sheath edge are uniform.

The real question is thus to determine how to produce homogeneous plasmas or uniform fluxes of charged species. Solving this problem in the most general case (collisional *and* non-collisional plasmas, volume ionization and recombination), implies considering the continuity equation in its complete form

$$\nabla \cdot n\mathbf{v} = \mathbf{v} \cdot \nabla n + n\nabla \cdot \mathbf{v} = S_p, \tag{9.30}$$

where

$$S_p = S_i - S_r \tag{9.31}$$

is the net result (Sec. 2.2.3) between the ion creation rate S_i (ionization) and the ion loss rate S_r (ion recombination) in the volume of the plasma.

When $S_p = 0$, obtaining a homogeneous plasma generally requires (Sec. 9.2.1) a directed plasma velocity $\mathbf{v} = 0$. Indeed, if $\mathbf{v} \neq 0$, the plasma velocity (obtained by integration of $\nabla \cdot \mathbf{v} = 0$) is necessarily a function of spatial coordinates[†] and the plasma cannot be homogeneous, since eqn. (9.15) for collisionless plasmas or eqn. (9.23) for collisional plasmas indicates that any variation in the plasma velocity induces a variation in plasma density.

When $S_p \neq 0$, eqn. (9.30) cannot lead to solutions that are independent of coordinates, i.e. we cannot set $n(\mathbf{r})v(\mathbf{r}) = n_h v_h$ in eqn. (9.30). As a result, we see that for $S_p \neq 0$: i) homogeneous plasmas cannot be obtained and ii) the plasma flux cannot be independent of the spatial coordinates. Nonetheless, a uniform plasma-surface interaction could be achieved by using specific geometries or symmetries such as to obtain uniform fluxes onto given surfaces in contact with the plasma. The general results of the present section are now applied to infinite, semi-infinite and bounded plasmas.

9.3.2. Homogeneity in infinite plasmas. The notion of illimited homogeneous plasmas has often been used as a simple approach to theoretical problems. This is the case with Langmuir probes which are assumed not to perturb the homogeneous, illimited isotropic plasma to be characterized. This simplifying description nonetheless implies restrictive physical conditions. To determine them, we turn to the time-dependent equation of continuity

$$\frac{\partial n}{\partial t} + \nabla \cdot n\mathbf{v} = S_p(n,t) . \tag{9.32}$$

For an illimited, homogeneous isotropic plasma, since then $n(\mathbf{r}) = n_h$, $\mathbf{v} = 0$, it reduces to

---

[†] Except in the trivial case of a one-dimensional configuration.

$$\frac{\partial n}{\partial t} = S_p(n, t).$$
(9.33)

Thus, under steady-state conditions, an illimited homogeneous plasma requires

$$S_p \equiv S_i - S_r = 0.$$
(9.34)

We conclude that plasma homogeneity can then be achieved in two cases: i) there is equilibrium between ionization and recombination rates ($S_i = S_r$); ii) plasma is produced at infinity and its volume recombination rate is negligible ($S_i = S_r = 0$). However, as only plasma sources with finite dimensions are conceivable, we pass on to more realistic situtations.

9.3.3. Homogeneity in semi-infinite plasmas. This refers to infinite plasmas produced from plasma sources of finite dimensions. As a first example, we consider a plasma expanding from a spherical source. We assume that ionization and ion recombination rates outside the plasma source are negligible ($S_i = S_r = 0$) and that the plasma flux is purely radial. In spherical coordinates since then $\partial/\partial\phi = 0$, $\partial/\partial\chi = 0$, eqn. (9.10) reduces to

$$\frac{1}{r^2} \frac{\partial}{\partial r}\left(r^2 nv\right) = 0.$$
(9.35)

Integration yields

$$n(r)v(r) = n_h \, v_h \left(\frac{R}{r}\right)^2,$$
(9.36)

where n_h and v_h are the plasma density and the radial plasma velocity at the edge of the plasma source ($r = R$). Although we obtain an inhomogeneous plasma, a uniform plasma-surface interaction can still be achieved on "r = constant" surfaces.

When the plasma source is made up of only a portion of a sphere (Fig. 9.2), the plasma flux is generally not purely radial. However, in the absence of collisions, provided that the plasma expands radially at the source exit, the plasma flux still varies according to eqn. (9.36). An example of this situation is the expansion of a plasma generated by a neutralized ion beam (a few tens of eVs). The divergence of the beam at the extraction electrode is small and is given by[2]

$$\chi \approx \tan \chi = \left(\frac{k_B T_i}{e V_0}\right)^{1/2},$$
(9.37)

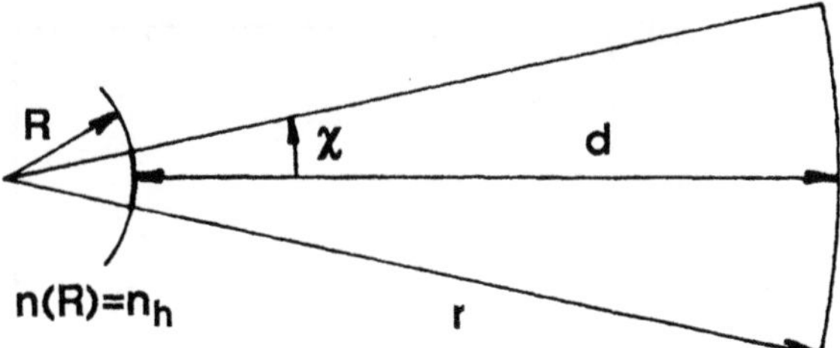

Fig. 9.2. Sketch characterizing a plasma expanding from a limited portion of a spherical source where χ is the polar angle defining the source aperture and d is the distance from the source exit (r = R).

where V_0 is the extraction voltage. In such a case, provided that the initial ion velocity v_0 is much greater than the ion acoustic velocity c_i, the beam velocity $v(r)$ can be assumed (eqn. (9.15)) to remain practically constant ($v \approx v_0$) during the expansion of the plasma. The density variation as a function of the distance d from the ion source (eqn. (9.36)) is given by

$$n = n_0 \frac{R^2}{(R+d)^2}. \tag{9.38}$$

Far from the source, the plasma density varies as the inverse of the square of the distance d from the ion source, as shown from experiment in Fig. 9.3.

The two examples considered above illustrate the fact that the free spatial expansion ($S_i = 0$) of a plasma from a finite, localized source cannot lead to a homogeneous plasma. Homogeneity can only be achieved with a one-dimensional uniform plasma source, assuming no volume recombination ($S_r = 0$) in the expanding plasma. In this particular case, the flux conservation yields

$$nv = n_h v_h, \tag{9.39}$$

i.e. we can have a homogeneous plasma with a non-zero directed velocity. However, since the infinite one-dimensional plasma cannot adequately approximate any realistic case, we examine below bounded plasmas.

9.3.4. Homogeneity in bounded plasmas. This case takes into account the finite size of laboratory plasma chambers. The problem of a uniform plasma-surface interaction in these plasmas will be considered on two examples, chosen for their

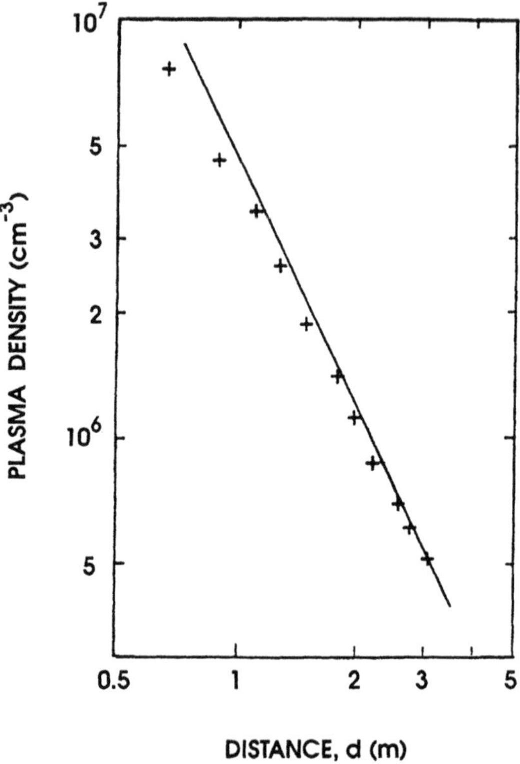

DISTANCE, d (m)

Fig. 9.3. Observed plasma density as a function of distance d from an aperture-limited source as in Fig. 9.2. Experimental conditions: argon discharge, expansion chamber pressure 10^{-5} torr, ion energy between 10 and 20 eV (from ref. 2).

prevalence. In both examples, we deal with an expanding plasma where the ionization and ion recombination rates are assumed to be negligible ($S_i = S_r = 0$).

As the first example, we consider a plasma expanding into a volume that is completely enclosed by a so-called peripheral plasma source of uniform density $n(R) = n_h$ (Fig. 9.4a).[†] Note that under steady-state conditions, and in the absence of volume recombination losses ($S_r = 0$), the *net plasma flux* entering the enclosed volume must be zero to avoid any infinite increase in plasma density. Equation (9.35), for example, shows that otherwise when $r \rightarrow 0$, then $n \rightarrow \infty$. This zero flux condition is consistent with the assumption that the density in an expanding plasma cannot exceed the density n_h in the source. Thus, the only physical solution which satisfies

---

[†] Such a peripheral plasma source corresponds to the case of DECR plasmas (see Chap. 14).

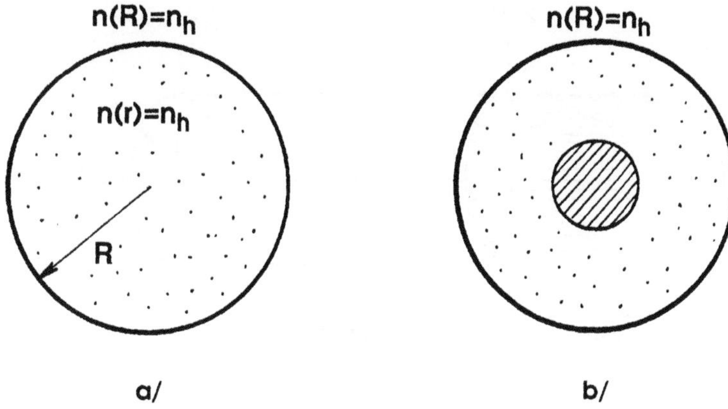

Fig. 9.4. Sketch showing a diffusion plasma generated at the periphery r = R with a density n(R) = n_h: a) diffusion in a free volume. The density of the enclosed plasma when S_p = 0 is then also n_h; b) case where a convex-shaped object of radius R_s is introduced (Appendix 9.1).

eqn. (9.10) and the boundary conditions n(R) = n_h whatever the geometry (excluding the one-dimensional case, which cannot yield a bounded plasma), is an isotropic homogeneous plasma ($\mathbf{v}(r) = 0$, n = n_h). We see that, in contrast to the case of plasma expansion from a limited aperture source (Sec. 9.3.3) where density decreases and plasma velocity is non-zero, a plasma (assuming S_p = 0) entirely confined within a volume surrounded by a peripheral plasma source is isotropic and perfectly homogeneous. However, if the volume of the expanding plasma is not entirely enclosed by the peripheral plasma source, plasma leaks out of this volume, and homogeneity is lost. In the same way, the introduction of an object into such a plasma (Fig. 9.4b) perturbs the plasma and homogeneity is again lost, except in the two limiting cases of small and large objects, as compared to the plasma mean free path and to the volume of the plasma respectively (see Appendix 9.1).

Our second example applies to many experimental devices:[3-5] a collisional plasma diffuses into a cylindrical chamber from a source located at one end (Fig. 9.5), and diffusion is assumed to be ambipolar. Under steady-state conditions and assuming S_p = 0, the expanding plasma obeys eqns. (9.10) and (9.23). Combining these equations, we see that plasma density is governed by

$$\nabla^2 n = 0. \tag{9.40}$$

Taking into account the existing cylindrical symmetry ($\partial/\partial\chi = 0$), eqn. (9.40) reduces to

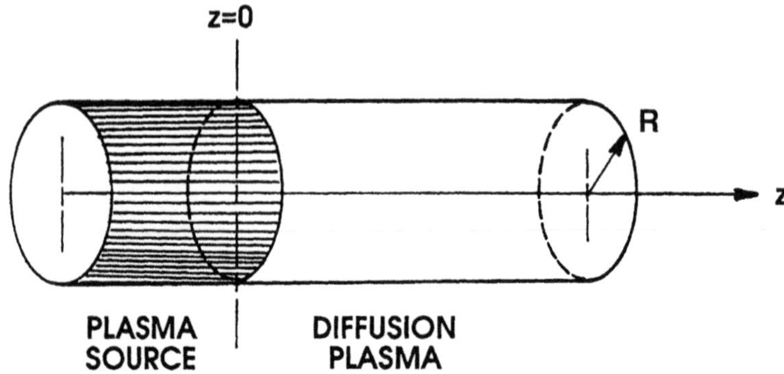

Fig. 9.5. Idealized configuration used to analyze plasma diffusion in a cylindrical chamber of radius R, assuming a plasma source of density $n_0 J_0(r/\lambda_a)$ at $z \leq 0$.

$$\frac{1}{r} \frac{\partial}{\partial r}\left(r \frac{\partial n}{\partial r}\right) + \frac{\partial^2 n}{\partial z^2} = 0, \tag{9.41}$$

where r and z are the radial and axial coordinates. The plasma source of radius R is located at $z = 0$, and the boundary conditions are

$$n(z=0, r) = n_0 J_0\left(\frac{r}{\lambda_a}\right) \tag{9.42}$$

and

$$v_z(z=0) = c_i. \tag{9.43}$$

We could show that the latter condition ensures flux conservation at the walls of the diffusion chamber. Then, the spatial evolution of the plasma density is given by

$$n = n_0 J_0\left(\frac{r}{\lambda_a}\right) \exp\left(-\frac{z}{\lambda_a}\right), \tag{9.44}$$

where J_0 is the zero-order Bessel function of the first kind and λ_a is the *ambipolar mean free path* defined as

$$\lambda_a = \frac{D_a}{c_i}. \tag{9.45}$$

This characteristic length is proportional to the ion mean free path λ_{in} (see eqn. (11.83) in Appendix 11.3) since we have

$$\lambda_a \approx \lambda_{in} \frac{c_i}{v_{T_i}},\qquad\qquad(9.46)$$

where v_{T_i} is the average thermal velocity of ions. The ion density exponential decrease along z predicted by eqn. (9.44) has been experimentally observed[4] as shown in Fig. 9.6, curve a. Figure 9.7 further shows the influence of the discharge pressure on plasma expansion.[3]

The previous example illustrates the difficulty of achieving uniform plasma-surface interaction under any given geometry. In fact, uniformity imposes in this case the use of

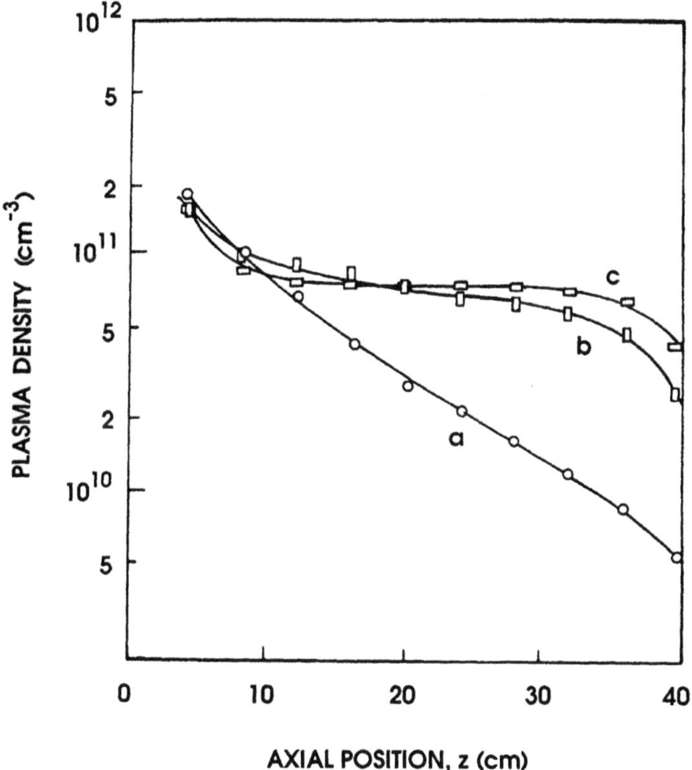

Fig. 9.6. Observed axial variation of plasma density in the diffusion region of a cylindrical chamber. The system configuration is similar to that in Fig. 9.5 with provision for multipolar magnetic confinement. The gas is argon at 0.5 mtorr: a) no magnets; b) radial confinement (side magnets only); c) radial confinement with side magnets and magnets added at the cylinder end opposite to the source (from ref. 4).

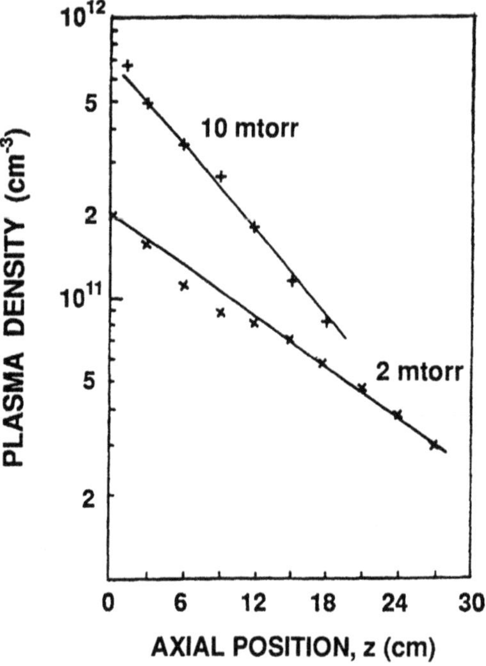

Fig. 9.7. Observed axial variation of plasma density in the diffusion region of a cylindrical chamber. The system configuration is similar to that in Fig. 9.5. The gas is argon and no magnetic confinement is used (from ref. 3).

specific symmetries or geometries for both the plasma source and the substrate surface. As we shall see, one solution to the present non-uniformity problem is to modify the boundary conditions, as suggested by our first example.

9.4. Plasma homogeneity through multipolar magnetic field confinement: experimental evidence

A way of modifying boundary conditions is to use magnetic field confinement, thereby reducing plasma loss on the chamber walls. The most common method is to magnetize the entire plasma volume. Usual configurations of this kind are the magnetic bottles found for example in multicharged ion beam systems, and the torus geometry of tokamaks used in fusion research. However, the bulk and cost of such devices are generally not really appealing for industrial applications. In addition, the anisotropy[†]

[†] A plasma can be homogeneous but nonetheless anisotropic, i.e. it exhibits, at any given point in space, mean particle velocities whose values depend on the direction considered with respect to the magnetic field lines.

imposed to the plasma by a generally non uniform magnetic field cannot lead to a uniform plasma-surface interaction. A more recent and innovative method is to confine plasma using multipolar magnetic field structures. This type of peripheral magnetic confinement, extensively described in Chap.10, offers the advantage of keeping the inner volume free from magnetic field, thus ensuring isotropy of the plasma. In the present section, we illustrate experimentally the influence of multipolar confinement in the two specific cases of Sec. 9.3 where the plasma is inhomogeneous in absence of confinement.

Figure 9.8 shows the improved homogeneity obtained in the neutralized ion beam example (Sec. 9.3.3):[2] the radial decrease in density is much less pronounced than without multipolar confinement (full line). In the case of a plasma diffusing into a cylindrical chamber (Sec. 9.3.4), Fig. 9.6 shows the improvement in plasma homogeneity and increase in density, obtained first when using magnets only along the

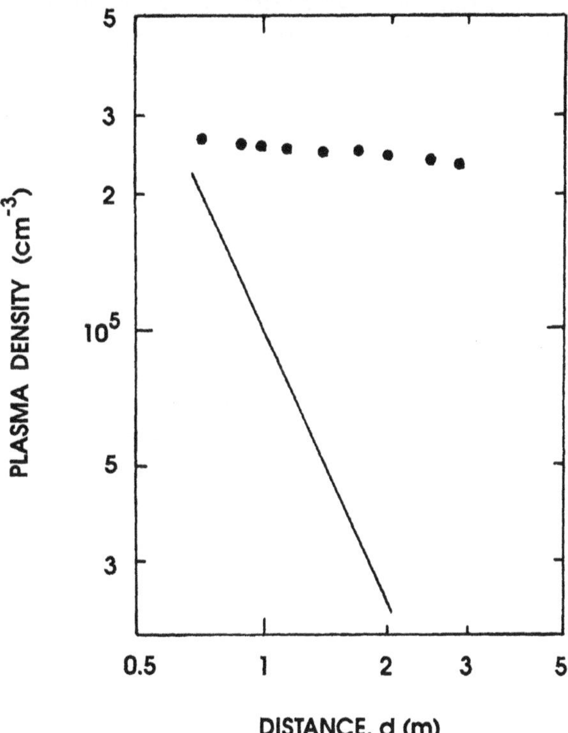

Fig. 9.8. Plasma density as a function of the distance d from an aperture-limited source as in Fig. 9.2. Experimental conditions: argon discharge, expansion chamber pressure 10^{-5} torr. The experimental values are obtained with multipolar magnetic confinement applied to the expansion chamber. The full line of slope -2 corresponds (see Fig. 9.3) to diffusion without magnetic confinement (from ref. 2).

cylinder generatrix (curve b),[4] and then when further adding magnets at the chamber extremity (curve c).[3,4]

The discharge conditions common to these two plasmas are: i) $S_i \approx 0$ (no energetic electrons); ii) low gas pressure (of the order of, or less than 1 mtorr), iii) a gas where the volume recombination rate can be assumed negligible ($S_r \approx 0$). In general, as we shall see in the next section, the condition $S_p = 0$ is required to have a homogeneous plasma.

9.5. Limits of plasma confinement

In high density plasmas, volume recombination of positive ions with electrons and negative ions (in short *ion recombination*) is probably the most prejudicial phenomenon that prevents achieving homogeneous plasmas because of the density gradients it introduces. Although generally negligible in plasmas generated in atomic gases or vapors (e.g. He, Ar, Cs, Hg), ion volume recombination is indeed likely to become a significant drawback for homogeneity in high density plasmas produced in molecular gases. We examine this matter by first listing the various charged particle recombination mechanisms.

Atomic ions. Atomic ion recombination can occur in three different ways, according to the domain of density and pressure considered:[6]

i) radiative recombination

$$A^+ + e^- \rightarrow A + \text{photon}, \tag{9.47}$$

where A^+ is an ionized A atom and e^- symbolizes the electron with which it collides. This reaction has a very low rate coefficient α_r, of the order of 10^{-12} cm^3s^{-1}. For example, $\alpha_r \approx 5 \times 10^{-12}$ cm^3s^{-1} for H^+ ions;[7]

ii) three-body recombination

$$e^- + A^+ + M \rightarrow A + M, \tag{9.48}$$

where M is the chaperon species. This process is only efficient in high pressure plasmas;

iii) collisional radiative recombination

$$e^- + e^- + A^+ \rightarrow A^* + e^- \rightarrow A + e^- + photon, \tag{9.49}$$

where A^* means an excited A atom. Such a recombination mechanism becomes important for plasma densities exceeding 10^{14} cm^{-3};

iv) dissociative recombination

$$A + A^+ \rightarrow A_2^+ \tag{9.50}$$

$$A_2^+ + e^- \rightarrow A_2 \rightarrow A + A. \tag{9.51}$$

The first three volume recombination mechanisms are generally negligible with respect to the fourth one in low pressure plasmas.[8] The latter mechanism is in turn negligible compared to diffusion losses for pressures below a few torr-cm.

Molecular ions. In molecular plasmas (often termed reactive plasmas), two mechanisms of ion volume recombination are known to be very efficient:

i) dissociative recombination

$$AB^+ + e^- \rightarrow A + B, \tag{9.52}$$

where AB^+ is an ionized AB molecule.

ii) mutual neutralization, which occurs in negative ion-rich plasmas

$$A^+ + B^- \rightarrow A + B. \tag{9.53}$$

These two mechanisms have rate coefficients of the order of 10^{-7} cm^{3}s^{-1}. For example, the rate coefficients for dissociative recombination of O_2^+ and N_2^+ are 2×10^{-7}cm^{3}s^{-1} and 2.7×10^{-7} cm^{3}s^{-1} respectively,[9,10] while the rate coefficients for the mutual neutralization of O_2^+ or O^+ with O_2^- or O^- are all of the order of 10^{-7} cm^{3}s^{-1}.[11]

In order to estimate the importance of the inhomogeneity effects introduced by ion volume recombination, we need to calculate the corresponding density profile for a plasma assumed to be homogeneous in the absence of such a recombination. For this purpose, we come back to the one-dimensional case (Sec. 9.3.3) where a plane source (located at $x = 0$) produces a uniform plasma ($n(x) = n_h$). In a collisional plasma where molecular ions dominate over atomic ions and where ion volume recombination losses dominate over ionization rate ($S_r \gg S_i$), eqns. (9.23) and (9.30) lead to

$$\nabla^2 n = \frac{S_r}{D_a} = \frac{\alpha_r}{D_a}\, n^2. \tag{9.54}$$

The corresponding spatial density profile is obtained by integration and we have

$$n = n_h \left[1 + \left(\frac{n_h\,\alpha_r}{6\,D_a}\right)^{1/2} x \right]^{-2}. \tag{9.55}$$

This equation shows the spatial inhomogeneity introduced by the rate coefficient α_r. More specifically, plasma density is decreased by a factor of 4 at a distance

$$x_{1/4} = \left(\frac{6\,D_a}{n_h\,\alpha_r}\right)^{1/2}. \tag{9.56}$$

For example, taking $\alpha_r = 10^{-7}$ cm^{3}s^{-1} in a molecular plasma and $D_a \approx 6 \times 10^5$ cm^{2}s^{-1}, we obtain $x_{1/4} = 60$ cm for $n_h = 10^{10}$ cm^{-3} and $x_{1/4} = 6$ cm only for $n_h = 10^{12}$ cm^{-3}. In atomic gas plasmas, where the rate coefficient $\alpha_r \approx 10^{-11}$ cm^{3}s^{-1} is many orders of magnitude smaller, ion volume recombination can be neglected since we find $x_{1/4} \approx 6$ m for $n_h = 10^{12}$ cm^{-3}. These results clearly show that it is more difficult to achieve homogeneity in molecular gas (reactive) plasmas than in atomic gas plasmas, and that this difficulty increases with increasing plasma density.

9.6. Conclusion

We have investigated the problem of plasma homogeneity and of uniform plasma-surface interaction. Our theoretical approach, illustrated and corroborated by a few selected examples, has shown that the conditions ensuring plasma homogeneity are rather restrictive. We have also seen that multipolar magnetic field confinement can improve significantly the homogeneity of diffusion plasmas. Nevertheless, we must conclude that in high density molecular gas plasmas, ion recombination will always be a major obstacle on the way to perfect homogeneity.

The confinement mechanisms of low pressure plasmas using multipolar magnetic fields are certainly worth examining in more detail. This is the subject of the five following chapters where we consider various plasma sources: hot cathode discharges (Chaps. 10 to 12) and microwave sustained plasmas (Chaps. 13 and 14).

APPENDIX 9.1

PERTURBATION OF A PLASMA BY A SUBSTRATE

We consider a plasma enclosed within a spherical ionization source of radius R (Fig. 9.4a) and of uniform density

$$n(R) = n_h. \tag{9.57}$$

When the ionization and ion volume recombination rates are assumed to be negligible ($S_i = S_r \approx 0$), the expanding plasma in the closed volume is isotropic and homogeneous since $\mathbf{v(r)} = 0$, $n(\mathbf{r}) = n_h$ (Sec. 9.3.4).

Let us introduce a spherical substrate of radius R_s into the plasma, in a concentric position with respect to the plasma source (Fig. 9.4b). We consider a collisional plasma and we assume that diffusion of the plasma is ambipolar. Under steady-state conditions, plasma diffusion obeys eqns. (9.10) and (9.23). Taking into account the symmetry of the problem, the plasma parameters only depend on the radial coordinate r. Thus, eqn. (9.40), which governs the plasma density, reduces to

$$\frac{1}{r^2}\frac{\partial}{\partial r} r^2 \frac{\partial n}{\partial r} = 0. \tag{9.58}$$

The boundary conditions are constituted by eqn. (9.57) and by the assumption of ion flux conservation at $r = R_s$ (Sec. 9.2.3). The latter condition requires the equality of the ion flux through the plasma sheath (eqn. (9.29)) with that of the plasma flux at the sheath edge (eqn. (9.23)). It yields

$$n(R_s)\, c_i = D_a \left(\frac{\partial n}{\partial r}\right)_{r=R_s}. \tag{9.59}$$

The solution of eqn. (9.58) is thus

$$n = n_h\left[1 - \frac{R\,R_s^2}{\lambda_a R + R_s (R - R_s)}\left(\frac{1}{r} - \frac{1}{R}\right)\right], \tag{9.60}$$

and the plasma density at the substrate ($r = R_s$) is given by

$$n(R_s) = n_h \frac{\lambda_a R}{\lambda_a R + R_s (R - R_s)} , \qquad (9.61)$$

where λ_a is the ambipolar mean free path defined in eqns. (9.45) and (9.46). Equations (9.60) and (9.61) show the density distribution $n(r)$ and the corresponding value $n(R_s)$ at $r = R_s$: the plasma density decreases with decreasing r, down to $n(R_s)$ at the substrate. The study of the different limiting cases will now prove very instructive.

The case of the infinite plasma can be obtained by setting R to infinity in eqns. (9.60) and (9.61). The radial variation of plasma density, and the density at the substrate surface are then

$$n = n_h \left[1 - \frac{R_s^2}{r(\lambda_a + R_s)} \right], \qquad (9.62)$$

$$n(R_s) = n_h \frac{\lambda_a}{\lambda_a + R_s} \qquad (9.63)$$

respectively. Equations (9.62) and (9.63) show immediately that the introduction of an object in a homogeneous infinite plasma (Sec. 9.3.2) perturbs the plasma unless $\lambda_a \gg R_s$.[†]

The next case is that of a finite size plasma where the radius of the substrate is varied. The influence of a large spherical substrate, as compared to the volume of the plasma, is obtained by letting R_s tend towards R. The plasma density, including that at the substrate surface, is then constant and equal to n_h (eqn. (9.60)), whatever the plasma condition (reminder: $R_s < r < R$). In fact, this result could be expected by considering this situation to be equivalent to the one-dimensional case.

As we have seen for the infinite plasma, the case $R_s \ll \lambda_a$ for a finite size plasma also leads to an homogeneous plasma of constant density $n(r) = n_h$. This situation applies to spherical Langmuir probes for example. More generally, this result shows that the perturbation brought to the plasma by the introduction of a substrate decreases when λ_a increases, i.e. when the plasma pressure decreases. The lower the pressure, the smaller the perturbation of the plasma.

Equation (9.60) shows that the perturbation of the plasma (in the geometry adopted) is maximum for

---

[†] In the case of a plasma with a non zero ion creation rate $(S_i \neq 0)$, perturbation of the plasma density becomes significant when the size of the object is larger than the plasma mean free path.[12]

$$R = 2R_s. \tag{9.64}$$

The density at the substrate surface is then at its minimum value as eqn. (9.61) yields

$$n(R_s) = n_h \frac{2\lambda_a}{2\lambda_a + R_s}. \tag{9.65}$$

Clearly, as in the previous examples, $n \approx n_h$ when $\lambda_a >> R_s$.

The different limiting cases investigated have shown that the lower the plasma pressure, the larger the ambipolar mean free path λ_a, and the more homogeneous the diffusion plasma within an enclosed volume.

References

[1] D. Bohm, *Characteristics of Electrical Discharges in Magnetic Fields*, A. Guthrie and R. K. Wakerling eds. (Mc Graw Hill, New York, 1949).

[2] Y. Arnal, Engineer Doctor Thesis, Université d'Orléans, France (1977), unpublished.

[3] C. Pomot, B. Mahi, B. Petit, Y. Arnal and J. Pelletier, J. Vac. Sci. Technol. **B4**, 1 (1986).

[4] R.W. Boswell, A.J. Perry and M. Emanl, J. Vac. Sci. Technol. **A7**, 3345 (1989).

[5] C. Charles, R.W. Boswell, A. Bouchoule, C. Laure and P. Ranson, J. Vac. Sci. Technol. **A9**, 661 (1991).

[6] J.B. Hasted, *Physics of Atomic Collisions*, J.B. Hasted ed. (Butterworths, London, 1972).

[7] D.R. Bates and A. Dalgarno, *Atomic and Molecular Processes*, D.R. Bates ed. (Academic, New York, 1962).

[8] A.B. Sá, Doctoral Thesis, Technical University of Lisbon, Portugal (1988), unpublished; A.B. Sá and C.M. Ferreira, ESCAMPIG 1988, Europhysics Conf. Abstracts **12H**, 191 (1988).

[9] E. Alge, N.G. Adams and D. Smith, J. Phys. **B16**, 1433 (1983).

[10] W.H. Kasner and M.A. Biondi, Phys. Rev. **137**, A317 (1965).

[11] W.H. Aberth and J.R. Peterson, Phys. Rev. **A1**, 158 (1970).

[12] J.F. Waymouth, Phys. Fluids 7, 1843 (1964).

CHAPTER 10

DISCHARGES CONFINED BY MULTIPOLAR MAGNETIC FIELDS *

10.1. Introduction

Multipolar magnetic field confinement of electrically charged particles was suggested two decades ago as a means of providing homogeneous plasmas of high ion density in a laboratory. The initial idea originated within the fusion research community where quadrupole magnetic field confinement served to produce energetic ion beams.[1,2] Nowadays, it has become a center of interest for scientists engaged in uniform surface treatment by plasmas.

This chapter is the first in a series of four which all relate to multipolar magnetic field confinement. This starting chapter is intended to get the reader progressively acquainted with the basic experimental aspects and properties of these discharges, termed in short multipolar discharges. We first review the pioneering investigations in that field and compare several magnetic field arrangements. Then, we describe the different regimes of the filament-excited multipolar discharge, the first multipolar plasma source to be developed. Next, we discuss the various aspects of the modeling of hot cathode discharges. We first verify the impossibility of confining electrons by using electrostatic confinement alone. Then we turn to an ambipolar diffusion model where the ionization by energetic electrons (the so-called primary electrons) is taken into account. Finally, we examine the influence of these electrons on the observed enhancement of plasma characteristics related to multipolar magnetic confinement.

10.2. Pioneering works

10.2.1. The spherical multipole: the first device. By 1965, different so-called minimum-B field traps (cusp-type and mirror-type[†]) were widely used to confine hot plasmas. In these configurations, the magnetic field intensity increases outwards from the confined plasma. Because they are open systems, they suffer at the ends from important losses in charged particles. Closing the magnetic system into a torus was not

---

† These are both minimum-B field configurations since there is an increasing magnetic field at the open ends to ensure reflection of charged particles. In the mirror-type system, however the whole plasma volume is submitted to a magnetic field while it is not necessarily the case with cusp-type devices.

* **Presented by R. Burke and J. Pelletier**

initially considered to be an attractive solution, as the toroidal field alone does not provide a "true" minimum - B field.

Since the particles that escape these systems are those with velocities parallel to the magnetic field lines, Sadowski[3-6] suggested a new magnetic field topology to reduce the corresponding loss cones. Termed the spherical, high-order multipolar magnetic field, it is realized by placing magnetic dipoles on a sphere's surface in a given symmetrical way. The magnets that surround the spherical chamber are reminiscent of quills on a porcupine or prickles on a cactus. Sadowski preferred the latter mimic and dubbed his research project "Kaktus".

Spherical symmetry in locating the magnets is obtained by inscribing a regular polyhedron in the sphere. Magnets of a given polarity are placed at the tips of the polyhedron, magnets of the other polarity being placed at the center of the polyhedron faces on the sphere. In the Kaktus project, a good compromise between the volume of the chamber (6 liter) and the space taken up by the electromagnets is the icosahedron geometry: 20 faces and 12 tips, as shown in Fig. 10.1. Twelve north poles are located at the apices of the icosahedron, and twenty south poles at the center of the faces. Losses through the thirty narrow loss cones, symmetrically situated on the sphere's surface, are significantly lower than the losses through the linear or circular loss gaps of

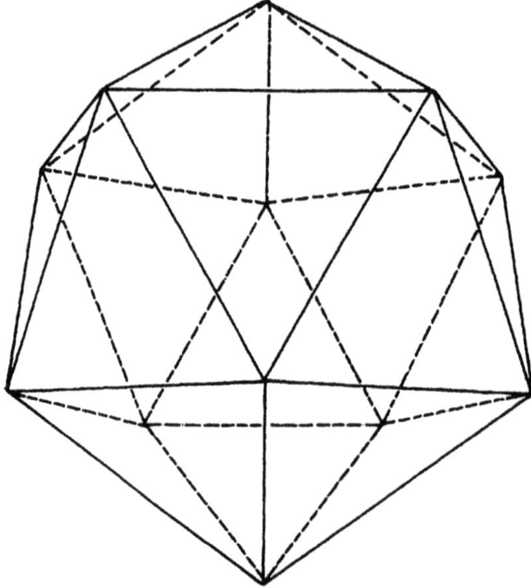

Fig. 10.1. The icosahedron: twenty sides and twelve apices. The north poles are located at the apices, and the south poles at the center of each face.

previous minimum-B designs. As a result, plasma containment time was increased threefold over a comparable conventional cusp trap.[5]

The spherical, high-order multipolar magnetic field configuration was ultimately abandoned as a confinement geometry for fusion, because fusion research at high energy progressed along different directions. Nevertheless, the important result obtained with this first multipolar device, a threefold extension of the plasma containment time due to decreased cusp losses, was still present in the minds when a few years later Limpaecher and MacKenzie took a closer look at the characteristics of multipolar plasmas.

10.2.2. Multipolar plasmas: the first detailed study

Limpaecher and MacKenzie[7] applied the multipolar confinement principle to obtain dense, large-volume, quiescent and, in particular, homogeneous plasmas of interest for research on plasma physics. Since it was no longer the intention to confine ions for fusion purposes, small permanent magnets (field strength of the order of a kgauss) could then be used to create the multipolar magnetic field and confine the plasma. With regard to cost and convenience, electromagnets obviously cannot compete with permanent magnets.

For more than a decade following their initial work, the excitation of multipolar plasmas was almost exclusively ensured by electron emission from one or more heated filaments (cathode assembly), which are negatively biased with respect to the multipolar magnetic structure and the chamber wall (anode). Even today, one can still find filament-excited multipolar discharges being used in some applications, such as the generation of H^- beams.[8] A thorough description of the filament-type plasma excitation follows in the fourth section of this chapter.

The first multipolar plasma chamber, used by Limpaecher and MacKenzie,[7] was a small (1 liter), rectangular parallelepiped, fitted with a checkerboard configuration of magnets, as shown in Fig. 10.2. The improvement in plasma characteristics over an identical unmagnetized structure was so dramatic that a cylindrical 86 liter device was immediately commissioned. The new magnet arrangement consisted of rings of alternating polarity, perpendicular to the cylinder axis, as described in Fig. 10.3. Other magnetic field configurations will be discussed in Sec. 10.3, but first let us list the characteristics of Limpaecher and MacKenzie's multipolar plasmas.

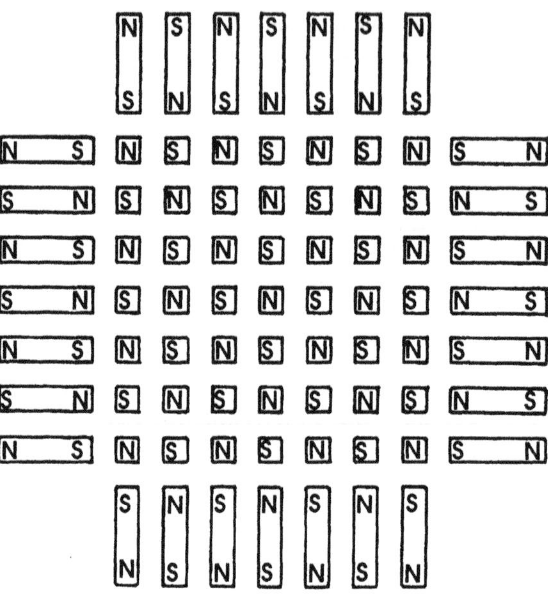

Fig. 10.2. End view of a rectangular parallelepiped chamber wall with a checkerboard arrangement of north and south poles (from ref. 7).

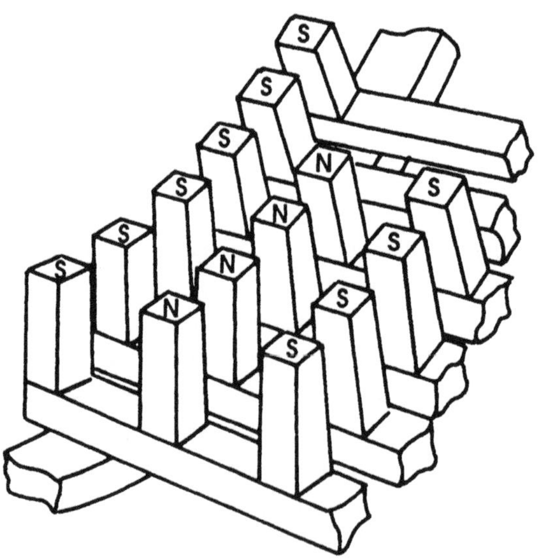

Fig. 10.3. Ring arrangement of north and south poles around a cylindrical chamber (from ref. 7).

Unmagnetized plasma. Most of the plasma volume yielded by multipolar magnetic field confinement is free from magnetic field, since the field generated by the small permanent magnets drops rapidly with distance from the magnets (see Appendix 11.4).

Enhanced plasma density. This is the most spectacular characteristic of multipolar plasmas. Compared to the same device without the multipolar magnetic field, and for a given power level of filament emission, the magnetic confinement can increase the plasma density by as much as two orders of magnitude. Figure 10.4 shows the electron current (proportional to plasma density) collected from an argon plasma at the center of the aforementioned parallelepiped chamber, as a function of pressure.[7] The magnetically contained plasma is much denser than the uncontained plasma, especially at low pressure. Notice also that it is feasible to strike the magnetically contained plasma at much lower pressures than in the absence of the multipolar field. At still lower pressures, however, one eventually reaches a point where insufficient

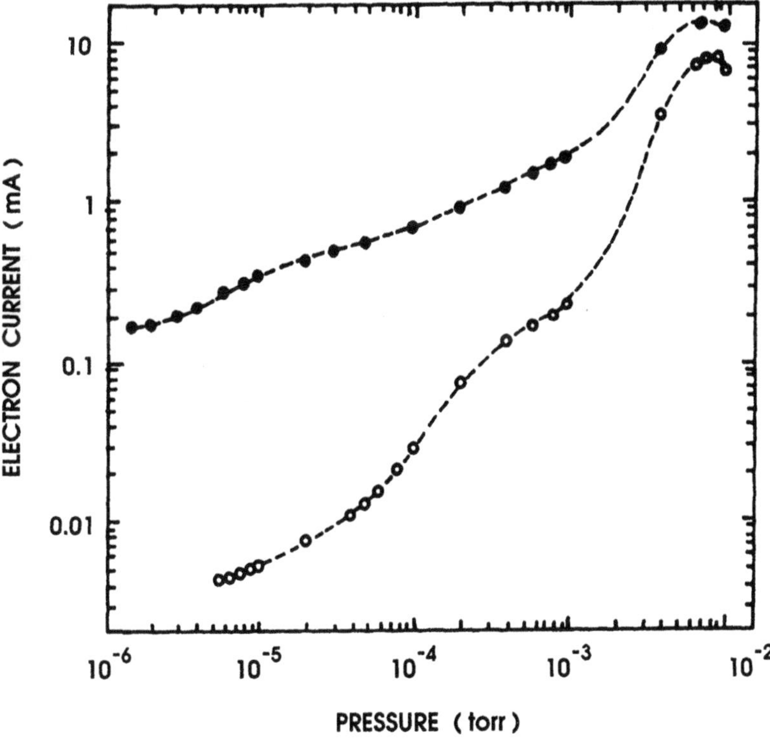

Fig. 10.4. Electron current (proportional to plasma density) as a function of pressure, for an argon plasma in the parallelepiped chamber with (closed circles) and without (open circles) magnetic confinement (from ref. 7).

neutral species are available for ionization. In the high pressure range of operation, where collisions become important, the two curves approach each other with a density about twice that of the plasma without magnets. At still higher pressures, plasma is no longer obtained due to excessive collisions preventing primary electrons to be collected at the chamber wall.

Homogeneity. The multipolar confinement yields homogeneous plasmas. Figure 10.5 shows the spatial density profile in the cylindrical chamber of Limpaecher and MacKenzie.[7] At low pressures (lower traces), the profile is extremely uniform, to better than 1%, in the field-free region inside the containment structure. When collisions become important (pressures above 10^{-4} torr), density gradients appear towards the filaments.

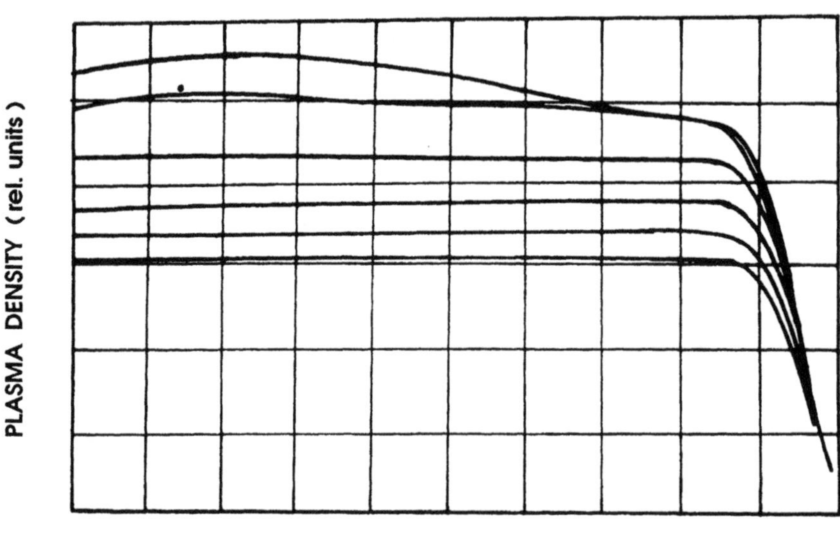

Fig. 10.5. Plasma density profile in an argon multipolar plasma, as measured on the axis of Limpaecher and MacKenzie's cylindrical chamber. Lower trace at 5×10^{-6} torr, and successive traces up at 10^{-5}, 5×10^{-5}, 10^{-4}, 10^{-3}, and 2×10^{-3} torr. The ordinates are not a direct measure of plasma density since the gain was changed to keep the traces on the screen. The density peaks at the center of the filament array region (left-hand side of the figure). A sharp density decrease occurs inside the multipolar field region and density drops to zero at the wall (right-hand side of the figure) (from ref. 7).

Quiescent plasma. The observed relative level of density fluctuations is very low, of the order of 10^{-4}, and increases in the vicinity of the magnets.[7] The spectrum of these fluctuations lies predominantly in the range below the ion-plasma frequency.

Plasma potential. The plasma potential measured is slightly positive,[7] a few volts with respect to the chamber wall, and is nearly independent of pressure, except at the lowest pressures. In the regime termed below the "excessively low pressure regime", it drops abruptly below the ground potential value: the ion density then decreases while the current of primary electrons remains constant, these electrons imposing their own potential.

Low electron temperature. The electron temperature decreases with increasing pressure, from approximately 7 eV to 2 eV over the pressure range examined by Limpaecher and MacKenzie, which is given in Fig. 10.4. At low pressure, the high energy electrons emitted by the filaments can be identified in the tail of the measured energy distribution function of electrons.

Large-volume plasma. Plasma of large dimensions can be produced. The second machine of Limpaecher and MacKenzie yielded 86 liters of plasma. To our knowledge, the largest multipolar plasma volume (5.5 m^3) is that comprising 6000 small magnets (4 x 2 x 1 cm) at the "Laboratoire de Physique et Chimie de l'Environnement" in Orléans.[9]

10.3 Comparison of multipolar magnetic field arrangements

10.3.1. Experimental results. In the checkerboard configuration (Fig. 10.6a) encountered in the parallelepiped chamber of Limpaecher and MacKenzie[7] (Fig. 10.2), each magnet determines a point cusp where charged particles can escape. In addition, at the center of each group of four such magnets (Fig. 10.6a), there is a zero-field point through which particles can also escape. A linear arrangement of alternating rows of like and unlike poles (Fig. 10.6b) was thought to reduce the latter losses. Limpaecher and MacKenzie[7] used it on the side wall of their cylindrical chamber (Fig. 10.3).

Later, Leung et al.[10] showed (Fig. 10.7) that plasma density is higher with broken-line cusps (Fig. 10.6b) than with point cusps (Fig. 10.6a), but obtained the highest density with the full-line cusp geometry (Figs. 10.6c, 10.6d), which eliminates the spacing between like poles. It should be noted, however, that in the latter case, the rows of identical magnets were placed not as rings circling around the cylinder (Fig. 10.6c) but parallel to its axis (Fig. 10.6d). We also note from Fig. 10.7 the much lower density obtained when the magnets are removed. From then on, a conventional multipolar

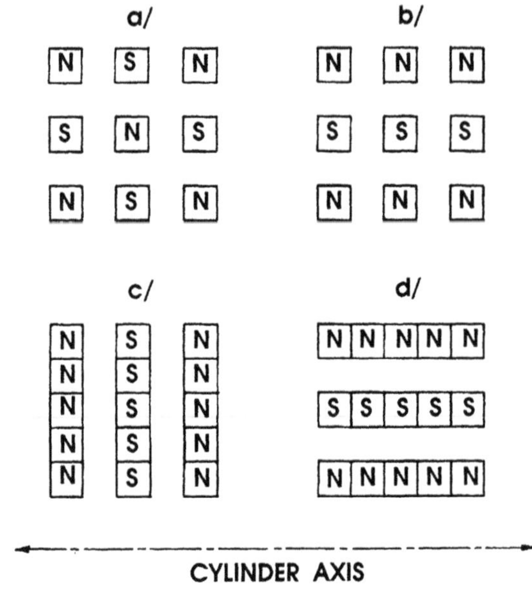

Fig. 10.6. Four magnet arrangements: (a) checkerboard; (b) broken-line cusp; (c) full-line cusp perpendicular to cylinder axis; (d) full-line cusp parallel to cylinder axis.

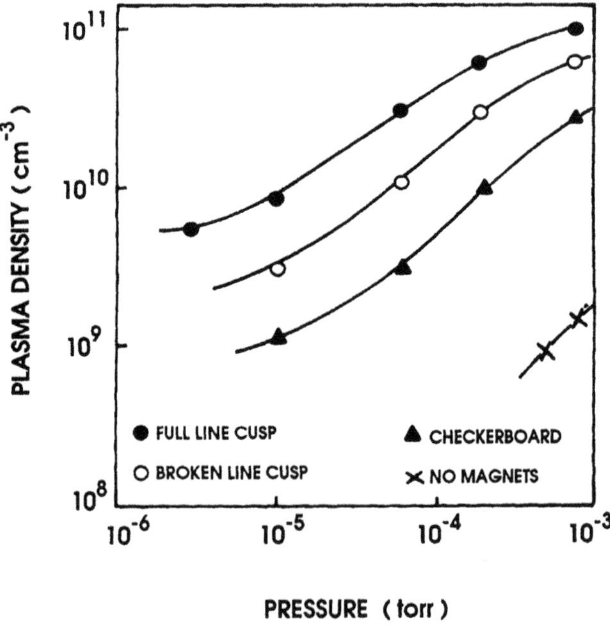

Fig. 10.7. Plasma density as a function of pressure, for three multipolar magnet arrangements and also without magnets (from ref. 10).

plasma chamber emerged, which is a cylinder covered with an even number of alternating pole magnets distributed parallel to the cylinder axis, which can be terminated with various magnetic field configurations. In the above work, the magnets were mounted externally, and nonmagnetic materials had to be used for the chamber walls. However, the magnets can also be mounted internally, in particular when diagnostic means impose the use of portholes.

Leung et al.[10] also measured the plasma density as a function of the number of magnet rows, parallel to the cylinder axis, as shown in Fig. 10.8. The spacing between magnet rows exhibits an optimum value. When the number of magnet rows is large, most of the plasma loss occurs at the cusps, i.e. along the pole magnets. Decreasing the number of rows increases magnet separation, which reduces total cusp area and hence plasma loss. However, when the number of rows is decreased even further, their spacing increases, resulting in low-field regions between the magnets; additional cusps then tend to form in between the magnets on the chamber wall and amplify plasma loss. The optimum position occurs when the two competing loss processes balance. This aspect of plasma confinement is to be more thoroughly discussed in Chaps. 11 and 12.

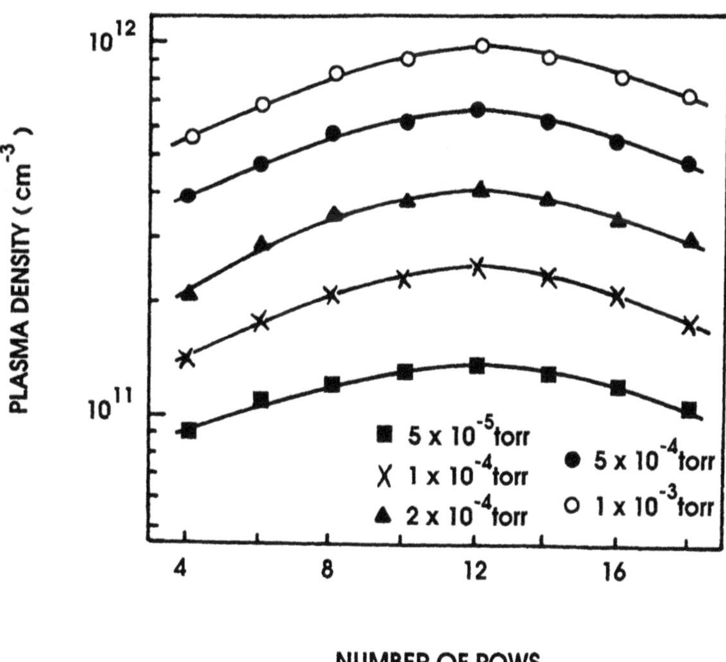

Fig. 10.8. Plasma density as a function of the number of magnet rows, at several pressures in an argon plasma (from ref. 10).

10.3.2. Interest of closed magnetic field configurations. As discussed above, the crucial point in magnet configuration is the perfect closing of the magnetic field structure. One way of achieving this is to have a unique magnetic polarity at the perimeter separating two surfaces of the magnetic structure. Examples follow that illustrate this concept. Three possible magnet arrangements for either cylindrical or parallelepiped chambers are shown in Fig. 10.9. These are the ring, the racetrack and the comb structures. We see that given these three arrangements, the ring geometry on the end walls is a good solution for closing the magnetic field structures of cylindrical chambers.

10.4. Operation of the filament-excited discharge

10.4.1. Principle. In a hot cathode discharge, as shown in Fig. 10.10, the plasma is generated by the so-called *primary electrons*, which are emitted by a thermionic filament (discharge current I_d)[†] and energized by the negative bias (discharge voltage V_d) of the filament with respect to the wall. When sufficiently energetic, these electrons undergo inelastic collisions with the neutral atoms or molecules. The result is excitation, dissociation and ionization of the gas. Finally, these electrons are collected by the wall of the discharge chamber providing a return current. Thus, plasma generation by primary electrons requires a metal wall (or at least a metal electrode within the plasma) so as to collect the DC electron current emitted by the negatively biased hot filament(s).

10.4.2. Striking and sustaining the discharge. In the absence of plasma, considering the large distance between the filament and the reactor wall, typically several centimeters, only few electrons can be extracted due to the Child-Langmuir space-charge limitation. Therefore, a rather high discharge voltage V_d is required to reach gas breakdown. Once the plasma is ignited, the emitted electrons are accelerated from the filament towards the plasma (at potential V_p) with energy $e(V_p - V_d)$ through the plasma sheath. This sheath being only a few Debye lengths thick, the space charge limitation is then negligible. Thus, after plasma ignition, the discharge voltage V_d can be significantly reduced according to the plasma density required. However, V_d must be kept high enough for the primary electrons to ionize the gas either by direct ionization or in two steps through metastable states.

[†] Due to the small filament area, the ion current accelerated toward it is negligible with respect to the thermoelectron current. Hence, the discharge current I_d can be taken equal to the current of primary electrons.

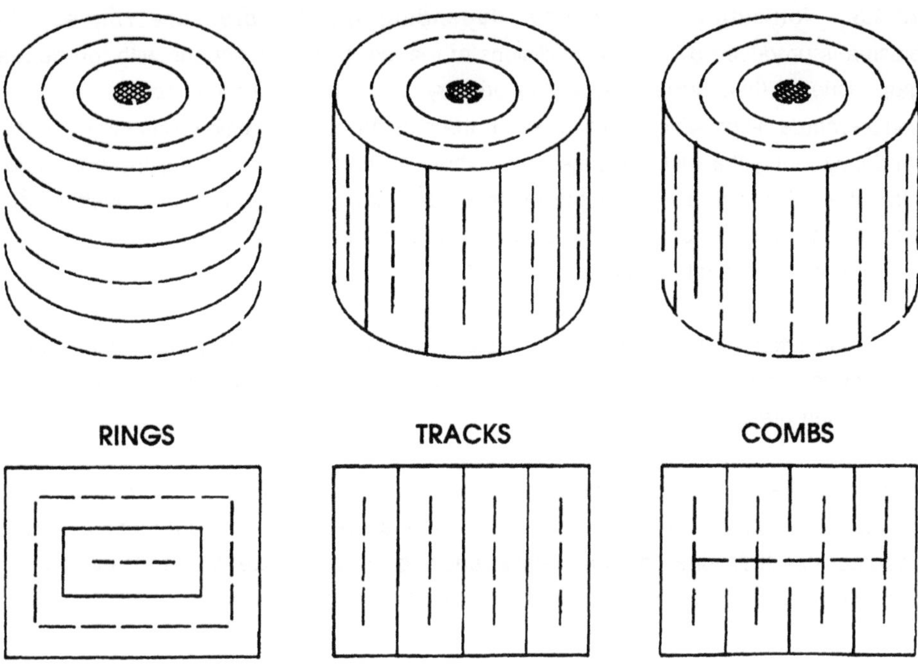

Fig. 10.9. Three possible magnet arrangements for cylindrical chambers and rectangular surfaces providing completely closed magnetic structures. Full and dashed lines schematize north and south magnetic polarities.

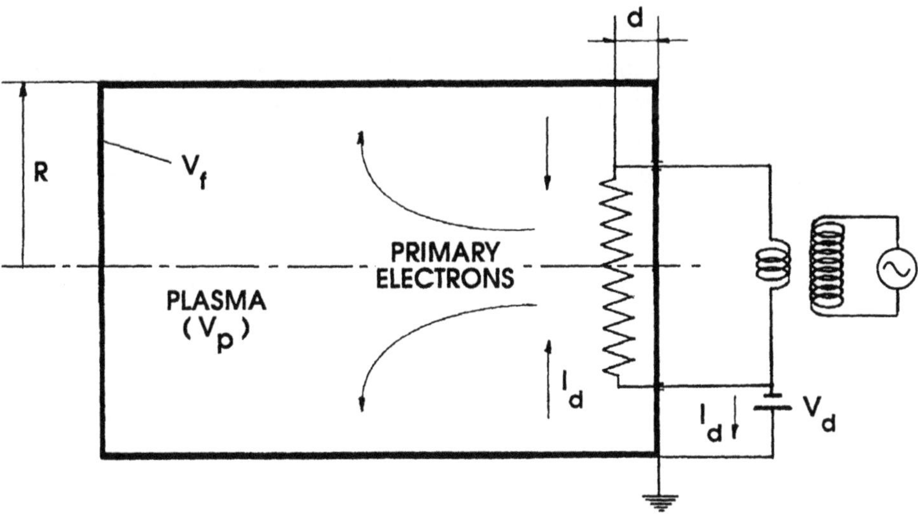

Fig. 10.10. A hot cathode discharge system where I_d is the discharge current, V_d the discharge voltage, V_p the plasma potential, V_f the floating potential, R the chamber radius and d the shortest filament-to-wall distance.

10.4.3.   Ionization mechanisms depending on the pressure range. The plasma is produced by inelastic collisions of the fast primary electrons with the neutral gas. Due to their large energy, the primary electrons are barely perturbed by the space-charge fields which develop within the plasma, and inelastic collisions will occur randomly along the electron trajectory. Thus, in the field-free region, the production rate of electrons and ions of the plasma does not depend on spatial position and plasma density. In contrast, the ionization rate in the magnetic field region may vary considerably with spatial position (Chap. 12).

The ionization of the gas leads to the slowing down of the primary electrons at a rate that depends on gas pressure. Thus the discharge will behave differently according to the working pressure range.

Low pressure range. The mean free path of primary electrons is larger than the chamber dimensions (see Appendix 11.1). On the average, these energetic electrons undergo less than one collision before reaching the discharge wall and, even when assuming one ionizing collision over that run, the energy loss is too small to convert these electrons into plasma thermal electrons ($k_B T_e \approx 1$ eV). Therefore, it is a good approximation to consider the primary electron population as totally independent of the plasma electron population. Further on, these two populations will thus often be considered as separate entities, in particular with respect to the action of the multipolar magnetic confinement.

High pressure range. The mean free path of the primary electrons is shorter than the chamber dimensions, i.e. they suffer several ionizing collisions. As a result, these electrons convert into plasma electrons before they reach the discharge wall. Clearly then, primary and plasma electrons do not constitute two different populations, and the action of the multipolar magnetic field is not to confine primary electrons but only to provide plasma confinement.

Excessively low pressure range. The mean free path of the fast primary electrons is now much larger than the chamber dimensions. Ionizing collisions are rare and isolated events. No collective effects appear in the discharge (the Debye length is of the order of the chamber dimensions) and thus a true plasma is not achieved.

Excessively high pressure range. The mean free path of the fast primary electrons is much smaller than the chamber dimensions. Collisions are so numerous that the plasma electrons (including the primary electrons) cannot even reach the discharge wall to provide a return current. The discharge is thus quenched, only an electron cloud being formed close to the filament.

Among the pressure ranges considered above, the low pressure range is the one of interest since applying multipolar magnetic confinement then yields approximately a tenfold increase in plasma density and an excellent plasma homogeneity (see Sec. 10.2.2).

10.4.4. Conclusions on filament-excited discharges. The main point from the above discussion is that a discharge, excited in the so-called low pressure regime by a thermionic filament with or without magnetic field confinement, contains two independent electron populations, the plasma electrons and the primary electrons, the latter producing the plasma. As a result, the characteristics of these DC plasmas, though they are sustained by ionization via the primary electrons, can be modeled without considering the population of the primary electrons. Also, as will prove useful later on in discharge modeling, the discharge current I_d (of primary electrons) can be simply superposed onto the currents of plasma electrons and ions.

10.5. Electrostatic confinement of a homogeneous discharge

In this section, we examine the possibility of using electrostatic confinement to increase plasma density. We investigate such a mechanism in a discharge excited by primary electrons. For the clarity of exposition, we first consider the simple case of a discharge enclosed in a chamber with equipotential walls and free of magnetic field. Afterwards, we turn to the case of the so-called *composite discharge*, where some parts of the chamber wall are at different potentials.

10.5.1. Electrostatic confinement in an equipotential conducting chamber. The plasma is generated by the energetic primary electrons emitted by a hot cathode biased negatively with respect to the chamber wall, as depicted in Fig. 10.10. We assume a homogeneous plasma, free of negative ions, filling completely the equipotential chamber. Recall that the population of primary electrons is independent of the population of plasma electrons. Since the production rates of plasma electrons and ions in the discharge volume are the same, plasma neutrality ($n_i = n_e = n$) requires that under steady state, the fluxes of plasma electrons and ions on the discharge wall be equal and thus

$$I_e + I_i = 0. \tag{10.1}$$

The total current I collected on the chamber wall also comprises primary electrons, and we have

$$I \equiv I_e + I_i + I_d = I_d. \tag{10.2}$$

Consider now the discharge wall as a large plane probe of area S. The currents of plasma electrons, I_e, and ions, I_i, collected through the plasma sheath as a function of the wall potential V with respect to the plasma potential V_p, are given in Sec. 9.2.3. We briefly recall these results. The saturation electron current collected in the case of an attractive potential $(V > V_p)$ is

$$I_{es} = - en\, S \left(\frac{k_B T_e}{2\pi m_e}\right)^{1/2}. \tag{10.3}$$

Here e is the absolute value of the elementary charge, k_B the Boltzmann constant, T_e the electron temperature, and m_e the electron mass. When the potential is repulsive $(V < V_p)$, the electron current decreases exponentially according to

$$I_e = - en\, S \left(\frac{k_B T_e}{2\pi m_e}\right)^{1/2} \exp\left[\frac{e(V - V_p)}{k_B T_e}\right]. \tag{10.4}$$

Under the same potential $(V < V_p)$, which is attractive for ions, the saturation ion current is given by

$$I_{is} = en\, S \left(\frac{k_B T_e}{m_i}\right)^{1/2}, \tag{10.5}$$

where m_i is the ion mass. From eqns. (10.4) and (10.5), it follows that eqn. (10.1) is satisfied for a plasma potential V_p positive with respect to the wall potential V.[†] More precisely, this potential V settles at the value

$$V \equiv V_f = V_p - \frac{k_B T_e}{2e} \log_e\left(\frac{m_i}{2\pi m_e}\right), \tag{10.6}$$

which is a few times $k_B T_e/e$ below the plasma potential and is known as the *floating potential* V_f. Note that the condition $V_f < V_p$ ensures the existence of the plasma, since it prevents all the electrons from escaping the discharge.[††] We term this mechanism *electrostatic confinement* of the plasma electrons. It results from the plasma potential adjusting at the value for which the current of plasma electrons exactly balances the ion saturation current, in order to maintain plasma neutrality.

[†] We neglect the effect of the negative space charge due to primary electrons, which makes the plasma potential slightly more negative.

[††] In the case where we would have $V_f > V_p$, the electrons owing to their greater mobility as compared to ions, would be swept out from the discharge and no plasma could subsist.

Using, in a first step, the assumption of a homogeneous plasma produced in an equipotential chamber, we have verified that the plasma potential is a few times $k_B T_e/e$ above the wall potential, ensuring the electrostatic confinement of the plasma. It is therefore worth examining whether we could improve such an electrostatic confinement, by applying different potentials on different parts of the chamber wall.

10.5.2. Electrostatic confinement in a composite reactor.

To investigate the possibility of using electrostatic confinement to increase plasma density, we consider a homogeneous plasma in a chamber that consists of two walls with areas S_1 and S_2, biased at potentials V_1 and V_2 respectively, such that $V_2 - V_1 >> k_B T_e/e$. Referring to the case of the equipotential chamber, we can expect the plasma potential to settle at a positive value above the potential of one or both chamber walls.

Case where $V_1 < V_p < V_2$. It is shown schematically in Fig. 10.11. Since $V_2 > V_p$, positive ions are repelled from the wall at V_2. As a result, the current I_2 collected by that wall is the electron saturation current

$$I_2 = -enS_2 \left(\frac{k_B T_e}{2\pi\, m_e}\right)^{1/2}. \tag{10.7}$$

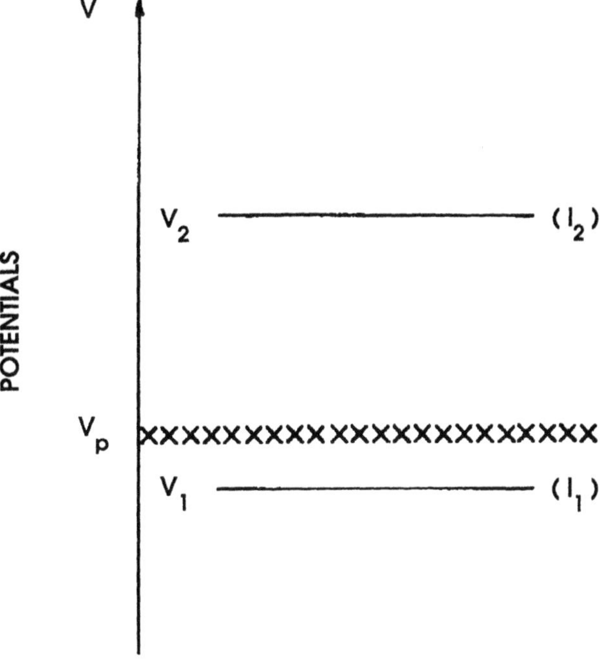

Fig. 10.11. Potential distribution on the walls of a composite reactor in the case $V_1 < V_p < V_2$.

The current I_1 collected on the wall biased at potential V_1 is the sum of the ion saturation current and of the electron current in the repulsive potential domain ($V_1 < V_p$), namely

$$I_1 = en\,S_1 \left(\frac{k_B T_e}{m_i}\right)^{1/2} - en\,S_1 \left(\frac{k_B T_e}{2\pi\,m_e}\right)^{1/2} \exp\left[\frac{e\,(V_1 - V_p)}{k_B T_e}\right].$$

(10.8)

Plasma neutrality requires that

$$I_1 + I_2 = 0,$$

(10.9)

which yields

$$V_p - V_1 = -\frac{k_B T_e}{e}\,\log_e\left[\left(\frac{2\pi m_e}{m_i}\right)^{1/2} \frac{S_2}{S_1}\right].$$

(10.10)

Clearly, the above potential distribution ($V_p - V_1 > 0$) is possible only if the condition

$$\frac{S_2}{S_1} < \left(\frac{2\pi m_e}{m_i}\right)^{1/2} \ll 1$$

(10.11)

is satisfied. Such a condition corresponds to the case of a small Langmuir probe ($S_2 \ll S_1$), which can be assumed not to perturb the plasma. In reality, eqn. (10.10) reduces to eqn. (10.6): the plasma characteristics are not modified as compared to the case of the equipotential reactor and there is no enhanced electrostatic confinement.

Case where $V_1 < V_2 < V_p$. Since the potential distribution considered in the preceding paragraph does not apply to approximately equal wall areas, we turn to the case $V_1 < V_2 < V_p$, depicted schematically in Fig. 10.12. The current I_2 drawn by the wall at potential V_2 is the sum of the ion saturation current and the electron current in the repulsive potential domain ($V_2 < V_p$)

$$I_2 = en\,S_2 \left(\frac{k_B T_e}{m_i}\right)^{1/2} - en\,S_2 \left(\frac{k_B T_e}{2\pi m_e}\right)^{1/2} \exp\left[\frac{e\,(V_2 - V_p)}{k_B T_e}\right].$$

(10.12)

Since we have assumed $V_2 - V_1 \gg k_B T_e/e$, the electron current collected on the other wall (at potential V_1) is negligible with respect to the ion saturation current and we have

$$I_1 = en\,S_1 \left(\frac{k_B T_e}{m_i}\right)^{1/2}.$$

(10.13)

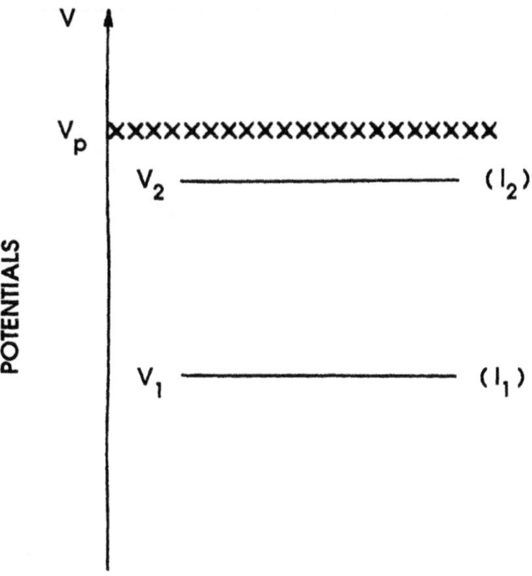

Fig. 10.12. Potential distribution on the walls of a composite reactor in the case $V_1 < V_2 < V_p$.

Again, plasma neutrality (eqn. (10.9)) leads to the plasma potential

$$V_p - V_2 = \frac{k_B T_e}{e}\left[\log_e\left(\frac{m_i}{2\pi m_e}\right)^{1/2} - \log_e\left(1 + \frac{S_1}{S_2}\right)\right].$$

(10.14)

The above potential distribution ($V_p - V_2 > 0$) is possible only if the condition

$$\frac{S_1}{S_2} < \left(\frac{m_i}{2\pi m_e}\right)^{1/2} - 1 \approx \left(\frac{m_i}{2\pi m_e}\right)^{1/2}$$

(10.15)

is satisfied. This condition is the reverse of condition (10.11) and corresponds to a large electrode in a plasma, hence to the true potential distribution in a composite reactor. The fact that in this case the plasma potential is always more positive than the potential of the most positive electrode (or chamber wall) in contact with the plasma, prevents electrons to be all collected by the walls. This situation thus ensures electrostatic confinement of the plasma electrons. As in the case of the equipotential chamber (Sec. 10.5.1), the value of the plasma potential adjusts so that the current of plasma electrons exactly balances the saturation ion current to ensure plasma neutrality.

Experimental confirmation[11] of the above discussion and initial assumptions can be deduced from Fig. 10.13 where a composite reactor is operated in the $V_1 < V_2 < V_p$

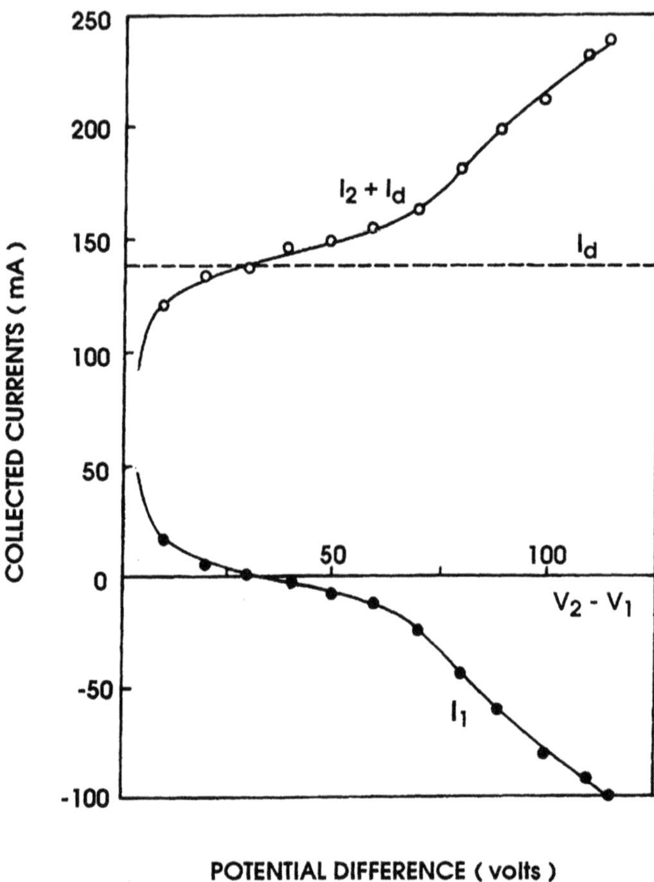

POTENTIAL DIFFERENCE (volts)

Fig. 10.13. Plasma currents I_1 and I_2 collected on the two walls of a composite reactor as a function of the difference V_2 - V_1 of the potentials applied. The plasma is sustained in argon at 5×10^{-4} torr by a hot cathode supplying the discharge current I_d. The discharge current is superposed on the plasma current of the most positive wall.

regime. We see that the contribution of plasma current to each wall is symmetrical but shifted by an amount I_d equal to the discharge current due to the primary electrons. The latter current, which can be measured independently (see Fig. 10.10), is preferentially collected on the most positive wall. This result also demonstrates the fact that the I_d contribution is superposed independently on the plasma currents, a feature which cannot be evidenced in an equipotential chamber. A similar behavior has been observed in a composite reactor submitted to multipolar confinement.[9]

10.5.3. Conclusion on electrostatic confinement.
Comparing the equipotential and composite chamber systems, we see from eqns. (10.5), (10.12), and (10.13) that

they both draw the same ion saturation current densities, whatever the wall potential values.   Therefore, when the equipotential chamber wall area equals that of the composite chamber, namely

$$S = S_1 + S_2,$$ \hfill (10.16)

the corresponding total ion saturation currents collected on these walls are also equal. This means that in a composite reactor, electrostatic confinement alone (without magnetic field) is totally inefficient to increase plasma density.   However, the results obtained will be useful for the understanding of potential distributions in a plasma and will guide us in determining substrate biasing, as discussed in Sec. 13.4.

10.6. Modeling of a hot cathode discharge

10.6.1.   Assumptions.   In Sec. 10.5, we have assumed that the plasma is homogeneous, i.e. the plasma density n and the plasma potential V_p are constant over the entire volume of the plasma. As this hypothesis is not valid in most discharges, we propose below a more realistic description of these hot cathode discharges.   The primary electrons are now assumed to produce plasma electrons and ions at an (equal) ionization rate S_i independent of position or plasma density. It is also assumed that the plasma is free of negative ions, and that volume ion recombination is negligible (Sec. 9.5).   The ion temperature is taken to be much lower than the electron temperature $(T_e \gg T_i)$.

10.6.2. Balance of charged particles.   Under steady-state conditions, particle conservation implies

$$\nabla \cdot n\mathbf{v}_e = \nabla \cdot n\mathbf{v}_i = S_i,$$ \hfill (10.17)

where $\mathbf{v}_e$ and $\mathbf{v}_i$ are the mean electron and ion directed velocities due to ∇n and $\mathbf{E}$. If V is the volume of plasma enclosed within a closed surface of area S, and $\mathbf{dS}$ is a vector normal to the surface and directed outwards, then from Ostrogradski's theorem

$$\int_V \nabla \cdot n\mathbf{v}_i \, dV = \int_S n\mathbf{v}_i \cdot \mathbf{dS},$$ \hfill (10.18)

and from eqn. (10.17),

$$\int_V S_i \, dV = \int_S n\mathbf{v}_i \cdot \mathbf{dS} = \int_S n\mathbf{v}_e \cdot \mathbf{dS} \ . \tag{10.19}$$

The latter equation shows that the ions and electrons created in the volume of the discharge (left-hand side) all flow to the walls (right-hand side). The collected current intensity is obtained by integration of eqn. (10.19),

$$I_i = - \ I_e = eS_i V \ . \tag{10.20}$$

To obtain the plasma density, we further assume that the ion flux is constant over the surface of the chamber wall and normal to it, namely $I_i = J_i S$. Since the wall potential is lower than V_p, the ion current density on the wall is the ion saturation current density J_{is} through the sheath

$$I_{is} = J_{is} \ S = n_s e S \left(\frac{k_B T_e}{m_i}\right)^{1/2} . \tag{10.21}$$

With eqn. (10.20), it yields the plasma density n_s at the sheath edge,

$$n_s = \frac{S_i V}{c_i S} , \tag{10.22}$$

where c_i is the ion-acoustic velocity (eqn. (9.16)). Equation (10.22) is a relation between the parameters influencing the charged particle balance. In particular, keeping V and S_i constant and increasing the area S (for example by introducing a collecting electrode in the plasma volume) induces a correlative decrease in the plasma density, as was experimentally demonstrated.[7] Plasma density can be increased by increasing S_i, i.e. for example, by increasing the primary electron current I_d.

Although the approach followed in this section suggests several means to increase the plasma density, it only yields n_s, the density value at the sheath edge, and thus gives no further indication on the plasma distribution inside the discharge. An even more detailed description of the discharge is thus required.

10.6.3. Ambipolar diffusion model. We consider a cylindrical discharge of infinite length and radius R. We retain the assumptions made in Secs. 10.6.1 and 10.6.2 and further consider that the mean free path of the primary electrons is much larger than R whereas the mean free path of the ions is assumed to be much smaller than R. From our previous discussions, supported by experimental results (Fig. 10.13), the primary and plasma electron populations can be considered distinct. We can thus assume the ambipolar diffusion model to apply only to the plasma electrons and to the ions. The complete set of equations that describe the plasma is then (Sec. 9.2.2)

$$\nabla \cdot n\mathbf{v} = S_i, \tag{10.23}$$

$$n\mathbf{v} = - D_a \nabla n, \tag{10.24}$$

$$\mathbf{E} = - \nabla V = - \frac{k_B T_e}{e} \frac{\nabla n}{n}, \tag{10.25}$$

where D_a is the ambipolar diffusion coefficient, $\mathbf{v}$ is the mean directed velocity of the ions and of the plasma electrons, $\mathbf{E}$ is the electric field, and V is the corresponding potential.

Due to the cylindrical symmetry of the problem, the plasma parameters exhibit only components depending on r. Integration of eqn. (10.23) leads to

$$nv = \frac{S_i}{2} r. \tag{10.26}$$

Substituting in eqn. (10.24) and integrating, we obtain the radial profile of plasma density

$$n(r) = n(0) - \frac{S_i}{4D_a} r^2, \tag{10.27}$$

where $n(0)$ is the density at the axis of the discharge.

To determine $n(0)$, we first assume a zero plasma density at the discharge wall ($r = R$), which leads to the classical parabolic profile

$$n(r) = \frac{S_i}{4D_a} \left(R^2 - r^2 \right). \tag{10.28}$$

This simple boundary condition provides us with the general shape of the density profile. However, it is not satisfactory since it leads to the divergence of the electric field and of the potential at the wall. A more physical boundary condition is to assume, as in Sec. 10.6.2, that the plasma flux at the sheath edge and the ion saturation flux through the sheath are equal (Bohm criterion). Neglecting the sheath thickness with respect to R, we obtain from eqn. (10.26)

$$n(R) \left(\frac{k_B T_e}{m_i}\right)^{1/2} = n(R)\, v(R) = \frac{S_i}{2} R. \tag{10.29}$$

The profile of the plasma density then finally takes the parabolic form shown in Fig. 10.14, namely

$$n(r) = \frac{S_i}{4 D_a}\left(R^2 + 2R\,\lambda_a - r^2\right), \tag{10.30}$$

where $\lambda_a = D_a/c_i$ is the ambipolar mean free path previously defined in eqn. (9.45).

Finally, as far as the plasma potential is concerned, integrating eqn. (10.25), we find

$$V_p(0) - V_p(R) = \frac{k_B T_e}{e} \log_e\left(1 + \frac{R}{2\lambda_a}\right). \tag{10.31}$$

This relation shows that the plasma potential $V_p(0)$ at the axis is a few times $k_B T_e/e$ above the plasma potential at the sheath edge. It also yields that $V_p(0) - V_p(R)$ depends on pressure through λ_a but is independent of S_i. The correlative space-charge electric field that develops inside the discharge and which is directed towards the wall (see Fig. 10.14), ensures the evacuation of the charged particles produced in the volume ($S_i \neq 0$).

10.6.4. Interest of introducing multipolar magnetic confinement. Equation (10.22) suggests that the plasma density can be enhanced by increasing the ion production rate S_i or by decreasing the collecting area S of the discharge wall. Recall however that a reduction of S cannot be achieved by reducing the effective collecting area: any attempt to bias surfaces independently in the chamber so as to reduce the collecting area is ineffective, since the plasma potential shifts accordingly. The ambipolar diffusion model that we just presented has shown that electrostatic confinement does not lead to a homogeneous plasma (Sec. 10.6.3). On the other hand, we know from experiment (Sec. 10.2.2) that multipolar magnetic confinement is very efficient in providing a homogeneous plasma with a density of up to two orders of magnitude higher than that obtained without magnetic confinement. The objective of the next section is thus to elucidate the discharge confinement mechanisms in a multipolar magnetic structure. For that purpose, we shall distinguish the effects of the multipolar magnetic field confinement on the plasma (electrons and ions) from those on the primary electrons (which produce the plasma).

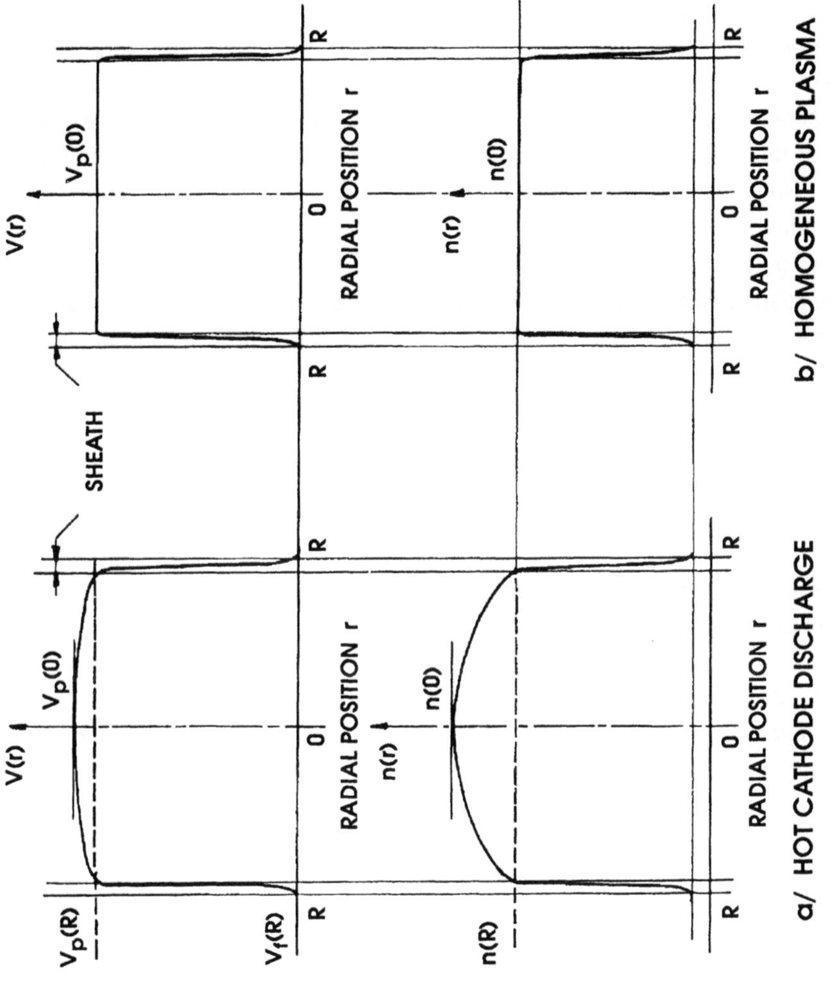

Fig. 10.14. Radial profiles of potential and plasma density in the case of the ambipolar diffusion model (a), and for comparison, in a homogeneous discharge (Sec. 10.5) (b), assuming in both cases the Bohm criterion to apply at the sheath edge.

10.7. Effects of multipolar magnetic confinement

10.7.1. Confinement of the plasma. The degree of plasma confinement can be measured in terms of plasma density enhancement or equivalently in terms of plasma lifetime enhancement. Such measurements must avoid the contribution from the primary electrons when they respond to the applied magnetic field. The increase of plasma lifetime due to multipolar magnetic confinement has been estimated experimentally to be approximately a factor of 2 to 3.[5,12] This result can be related to the increase in plasma density by a factor of two which is observed in the high pressure range (Sec. 10.4.3) of a hot cathode discharge (see Fig. 10.4). Recall that in this pressure range, the primary electrons become plasma electrons before they reach the wall, so that the effect of the magnetic confinement on the primary electrons may be neglected. Hence, in the high pressure range, the density enhancement results only from the confinement of the plasma. The relatively low efficiency of plasma confinement by the multipolar magnetic field can be understood from the iso-density lines in the cusp region, as shown in Fig. 10.15. The confinement of ions compared to that of primary electrons is less effective, because of a much larger leaking of the ions to the wall at the pole magnets (cusps). Clearly, the wider this leaking region (hereafter referred to as the *leak width*), the lower the efficiency of confinement.

Another important feature of plasma confinement inside a multipolar magnetic field configuration is the presence, under certain plasma conditions, of positive *potential hills* between the cusps, as shown in Figs. 10.15c and 10.15d. The development of such space-charge electrostatic fields (see Chap. 11) can play an important role in the confinement of plasma by reducing the ion leaks at the cusps.[13]

Finally, as far as plasma homogeneity is concerned, experiments showed[7,14] that the multipolar magnetic confinement significantly and progressively improves plasma homogeneity as pressure is decreased (see Fig. 10.5). This increased homogeneity cannot be explained by models assuming uniform volume ionization (Sec. 9.3.1) in collisional (Sec. 10.6.3) or collisionless hot cathode discharges.[14] We are thus compelled to explore the role of the primary electrons to account for the uniformity and increased density in multipolar discharges.

10.7.2. Confinement of the primary electrons. The degree of confinement of the primary electrons can be measured, in a similar way as for plasma confinement, i.e. in terms of plasma density enhancement or, equivalently, in terms of increased lifetime or mean free paths, provided that the plasma confinement contribution can be separated from it. Figure 10.16 shows the increase in the mean free path of the primary electrons when the multipolar magnetic field confinement is added.[12] It corresponds to

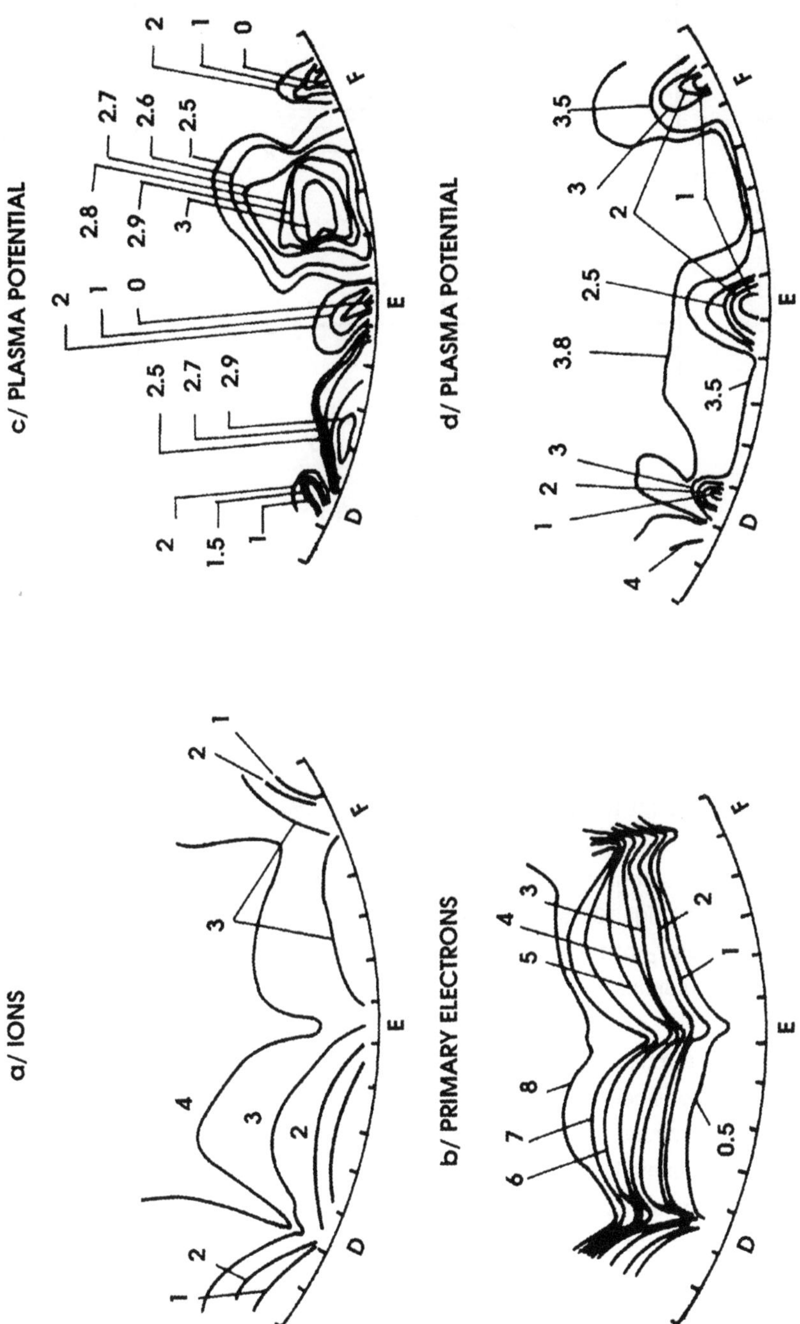

Fig. 10.15. Equi-ion, equi-primary electron, and equi-plasma potential contours in the neighborhood of three cusps labelled D, E, and F: (a) equi-ion density contours; (b) equi-primary electron density contours; (c) and (d) equi-plasma potential contours labelled in volt with respect to the chamber walls (from ref. 13).

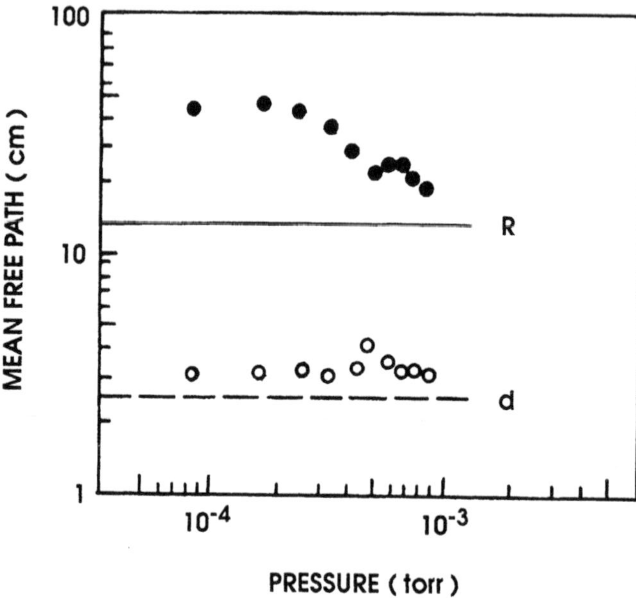

Fig. 10.16. Observed mean free path of the primary electrons as a function of argon pressure, with (closed circles) and without multipolar magnetic confinement (open circles). The solid line represents R, the radius of the chamber, and the dashed line corresponds to d, the shortest filament-to-wall distance (see Fig. 10.10) (from ref. 12). Without magnetic confinement, most of the discharge current is collected on the wall nearest to the filament.

an increase of plasma density by a factor of ten. Thus, the confinement effect of the multipolar magnetic field on the primary electrons is much more pronounced than on the plasma itself. This is correlated with a leak width at the cusps narrower for primary electrons than for ions (Fig. 10.15). The leak width of the primary electrons is of the order of the gyroradius r_p of these electrons at the cusp,[13] as expected from single particle trajectory calculations.

The efficiency of the multipolar magnetic field confinement on the primary electrons can be visualized by considering Figs. 10.17 and 10.18. The photograph in Fig. 10.17 shows successive reflections of a primary electron beam as a result of the multipolar magnetic field.[12] No plasma is produced during this experiment, but an argon pressure of 5×10^{-4} torr is maintained to track the electron beam trajectory by recording the light emission following inelastic collisions with argon atoms. Several reflections are observed, except when the electron gun is pointed exactly at a magnet pole (cusp), as expected from theory (see Chap. 12). Photograph 10.18 is obtained by rotating the gun around the axis of the cylindrical chamber during exposure time. The number of

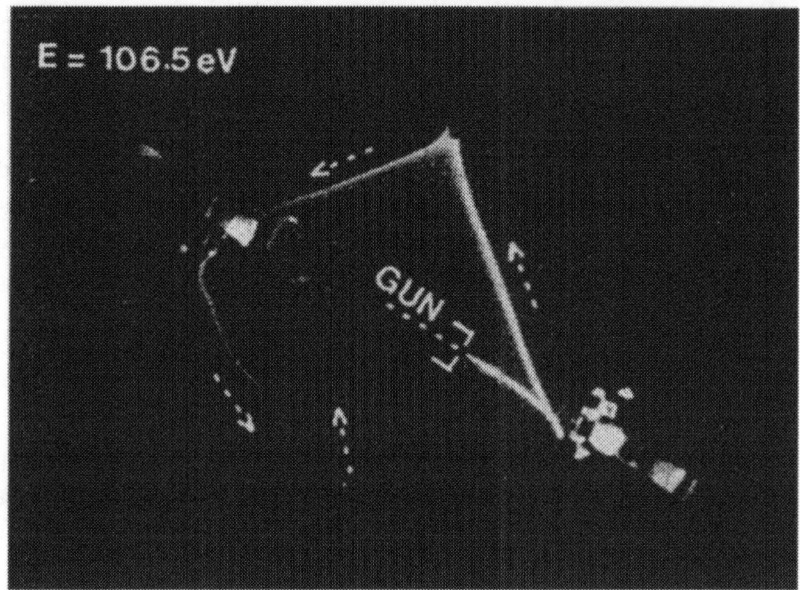

Fig. 10.17. Photograph of a primary electron trajectory observed in an argon filled chamber, showing successive reflections from the multipolar magnetic field (from ref. 12).

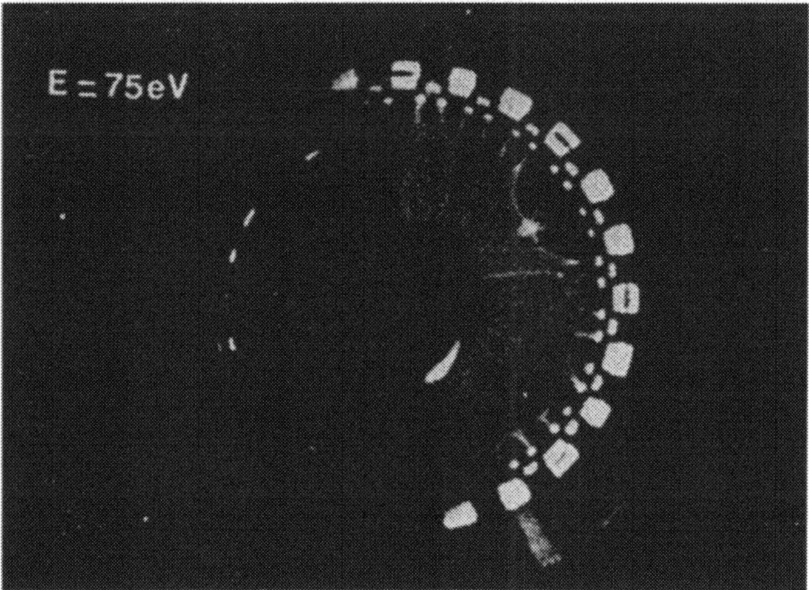

Fig. 10.18. Envelope of the trajectories (gray area) obtained at a given beam energy by rotating the electron gun in the plane of the figure during exposure time (from ref. 12).

photons emitted along the various trajectories of these electrons lead to distinguish three regions of unequal light intensity: i) the central region, which is free from magnetic field, exhibits a uniform light emission; ii) peripheral dark regions located between the cusps and the wall, i.e. between successive magnet poles; iii) a peripheral bright region which separates the central region from the dark regions. This bright emission region seems to link each magnet pole to the adjacent one via lobes (see Fig. 14.1) that follow the magnetic field lines. This region of enhanced light emission appears to be the location where primary electrons interact with the magnetic field and where some of them are trapped by the magnetic field lines whereas the dark regions between these field lines and the wall seem to be locations from which primary electrons are excluded.

10.8. Conclusion

We have seen that multipolar discharges excited by primary energetic electrons and operated at low pressure, yield a large volume of homogeneous, high density plasma. The confinement effect of the multipolar magnetic field, although weak on the plasma itself, increases considerably the path of the primary electrons in the chamber before they are collected on the walls, resulting in increased ionization. These electrons behave as a population distinct from the plasma electrons. The models developed in this chapter use elementary assumptions that cannot explain all of the observed experimental features. A more specific theoretical treatment is required to better model the plasma confinement mechanisms in a multipolar magnetic field. For this reason, we review in the next chapter the collisional diffusion model developed by Koch and Matthieussent,[16,17] which explains most of the experimental observations, excluding plasma homogeneity. Then, in Chap. 12, we summarize the results obtained by Gauthereau and Matthieussent[18,19] on the role of primary electrons in providing plasma homogeneity.

References

[1]   E.D. Courant, M.S. Livingston and H.S. Snyder, Phys. Rev. **88**, 1190 (1952).

[2]   J.P. Blewett, Phys. Rev. **88**, 1197 (1952).

[3]   M. Sadowski, Phys. Lett. **25A**, 695 (1967).

[4]   M. Sadowski, Phys. Lett. **27A**, 435 (1968).

[5]   M. Sadowski, Phys. Lett. **28A**, 626 (1969).

[6]   M. Sadowski, Rev. Sci. Instrum. **40**, 1545 (1969).

[7]   R. Limpaecher and K.R. MacKenzie, Rev. Sci. Instrum. **44**, 726 (1973).

[8]   K.N. Leung, C.F.A. van Os and W.B. Kunkel, Appl. Phys. Lett. **58**, 1467 (1991).

[9]   J. Pelletier, Y. Arnal, R. Debrie, L. Pomathiod and J.C. Rifflet, Rev. Sci. Instrum. **55**, 1636 (1984).

[10]   K.N. Leung, T.K. Samec and A. Lamm, Phys. Lett. **51A**, 490 (1975).

[11]   E. Nicolopoulou, thèse 3ème cycle, Université Paris-XI, Orsay (1976).

[12]   J.M. Buzzi, J. Snow and J.L. Hirshfield, Phys. Lett. **54A**, 344 (1975).

[13]   N. Hershkowitz, J.R. Smith and J.R. Dekock, Plasma Phys. **21**, 823 (1979).

[14]   C. Gauthereau and G. Matthieussent, Phys. Lett. **A102**, 231 (1984).

[15]   K.N. Leung, N. Hershkowitz and K.R. MacKenzie, Phys. Fluids **19**, 1045 (1976).

[16]   C. Koch and G. Matthieussent, J. Physique **43**, 67 (1982).

[17]   C. Koch and G. Matthieussent, Phys. Fluids, **26**, 545 (1983).

[18]   C. Gauthereau and G. Matthieussent, J. Physique **45**, 1113 (1984).

[19]   C. Gauthereau and G. Matthieussent, Phys. Lett. **A121**, 342 (1987).

CHAPTER 11

AMBIPOLAR DIFFUSION MODEL OF MULTIPOLAR PLASMAS *

11.1. Introduction

Since the experiments of Limpaecher and MacKenzie[1] on discharges confined by multipolar magnetic field, many other devices also using multipolar field configurations[2-4] have been constructed. Compared with the same chamber used without magnetic field, and for a given discharge power level, the multipolar magnetic confinement was shown to increase plasma density by as much as two orders of magnitude and also to improve plasma homogeneity.[1] Because of the high ionization efficiency of the primary electrons (emitted by negatively biased filaments), which are bouncing back between the multipolar magnetic field regions, these devices can be operated at gas pressures as low as 10^{-5} torr where classical discharges are inefficient. However, when the mean free path of primary electrons becomes smaller than the characteristic dimensions of the device, these multipolar systems behave almost as if there was no magnetic confinement: no real improvement in plasma density and homogeneity is found by adding multipolar confinement.

Experiments have confirmed that the main role played by the multipolar magnetic field is to confine the primary ionizing electrons.[5,6] From the observation of potential hills (Sec. 10.7.1) between regions of high magnetic field intensity located at the magnet poles, the so-called cusp regions,[7,8] it has been inferred that space charge electrostatic fields can play an important role in plasma confinement.[7] However, ion (plasma) lifetime measurements show that plasma confinement is very weak compared to the confinement of primary electrons acting as a distinct population from plasma electrons.[5] The leak widths (Sec. 10.7.1) at the cusps have been measured for primary electrons and plasma particles.[9] As the variation of energy of primary electrons due to the space charge electric field induced by the plasma is negligible compared to their actual value, the motion of these electrons is that of a single particle in a magnetic field, and the leak width at the cusp is comparable to their gyroradius r_p. Plasma electrons and ions being coupled through the space charge electric field, a common leak width is measured, that lies between the ion and the hybrid gyroradius r_h. The latter is the geometric mean of r_i and r_e, the ion and electron gyroradius respectively.

---

* **Presented by G. Matthieussent and J. Pelletier**

At the present time, most of these experimental results are not fully understood and there is no theoretical self-consistent treatment. In the case of high-beta plasmas (beta is the ratio of the plasma kinetic pressure to the magnetic pressure), several analytical models proposed an effective magnetic sheath that leads to the plasma leak width at a cusp.[10-13] These are for example the Rosenbluth sheath and the models based on either the ion-acoustic instability or the cross-field drift instability, and also a visco-resistive sheath model.

In this chapter, a collisional diffusion model[14-15] of a low-beta plasma in a two-dimensional magnetic field is presented, which assumes the joint diffusion of electrons and ions in the plane of the magnetic field. The distribution of plasma density comes from a diffusion type equation which depends locally on the parameter $\lambda_e \lambda_{in}/r_i r_e$, where λ_e and λ_{in} are the electron and ion mean free paths respectively. This parameter characterizes plasma diffusion perpendicularly to the magnetic field.

This model is then applied to multipolar discharges where the mean free paths of plasma electrons and ions are much less than the characteristic dimensions of the device, while the mean free path of the primary electrons is of the order of, or greater than these dimensions. These conditions correspond to gas pressures and plasma densities of commonly used multipolar devices:[1-5] discharge radius of several tens of centimeters, argon pressure of the order of 10^{-3} torr, plasma density greater than 10^{10} cm^{-3}, electron and ion temperature of approximately 1 and 0.2 eV respectively (Appendix 11.1).

Spatial regions with different regimes of diffusion can be identified from qualitative arguments based on the above model and on the magnetic field configuration. The cusp leak width is shown to be proportional to the hybrid gyroradius.[11] Plasma density and potential distributions are then obtained by numerically solving the diffusion equation for a multipolar device with $2m$ ($2m = 56$) rows of magnets

11.2. Collisional diffusion in a two-dimensional magnetic field

We consider a partially ionized plasma, produced by collisions of primary electrons with neutral gas atoms, which diffuses in a region submitted to a magnetic field $\mathbf{B}_0(x,y)$, lying in the (x,y) plane and invariant under translation along z. We assume that ionization yields an equal number of plasma electrons and ions and that this production rate S_i does not depend on position and plasma density. Based on these simple hypotheses, we arrive at a collisional diffusion model for a low-beta plasma.

11.2.1. Ion and plasma electron transport coefficients.
Over the range of plasma parameters considered, electron-ion Coulomb collisions of frequency v_{ei} cannot be ignored compared to electron-neutral or ion-neutral collisions of frequency v_{en} and v_{in} respectively (see Appendix 11.1). When the mean free paths of electrons and ions are less than the characteristic dimensions of the plasma, then, assuming quasineutrality and neglecting second order terms, the plasma is described, under steady-state conditions, by the following fluid equations for electrons and ions (subscripts e and i respectively)[16]

$$- k_B T_e \, \nabla n - ne\mathbf{E} - ne\mathbf{v}_e \times \mathbf{B}_0 - nm_e v_{en} \, \mathbf{v}_e - nm_e \, v_{ei} \, (\mathbf{v}_e - \mathbf{v}_i) = 0, \tag{11.1}$$

$$- k_B T_i \, \nabla n + ne\mathbf{E} + ne\mathbf{v}_i \times \mathbf{B}_0 - nm_i \, v_{in} \, \mathbf{v}_i + nm_e \, v_{ei} \, (\mathbf{v}_e - \mathbf{v}_i) = 0, \tag{11.2}$$

$$\nabla \cdot n\mathbf{v}_e = \nabla \cdot n\mathbf{v}_i = S_i, \tag{11.3}$$

where the velocities are functions of x, y and z but all other quantities depend on x and y only; $\mathbf{E}$ is the space charge electric field and similarly to ∇n, it lies in the (x,y) plane of the magnetic field. The electron charge is -e; m_e and m_i, $\mathbf{v}_e$ and $\mathbf{v}_i$ stand for the mass and velocity of electrons and ions respectively.

The momentum equations (11.1) and (11.2) are coupled through electron-ion collisions. They are solved in Appendix 11.2 in the (x,y) plane of the magnetic field, along the field (subscript $\parallel$), in a direction perpendicular to it (subscript $\perp$), and along the z axis. In the (x,y) plane, the electron and ion fluxes parallel and perpendicular to the magnetic field are respectively

$$n\mathbf{v}_{e\parallel} = - D_{e\parallel} \, \nabla_{\parallel} n - n\mu_{e\parallel} \, \mathbf{E}_{\parallel}, \tag{11.4}$$

$$n\mathbf{v}_{i\parallel} = - D_{i\parallel} \, \nabla_{\parallel} n + n\mu_{i\parallel} \, \mathbf{E}_{\parallel}, \tag{11.5}$$

$$n\mathbf{v}_{e\perp} = - D_{e\perp} \, \nabla_{\perp} n - n\mu_{e\perp} \, \mathbf{E}_{\perp}, \tag{11.6}$$

$$n\mathbf{v}_{i\perp} = - D_{i\perp} \, \nabla_{\perp} n + n\mu_{i\perp} \, \mathbf{E}_{\perp}, \tag{11.7}$$

where $D_{e,i}$ and $\mu_{e,i}$ are the free diffusion coefficients and mobilities respectively. These coefficients are explicitly calculated in Appendix 11.2.

Clearly, mobilities and diffusion coefficients parallel to the magnetic field are not modified by the magnetic field whereas, in the direction perpendicular to the field, these transport coefficients are reduced (Appendix 11.2). Provided the plasma parameters and the local value of the magnetic field satisfy the inequality[17]

$$\gamma_c B_0^2(x,y) > 1, \tag{11.8}$$

with

$$\gamma_c = e^2/m_e\, m_i\, \nu_{in}\, \nu_e\,, \tag{11.9}$$

$$\nu_e = \nu_{en} + \nu_{ei}\left(1 + \frac{m_e \nu_{en}}{m_i \nu_{in}}\right), \tag{11.10}$$

where ν_e is an equivalent collision frequency for electrons, associated to the mean free path λ_e, we find

$$\mu_{e\perp} < \mu_{i\perp}. \tag{11.11}$$

Furthermore, provided that

$$\gamma_c B_0^2(x,y) > \frac{m_i \nu_{in} T_e - m_e \nu_{en} T_i}{m_i \nu_{in} T_i - m_e \nu_{en} T_e}\,, \tag{11.12}$$

we find

$$D_{e\perp} < D_{i\perp}. \tag{11.13}$$

To further analyze these results, we note that the dimensionless quantity $\gamma_c B_0^2(x,y)$ can be written either as

$$\gamma_c B_0^2(x,y) = \frac{\omega_{ci}\omega_{ce}}{\nu_{in}\nu_e} \tag{11.14}$$

or

$$\gamma_c B_0^2(x,y) = \frac{\lambda_e \lambda_{in}}{r_e r_i}\,, \tag{11.15}$$

where ω_{ce} and ω_{ci} are the electron and ion cyclotron frequencies respectively. Due to the heavier mass of ions relatively to electrons, mobilities and diffusion coefficients across the magnetic field practically satisfy the above inequalities (11.11) and (11.13), provided

$$\omega_{ce}/v_e > 1, \qquad \omega_{ci}/v_{in} \lesssim 1. \tag{11.16}$$

In magnetic field free plasmas, electron transport coefficients are much larger than those for ions (Appendix 11.2). In contrast, in strongly magnetized plasmas, these coefficients in the direction perpendicular to $\mathbf{B}_0$ become smaller than those of ions. In conclusion, electrons are more strongly tied to the magnetic field lines than ions.

11.2.2.   Plasma density and potential under ambipolar diffusion.

As in the case of ambipolar diffusion without magnetic field, quasineutrality is assumed and an assumption about electron and ion fluxes is needed that must be compatible with particle conservation (congruence approximation, see Sec. 6.2.2). In the present two-dimensional geometry, we assume thus that in the plane of the magnetic field, electron and ion fluxes are equal

$$nv_{e\parallel} = nv_{i\parallel}, \tag{11.17}$$

$$nv_{e\perp} = nv_{i\perp}. \tag{11.18}$$

Taking the divergence of the previous equalities, we verify that ion and electron production rates are equal. The congruence approximation will be examined further based on experimental results. With the help of eqns. (11.17) and (11.18), the electric field can be eliminated from eqns. (11.4) to (11.7). This allows us to express ion fluxes in the plane of the magnetic field, in terms of density gradients as

$$nv_{i\parallel} = -D_\parallel \nabla_\parallel n, \tag{11.19}$$

$$nv_{i\perp} = -D_\perp \nabla_\perp n, \tag{11.20}$$

where the plasma diffusion coefficients

$$D_\parallel = \frac{\mu_{e\parallel}D_{i\parallel} + \mu_{i\parallel}D_{e\parallel}}{\mu_{i\parallel} + \mu_{e\parallel}} \tag{11.21}$$

and

$$D_\perp = \frac{\mu_{e\perp}D_{i\perp} + \mu_{i\perp}D_{e\perp}}{\mu_{i\perp} + \mu_{e\perp}} \tag{11.22}$$

are explicitly expressed in Appendix 11.3. Taking the divergence of the flux, and introducing the conservation of ions as expressed by eqn. (11.3), we obtain a differential equation for the plasma density

$$\nabla \cdot \left(D_{\parallel} \nabla_{\parallel} n + D_{\perp} \nabla_{\perp} n \right) + S_i = 0 \tag{11.23}$$

with[†]

$$D_{\perp} = \frac{D_{\parallel}}{1 + \gamma_c B_0^2(x,y)}, \tag{11.24}$$

where $D_{\parallel}$ does not depend on B_0, as shown in Appendix 11.3.

Considering eqns. (11.4) to (11.7), (11.17) and (11.18), we can obtain the space-charge electric field in terms of the density gradients, namely

$$\mathbf{E} = -\frac{D_{e\parallel} - D_{i\parallel}}{\mu_{e\parallel} + \mu_{i\parallel}} \frac{\nabla_{\parallel} n}{n} - \frac{D_{e\perp} - D_{i\perp}}{\mu_{e\perp} + \mu_{i\perp}} \frac{\nabla_{\perp} n}{n}. \tag{11.25}$$

The electric field can also be written in terms of the diffusion temperatures $T_{\parallel}$, $T_{\perp}'$ and $T_{\perp}''$ given in Appendix 11.3, as

$$\mathbf{E} = -\frac{k_B T_{\parallel}}{e} \frac{\nabla_{\parallel} n}{n} - \left[\frac{k_B T_{\perp}'}{e} - \gamma_c B_0^2 \frac{k_B T_{\perp}''}{e} \right] \frac{1}{1 + \gamma_c B_0^2} \frac{\nabla_{\perp} n}{n}, \tag{11.26}$$

where

$$T_{\parallel} = T_{\perp}' \approx T_e, \tag{11.27}$$

$$T_{\perp}'' \approx T_i. \tag{11.28}$$

The above relations show that the dimensionless quantity $\gamma_c B_0^2(x,y)$ characterizes locally the plasma diffusion and the direction of the electric field with respect to the density gradients. From the diffusion eqns. (11.19), (11.20) and (11.24), one can see that in regions where

---

[†] The expression for $D_{\perp}$ (eqn. (11.24)) does not fundamentally differ from those (eqns. (6.7) and (6.12)) derived in Chap. 6.

$$\gamma_c B_0^2(x,y) \ll 1, \tag{11.29}$$

the effect of the magnetic field may be ignored and the diffusion is isotropic. Also, as in a magnetic field free plasma, the electric field is directed opposite to the density gradient. Furthermore, electrons diffuse faster than ions, and an electric field develops that accelerates ions towards lower density regions to maintain quasineutrality.

Conversely,[17] in regions where

$$\gamma_c B_0^2(x,y) > 1, \tag{11.30}$$

perpendicular diffusion is weakened, as shown by expressions (11.20) and (11.24), except for strong density gradients perpendicular to $\mathbf{B}_0$. Indeed, while the electric field component parallel to $\mathbf{B}_0$, $E_{\parallel}$, is always directed opposite to the corresponding component of the density gradient, the electric field component, in the plane of the magnetic field and perpendicular to it, $E_{\perp}$, as shown by eqn. (11.26), is in the direction of the density gradient in regions where

$$\gamma_c B_0^2(x,y) > T_e/T_i. \tag{11.31}$$

Such a reversal of the electric field is a consequence of a faster perpendicular diffusion of ions compared to that of electrons, as shown by eqn. (11.25), and leads to the acceleration of electrons.

These conclusions are similar to those obtained in Sec. 11.2.1 from single particle transport coefficients. Equation (11.23) has now to be solved for given boundary conditions. This leads to the two-dimensional dependence of density, from which one can compute the electric field and the potential as well as electron and ion fluxes. For electron and ion fluxes perpendicular to the magnetic field plane, we must also assume either sources and sinks of particles to be present at infinity or that such fluxes are closing on themselves as in experimental devices (see footnote in Appendix 11.2).

11.3. Application of the diffusion model to multipolar plasmas

The above collisional model is applied to plasmas confined in a multipolar magnetic field under the typical discharges parameters[7] mentioned previously ($n > 10^{11}\,\mathrm{cm^{-3}}$, $p \approx 10^{-3}\,\mathrm{torr}$, $T_e \approx 1$ eV, $T_i \approx 0.2$ eV). Under such conditions, the characteristic collision frequencies satisfy (see Appendix 11.1)

$$m_i \nu_{in} \gg m_e \nu_e , \tag{11.32}$$

$$v_{ei} \gg v_{en}. \tag{11.33}$$

We derive below an analytical expression for the dimensionless quantity $\gamma_c B_0^2(x,y)$, which characterizes locally the diffusion and the electric field. This requires to first look for the mathematical representation of the magnetic field $\mathbf{B}_0(x,y)$ in a multipolar configuration. Then, we discuss the validity of the model assumptions when applied to multipolar plasmas and, finally, we compare the qualitative results obtained with experiments.

11.3.1. Map of the multipolar magnetic field. Figure 11.1 shows the structure of the device considered. The plasma is contained inside a conductive cylinder of radius R centered on the Oz axis. The chamber is assumed to be of infinite extent along Oz and to have $2m$ rows of permanent, cylindrical magnets (radius a) of alternating polarity, parallel to Oz and equally spaced azimuthally, at a distance R_m from the Oz axis. In the case of Fig. 11.1 where the magnets are located outside the chamber, $R_m \approx R + a$. Such a configuration being independent of z, yields a two-dimensional magnetic field which only depends on the cylindrical coordinates (r, φ). Due to the $2m$-fold rotational symmetry of the configuration, the solution of the plasma equations can be limited to values of φ such that

$$0 \leq \varphi \leq \pi/2m. \tag{11.34}$$

In what follows, the coordinate system used is that of Fig. 11.1 with

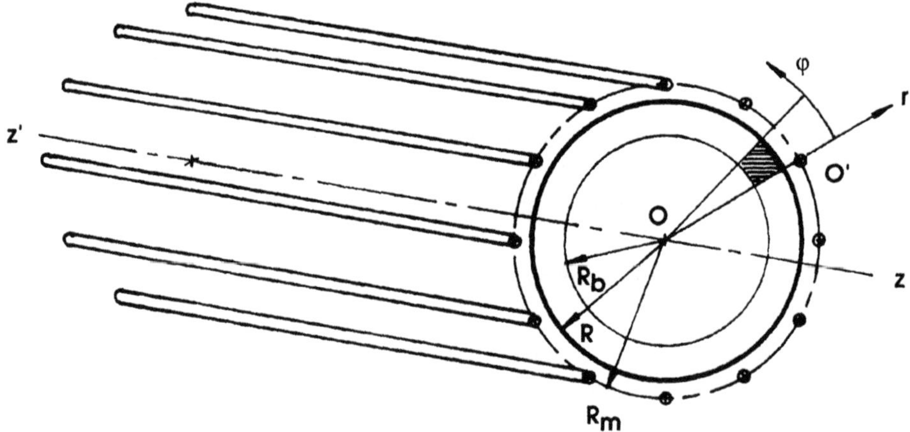

Fig. 11.1. Geometry of a multipolar discharge. The axes of the infinitely long cylindrical magnets are located at a radial distance $r = R_m$ from the discharge axis Oz and at azimuthal angles $\varphi = p\pi/m$. The radius of the discharge is R.

$$X = r/R_m, \qquad Y = m\,\varphi, \tag{11.35}$$

where X is the reduced distance measured from the axis Oz, and $Y = mR_m\varphi/R_m$ is an angle corresponding to the reduced distance measured along the circle of radius R_m. The axes of the magnets are then located at $X = 1$, $Y = p\pi$, with p an integer varying from 0 to $2m - 1$. Although all calculations are made using cylindrical geometry, the density and potential mappings are more simply plotted in rectangular coordinates O'X, O'Y, where the O'Y axis is a straight line going from the axis of one magnet to the axis of the next one, as shown in Fig. 11.2. This representation is strictly valid for a planar, infinite array of equally spaced magnets separated by a distance d, or in cylindrical geometry, provided the number of rows is very large in such a way that

$$\frac{m\,d}{\pi R_m} \approx 1. \tag{11.36}$$

The expressions of the magnetic fields in cylindrical and planar geometries are given in Appendix 11.4. The field lines (lines of constant λ_b), the lines perpendicular to the field (lines of constant μ_b), and the lines of constant magnetic field amplitude (b_b lines) are shown in Fig. 11.2, for a multipolar configuration with 56 rows of magnets.

In a multipolar magnetic field configuration produced by permanent magnets (as opposed to the picket fence configuration[†]), the cusps correspond to the regions where the field lines converge on the magnets. Along the lines $Y = p\pi$ which go from the magnet axis to the axis of the plasma ($\lambda_b = 0$ lines), B_0 decreases nearly exponentially (Appendix 11.4) on a scale length R_m/m (or d/π) and is thus confined to regions close to the magnets. Along the lines $Y = \pi/2 + p\pi$ going from the midpoint between magnets to the axis of the plasma ($\mu_b = 0$ lines), B_0 is equal to zero at $r = R_m$ and increases as r decreases to reach a maximum value $B_{0m}/2$ [††] at the point P (Fig. 11.2), which is a saddle point for $\mathbf{B}_0$. This point P is at a distance

$$\Delta R_m \approx 0.89\, R_m/m \tag{11.37}$$

[†] In the picket fence configuration,[6] the multipolar magnetic field is produced by current flowing with alternating polarity through equally spaced conductors. In this case, the cusps are located between conductors.

[††] As calculated in Appendix 11.4, $B_{0m} = B_{0s}\, m^2 a^2/R_m^2$, where B_{0s} is the magnetic field at the surface of a single magnet of radius a. The maximum value at the saddle point is strictly equal to $B_{0m}/2$ in the plane configuration. For the exact value in cylindrical geometry, see Appendix 11.4.

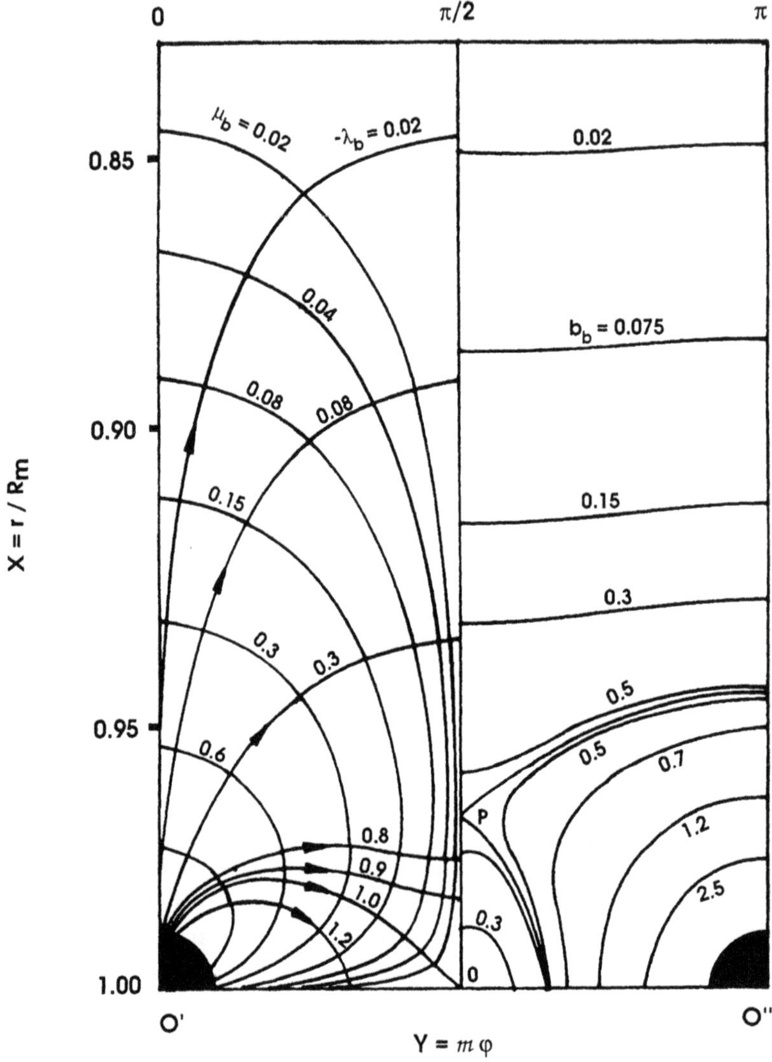

Fig. 11.2. Topology of the multipolar magnetic field in a cylindrical multipolar chamber. The axes of the infinitely long cylindrical magnets are located at $X = 1$, $Y = 0$ and $X = 1$, $Y = \pi$ (points O' and O''). $X = r/R_m$ is the reduced distance from the chamber axis and $Y = m\,\varphi$ is the product of the azimuthal angle φ and the number m of magnet pairs. "λ_b = constant" lines are field lines, "μ_b = constant" lines are orthogonal trajectories (perpendicular to the field lines) and "b_b = constant" lines are lines of constant magnetic field intensity. Point P is a saddle point for the magnetic field.

from the circle of radius R_m (Appendix 11.4). Then, beyond P, B_0 decreases exponentially towards the axis of the plasma, again with a scale length R_m/m, leaving most of the discharge volume without any significant magnetic field.

11.3.2. Limits of validity of the model. In order to solve the diffusion equation, the electron and ion production rate S_i has been assumed independent of position. This rate is proportional to the primary electron density since ionization results from collisions of primary electrons with the neutral gas. However, considering the initial energy of primary electrons (50 eV or above), their mean free path λ_{pi} (see Appendix 11.1) can be assumed much larger than the dimensions of the chamber for the pressure range considered ($\approx 10^{-3}$ torr). As a result, only very few of them undergo successive inelastic collisions before being collected by the wall. Thus, the slowing down of these electrons by excitation or ionization of the neutral gas is not enough for these electrons to be generally considered afterwards as plasma electrons for which $k_B T_e \approx 1$ eV. Thus, the motion of primary electrons can be studied (see Sec. 12.3) separetely from the joint (ambipolar) diffusion of ions and plasma electrons in the multipolar magnetic field.

Since the spatial variations of the plasma potential due to density gradients do not exceed a few volts, the motion of the energetic electrons (≈ 50 eV) is only determined by the magnetic field configuration $\mathbf{B}_0(x,y)$. The analytical model developed in Sec. 12.3 considers two invariants of the primary electron motion, namely the kinetic energy and the component of momentum along Oz. This allows to show that the primary electrons created in the magnetic field free region of the discharge will not be able to move beyond $\lambda_{b(max)} = 2\lambda_{b0}$ (eqn. (12.22)) lines, i.e. they cannot penetrate the peripheral regions between the cups (see Fig. 11.2). In the latter region, we thus take $S_i = 0$ while S_i is assumed constant in authorized regions.

The assumption of equal ion and plasma electron fluxes, as expressed by eqns. (11.17) and (11.18) can be considered as a good approximation for a cylindrical plasma provided it is magnetic field free (see Sec. 6.2.2). The derivation of the ion and electron transport coefficients in the present magnetic field configuration shows clearly that this assumption does not hold along Oz (Appendix 11.4). However, experimental results in a multipolar confined discharge, suggest that the congruence approximation can be used in the (x,y) plane of the magnetic field, since no charge separation is observed across the cusps.[9] Therefore, one can assume the joint diffusion of electrons and ions across the magnetic field.

11.3.3. Spatial diffusion characteristics, leak width and magnetic sheath thickness. From the configuration of the magnetic field and from the collisional diffusion model, some conclusions can be drawn on the diffusion characteristics in the different regions of a multipolar discharge, on the leak width of the plasma at the cusps as well as on the thickness of the so-called magnetic sheath.

Figure 11.3 shows that, in a multipolar discharge, one can distinguish three kinds of regions, each being characterized by specific diffusion regimes. These regions are differentiated according to the magnetic field configuration and to the local value of $\mathbf{B}_0(X,Y)$. Regions of the first type are such that

$$\gamma_c B_0^2(X,Y) < 1. \tag{11.38}$$

They can be considered as free from magnetic field since parallel and perpendicular diffusion coefficients are nearly equal. In such a case, the electric field is directed opposite to the density gradient. Two domains of this kind exist in multipolar

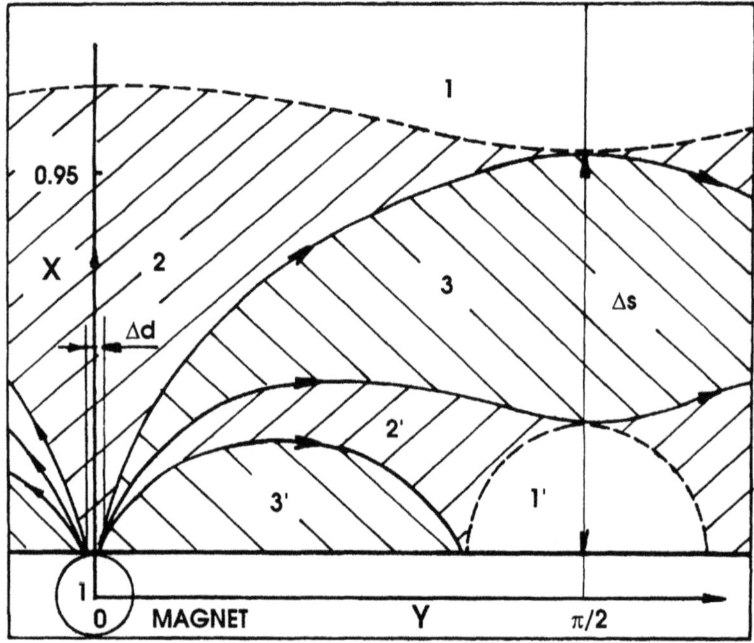

Fig. 11.3. Regions of different diffusion regimes in a multipolar discharge. Full lines are magnetic field lines, dashed lines are "$\gamma_c B_0^2 = 1$" lines, $X = r/R_m$ and $X = m\phi$ are reduced coordinates. Regions 1 and 1' are magnetic field free. In the shaded regions, diffusion in the direction perpendicular to the field lines is weakened. Regions of the second type (2 and 2') are linked to the first type regions (1 and 1') by parallel diffusion, while the third type regions (3 and 3') are not. Sheath thickness Δs and leak width Δd at the cusp are indicated.

discharges: domain 1 extends from the axis of the discharge to the $\gamma_c B_0^2 = 1$ boundary line while domain 1' extends from the midpoint between magnets to the near by $\gamma_c B_0^2 = 1$ boundary line. Regions of the second and third type, shown as shaded area in Fig. 11.3, are enclosed between the two $\gamma_c B_0^2 = 1$ boundary lines mentioned above and the chamber wall. The 2, 3, 2', and 3' domains are separated from each other by the three field lines that are tangent to the $\gamma_c B_0^2 = 1$ lines.

Regions 2 and 3 are such that $\gamma_c B_0^2 > 1$ and such that perpendicular diffusion is weakened compared with parallel diffusion. While electrons and ions can diffuse from regions of the first type to regions of the second type without crossing a magnetic field line (so-called parallel diffusion), diffusion from third type regions to other regions can only occur by crossing magnetic field lines (so-called perpendicular diffusion). This implies that the main flux of plasma towards the chamber wall occurs through region 2 linking the central part of the plasma to the cusp by parallel diffusion. In region 3, as a result of magnetic field symmetry with respect to $Y = \pi/2$, the magnetic field line is directed along $r = \text{const.}$ contours, hence the density gradients parallel to the magnetic field can be considered negligible close to $Y = \pi/2$. Thus, the density gradients are purely radial, i.e. perpendicular to $\mathbf{B}_0$ in that region. Then considering eqn. (11.26), provided that the strong field condition expressed by inequality (11.31) is verified, we see that the electric field is perpendicular to $\mathbf{B}_0$, and parallel and in the same direction as the density gradient. This explains the origin of the potential hills (Fig. 10.15) observed under certain discharge conditions.

Recall that the maximum magnetic field intensity along the $Y = \pi/2$ line is at the saddle point P. Then from Figs. 11.2 and 11.3, provided the magnetic field value $B_{0m}/2$ at the saddle-point pass satisfies

$$\gamma_c(B_{0m}/2)^2 > 1, \tag{11.39}$$

diffusion to the wall takes place mainly parallel to the magnetic field lines of region 2. When inequality (11.39) is reversed, that is when the magnet bars are too far from each other, domains 1 and 1' are connected (Fig. 11.3), and plasma diffusion also occurs between the cusps. This occurs for B_{0m} of the order of a few tens of gauss (see Appendix 11.2).

When inequality (11.39) is satisfied, a *magnetic sheath* of thickness Δs can be defined as the region of high intensity magnetic field limited by the chamber wall and by the far most $\gamma_c B_0^2 = 1$ line,[†] as shown in Fig. 11.3. The larger the sheath, the lower the

---

[†] The magnetic sheath thus defined allows one to determine approximately the outer limit of the useful volume for processing in a magnetic field free plasma.

perpendicular diffusion towards the wall.   Using expressions for the magnetic field given in Appendix 11.4, recalling $R_m - R \approx a$, the leak width Δd and sheath thickness Δs can be expressed respectively as

$$\Delta d = \frac{2\pi a^2}{d\left(\gamma_c B_{0m}^2\right)^{1/2}},$$

(11.40)

$$\Delta s = \frac{d}{2\pi}\log_e\left(4\gamma_c B_{0m}^2\right).$$

(11.41)

This implies that the number of magnet rows is such that the magnetic field can be approximated by that produced by a planar, infinite array of magnet bars.   The leak width at the cusp (eqn. (11.40)) is proportional to the ratio of the hybrid gyroradius r_h, calculated at the pass ($B_0 = B_{0m}/2$), to the square root of the product of the mean free paths of electrons and ions (see eqn. (11.15)).   This result agrees with the leak width obtained experimentally at intermediate and low pressure regimes.[4,9]

11.4.   Plasma density and potential profiles in multipolar discharges: numerical results.

The spatial variation of plasma density and potential in multipolar discharges is obtained by numerically solving the diffusion equation with the appropriate magnetic field configuration.   The density and potential variations found are similar to those mentioned in different experiments on multipolar devices.[7,8]

11.4.1.   Numerical procedure.   Before proceeding to the numerical solution of diffusion eqn. (11.23), two more assumptions are needed: i) since the diffusion equation is nonlinear because of the dependence of γ_c on density through the electron-ion collision frequency (Appendix 11.1), this parameter is calculated for a mean plasma density; ii) the production rate S_i is assumed constant in the central part of the discharge and equal to zero at the periphery between the cusps (see Sec. 11.3.2).

Taking into account the symmetry of the magnetic field, the diffusion equation can be integrated over a limited region of the chamber (Fig. 11.4) defined by

$$0 < m \, \varphi < \pi/2$$

(11.42)

with

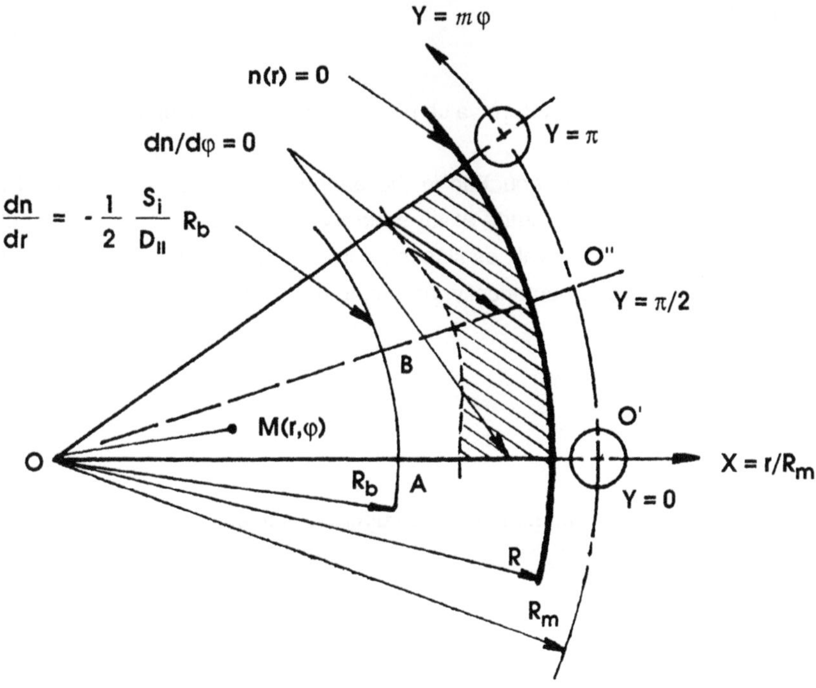

Fig. 11.4. Boundaries of the region where the diffusion equation is solved numerically. These are the chamber wall, the circle of radius R_b and the two radii defined by $Y = 0$ and $Y = \pi/2$. The conditions imposed at these boundaries are indicated and the shaded area corresponds to $\gamma_c B_0^2 > 1$.

$$\left(\frac{\partial n}{\partial \varphi}\right)_{\varphi=0} = 0, \qquad \left(\frac{\partial n}{\partial \varphi}\right)_{\varphi=\pi/2\,m} = 0, \qquad n(R) = 0, \tag{11.43}$$

as boundary conditions.

In the central part of the discharge, the plasma is magnetic field free and diffusion equation (eqn. (11.23)) reduces to

$$\nabla^2 n + \frac{S_i}{D_{\parallel}} = 0. \tag{11.44}$$

Neglecting any dependence of density on the azimuthal angle φ, the plasma density is obtained by integration of eqn. (11.44)

$$n(r) = n(0) - \frac{S_i}{4D_\parallel} r^2, \tag{11.45}$$

where $n(0)$ is the presently unknown value of the density at the center of the discharge.

Integration of the diffusion equation in the peripherical region (as opposed to the central part of the discharge) starts at an arbitrary radius value R_b chosen such that $\gamma_c B_0^2 \ll 1$, up to the chamber wall, as illustrated in Fig. 11.4. In this case, the diffusion equation is to be computed numerically, assuming the continuity of the density slope at R_b, namely

$$\left(\frac{\partial n}{\partial r}\right)_{r=R_b} = -\frac{S_i}{2D_\parallel} R_b. \tag{11.46}$$

The value $n(R_b)$ obtained from numerical calculations is used to determine $n(0)$ through eqn. (11.45),

$$n(0) = n(R_b) + \frac{S_i}{4D_\parallel} R_b^2. \tag{11.47}$$

The density profile obtained (analytically in the center, numerically near the wall) is independent of the choice of R_b (where both solutions are linked), provided the radius R_b is within the magnetic field free region.

To facilitate the numerical integration, we turn to the orthonormal system of (λ_b, μ_b) coordinates shown in Fig. 11.2, where the diffusion equation takes the form (Appendix 11.5)

$$\frac{\partial^2 \mathcal{N}}{\partial \mu_b^2} + \frac{\partial}{\partial \lambda_b}\left(\frac{1}{1 + \gamma_c \, b_b^2 \, B_{0m}^2} \frac{\partial \mathcal{N}}{\partial \lambda_b}\right) + \frac{1}{b_b^2} = 0, \tag{11.48}$$

where the reduced density $\mathcal{N}$ is defined as

$$\mathcal{N} = n\left(\frac{m^2 \, D_\parallel}{S_i \, R_m^2}\right). \tag{11.49}$$

The integration region in the (λ_b, μ_b) plane is delimited by the $\lambda_b = 0$, $\mu_b = 0$ lines, the wall and the circle of radius R_b. Provided that the inequality (Appendix 11.4)

$$\left(\frac{R_m}{R_b}\right)^m \gg 1 \qquad (11.50)$$

is satisfied, the transform of the circle R_b in the (r,φ) plane is, in the (λ_b,μ_b) plane, a circle of radius $\mathbb{R}_b$,

$$\mathbb{R}_b \approx 2\left(\frac{R_b}{R_m}\right)^m. \qquad (11.51)$$

To proceed with the numerical integration, we note (Appendix 11.4) that the value of $\mathbb{R}_b$ is small compared to the maximum value of λ_b and μ_b reached in the integration region. Therefore $\mathbb{R}_b$ can be chosen as the integration step. The boundary conditions (11.43) in the plasma become in the (λ_b,μ_b) coordinates

$$\left(\frac{\partial \mathfrak{N}}{\partial \lambda_b}\right)_{\lambda_b=0} = 0 \quad , \quad \left(\frac{\partial \mathfrak{N}}{\partial \mu_b}\right)_{\mu_b=0} = 0 \quad , \qquad (11.52)$$

and the continuity of the density slope given by eqn. (11.46) needs only to be taken at the points A ($\lambda_b = 0$, $\mu_b = \mathbb{R}_b$) and B ($\lambda_b = \mathbb{R}_b$, $\mu_b = 0$) shown in Fig. 11.4. The diffusion equation is then numerically computed by a Gauss-Seidel iteration procedure[18] on a mesh of 100 x 50 points. The accuracy of the method is tested by comparing the numerical solution without magnetic field to the analytical one given by eqn. (11.45); the accuracy is better than 3 % and the errors result mainly from the use of grid point approximations for the boundary conditions on the wall. To check our assumption that the plasma density in the magnetic field free region is independent of the azimuthal angle, the density values at $r = R_b$ for points A and B were calculated. The difference n(A) - n(B) is less than the accuracy of computation, supporting our initial assumption. Finally, from the density gradients obtained, the electric field is computed from eqn. (11.26), providing the reduced potential $\Phi = eV/k_BT_e$.

11.4.2. Spatial variations of plasma density and potential. The spatial variations of reduced density $\mathfrak{N}$ and potential Φ are computed for a cylindrical multipolar discharge with the following characteristics: number of magnet rows $2m = 56$, reduced wall radius $R/R_m = 0.984$, energy of primary electrons 50 eV. The parameter $\gamma_c B_{0m}^2$ is varied from 0 to 80. This corresponds to an argon plasma at a pressure of 10^{-3} torr with an electron density of 10^{12} cm^{-3}, to electron and ion temperatures of 1 eV and 0.2 eV respectively. Under such conditions, γ_c defined by eqn. (11.9) is 7×10^5 m^4 V^{-2} sec^{-2} (Appendix 11.1). Then, the corresponding value of B_{0m} is varied from 0 to 130 gauss (B_{0m} is twice the value at the saddle point of a multipolar discharge).

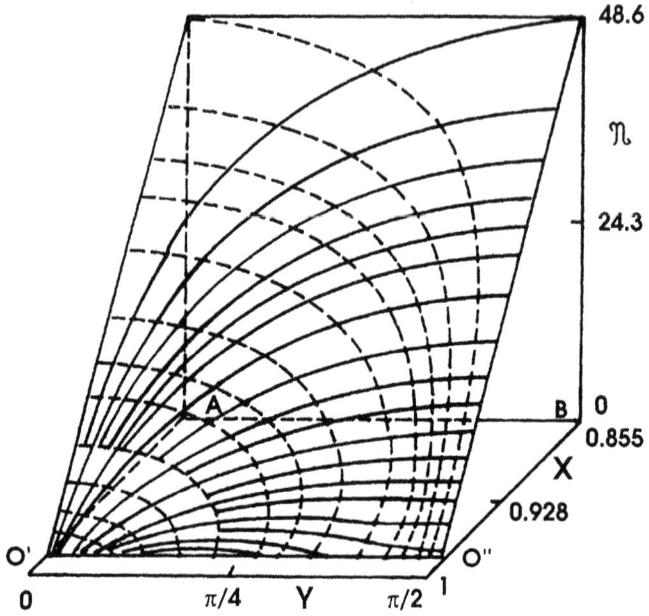

Fig. 11.5. Reduced density $\mathfrak{N}$ in the absence of magnetic field as plotted along what would be the field lines (full lines) and the orthogonal trajectories (dashed lines) in the presence of B_{0m} ($B_{0m} \rightarrow 0$). The coordinate system and points O', O", A and B refer to Fig. 11.4.

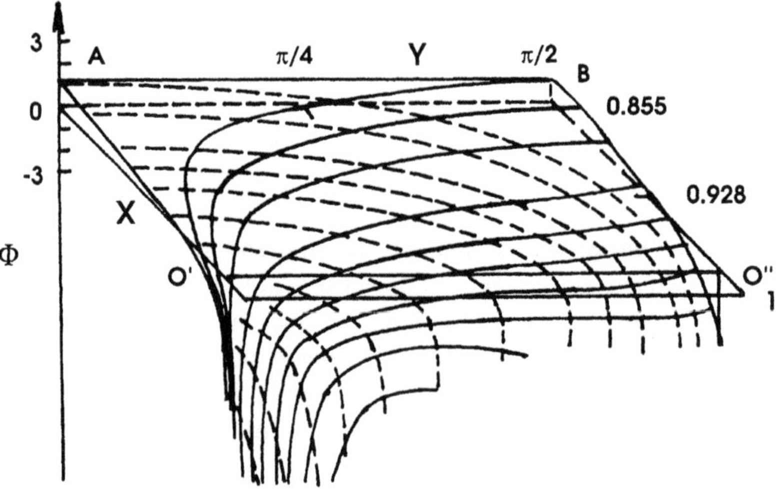

Fig. 11.6. Reduced potential Φ in the absence of magnetic field as plotted along what would be the field lines (full lines) and the orthogonal trajectories (dashed lines) in the presence of B_{0m} ($B_{0m} \rightarrow 0$). The coordinate system and points O', O", A and B refer to Fig. 11.4.

The coordinate system used for plotting the results is the same as in Figs. 11.1 and 11.4. The A, B, O', O" points refer to Fig. 11.4, the magnet being located at point O'. The reduced density and potential are plotted along the field lines (λ_b lines) and along the orthogonal trajectories (μ_b lines). The results obtained for the plasma density and potential are consistent with the expected theoretical behavior. However, as a result of the boundary conditions retained ($n(R) = 0$), the electric field diverges at the wall and the corresponding values of potential were discarded.

Figures. 11.5 and 11.6 show the reduced density and potential variations near the wall in the absence of magnetic field ($B_{0m} \to 0$) while Figs. 11.7 and 11.8 are plotted for $\gamma_c B_{0m}^2 = 80$. As can be seen by comparing density profiles with and without magnetic field, the effect of the magnetic field, for a given ionization rate S_i, is to increase the density everywhere in the discharge except in regions situated between the magnets (compare density profiles for $Y = \pi/2$). In these regions between magnets and close to the wall where field lines are mainly parallel to the wall, ions diffuse faster than electrons across the magnetic field towards the wall. Thus, as shown in Fig. 11.8, a reversal of the electric field occurs at these locations, generating potential hills (see Fig. 10.15 in Sec. 10.7.1).

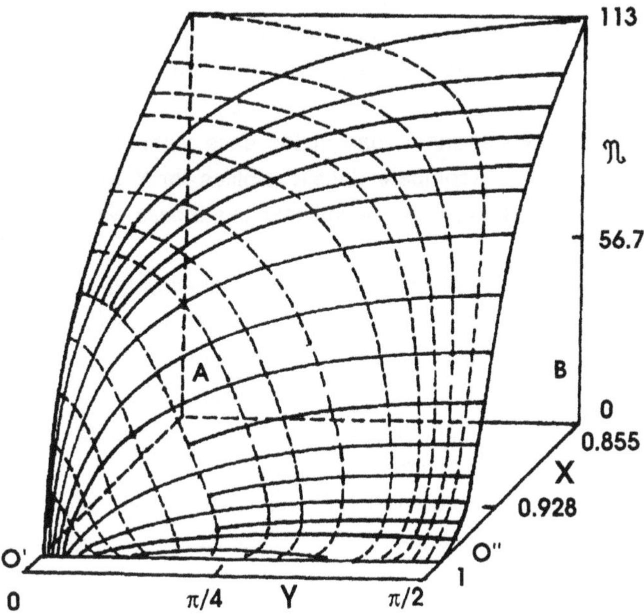

Fig. 11.7. Reduced density $\mathfrak{N}$ in a multipolar discharge for $m = 28$ and $\gamma_c B_{0m}^2 = 80$ plotted along the field lines (full lines) and the orhogonal trajectories (dashed lines). The coordinate system and points O', O", A and B refer to Fig. 11.4.

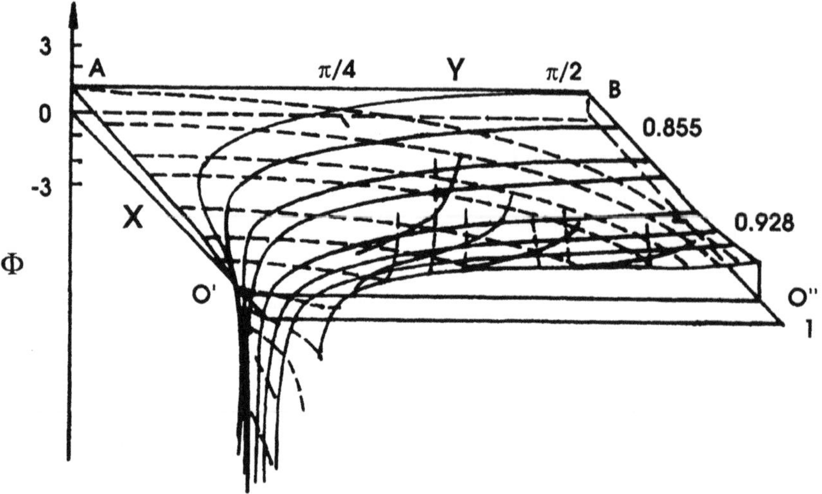

Fig. 11.8. Reduced density Φ in a multipolar discharge for $m = 28$ and $\gamma_c B_{0m}^2 = 80$ plotted along field lines (full lines) and orhogonal trajectories (dashed lines). The coordinate system and points O', O", A and B refer to Fig. 11.4.

Electrons and ions, which are produced mainly in the magnetic field free volume of the discharge according to the present model, are evacuated to the wall by parallel diffusion along the field lines, these lines forming a section which becomes narrower as particles move towards the wall. Hence in the cusp, the density gradient parallel to these field lines is much stronger than in the case without a magnetic field, and increases towards the wall (compare density profiles for $Y = 0$).[8] Correlatively, the electric field grows more rapidly near the wall than in the case without magnetic field.

Numerical computations also confirmed that the leak width and sheath thickness (Fig. 11.3) can be approximated by expressions (11.40) and (11.41), previously derived. The profiles and potential maps have been obtained experimentally for multipolar configurations[9] in a regime where the mean free paths of plasma particles are of the order of, or greater than, the characteristic length of the device whereas it was assumed shorter in the present model. It was found that, despite the fact that the two collisional regimes differ, the nature of the phenomena involved, such as space charge effects and the need to maintain quasineutrality, seems to have the same influence on diffusion and potential.

11.4.3. Plasma density at the axis. Figure 11.9 shows the ratio of plasma density $n_b(0)$ at the axis of a multipolar discharge to the density $n_0(0)$ of the same discharge without a magnetic field as a function of the parameter $\gamma_c B_{0m}^2$, assuming the

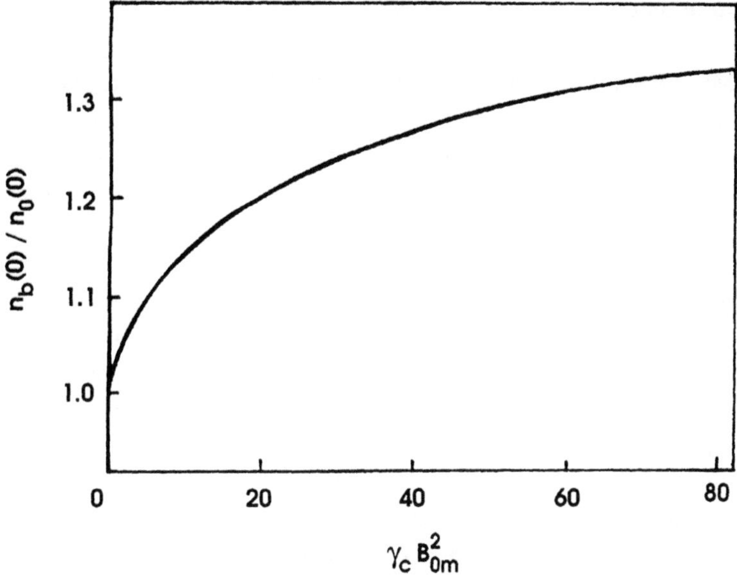

Fig. 11.9. Ratio of plasma density $n_b(0)$ at the axis of a multipolar discharge to the density $n_0(0)$ at the same position without the multipolar field as a function of $\gamma_c B_{0m}^2$, assuming the same ionization rate in both cases.

same ionization rate S_i. As a result of this assumption, this ratio is a mere indication of efficiency of plasma confinement. Since the model does not take into account the increase in primary electron density due to magnetic confinement, the true ionization rate S_i' in presence of multipolar confinement is larger than S_i. Thus the ratio of the plasma densities at the axis with and without magnetic field, $n_b'(0)/n_0(0)$ is given by

$$\frac{n_b'(0)}{n_0(0)} = \frac{S_i'}{S_i}\frac{n_b(0)}{n_0(0)},$$

(11.53)

where S_i'/S_i takes into account the increased ionization due to primary electron confinement while $n_b(0)/n_0(0)$ corresponds to the increase in plasma density due to plasma confinement.

The numerical results show that the efficiency of plasma confinement by the cusp magnetic field is weak due to the diffusion of particles along field lines. This result is in qualitative agreement with confinement time measurements that show that permanent magnets reduce the loss rate of plasma by a factor of approximately two (hence $n_b(0)/n_0(0) \approx 2$) while the increase in plasma density is mainly due to the confinement of primary electrons ($S_i'/S_i \approx 30$).[5]

11.5. Conclusion

We have presented a collisional diffusion model for a low-beta plasma under the assumptions of quasineutrality and of joint diffusion of electrons and ions in the plane of the magnetic field. This model was applied to high density multipolar discharges assuming the mean free path of the primary electrons to be larger than the characteristic diameter of the chamber, while the mean free paths of plasma electrons and ions are much smaller. Although most experiments were conducted in a collisionless regime, a good qualitative agreement was found between measured and computed density and potential profiles. This model accounts for the existence of positive potential hills between the cusps, and for the formation of a potential channel at the cusp through which diffusion takes place. The density profiles are comparable to measured profiles, steeper at the cusp than in-between. This diffusion model also confirms that the increase of plasma density due to the magnetic multipolar configuration is mainly due to the confinement of primary electrons. However, the model is unable to explain why plasma homogeneity is progressively improved when decreasing the operating pressure of the discharge. Again, primary electrons may be involved in this phenemenon, showing the necessity of a closer study of primary electrons within multipolar discharges. This is the object of the next chapter.

APPENDIX 11.1

MAIN PARAMETERS OF AN ARGON MULTIPOLAR DISCHARGE

Discharge conditions

Gas: argon

Pressure: $p = 10^{-3}$ torr

Density of neutral atoms: $N = 3.5 \times 10^{13}$ cm^{-3} at 300 K

Energy and velocity of particles

Electron temperature: $k_B T_e = 1$ eV ($T_e = 11560$ K)

Thermal electron velocity: $v_{T_e} = (8 \, k_B T_e / \pi m_e)^{1/2} = 6.7 \times 10^5$ m/s

Ion temperature: $k_B T_i = 0.2$ eV ($T_i = 2890$ K)

Thermal ion velocity: $v_{T_i} = (8 \, k_B T_i / \pi m_i)^{1/2} = 1.2 \times 10^3$ m/s

Energy of primary electrons: 50 eV $= 1/2 \, m \, v_p^2$

Velocity of primary electrons: $v_p = 4.2 \times 10^6$ m/s

Primary electron collisions with neutrals (10^{-3} torr)

Momentum transfer

Cross-section: Q_c (50 eV) $= 10^{-19}$ m^2 (ref. 19)[†]

Collision frequency: $v_{pn} = N \, v_p \, Q_c = 1.5 \times 10^7$ s^{-1}

Mean-free-path: $\lambda_{pn} = 1 \, / \, NQ_c = 0.30$ m

[†] For argon, the electron momentum transfer cross-section[19] is practically equal to the total elastic scattering cross-section.[20]

Ionization

Cross-section: Q_i (50 eV) = 2.5 x 10^{-20} m^2 (ref. 21)

Ionization frequency: $v_i = N\, v_p\, Q_i$ = 3.8 x 10^6 s^{-1}

Mean-free-path: λ_{pi} = 1 / NQ_i = 1.15 m

Collisions of plasma electrons with neutrals (10^{-3} torr)

Cross-section: Q_c (1 eV) = 1.4 x 10^{-20} m^2 (ref. 19)

Collision frequency: $v_{en} = N\, v_{T_e}\, Q_c$ = 3.3 x 10^5 s^{-1}

Mean-free-path: λ_{en} = 1 / NQ_c = 2.0 m

Electron mobility: $\mu_e = e/m_e v_{en}$ = 5.3 x 10^5 m^2 V^{-1} s^{-1}

Collisions of ions with neutrals (10^{-3} torr)

Ion mobility: $\mu_i = e/m_i v_{in}$ = 1.1 x 10^2 m^2 V^{-1} s^{-1} (ref. 22)

Collision frequency: v_{in} = 2.1 x 10^4 s^{-1}

Mean-free-path: $\lambda_{in} = v_{T_i} / v_{in}$ = 4.8 cm

Comparison between characteristic collision frequencies (10^{-3} torr)

According to Table 11.1, the following inequalities are satisfied

$$m_e\, v_{en} \ll m_i\, v_{in}, \tag{11.54}$$

$$\frac{v_{en}}{v_{ei}} \ll 1 \ll \frac{m_i v_{in}}{m_e v_{ei}} \tag{11.55}$$

for $10^{11} < n < 10^{12}$ cm^{-3}. Equation (11.55) shows that Coulomb collisions cannot be neglected with respect to electron-neutral collisions in a multipolar discharge at 10^{-3} torr in argon.

Table 11.1. Values of the main parameters in a multipolar argon discharge ($p = 10^{-3}$ torr) at two plasma densities.

Plasma parameters	Expressions	Plasma density	
		$n = 10^{10}$ cm^{-3}	$n = 10^{12}$ cm^{-3}
Electron plasma frequency	$\dfrac{\omega_{pe}}{2\pi} = \dfrac{1}{2\pi}\left(\dfrac{ne^2}{\varepsilon_0 m_e}\right)^{1/2}$	9×10^{8} s^{-1}	9×10^{9} s^{-1}
Debye length	$\lambda_D = \left(\dfrac{\varepsilon_0 k_B T_e}{ne^2}\right)^{1/2}$	74 µm	7.4 µm
$\Lambda_c = \lambda_D/\mathcal{P}_c$ ($\mathcal{P}_c$: critical impact parameter)	$\mathcal{P}_c = \dfrac{e^2}{12\,\pi\,\varepsilon_0\,k_B\,T_e}$	1.5×10^{5}	1.5×10^{4}
Coulomb logarithm	$\log_e \Lambda_c$	12	9.6
Electron-ion collision frequency	$\nu_{ei} = \dfrac{2\,\omega_{pe}}{\sqrt{2\pi}}\,\dfrac{\log_e \Lambda_c}{\Lambda_c}$	3.6×10^{5} s^{-1}	2.9×10^{7} s^{-1}
Electron-electron collision frequency	$\nu_{ee} \approx \dfrac{\nu_{ei}}{\sqrt{2}}$	2.5×10^{5} s^{-1}	2.0×10^{7} s^{-1}
Electron-mean-free path (Coulomb collisions)	$\lambda_{e(i+e)} = \dfrac{v_{T_e}}{\nu_{ei} + \nu_{ee}}$	1.1 m	1.4 cm
ν_{en}/ν_{ei}		~ 1	10^{-2}
$m_i \nu_{in}/m_e \nu_{ei}$		5000	50
ν_e	$\nu_e \approx \nu_{en} + \nu_{ei}$	6.9×10^{5} s^{-1}	2.9×10^{7} s^{-1}
γ_c	$\gamma_c = \dfrac{e^2}{m_e m_i \nu_{in} \nu_e}$	0.3×10^{8} m^{4}V^{-2}s^{-2}	0.7×10^{6} m^{4}V^{-2}s^{-2}
$\gamma_c B_0^2 = 1$	$B_0 = \gamma_c^{-1/2}$	1.8 gauss	12 gauss

APPENDIX 11.2

**MOBILITIES AND DIFFUSION COEFFICIENTS FOR ELECTRONS
AND IONS IN A MAGNETIC FIELD**

Expressions for the transport coefficients

We derive below the expressions used in Chap. 11 for the transport coefficients, assuming a magnetic field $\mathbf{B}_0(x,y)$, lying in the (x,y) plane and invariant under translation along z. Momentum equations (11.1) and (11.2) can be reduced to

$$\mathbf{C}_e - \frac{e}{m_e \nu_{ei}} \mathbf{v}_e \times \mathbf{B}_0 - (\mathbf{v}_e - \mathbf{v}_i) - \frac{\nu_{en}}{\nu_{ei}} \mathbf{v}_e = 0, \tag{11.56}$$

$$\mathbf{C}_i + \frac{e}{m_e \nu_{ei}} \mathbf{v}_i \times \mathbf{B}_0 + (\mathbf{v}_e - \mathbf{v}_i) - \frac{m_i \nu_{in}}{m_e \nu_{ei}} \mathbf{v}_i = 0, \tag{11.57}$$

where

$$\mathbf{C}_e = \left(- k_B T_e \, \nabla n - en \, \mathbf{E}\right) / n m_e \nu_{ei}, \tag{11.58}$$

$$\mathbf{C}_i = \left(- k_B T_i \, \nabla n + en \, \mathbf{E}\right) / n m_e \nu_{ei}. \tag{11.59}$$

These equations are solved in the (x,y) plane of the magnetic field, parallel to the field (subscript $\|$), perpendicular to it (subscript $\perp$), and along the z-axis (subscript z). Assuming

$$E_z = 0, \qquad \nabla_z n = 0, \tag{11.60}$$

the projection of eqns. (11.56) and (11.57) along the different axes yields for electrons

$$\mathbf{C}_{e\|} + v_{i\|} - \left(1 + \frac{\nu_{en}}{\nu_{ei}}\right) v_{e\|} = 0, \tag{11.61}$$

$$\mathbf{C}_{e\perp} + v_{i\perp} - \left(1 + \frac{\nu_{en}}{\nu_{ei}}\right) v_{e\perp} - \frac{eB_0}{m_e \nu_{ei}} v_{ez} = 0, \tag{11.62}$$

$$v_{iz} - \left(1 + \frac{\nu_{en}}{\nu_{ei}}\right) v_{ez} + \frac{eB_0}{m_e \nu_{ei}} v_{e\perp} = 0, \tag{11.63}$$

and for ions

$$\mathfrak{C}_{i\parallel} + v_{e\parallel} - \left(1 + \frac{m_i v_{in}}{m_e v_{ei}}\right) v_{i\parallel} = 0, \tag{11.64}$$

$$\mathfrak{C}_{i\perp} + v_{e\perp} - \left(1 + \frac{m_i v_{in}}{m_e v_{ei}}\right) v_{i\perp} + \frac{e B_0}{m_e v_{ei}} v_{iz} = 0, \tag{11.65}$$

$$v_{ez} - \left(1 + \frac{m_i v_{in}}{m_e v_{ei}}\right) v_{iz} - \frac{e B_0}{m_e v_{ei}} v_{i\perp} = 0. \tag{11.66}$$

To obtain the mobility and diffusion coefficients in the parallel direction, we solve eqns. (11.61) and (11.64) for $v_{e\parallel}$ and $v_{i\parallel}$. The mobilities and diffusion coefficients parallel to the magnetic field are then deduced from eqns. (11.4) and (11.5)

$$\mu_{e\parallel} = \frac{e}{m_e v_e}, \tag{11.67}$$

$$D_{e\parallel} = \frac{k_B T_e + \dfrac{m_e v_{ei}}{m_i v_{in}}(k_B T_e + k_B T_i)}{m_e v_e}, \tag{11.68}$$

$$\mu_{i\parallel} = \frac{e}{m_i v_{in} v_e / v_{en}}, \tag{11.69}$$

$$D_{i\parallel} = \frac{k_B T_i + \dfrac{v_{ei}}{v_{en}}(k_B T_e + k_B T_i)}{m_i v_{in} v_e / v_{en}}, \tag{11.70}$$

where v_e is defined by eqn. (11.10).

To obtain the mobility and diffusion coefficients in the perpendicular direction, we solve the system of eqns. (11.62), (11.63), (11.65) and (11.66), which yield $v_{e\perp}$, $v_{i\perp}$, v_{ez} and v_{iz} where v_{ez} and v_{iz} express the drift velocities of electrons and ions along the z axis. The mobilities and diffusion coefficients in the plane of the magnetic field and perpendicular to it[†] are then deduced from eqns. (11.6) and (11.7)

$$\mu_{e\perp} = \frac{\mu_{e\parallel}}{\Xi}\left(1 + \frac{m_e v_{en}}{m_i v_{in}} \gamma_c B_0^2\right), \tag{11.71}$$

$$\mu_{i\perp} = \frac{\mu_{i\parallel}}{\Xi}\left(1 + \frac{m_i v_{in}}{m_e v_{en}} \gamma_c B_0^2\right), \tag{11.72}$$

[†] The transport coefficients along the Oz axis can be also calculated in the same way. The electron and ion mobilities have opposite sign, indicating opposite drift velocities of electrons and ions along the Oz axis.

$$D_{e\perp} = \frac{\mu_{e\parallel}}{e\Xi}\left[k_BT_e\left(1 + \frac{m_e\nu_{en}}{m_i\nu_{in}}\gamma_c B_0^2\right) + \frac{m_e\nu_{ei}}{m_i\nu_{in}}(k_BT_e + k_BT_i)\left(1 + \gamma_c B_0^2\right)\right], \tag{11.73}$$

$$D_{i\perp} = \frac{\mu_{i\parallel}}{e\Xi}\left[k_BT_i\left(1 + \frac{m_i\nu_{in}}{m_e\nu_{en}}\gamma_c B_0^2\right) + \frac{\nu_{ei}}{\nu_{en}}(k_BT_e + k_BT_i)\left(1 + \gamma_c B_0^2\right)\right], \tag{11.74}$$

with

$$\Xi = \left(1 + \gamma_c B_0^2\right)^2 + \frac{\gamma_c^2 B_0^2}{e^2}\left(m_i\nu_{in} - m_e\nu_{en}\right)^2 \tag{11.75}$$

where γ_c is defined by equation (11.9).

Comparison of transport coefficients

In the direction parallel to the magnetic field lines, mobilities and diffusion coefficients are the same as in magnetic field free plasma and are much greater for electrons than for ions

$$\frac{\mu_{e\parallel}}{\mu_{i\parallel}} = \frac{m_i\nu_{in}}{m_e\nu_{en}} \gg 1, \tag{11.76}$$

$$\frac{D_{e\parallel}}{D_{i\parallel}} \approx \frac{m_i\nu_{in}}{m_e\nu_{ei}} \gg 1. \tag{11.77}$$

In the direction perpendicular to the magnetic field lines, the transport coefficients in weak magnetic fields ($\gamma_c B_0^2 \ll 1$) remain practically the same as in magnetic field free plasmas. In large magnetic fields ($\gamma_c B_0^2 \gg 1$), mobilities and diffusion coefficients, which decrease when the magnetic field increases, become much greater for ions than for electrons

$$\frac{\mu_{e\perp}}{\mu_{i\perp}} \approx \frac{m_e\nu_{en}}{m_i\nu_{in}} \ll 1, \tag{11.78}$$

$$\frac{D_{e\perp}}{D_{i\perp}} \approx \frac{m_e\nu_{ei}}{m_i\nu_{in}} \ll 1. \tag{11.79}$$

Ion and electron diffusion coefficients are equal ($D_{i\perp} = D_{e\perp}$) for

$$\gamma_c B_0^2 = \frac{m_i\nu_{in}T_e - m_e\nu_{en}T_i}{m_i\nu_{in}T_i - m_e\nu_{en}T_e} \approx \frac{T_e}{T_i}, \tag{11.80}$$

while ion and electron mobilities are equal ($\mu_{i\perp} = \mu_{e\perp}$) when

$$\gamma_c B_0^2 = 1. \tag{11.81}$$

The latter condition is realized in the plasma ($n = 10^{12}$ cm^{-3}) defined in Appendix 11.1, when

$$B_0 = 12 \text{ gauss.} \tag{11.82}$$

APPENDIX 11.3

DIFFUSION COEFFICIENTS OF A PLASMA
SUBMITTED TO A MAGNETIC FIELD

The plasma diffusion coefficients appearing in eqns. (11.21) and (11.22), using electron and ion transport coefficient from Appendix 11.2, can be expressed as

$$D_{\parallel} = \frac{k_B T_e + k_B T_i}{m_i \nu_{in} + m_e \nu_{en}},$$ (11.83)

$$D_{\perp} = \frac{D_{\parallel}}{1 + \gamma_c B_0^2}.$$ (11.84)

The expressions of the different diffusion temperatures $T_{\parallel}$, $T'_{\perp}$ and $T''_{\perp}$ appearing in eqn. (11.26) are obtained from eqns. (11.25) and (11.26) for the electric field using electron and ion transport coefficients in Appendix 11.2,

$$T_{\parallel} = \frac{m_i \nu_{in} T_e - m_e \nu_{en} T_i}{m_i \nu_{in} + m_e \nu_{en}} \approx T_e,$$ (11.85)

$$T'_{\perp} = \frac{m_i \nu_{in} T_e - m_e \nu_{en} T_i}{m_i \nu_{in} + m_e \nu_{en}} \approx T_e,$$ (11.86)

$$T''_{\perp} = \frac{m_i \nu_{in} T_i - m_e \nu_{en} T_e}{m_i \nu_{in} + m_e \nu_{en}} \approx T_i.$$ (11.87)

APPENDIX 11.4

MAGNETIC POTENTIALS AND FIELDS IN A MULTIPOLAR CONFIGURATION

Geometry

We start by considering a multipolar magnetic field configuration of cylindrical symmetry, which results from $2m$ rows of permanent cylindrical magnets (Fig. 11.1) parallel to the axis and radially equally spaced. These rows have radially uniform magnetization with alternate polarity. They are assumed to be of infinite axial extent, producing a two-dimensional magnetic field perpendicular to the axis.

Background

Outside the magnets, the static magnetic induction $\mathbf{B}_0$ can be indifferently derived from the vector potential $\mathbf{A}$ or from the magnetic scalar potential $\mathcal{U}$

$$\mathbf{B}_0 = \nabla \times \mathbf{A}, \tag{11.88}$$

$$\mathbf{B}_0 = - \nabla \, \mathcal{U}. \tag{11.89}$$

In the particular case of the infinitely long multipolar configuration defined above, $\partial/\partial z = 0$ and the magnetic field $\mathbf{B}_0$ has no component along $0z$. We use cylindrical coordinates (r, φ, z) where r is the distance from the $0z$ axis and φ is the azimuthal angle. Then the vector potential components are

$$A_r = 0, \tag{11.90}$$

$$A_\varphi = 0, \tag{11.91}$$

$$A_z = A_z \, (r,\varphi) \tag{11.92}$$

and the relationship between $\mathbf{A}$, $\mathbf{B}_0$ and $\mathcal{U}$ is given by

$$(\mathbf{B}_0)_r = \frac{1}{r} \frac{\partial A_z}{\partial \varphi} = - \frac{\partial \mathcal{U}}{\partial r}, \tag{11.93}$$

$$(\mathbf{B}_0)_\varphi = - \frac{\partial A_z}{\partial r} = - \frac{1}{r} \frac{\partial \mathcal{U}}{\partial \varphi}, \tag{11.94}$$

$$(\mathbf{B_0})_z = 0. \tag{11.95}$$

Although calculation of the $\mathbf{B_0}$ field can be carried out either from $\mathbf{A}$ or from u, the knowledge of both A_z and u is of practical interest: the "A_z = constant" lines and "u = constant" lines are the field lines and the lines perpendicular to the field lines respectively. Since plasma diffusion occurs quite differently in directions parallel and perpendicular to the field lines, i.e. along the "A_z = constant" lines or along the "u = constant" lines, these lines will serve as the basis of a system that facilitates the numerical resolution of the diffusion equation (see Appendix 11.5).

Symmetries of the multipolar magnetic field

In the multipole sector presented in Fig. 11.10, the coordinates of points M_0, M_1, M_2 and M_3 are (r, φ), $(r, -\varphi)$, $(r, \varphi + \pi/m)$ and $(r, -\varphi + \pi/m)$ respectively. The relations between the different B_0 components are given by

$$(\mathbf{B_0})_r\,(r, -\varphi) = (\mathbf{B_0})_r\,(r, \varphi), \tag{11.96}$$

$$(\mathbf{B_0})_\varphi\,(r, -\varphi) = -(\mathbf{B_0})_\varphi\,(r, \varphi), \tag{11.97}$$

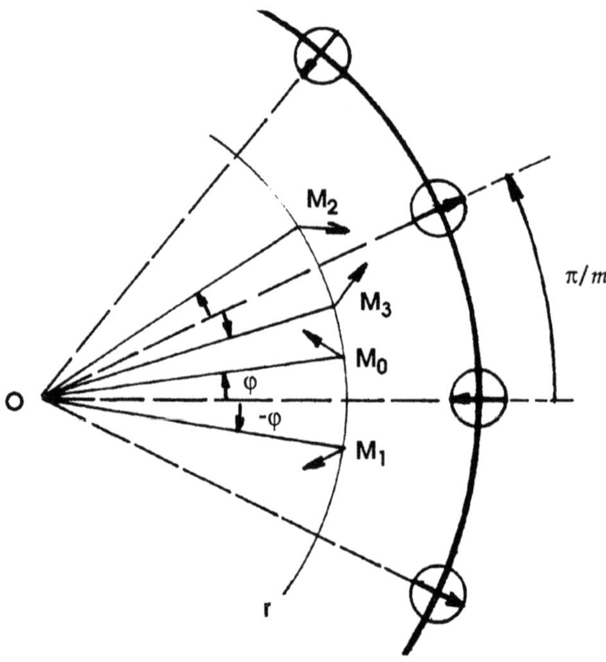

Fig. 11.10. Symmetries of the multipolar magnetic field.

$$(\mathbf{B}_0)_r\left(r,\ \varphi + \tfrac{\pi}{m}\right) = -\,(\mathbf{B}_0)_r(r,\ \varphi), \tag{11.98}$$

$$(\mathbf{B}_0)_\varphi\left(r,\ \varphi + \tfrac{\pi}{m}\right) = -\,(\mathbf{B}_0)_\varphi(r,\ \varphi), \tag{11.99}$$

$$(\mathbf{B}_0)_r\left(r,\ -\varphi + \tfrac{\pi}{m}\right) = -\,(\mathbf{B}_0)_r(r,\ \varphi), \tag{11.100}$$

$$(\mathbf{B}_0)_\varphi\left(r,\ -\varphi + \tfrac{\pi}{m}\right) = (\mathbf{B}_0)_\varphi(r,\ \varphi). \tag{11.101}$$

Such a symmetry shows that the study of the multipolar magnetic field may be restricted to a $0 < \varphi < \pi/2m$ sector.

Magnetic scalar potential from a single row

The magnetic scalar potential $\mathcal{U}$ generated at position $\mathbf{r}$ by a magnet row is given by

$$\mathcal{U} = -\,\frac{\mu_0}{4\pi}\int_{\mathbb{V}} \mathbf{M}\cdot\nabla\ \frac{1}{|\mathbf{r}-\mathbf{r}'|}\,d\mathbb{V}, \tag{11.102}$$

where μ_0 is the vacuum permeability and $\mathbf{r}'$ defines the position of each elementary magnetic moment $\mathbf{M}\,d\mathbb{V}$ in the volume $\mathbb{V}$ of the magnet. Assuming a uniform magnetization density $\mathbf{M}$, eqn. (11.102) can be rewritten according to

$$\mathcal{U} = \frac{\mu_0\mathbf{M}}{4\pi}\cdot\int_{\mathbb{V}} \frac{\mathbf{r}-\mathbf{r}'}{|\mathbf{r}-\mathbf{r}'|^3}\,d\mathbb{V}. \tag{11.103}$$

The integral in eqn. (11.103) is proportional to the electric field $\mathbf{E}$ created at position $\mathbf{r}$ by a uniform volumic electric charge $\mathbb{Q}$ in the same volume $\mathbb{V}$ since

$$\mathbf{E} = \frac{\mathbb{Q}}{4\pi\varepsilon_0}\int_{\mathbb{V}} \frac{\mathbf{r}-\mathbf{r}'}{|\mathbf{r}-\mathbf{r}'|^3}\,d\mathbb{V}. \tag{11.104}$$

An infinitely long cylinder of radius a (a is the radius of an elementary magnet) with uniform volumic electric charge $\mathbb{Q}$ produces a purely radial electric field. Its value at position $\mathbf{r}$ from the (magnet) cylinder axis is obtained by application of Gauss's theorem, yielding

$$\mathbf{E} = \frac{\mathbb{Q}}{\varepsilon_0}\,a^2\,\frac{\mathbf{r}}{r^2}. \tag{11.105}$$

With the help of eqns. (11.104) and (11.105), we find that the magnetic scalar potential produced at position **r** from the axis of an infinitely long cylinder of radius a (Fig. 11.11) and uniform magnetization density **M** is thus[†]

$$\mathcal{U} = \mu_0 \frac{a^2}{r^2}\, \mathbf{M} \cdot \mathbf{r} = \frac{a^2}{r^2}\, \mathbf{J} \cdot \mathbf{r},\tag{11.106}$$

where $\mathbf{J} = \mu_0\, \mathbf{M}$ is the magnetization intensity.

Using the cylindrical coordinates (r, ϕ) for a single magnet (see Fig. 11.11), the magnetic scalar potential produced by a single row becomes

$$\mathcal{U}(r, \phi) = \mathbf{J}\, a^2 \frac{\sin \phi}{r}.\tag{11.107}$$

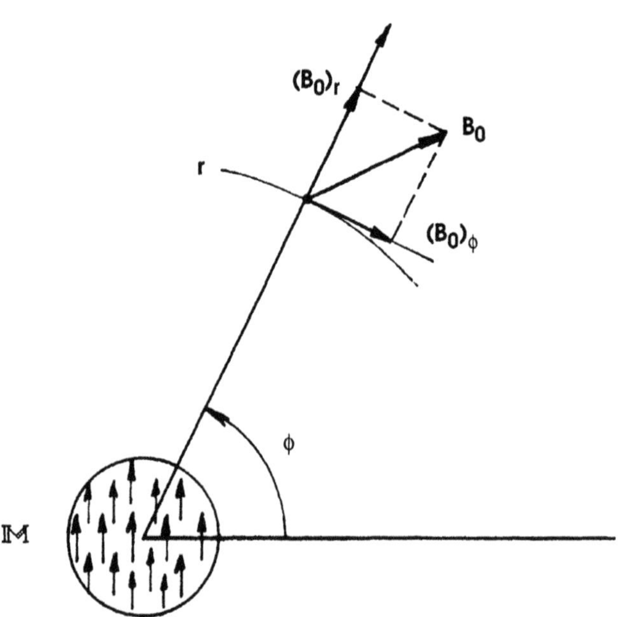

Fig. 11.11. Magnetic field created by a single, infinitely long cylindrical magnet row of uniform magnetization density **M**.

[†] For any other magnet shape, eqn. (11.106) is valid only for r >> a.

The B_0 components deduced from eqns. (11.93) and (11.94) are

$$(\mathbf{B}_0)_r = B_{0s}\,\frac{a^2}{r^2}\,\sin\phi,\tag{11.108}$$

$$(\mathbf{B}_0)_\phi = -\,B_{0s}\,\frac{a^2}{r^2}\,\cos\phi,\tag{11.109}$$

where $B_{0s} = \mathbb{J}$ is the magnetic field at the surface of a single magnet. Then the intensity B_0 of the magnetic field is independent of the azimuthal angle ϕ

$$B_0 = B_{0s}\,\frac{a^2}{r^2}.\tag{11.110}$$

Magnetic scalar potential created by a multipolar structure

The magnetic scalar potential $\mathcal{U}(r,\varphi)$ created at point $M(r,\varphi)$ by the $2m$ magnet rows with alternate polarity of the multipolar structure shown in Fig. 11.12 is

$$\mathcal{U}(r,\varphi) = \sum_{p=0}^{2m-1} (-1)^p\,\mathcal{U}_p\!\left(r_p,\phi_p\right),\tag{11.111}$$

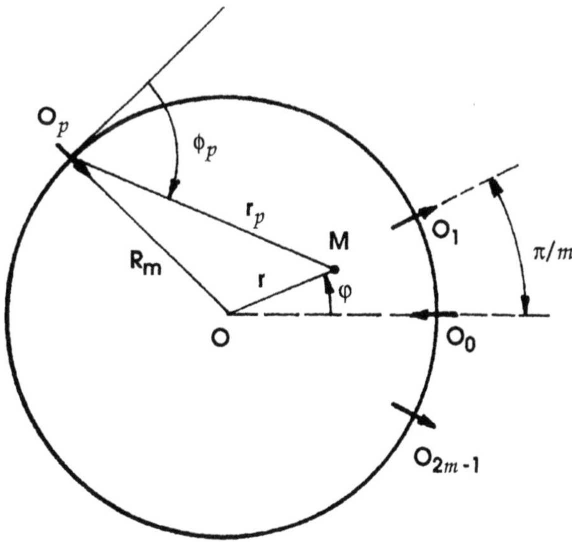

Fig. 11.12. Coordinates involved in the derivation of the multipolar magnetic field. The subscript p refers to the p-th magnet.

$$\mathcal{U}(r, \varphi) = B_{0s}\, a^2 \sum_{p=0}^{2m-1} (-1)^p\, \frac{\sin \phi_p}{r_p}\,. \tag{11.112}$$

Considering the (OMO_p) triangle, we obtain

$$r_p^2 = R_m^2 + r^2 - 2\, R_m\, r\, \cos\!\left(p\,\frac{\pi}{m} - \varphi\right), \tag{11.113}$$

$$r^2 = R_m^2 + r_p^2 - 2\, R_m\, r_p\, \sin \phi_p\,. \tag{11.114}$$

Equation (11.114) gives

$$\frac{\sin \phi_p}{r_p} = \frac{1}{2\, R_m} + \frac{R_m^2 - r^2}{2\, R_m\, r_p^2}\,. \tag{11.115}$$

Substituting back eqn. (11.115) into eqn. (11.112), the scalar potential becomes

$$\mathcal{U} = B_{0s}\, a^2 \left[\sum_{p=0}^{2m-1} \frac{(-1)^p}{2\, R_m} + \sum_{p=0}^{2m-1} (-1)^p\, \frac{R_m^2 - r^2}{2\, R_m\, r_p^2} \right]. \tag{11.116}$$

where the first sum is zero. With the help of eqn. (11.113), $\mathcal{U}$ becomes

$$\mathcal{U} = B_{0s}\, a^2\, \frac{R_m^2 - r^2}{4\, R_m^2\, r} \sum_{p=0}^{2m-1} \frac{(-1)^p}{\dfrac{1}{2}\left(\dfrac{R_m}{r} + \dfrac{r}{R_m}\right) - \cos\!\left(p\,\dfrac{\pi}{m} - \varphi\right)}\,. \tag{11.117}$$

Taking as a new variable

$$\rho = \log_e \frac{R_m}{r}\,, \tag{11.118}$$

eqn. (11.117) becomes

$$\mathcal{U} = \frac{B_{0s}\, a^2}{2 R_m}\, \sinh \rho \sum_{p=0}^{2m-1} \frac{(-1)^{\,p}}{\cosh \rho - \cos\!\left(p\,\dfrac{\pi}{m} - \varphi\right)}\,. \tag{11.119}$$

Using the expansion [23]

$$\sum_{k=0}^{q-1} \frac{1}{\cosh \rho - \cos \left(2\, k\frac{\pi}{q} - \varphi\right)} = \frac{\sinh q\rho}{\sinh \rho}\, \frac{q}{\cosh q\rho - \cos q\varphi}, \tag{11.120}$$

eqn. (11.119) becomes, after separating of the positive $(p = 2k)$ and negative $(p = 2k + 1)$ terms,

$$u = \frac{B_{0s}\, a^2}{2\, R_m}\, m \sinh m\rho \left[\frac{1}{\cosh m\rho - \cos m\varphi} - \frac{1}{\cosh m\rho - \cos m\left(\varphi - \frac{\pi}{m}\right)}\right], \tag{11.121}$$

i.e. finally

$$u(m\rho,\, m\varphi) = \frac{B_{0s}\, m\, a^2}{R_m}\, \frac{\sinh m\rho \,\cos m\varphi}{\cosh^2 m\rho - \cos^2 m\varphi}. \tag{11.122}$$

Components of the multipolar magnetic field in the $(m\rho,\, m\varphi)$ coordinate system

Writing the multipolar magnetic field as

$$\mathbf{B}_0 = B_{0m}\, \mathbf{b_b}, \tag{11.123}$$

where $\mathbf{b_b}$ is the reduced magnetic field and

$$B_{0m} \equiv B_{0s}\, m^2\, \frac{a^2}{R_m^2}. \tag{11.124}$$

The multipolar magnetic scalar potential (eqn. (11.122)) then becomes

$$u = B_{0m}\, \frac{R_m}{m}\, \mu_b, \tag{11.125}$$

where μ_b is the reduced magnetic scalar potential

$$\mu_b = \frac{\sinh m\rho \,\cos m\varphi}{\cosh^2 m\rho - \cos^2 m\varphi}. \tag{11.126}$$

The reduced magnetic field $\mathbf{b_b}$ is then given by

$$\mathbf{b_b} = -\frac{R_m}{m}\, \nabla\, \mu_b. \tag{11.127}$$

In the $(m\rho, m\varphi)$ coordinate system, the radial and azimuthal components of the reduced multipolar magnetic field are expressed by

$$(\mathbf{b}_b)_r = \frac{R_m}{r} \frac{\partial \mu_b}{\partial (m\rho)} , \tag{11.128}$$

$$(\mathbf{b}_b)_\varphi = - \frac{R_m}{r} \frac{\partial \mu_b}{\partial (m\varphi)} , \tag{11.129}$$

whose full expressions are

$$(\mathbf{b}_b)_r = - \frac{R_m}{r} \cosh m\rho \cos m\varphi \; \frac{\sinh^2 m\rho - \sin^2 m\varphi}{\left(\cosh^2 m\rho - \cos^2 m\varphi \right)^2} , \tag{11.130}$$

$$(\mathbf{b}_b)_\varphi = \frac{R_m}{r} \sinh m\rho \sin m\varphi \; \frac{\cosh^2 m\rho + \cos^2 m\varphi}{\left(\cosh^2 m\rho - \cos^2 m\varphi \right)^2} . \tag{11.131}$$

Multipolar magnetic vector potential

Setting $\mathbf{A} = B_{0m} R_m \boldsymbol{\lambda}_b / m$, the reduced magnetic field $\mathbf{b}_b$ may also be given by

$$\mathbf{b}_b = \frac{R_m}{m} \nabla \times \boldsymbol{\lambda}_b , \tag{11.132}$$

where $\boldsymbol{\lambda}_b$ is the reduced vector potential. In the $(m\rho, m\varphi)$ coordinate system, the radial and azimuthal components of the magnetic field are then expressed by

$$(\mathbf{b}_b)_r = \frac{R_m}{r} \frac{\partial \lambda_b}{\partial (m\varphi)} , \tag{11.133}$$

$$(\mathbf{b}_b)_\varphi = \frac{R_m}{r} \frac{\partial \lambda_b}{\partial (m\rho)} . \tag{11.134}$$

It is easy to verify that the solution

$$(\boldsymbol{\lambda}_b)_z = - \frac{\cosh m\rho \sin m\varphi}{\cosh^2 m\rho - \cos^2 m\varphi} \tag{11.135}$$

combined with eqns. (11.133) and (11.134), yields again expressions (11.130) and (11.131).

Intensity of the multipolar magnetic field

The intensity b_b of the reduced magnetic field is

$$b_b^2 = (b_b)_r^2 + (b_b)_\varphi^2.\tag{11.136}$$

According to eqns. (11.130) and (11.131), the intensity b_b may be expressed as

$$b_b = \frac{\left(\cosh^2 m\rho - \sin^2 m\varphi\right)^{1/2}}{\cosh^2 m\rho - \cos^2 m\varphi}\,\exp\rho\,.\tag{11.137}$$

It can also be useful to calculate b_b directly from λ_b and μ_b. Equations (11.126) and (11.135) yield

$$\sinh^2 m\rho = \frac{1 - \lambda_b^2 - \mu_b^2 + \left[\left(\lambda_b^2 + \mu_b^2 - 1\right)^2 + 4\,\mu_b^2\right]^{1/2}}{2\left(\lambda_b^2 + \mu_b^2\right)},\tag{11.138}$$

$$\cosh^2 m\rho = \frac{1 + \lambda_b^2 + \mu_b^2 + \left[\left(\lambda_b^2 + \mu_b^2 - 1\right)^2 + 4\,\mu_b^2\right]^{1/2}}{2\left(\lambda_b^2 + \mu_b^2\right)},\tag{11.139}$$

$$\sin^2 m\varphi = \frac{\lambda_b^2 + \mu_b^2 + 1 - \left[\left(\lambda_b^2 + \mu_b^2 - 1\right)^2 + 4\,\mu_b^2\right]^{1/2}}{2\left(\lambda_b^2 + \mu_b^2\right)},\tag{11.140}$$

$$\cos^2 m\varphi = \frac{\lambda_b^2 + \mu_b^2 - 1 + \left[\left(\lambda_b^2 + \mu_b^2 - 1\right)^2 + 4\,\mu_b^2\right]^{1/2}}{2\left(\lambda_b^2 + \mu_b^2\right)}.\tag{11.141}$$

The resulting expression for b_b using eqn. (11.137) is

$$b_b^2 = \frac{R_m^2}{r^2}\left(\lambda_b^2 + \mu_b^2\right)\left[\left(\lambda_b^2 + \mu_b^2 - 1\right)^2 + 4\,\mu_b^2\right]^{1/2}.\tag{11.142}$$

Equation of a circle of radius R_0 in the (λ_b, μ_b) coordinate system

The equation of a circle of radius R_0 is simply $r = R_0$ ($\rho = \rho_0$). Using the (λ_b, μ_b) coordinates, the circle equation may be derived, for example, from (eqn. (11.139)),

$$\frac{\lambda_b^2 + \mu_b^2 + 1 + \left[\left(\lambda_b^2 + \mu_b^2 - 1\right)^2 + 4\mu_b^2\right]^{1/2}}{2\left(\lambda_b^2 + \mu_b^2\right)} = \frac{1}{\mathbb{R}_0^2} \tag{11.143}$$

with

$$\frac{1}{\mathbb{R}_0} = \cosh m\rho_0. \tag{11.144}$$

Setting

$$\lambda_b^2 + \mu_b^2 = \xi_b^2 \tag{11.145}$$

with $\lambda_b = \xi_b \cos\phi$ and $\mu_b = \xi_b \sin\phi$, eqn. (11.143) becomes

$$\xi_b^2 = \mathbb{R}_0^2 \, \cos^2\phi + \frac{\mathbb{R}_0^2}{1 - \mathbb{R}_0^2} \sin^2\phi \, . \tag{11.146}$$

It should now be noted that $\mathbb{R}_0 \ll 1$ when $m\rho_0 \gg 1$, i.e.

$$\left(\frac{R_0}{R_m}\right)^m \ll 1. \tag{11.147}$$

Equation (11.146) then reduces to

$$\lambda_b^2 + \mu_b^2 = \mathbb{R}_0^2 \, , \tag{11.148}$$

which is also, in the (λ_b, μ_b) coordinate system, a circle of radius

$$\mathbb{R}_0 \approx 2\left(\frac{R_0}{R_m}\right)^m. \tag{11.149}$$

Provided inequality (11.147) is verified we have, $\mathbb{R}_0 \ll 1$ and, according to eqn. (11.148), we also have $\lambda_b \ll 1$ and $\mu_b \ll 1$. Equation (11.142) comes to

$$b_b^2 \approx \frac{R_m^2}{r^2} \left(\lambda_b^2 + \mu_b^2\right), \tag{11.150}$$

which shows that, for $\lambda_b, \mu_b \ll 1$, "b_b = constant" lines are circles in both (λ_b, μ_b) and (r, φ) coordinate systems.

Multipolar magnetic field produced by a planar array of magnets

Let us now consider a planar array of infinitely long cylindrical magnets equally separated by a distance d. The magnetic field configuration can be deduced from the above results by assuming an infinite number of magnets set up on a cylinder of infinite radius such that

$$2\pi R_m = 2m\, d. \tag{11.151}$$

If $y = yR_m/m$ is the abcissa (see Sec. 11.3.1) along the infinite circle,

$$\varphi = \frac{y}{R_m} \tag{11.152}$$

and, according to eqn. (11.151),

$$m\,\varphi = \frac{\pi\, y}{d}. \tag{11.153}$$

If $x = R_m - r$ is the ordinate along a radius whose origin lies on the circle (note that x is different from X in Sec. 11.3.1)

$$m\,\rho = -\, m \log_e \left(1 - \frac{x}{R_m}\right) \tag{11.154}$$

and, according to eqn. (11.151),

$$m\,\rho = \frac{\pi\, x}{d}. \tag{11.155}$$

Thus the configuration produced by a planar array of magnets is obtained by replacing $m\rho$ by $\pi x/d$, $m\varphi$ by $\pi y/d$, m/R_m by π/d and R_m/r by 1 in the different expressions. We find

$$\lambda_b\,(x, y) = - \frac{\cosh \dfrac{\pi x}{d}\ \sin \dfrac{\pi y}{d}}{\cosh^2 \dfrac{\pi x}{d}\ -\ \cos^2 \dfrac{\pi y}{d}}, \tag{11.156}$$

$$\mu_b\,(x, y) = \frac{\sinh \dfrac{\pi x}{d}\ \cos \dfrac{\pi y}{d}}{\cosh^2 \dfrac{\pi x}{d} - \cos^2 \dfrac{\pi y}{d}}, \tag{11.157}$$

$$b_b(x, y) = \frac{\left(\cosh^2 \frac{\pi x}{d} - \sin^2 \frac{\pi y}{d}\right)^{1/2}}{\cosh^2 \frac{\pi x}{d} - \cos^2 \frac{\pi y}{d}} \quad . \tag{11.158}$$

Topograghy of the magnetic field

The field lines defined as "λ_b = constant" intensity, the orthogonal trajectories defined as lines of "μ_b = constant" value, and the b_b lines of constant magnetic field intensity, are shown in Fig. 11.2 in the $X = r/R_m$ and $Y = m\varphi$ coordinate system. The *cusps*, where the field lines converge, are localized in front of the magnet poles. Between the cusps, along the $Y = \pi/2$ radial lines, the magnetic field intensity reaches a maximum value at a point P which is a saddle point for $\mathbf{B_0}$. The derivative $\partial b_b/\partial \rho$ from eqn. (11.137) is zero for $m\varphi = \pi/2$ when

$$\cosh m\varphi \sinh m\varphi = m\,(\sinh^2 m\varphi - 1) \tag{11.159}$$

or when

$$(m^2 - 1)\sinh^4 m\varphi - (2\,m^2 + 1)\sinh^2 m\varphi + m^2 = 0 \; . \tag{11.160}$$

Assuming a large number of magnet rows ($m^2 \gg 1$), the maximum intensity of the magnetic field at the saddle point is obtained for

$$\sinh m\varphi = 1, \tag{11.161}$$

i.e. for $m\varphi = 0.89$.

The intensity of the magnetic field at this point, which is strictly equal to

$$b_b = 1/2 \qquad\qquad (B_0 = B_{0m}/2) \tag{11.162}$$

in the planar configuration, takes a slightly higher value in the cylindrical configuration, namely

$$b_b = \frac{R_m}{2r} \approx \frac{1}{2} + \frac{0.89}{2m} \; . \tag{11.163}$$

The intensity of the magnetic field as a function of the distance $x = R_m - r$ from the magnets is expressed by eqn. (11.158). When $\cosh(\pi x/d) \gg 1$, eqn. (11.158) reduces to

$$b_b(x,y) \approx 2 \exp\left(-\frac{\pi x}{d}\right), \tag{11.164}$$

the magnetic field intensity then being independent of coordinate y (or of azimuthal angle φ) and decreasing exponentially over a scale length d/π (or R_m/m when considering eqn. (11.151)).

APPENDIX 11.5

DIFFUSION EQUATION IN THE (λ_b, μ_b) COORDINATE SYSTEM

The diffusion equation of the plasma in the magnetic field region (eqn. (11.23)) can be written in the (x,y) coordinate system

$$\nabla \cdot \left[\nabla_\parallel n + \frac{1}{1 + \gamma_c B_0^2} \nabla_\perp n \right] + \frac{S_i}{D_\parallel} = 0. \tag{11.165}$$

However, it is easier to solve numerically this equation in the (λ_b, μ_b) coordinate system since diffusion is markedly different along the field lines (λ_b = constant lines) and perpendicularly to the field lines, along orthogonal trajectories (μ_b = constant lines).

The relationship between $\pmb{\lambda}_b$, μ_b and $\mathbf{b}_b$ (eqns. (11.123), (11.127) and (11.132)) can be rewritten according to

$$\frac{m}{R_m} \mathbf{b}_b = - \nabla \mu_b = \nabla \times \pmb{\lambda}_b, \tag{11.166}$$

$$\frac{m}{R_m} (\mathbf{b}_b)_x = - \frac{\partial \mu_b}{\partial x} = \frac{\partial \lambda_b}{\partial y}, \tag{11.167}$$

$$\frac{m}{R_m} (\mathbf{b}_b)_y = - \frac{\partial \mu_b}{\partial y} = - \frac{\partial \lambda_b}{\partial x}, \tag{11.168}$$

with

$$\frac{m^2}{R_m^2} b_b^2 = \left(\frac{\partial \mu_b}{\partial x} \right)^2 + \left(\frac{\partial \mu_b}{\partial y} \right)^2 = \left(\frac{\partial \lambda_b}{\partial x} \right)^2 + \left(\frac{\partial \lambda_b}{\partial y} \right)^2. \tag{11.169}$$

To transform eqn. (11.165) into the (λ_b, μ_b) orthogonal coordinate system, we must express unit vectors $\mathbf{e}_\mu$ and $\mathbf{e}_\lambda$, defined in Fig. 11.13, in the (x, y) coordinate system

$$\frac{m|\mathbf{b}_b|}{R_m} \mathbf{e}_\mu = - \frac{\partial \mu_b}{\partial x} \mathbf{e}_x - \frac{\partial \mu_b}{\partial y} \mathbf{e}_y = \frac{\partial \lambda_b}{\partial y} \mathbf{e}_x - \frac{\partial \lambda_b}{\partial x} \mathbf{e}_y, \tag{11.170}$$

$$\frac{m|\mathbf{b}_b|}{R_m} \mathbf{e}_\lambda = \frac{\partial \lambda_b}{\partial x} \mathbf{e}_x + \frac{\partial \lambda_b}{\partial y} \mathbf{e}_y = \frac{\partial \mu_b}{\partial y} \mathbf{e}_x - \frac{\partial \mu_b}{\partial x} \mathbf{e}_y. \tag{11.171}$$

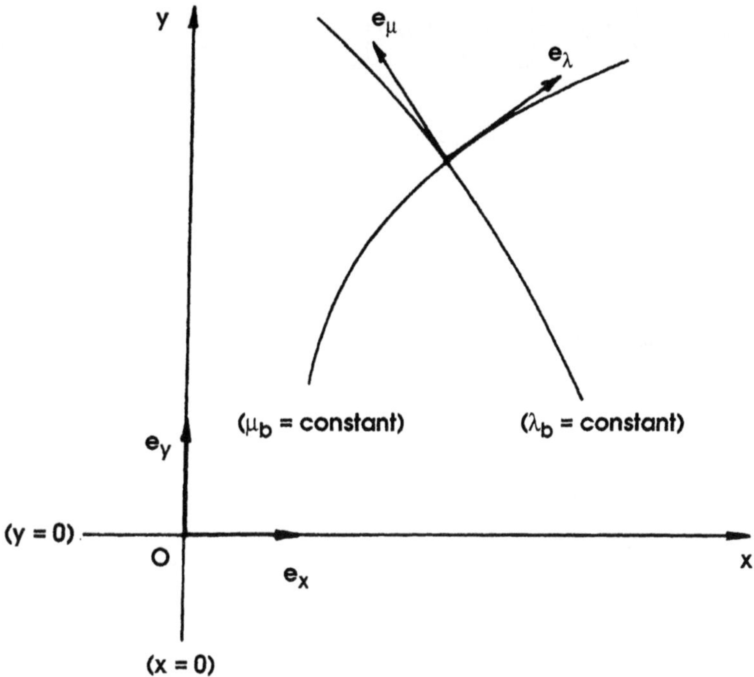

Fig. 11.13. Diagram of the two orthonormal coordinate systems used to derive and solve numerically the diffusion equation of the plasma in a magnetic field. The "λ_b = constant" lines are field lines, and the "μ_b = constant" lines are orthogonal trajectories.

We can now derive the different parts of eqn. (11.165). With the help of eqn. (11.170), we obtain along the field lines

$$\nabla_{\parallel} n = \left(\frac{\partial n}{\partial x}\right)_{\lambda_b} e_x + \left(\frac{\partial n}{\partial y}\right)_{\lambda_b} e_y = -\frac{m|b_b|}{R_m}\frac{\partial n}{\partial \mu_b} e_\mu. \tag{11.172}$$

Using eqn. (11.171), we obtain along the orthogonal trajectories

$$\nabla_{\perp} n = \left(\frac{\partial n}{\partial x}\right)_{\mu_b} e_x + \left(\frac{\partial n}{\partial y}\right)_{\mu_b} e_y = \frac{m|b_b|}{R_m}\frac{\partial n}{\partial \lambda_b} e_\lambda. \tag{11.173}$$

To reach the next step of the derivation, we calculate the divergence along the field lines of the expression

$$\nabla \cdot \left[\mathcal{F}(\lambda_b, \mu_b)\, \mathbf{e}_\mu \right] = \left[\frac{\partial}{\partial x}\, (\mathbf{e}_\mu)_x\, \mathcal{F} \right]_{\lambda_b} + \left[\frac{\partial}{\partial y}\, (\mathbf{e}_\mu)_y\, \mathcal{F} \right]_{\lambda_b} .$$

(11.174)

Considering eqns. (11.167), (11.168) and (11.170), eqn. (11.174) becomes

$$\frac{\partial}{\partial x} \left[-\frac{\mathcal{F} R_m}{m\,|b_b|}\frac{\partial \mu_b}{\partial x} \right] + \frac{\partial}{\partial y} \left[-\frac{\mathcal{F} R_m}{m\,|b_b|}\frac{\partial \mu_b}{\partial y} \right] = $$

$$-\frac{\mathcal{F} R_m}{m\,|b_b|}\left[-\frac{\partial^2 \lambda_b}{\partial x \partial y} + \frac{\partial^2 \lambda_b}{\partial x \partial y} \right] + \left[\left(\frac{\partial \mu_b}{\partial x}\right)^2 + \left(\frac{\partial \mu_b}{\partial y}\right)^2 \right] \frac{\partial}{\partial \mu_b}\left(-\frac{\mathcal{F} R_m}{m\,|b_b|} \right).$$

(11.175)

Taking into account eqn. (11.169), we finally obtain

$$\nabla \cdot \left[\mathcal{F}(\lambda_b, \mu_b)\, \mathbf{e}_\mu \right] = \frac{m^2\, b_b^2}{R_m^2}\frac{\partial}{\partial \mu_b}\left(-\frac{\mathcal{F} R_m}{m\,|b_b|} \right).$$

(11.176)

In the same manner, we can obtain

$$\nabla \cdot \left[\mathcal{F}(\lambda_b, \mu_b)\, \mathbf{e}_\lambda \right] = \frac{m^2 b_b^2}{R m^2}\frac{\partial}{\partial \lambda_b}\left(-\frac{\mathcal{F} R_m}{m\,|b_b|} \right).$$

(11.177)

Thus, using eqns. (11.172) and (11.176), we obtain

$$\nabla \cdot \left(\nabla_\parallel n \right) = \frac{m^2\, b_b^2}{R_m^2}\frac{\partial^2 n}{\partial \mu_b^2} .$$

(11.178)

Using eqns. (11.165), (11.173) and (11.177), we also obtain

$$\nabla \cdot \left(\frac{1}{1 + \gamma_c B_0^2}\, \nabla_\perp n \right) = \frac{m^2 b_b^2}{R_m^2}\frac{\partial}{\partial \lambda_b}\left(\frac{1}{1 + \gamma_c B_{0m}^2 b_b^2}\frac{\partial n}{\partial \lambda_b} \right).$$

(11.179)

Finally, in the (λ_b, μ_b) coordinate system, diffusion equation (11.165) takes the form

$$\frac{m^2 D_\parallel}{S_i R_m^2}\left[\frac{\partial^2 n}{\partial \mu_b^2} + \frac{\partial}{\partial \lambda_b}\left(\frac{1}{1 + \gamma_c b_b^2 B_{0m}^2}\frac{\partial n}{\partial \lambda_b} \right) \right] + \frac{1}{b_b^2} = 0.$$

(11.180)

References

[1]   R. Limpaecher and K.R. MacKenzie, Rev. Sci. Instrum. **44**, 726 (1973).

[2]   W.F. Divergilio, A.Y. Wong, H.C. Kim and V.C. Lee, Phys. Rev. Lett. **38**, 541 (1977).

[3]   T.D. Mantei and G. Matthieussent, Phys. Fluids **20**,1005 (1977).

[4]   N. Hershkowitz, K.N. Leung and T. Romesser, Phys. Rev. Lett. **35**, 277 (1975).

[5]   K.N. Leung, T.K. Samec and A. Lamm, Phys. Lett. **51A**, 490 (1975); T.K. Samec, Ph. D. thesis, University of California, 1976.

[6]   J.M. Buzzi, J. Snow and J.L. Hirshfield, Phys. Lett. **54A**, 344 (1975).

[7]   N. Hershkowitz, J.R. Smith and H. Kozima, Phys. Fluids **22**, 122 (1979).

[8]   N. Hershkowitz, J.R. Smith and F.R. De Kock, Plasma Phys. **21**, 823 (1979).

[9]   K.N. Leung, N. Hershkowitz and K.R. MacKenzie, Phys. Fluids **19**, 1045 (1976).

[10]   I.J. Spalding, in *Advances in Plasma Physics*, A. Simon and W.B. Thompson eds. (Interscience, New York, 1971) Vol. 4, p. 79.

[11]   M.G. Haines, Nucl. Fusion **17**, 811 (1977).

[12]   N. Hershkowitz, J.R. De Kock and C. Chan, Nucl. Fusion **20**, 695 (1980).

[13]   A.J. Marcus, G. Knorr and G. Joyce, Plasma Phys. **22**, 1015 (1980).

[14]   C. Koch and G. Matthieussent, J. Physique **43**, 67 (1982).

[15]   C. Koch and G. Matthieussent, Phys. Fluids **26**, 545 (1983).

[16]   S.I. Braginski, in *Review of Plasma Physics*, M.A. Leontovich eds. (Consultants Bureau, New York, 1965) Vol. 1, p. 284.

[17]   A.V. Gurevich, *Nonlinear Phenomena in the Ionosphere* (Springer-Verlag, New York, 1978) pp. 29 and 221.

[18]   R.W. Hockney, in *Methods in Computational Physics* (Academic, New York, 1970) Vol. 9, p. 135.

[19]   E.W. McDaniel, *Atomic Collisions* (Wiley, New York, 1989) p. 211.

[20]   J. Ferch, B. Granitza, C. Masche and W. Raith, J. Phys. **B18**, 967 (1985).

[21]   K. Stephan, H. Helm and T.D. Marle, J. Chem. Phys. **73**, 3763 (1980).

[22]   E.C. Berty, *Proc. Vth Int. Conf. Phenom. Ionized Gases* (North-Holland, Amsterdam, 1962) vol. 1, p. 183.

[23]   E.R. Hansen, *A Table of Series and Product* (Prentice-Hall, Englewood Cliffs, NJ, 1975) p. 271.

CHAPTER 12

HOMOGENEITY IN MULTIPOLAR DISCHARGES: THE ROLE OF PRIMARY ELECTRONS *

12.1. Introduction

The ambipolar diffusion model for multipolar plasmas developed in the previous chapter has confirmed that the confinement of plasma by multipolar magnetic fields is weak, as already observed from lifetime measurements.[1-3] Recall that this led to suggest that the increase in plasma density observed when applying multipolar confinement rather results from an efficient confinement of the ionizing primary electrons. We also saw that the plasma homogeneity usually obtained in these low pressure plasmas could not be accounted for either by assuming a uniform ionization rate in the volume of the plasma, or by considering a significant decrease of the plasma leakage through the multipolar magnetic sheath (Chap. 11). The objective of the present chapter is to thoroughly study the role played by primary electrons in improving plasma homogeneity.

The chapter begins by a reexamination of the general hypotheses for discharges confined by a multipolar magnetic field, in order to point out that plasma homogeneity is improved when one assumes the ionization by primary electrons to be mainly localized near the wall of the discharge. This assumption is largely supported by the detailed study of the trajectories of primary electrons in a multipolar discharge that we are going to present. Then, using the two invariants in the motion of primary electrons, we can distinguish the particles trapped near the wall in the multipolar magnetic field, from the free particles reflected by the magnetic sheath. Finally, we propose two competing mechanisms that ensure a large enough ionization rate near the walls to account for the plasma homogeneity in low pressure multipolar discharges.

12.2. Plasma homogeneity in multipolar discharges: experimental evidence and theoretical aspects

12.2.1. Experimental results. The fact that plasmas confined by multipolar magnetic fields can be homogeneous was qualitatively demonstrated for the first time by Limpaecher and MacKenzie.[4] They observed that plasma homogeneity gradually improved with decreasing pressure, as shown in Fig. 10.5. More recently, in order to compare with theoretical results (which are presented in the next section), Gauthereau

---

* **Presented by J. Pelletier and G. Matthieussent**

and Matthieussent measured the radial density profile in a 1 meter diameter, 1 meter long multipolar discharge in argon as a function of pressure.[5] Their results are presented in Fig. 12.1 for three gas pressures, namely 5×10^{-4}, 10^{-3} and 8×10^{-3} torr, which correspond to primary electron mean free paths λ_{pi} of 2.3 m, 1.15 m and 0.15 m respectively (see Appendix 11.1 for λ_{pi} values). When λ_{pi} is larger than the characteristic dimensions of the discharge (here 1 meter), we say that the *primary electrons* are in a *collisionless regime.* At 5×10^{-4} torr, the lowest pressure examined, the ion mean free path λ_{in} for ion-neutral collisions is less than 10 cm, i.e. it is much shorter than the chamber dimensions. Thus in the whole pressure range investigated, the *plasma* can be considered as *collisional.*

Figure 12.1 shows that, at the lowest pressure (5×10^{-4} torr), i.e. in the collisionless regime for primary electrons, perfect plasma homogeneity is achieved in the volume of the discharge where it is free from magnetic field. In contrast, at the highest pressure (8×10^{-3} torr), i.e. in the collisional regime for primary electrons, the parabolic plasma density profile (see eqn. (10.27)) commonly obtained in non magnetically confined discharges is observed. Figure 12.1 also shows that the transition from a

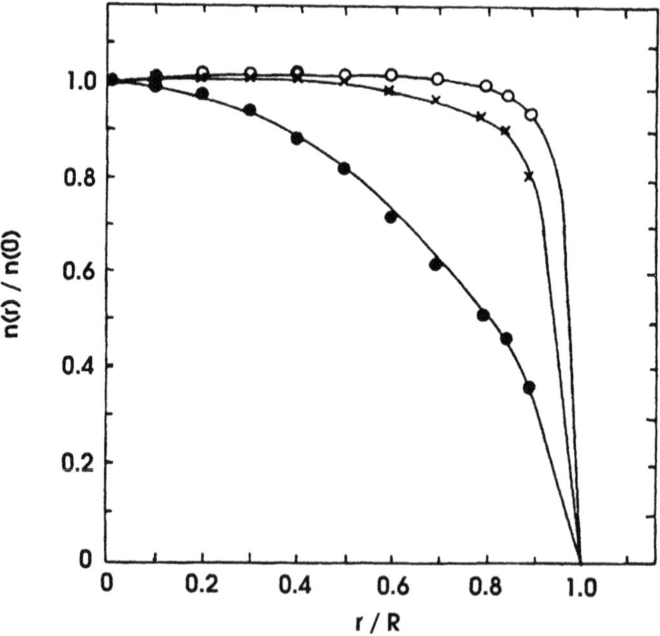

Fig. 12.1. Measured radial plasma density profile in a 1 meter long multipolar discharge of radius $R = 0.5$ m, at different argon pressures (o: 5×10^{-4} torr, x: 10^{-3} torr, •: 8×10^{-3} torr). The central density n(0) is close to 2×10^{11} cm^{-3} (from ref. 5).

homogeneous plasma to a plasma with a parabolic density profile occurs at an argon pressure of the order of 10^{-3} torr, i.e. at the onset of the collisional regime for primary electrons $(\lambda_{pi} \approx 2R)$.

These experimental results confirm the major role played by primary electrons in achieving plasma homogeneity in a multipolar discharge. Thus, bearing in mind that primary electrons produce the plasma and that the assumption of a constant ionization rate over the whole plasma volume does not lead to a homogeneous plasma (see Chap. 11), we reexamine the hypotheses which govern the particle balance in a multipolar discharge.

12.2.2. Theoretical aspects. We consider first a multipolar discharge with λ_{in} shorther than the radius of the chamber (collisional plasma). Under steady-state conditions and in the absence of ion volume recombination (Sec. 9.5), the plasma can be described by an ambipolar diffusion model obeying the following equations (see Sec. 6.2.2)

$$\nabla \cdot \boldsymbol{\varGamma} = S_i, \tag{12.1}$$

$$\boldsymbol{\varGamma} = -\underline{D} \cdot \nabla n, \tag{12.2}$$

where $\boldsymbol{\varGamma}$ is the ion (or electron) flux, $\underline{D}$ is the diffusion tensor, n is the plasma density and S_i is the ionization rate. Equations (12.1) and (12.2) mean that the particles produced in a given volume are evacuated outside this volume by diffusion (eqn. (12.1)) under the influence of density gradients (eqn. (12.2)). The diffusion tensor and the ionization rate are two parameters that depend on discharge conditions. As an example, consider the ambipolar diffusion of a plasma in a static magnetic field $\mathbf{B}_0$ (see Sec. 6.2.2). The components $D_{\parallel}$ and $D_{\perp}$ of the diffusion tensor (so-called parallel and perpendicular to the magnetic field coefficients respectively) are

$$D_{\parallel} \approx \frac{k_B T_e}{m_i \nu_{in}}, \tag{12.3}$$

$$D_{\perp} = \frac{D_{\parallel}}{1 + \gamma_c B_0^2}, \tag{12.4}$$

where T_e is the electron temperature, k_B the Boltzmann constant, m_i the ion mass, ν_{in} the ion-neutral collision frequency and γ_c the product of electron and ion mobilities in absence of magnetic field (see eqn. (6.7) or (6.12)). We see that these diffusion coefficients both vary with gas pressure through collision frequencies, while the diffusion coefficient perpendicular to the magnetic field can be reduced independently by increasing the intensity of the applied magnetic field $\mathbf{B_0}$.

In this chapter, we define a homogeneous plasma as a plasma where the density gradients are negligible in the central volume of the chamber, but may become significant near the walls. Thus, considering eqns. (12.1) and (12.2), plasma homogeneity can be improved in two ways: i) by localization of the density gradients close to the chamber walls: assuming a given flux $\mathit{\Gamma}$ in eqn. (12.2), this requires the decrease of the diffusion coefficients close to the chamber walls; ii) by localization of the ionization rate near the walls: the flux $\mathit{\Gamma}$ of charged particles is thus reduced in the central region of the discharge (eqn. (12.1)) together with the density gradients (eqn. (12.2)). However, close to the walls, the density gradient must increase to evacuate the charged particles (eqn. (12.2)) produced near the walls (eqn. (12.1)).

To test the above alternative, we consider an infinitely long cylindrical chamber of radius R. Assuming for simplicity a scalar diffusion coefficient, we obtain from eqns. (12.1) and (12.2)

$$\nabla^2 n(r) = -\frac{S_i}{D_a},\tag{12.5}$$

where r is the radial coordinate, and where the ionization rate and the ambipolar diffusion coefficient take the value S_i and D_a in the central region of the discharge ($r < R'$), and S_i' and D_a' near the walls ($R' < r < R$). The continuity of plasma density and flux at $r = R'$, and a zero plasma density at the wall, are the imposed boundary conditions. The analytical expression for the radial variation of plasma density is

$$n(r) = \frac{S_i'}{4D_a'}\left(R^2 - r^2\right) + \frac{R'^2}{2D_a'}\left(S_i - S_i'\right)\log_e\frac{R}{r}\tag{12.6}$$

in the peripheral region ($R' < r < R$), and

$$n(r) = \frac{S_i}{4D_a}\left(R'^2 - r^2\right) + \frac{S_i'}{4D_a'}\left(R^2 - R'^2\right) + \frac{R'^2}{2D_a'}\left(S_i - S_i'\right)\log_e\frac{R}{R'}\tag{12.7}$$

in the central region ($r < R'$). Note that the classical parabolic profile (eqn. (10.28)) obtained in cylindrical discharges under both constant diffusion coefficient and constant ionization rate, is recovered assuming either $R' \to R$ or $D_a = D_a'$ and $S_i = S_i'$ in eqns. (12.6) and (12.7).

To illustrate the first possibility of the above alternative for improved plasma homogeneity, we assume a constant ionization rate everywhere ($S_i = S_i'$) and a smaller diffusion coefficient in the peripheral region than in the central region of the cylindrical chamber ($D_a > D_a'$). Figure 12.2 shows, for R'/R = 0.8, the evolution of the radial profile of plasma density as a function of the ratio D_a/D_a'. We see that plasma homogeneity is improved when D_a/D_a' is increased, i.e. when D_a' is decreased. Equation (12.7) clearly shows that the plasma density in the central region is constant for $D_a' \ll D_a$.

To illustrate the second possibility, we assume everywhere a constant diffusion coefficient ($D_a = D_a'$) and an ionization rate larger in the peripheral region than in the central region of the chamber ($S_i' > S_i$).   Figure 12.3 shows, for R'/R = 0.8, the evolution of the radial profile of plasma density as a function of the ratio S_i'/S_i. We see that plasma homogeneity also improves when S_i'/S_i increases. Equation (12.7) yields a constant plasma density for $S_i' \gg S_i$.

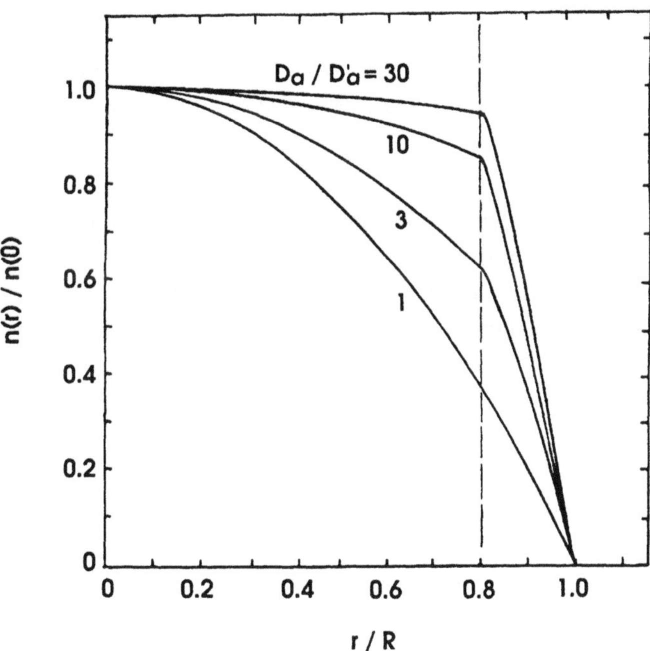

Fig. 12.2. Calculated radial profile n(r)/n(0) of the plasma density obtained at different ratios D_a/D_a' where D_a and D_a' are the ambipolar diffusion coefficients in the central region (r/R < 0.8) and near the walls (0.8 ≤ r/R ≤ 1) respectively (from ref. 5).

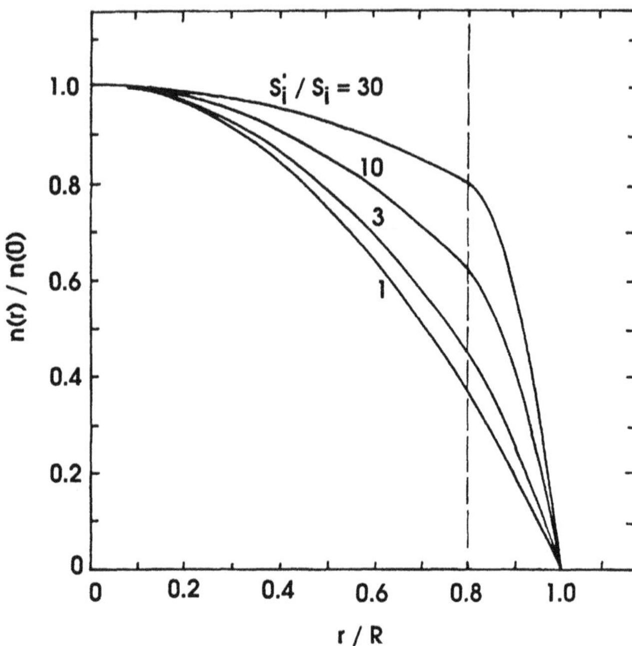

Fig. 12.3. Calculated radial profiles $n(r)/n(0)$ of the plasma density obtained at different ratios S'_i / S_i, where S_i and S'_i are the ionization rates in the central region ($r/R < 0.8$) and near the walls ($0.8 \leq r/R \leq 1$) respectively (from ref. 5).

From the above discussion, we conclude that plasma homogeneity can be accounted for by assuming either a decrease of plasma leakage towards the chamber walls or the localization of ionization in the peripheral region near the walls. In both cases, the effect considered must show up very strongly in order to lead to perfect homogeneity.

12.2.3. Necessity of peripheral ionization. The first method proposed to improve plasma homogeneity can be further examined using multipolar magnetic field confinement as an example. In such a configuration, we can assume that, in the peripheral magnetic sheath, D'_a is some kind of a spatial average on the components D_{mn} of the diffusion tensor in the region where $B_0 \neq 0$, D'_a being smaller than D_a in the magnetic field free region because, as shown by eqn. (12.4), plasma diffusion perpendicular to the magnetic field is reduced.

However, since ion lifetime measurements[1-3] indicate an increase by a factor of only 2 to 3 over a plasma without magnets, the multipolar magnetic field is clearly not very efficient in confining plasma. Because the lifetime of the plasma is closely dependent on the leakage of ions towards the walls, we cannot assume the average diffusion coefficient D_a' in the magnetic sheath to be less than about one third of its value D_a in the magnetic field free region. As a confirmation of this, we return to the numerical calculations of plasma diffusion across the multipolar magnetic sheath presented in Sec. 11.4. Assuming a constant ionization rate S_i, it is found that plasma density at the chamber axis is increased by a factor of about 1.5 only compared to the same discharge without magnetic confinement. This result corresponds approximately to the case $D_a/D_a' = 3$ since eqns. (10.28) and (12.7) then yield an increase of 1.72 in plasma density. Since Fig. 12.2 then shows a plasma density profile that is not uniform, we must admit that plasma confinement *alone* cannot explain the observed profiles. We must therefore turn to the second possibility of our initial discussion and consider the additional contribution of peripheral ionization to plasma homogeneity. This assertion is supported by Fig. 10.18, which is a front view photograph of a multipolar discharge showing that the emission of photons (produced by the species excited by primary electrons) is more intense near the edge of the chamber than inside it.

To prove that the above qualitative considerations are well founded, the results of a rigorous numerical calculation of plasma diffusion across the magnetic sheath are reported in Fig. 12.4. The calculation is similar to that performed in Sec. 11.4, but an enhanced ionization rate near the walls ($S_i'/S_i = 30$) has been added. The figure shows the radial profile of plasma density computed along a radius that passes though a cusp (full lines) and between two cusps (dashed line). The profiles obtained with $S_i'/S_i = 30$ definitely show a much more homogeneous plasma than with $S_i' = S_i$.

The whole question that remains is to understand how primary electrons can produce such an enhanced peripheral ionization. To find out about the mechanisms involved, we turn to a detailed study of the trajectories of primary electrons in the multipolar magnetic field.

12.3. Trajectories of primary electrons

12.3.1. General hypotheses. The general hypotheses adopted below correspond to the experimental conditions described in Sec. 12.2.1 where the plasma is homogeneous. Recall that the plasma is then collisional with an ion mean free path shorter than the characteristic dimensions of the discharge chamber, but with a primary electron mean free path for ionization much larger than these dimensions. Then, as discussed is Sec. 10.4.3, the primary electrons on the average undergo less than one inelastic ionizing collisions before being collected by the chamber wall. Clearly, these energetic primary electron ($W_{pe} \approx 50$ eV) are distinct from the population of plasma electrons with a thermal energy $k_B T_e$ of the order of a few eVs. As a result, primary

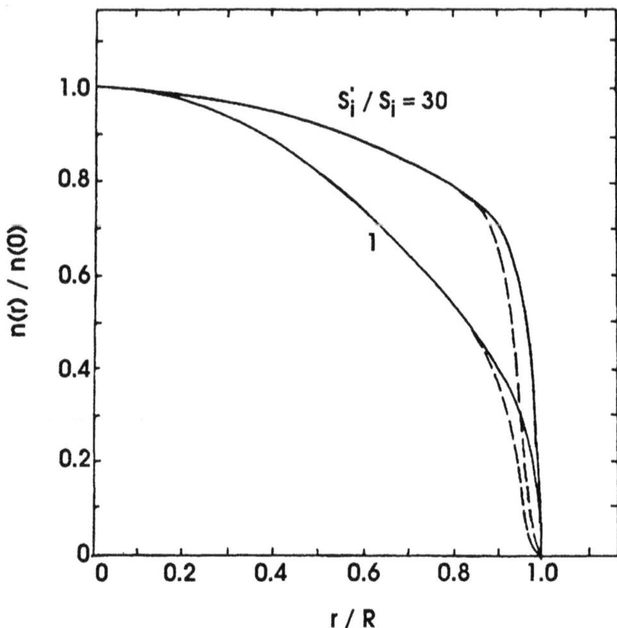

Fig. 12.4. Radial profile n(r)/n(0) of the plasma density obtained by solving numerically the diffusion equation in the magnetic field for different ratios S_i' / S_i of the ionization rate inside a multipolar discharge. The change in the ionization rate occurs at r/R = 0.8. The full lines are computed along a radius which passes through a cusp and the dashed lines between two cusps (from ref. 5).

electrons behave like free energetic particles in motion in external fields, which include the multipolar magnetic field and the space charge field of the plasma. However, as the variation of the potential inside the plasma does not usually exceed a few $k_B T_e/e$ (see eqn. (10.31)), the trajectory of energetic primary electrons is, in contrast to that of plasma electrons, weakly influenced by the space charge electric field. We neglect this effect in the following with respect to the action of the multipolar magnetic field on the primary electron trajectories.

12.3.2. Configuration of the multipolar magnetic field. For the sake of convenience, the cylindrical multipolar configuration is replaced by a planar structure. Such a simplification does not modify much the physical situation as compared to that of a cylindrical structure where the chamber radius is much larger than the distance between magnets. The magnetic field is produced by an array of infinitely long cylindrical magnets, parallel to the Oz axis (Fig. 12.5).

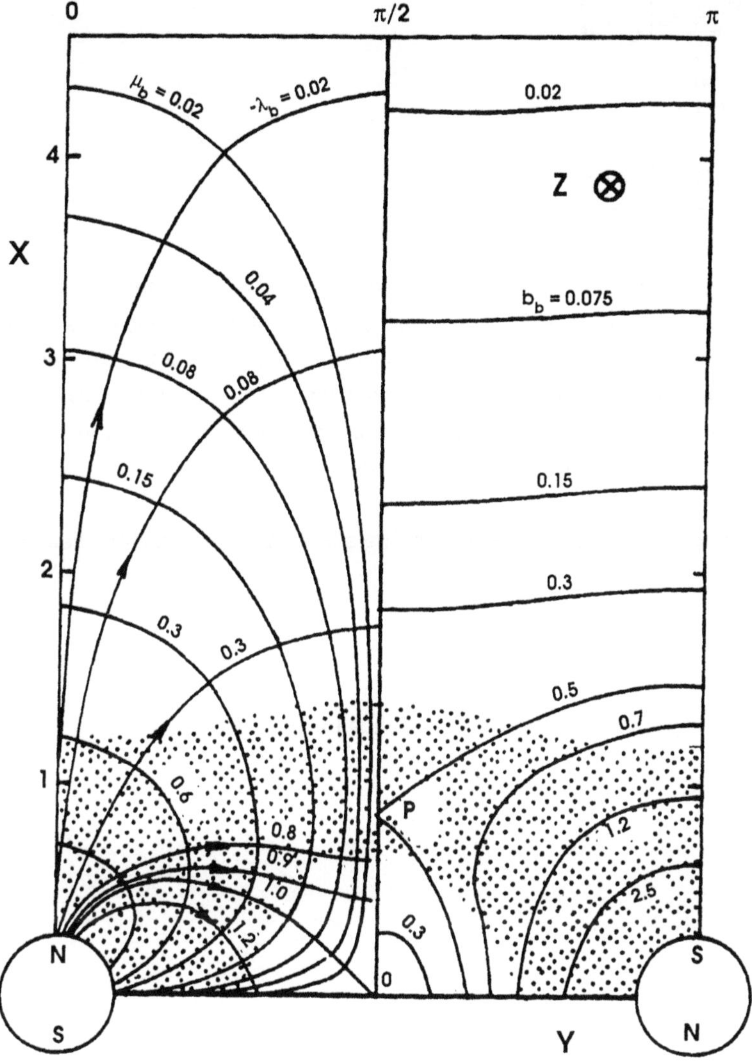

Fig. 12.5. Magnetic field configuration. The multipolar magnetic field is produced by a planar array of magnet rows parallel to the Oz axis. The shape of field lines ("λ_b = constant" lines), of orthogonal trajectories ("μ_b = constant" lines) and of "b_b = constant" lines is indicated. The shaded area corresponds to regions where the adiabaticity condition is fulfilled.

A two-dimensional magnetic field $\mathbf{B}_0(x,y)$ is obtained (Appendix 11.4), which derives from a magnetic vector potential $\mathbf{A}(x,y)$, parallel to the Oz axis,

$$A_z(x,y) = \frac{B_{0m}d}{\pi} \lambda_b\left(\frac{\pi x}{d}, \frac{\pi y}{d}\right) \tag{12.8}$$

with (eqn. (11.156))

$$\lambda_b(x,y) \equiv - \frac{\cosh X \sin Y}{\cosh^2 X - \cos^2 Y}, \tag{12.9}$$

where $X = \pi x/d$, $Y = \pi y/d$ are the reduced coordinates, d is the distance between magnet axes and B_{0m} is a value defined by eqn. (11.124), which characterizes the multipolar magnetic field intensity, and $\boldsymbol{\lambda}_b$ is the reduced vector potential. Note that "λ_b = constant" lines are the so-called *field lines*. In the numerical calculations that follow, we have $B_{0m} = 100$ gauss and $d = 6.2$ cm, which correspond to current values in multipolar devices using conventional magnets.

The magnetic vector potential **A** being parallel to Oz, the magnetic field **B**$_0$ derived from

$$\mathbf{B}_0 = \nabla \times \mathbf{A}, \tag{12.10}$$

lies in the (x,y) plane. Its components can thus be written according to

$$(\mathbf{B}_0)_x = \frac{B_{0m}d}{\pi} \frac{\partial \lambda_b}{\partial y}, \tag{12.11}$$

$$(\mathbf{B}_0)_y = - \frac{B_{0m}d}{\pi} \frac{\partial \lambda_b}{\partial x}. \tag{12.12}$$

Using the reduced coordinates X and Y, the expressions for the reduced magnetic scalar potential μ_b (eqn. (11.157)) and for the reduced magnetic field intensity b_b (eqn. (11.158)) come to

$$\mu_b(X,Y) = \frac{\sinh X \cos Y}{\cosh^2 X - \cos^2 Y}, \tag{12.13}$$

$$b_b(X,Y) = \frac{\left(\cosh^2 X - \sin^2 Y\right)^{1/2}}{\cosh^2 X - \cos^2 Y}. \tag{12.14}$$

Figure 12.5 shows the topography of the multipolar magnetic field in the case of a plane configuration (see Sec. 11.3.1 and Appendix 11.4). The field lines

("λ_b = constant" lines) converge on magnet poles to form the so-called *cusps*. Along the $Y = 0$ line, i.e. in front of the magnet, the intensity of the magnetic field decreases monotonically away from the magnet pole. Along the $Y = \pi/2$ line, i.e. between two magnets, B_0 is equal to zero at $X = 0$ and increases as X increases to reach a maximum value $B_{0m}/2$ at point P (Fig. 12.5) which is a saddle point for $\mathbf{B}_0$. This point is at a distance $X \approx 0.89$ from the plane defined by the magnet axes. Then beyond P, B_0 (i.e. b_b) becomes nearly independent of Y for $X \gtrsim 2$ (see Fig. 12.5) and decreases exponentially over a scale length d/π (Appendix 11.4).

12.3.3. Invariants in the motion of primary electrons.
The expressions for the multipolar magnetic field are too complicated to permit one integrating analytically the equations of motion of the primary electrons. However, using the invariants in the motion of the primary electrons allows one to circumvent this difficulty. The field geometry considered in Sec. 12.3.2 lends itself to an easy derivation of these invariants. In the absence of collisions, the *first invariant* is the kinetic energy W_{pe} of the primary electrons

$$W_{pe} = \frac{1}{2} m_e v_p^2 \tag{12.15}$$

with

$$v_p^2 = v_{px}^2 + v_{py}^2 + v_{pz}^2, \tag{12.16}$$

where m_e is the electron mass and v_{px}, v_{py} and v_{pz} are the components at time t of the primary electron velocity v_p. The second invariant is obtained from the evolution of the primary electron momentum along Oz

$$m_e \frac{dv_{pz}}{dt} = - e \left[v_{px} (\mathbf{B}_0)_y - v_{py} (\mathbf{B}_0)_x \right], \tag{12.17}$$

where e is the absolute value of the electron charge. Using eqns. (12.11) and (12.12), we obtain

$$m_e \frac{dv_{pz}}{dt} = \frac{e B_{0m} d}{\pi} \frac{d\lambda_b}{dt} \tag{12.18}$$

which, after integration, yields the *second invariant* in the motion

$$p_z = m_e v_{pz} - \frac{e B_{0m} d}{\pi} \lambda_b(x,y), \tag{12.19}$$

where p_z is the momentum along Oz of the primary electron.

We now derive the characteristics of the trajectories using these two invariants. Equation (12.16) shows that during the motion of the primary electron, its component v_{pz} evolves between $-v_p$ and v_p, and we have the condition

$$v_{pz}^2 \leq v_p^2. \tag{12.20}$$

Taking into account eqn. (12.19), this condition becomes

$$\left(\frac{p_z}{m_e} + v_p \frac{\lambda_b}{\lambda_{b0}} \right)^2 < v_p^2, \tag{12.21}$$

where

$$\lambda_{b0} = \frac{\pi m_e v_p}{e B_{0m} d} \tag{12.22}$$

characterizes the motion in the multipolar magnetic field of primary electrons of initial velocity v_p. For a 50 eV electron, we get $\lambda_{b0} = 0.12$ with the values previously adopted for B_{0m} and d. Equation (12.21) yields the minimum and maximum λ_b values of the field lines between which primary electrons of velocity v_p and momentum p_z can move, namely

$$\lambda_{bm} < \lambda_b < \lambda_{bM} \tag{12.23}$$

with

$$\lambda_{bm} = - \lambda_{b0} \left(\frac{p_z}{m_e v_p} + 1 \right), \tag{12.24}$$

$$\lambda_{bM} = - \lambda_{b0} \left(\frac{p_z}{m_e v_p} - 1 \right), \tag{12.25}$$

$$\lambda_{bM} - \lambda_{bm} = 2\lambda_{b0}. \tag{12.26}$$

Equation (12.23) thus indicates that the motion of a given primary electron of velocity v_p and momentum p_z is restricted to the region of the (x,y) plane bounded by the two field lines defined by eqns. (12.24) and (12.25).

Equations (12.24) and (12.25) suggest that different cases must be distinguished according to the value of p_z. If

$$\frac{|p_z|}{m_e v_p} > 1, \tag{12.27}$$

then λ_{bm} and λ_{bM} have the same sign, opposite to that of p_z. In this case, the motion of the primary electron is confined within the two adjacent field lines, λ_{bm} and λ_{bM}, as shown in Fig. 12.6 (unshaded area): the electron is trapped in the magnetic field and cannot escape to the magnetic field free volume of the multipolar discharge. In the same way, a primary electron coming from the magnetic field free volume of the discharge cannot get trapped without violation of the invariants.

In contrast, if now

$$\frac{|p_z|}{m_e v_p} < 1, \tag{12.28}$$

then λ_{bm} and λ_{bM} have opposite signs. This situation corresponds to free electrons moving in the magnetic field free region (unshaded area in Fig. 12.7). These electrons cannot reach the regions between the field lines and the discharge wall, as shown in Fig. 12.7. In this case, the volume accessible to primary electrons is open: electrons coming from infinity (zero field region) penetrate into the magnetic field region, are then mirrored and turned back to the magnetic field free part of the discharge. In this region, the momentum p_z of primary electrons reduces to $m_e v_{pz}$ and, according to eqn. (12.20), may vary in the interval

$$-m_e v_p < p_z < m_e v_p. \tag{12.29}$$

These free electrons, whatever their momentum, cannot penetrate in the doubly hatched region. Thus, the "$\lambda_b = 2\lambda_{b0}$" and "$\lambda_b = -2\lambda_{b0}$" field lines (see eqns. (12.24) and (12.25)) define the forbidden regions for electrons of velocity v_p coming from the field free region. These lines set the frontier separating free and trapped electrons of velocity v_p (Figs. 12.6 and 12.7).

A consequence of the above discussion is that the leaking of primary electrons to the wall can only occur at the cusps, i.e. at the magnet poles. The corresponding leak width can thus be defined by

$$\Delta d_p = 2y_0, \tag{12.30}$$

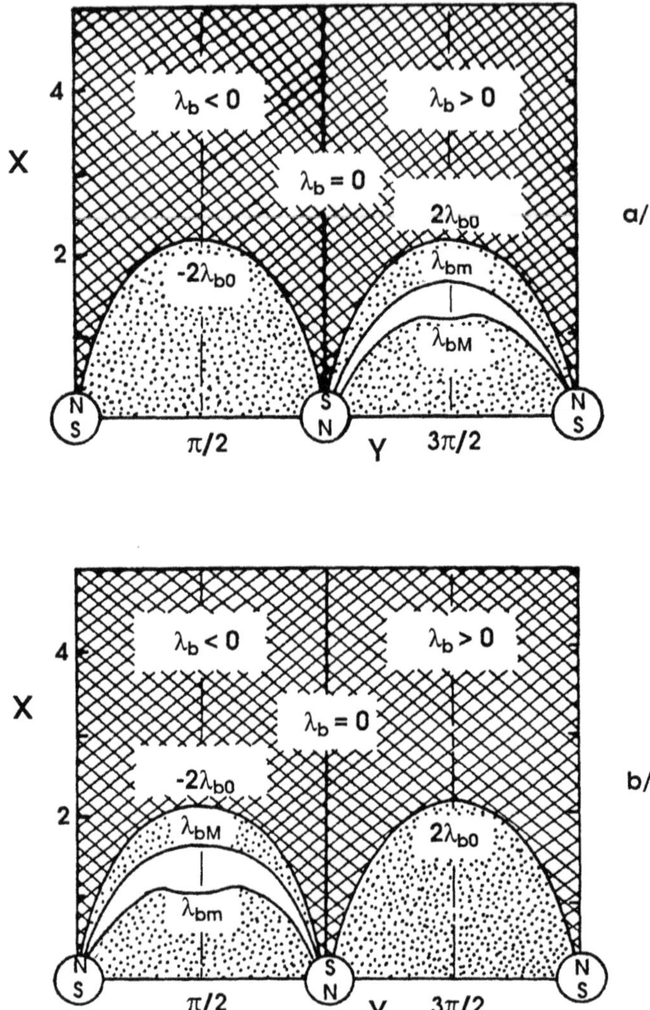

Fig. 12.6. Domain of the multipolar magnetic field (unshaded area) allowing the motion of trapped electrons of velocity v_p and momentum p_z in the case $|p_z| / m_e v_p > 1$: a) $p_z/m_e v_p < -1$; b) $p_z/m_e v_p > 1$.   Shaded domains are regions that are accessible to trapped electrons whatever their momentum p_z, while doubly hatched domains are not accessible to trapped electrons. The $\pm 2\lambda_{b0}$ field lines thus separate the domains of free and trapped electrons of given velocity v_p.

where the coordinates $\pm y_0$ are the intersection of the field lines $\lambda_b = \pm 2\lambda_{b0}$ with the plasma wall located at $x \approx a$ (a is the radius of the magnets) (see Fig. 12.7).

According to eqn. (12.9), y_0 is given by the expression

$$2\lambda_{b0} = \frac{\cosh(\pi a/d)\,\sin(\pi y_0/d)}{\cosh^2(\pi a/d) - \cos^2(\pi y_0/d)},$$ (12.31)

with λ_{b0} given by eqn. (12.22). Assuming $\pi a/d$, $\pi y_0/d \ll 1$, we get

$$\Delta d_p = r_{p0}\left(\frac{2\pi a}{d}\right)^2$$ (12.32)

which is proportional to the Larmor radius r_{p0} of the primary electrons

$$r_{p0} = \frac{eB_{0m}}{m_e v_p}$$ (12.33)

calculated at $B_0 = B_{0m}$.

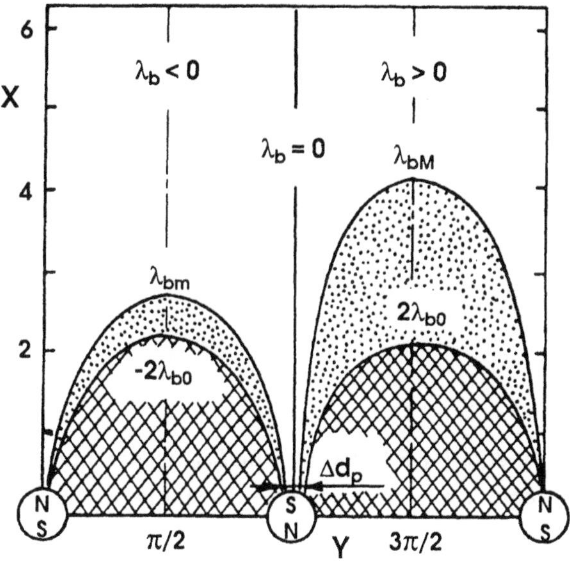

Fig. 12.7. Domain of the multipolar magnetic field (unshaded area) allowing the motion of free electrons of velocity v_p and momentum p_z, coming from zero field regions, in the case $|\,p_z\,|\,/\,m_e v_p < 1$. Shaded domains are regions that are not accessible to free electrons of velocity v_p and momentum p_z. Doubly hatched regions are forbidden to all free electrons of given velocity v_p, whatever their momentum p_z. The leak width Δd_p through which free electrons reach the wall is indicated. The $\pm\,2\lambda_{b0}$ field lines separate the domains of free and trapped electrons of velocity v_p.

After having determined the two invariants of the primary electron motion in a multipolar magnetic field, we consider the detailed description of the electron motion in the regions just defined by turning to the guiding center approximation, using the adiabatic invariants in the motion.

12.3.4. Electron motion characteristics: adiabatic approximation. When the Larmor radius r_p of a charged particle becomes small with respect to the scale length of the magnetic field inhomogeneity (see below), its motion can be considered as a perturbation of the helical motion that it would undergo in a uniform static magnetic field.[6-8] This approximation leads to the picture of a particle gyrating about a small radius circle whose center, the so-called *guiding center*, is slowly drifting. The smaller the radius of gyration compared to the spatial variation of the magnetic field, the more correct becomes the picture of a slowly drifting circular trajectory. The motion of the guiding center directly follows from the adiabatic invariants of charged particles in a slowly varying magnetic field. This approximation is now used to examine the trajectories of primary electrons in the different regions of a multipolar discharge.

Domain of validity of the adiabatic theory. The scale length of the magnetic field inhomogeneity is defined as $B_0 / |\nabla B_0|$, where B_0 is the magnitude of $\mathbf{B}_0$. The *adiabatic approximation* requires that r_p be much smaller than the scale length of the field inhomogeneity, namely

$$r_p \frac{|\nabla B_0|}{B_0} \ll 1, \tag{12.34}$$

which can be written as

$$\frac{\lambda_{b0}\, d\, |\nabla B_0|}{\pi\, b_b^2} \ll 1, \tag{12.35}$$

where λ_{b0} is given by eqn. (12.22) and $b_b = B_0/B_{0m}$ comes from eqn. (11.158). We delimit below the regions of the multipolar discharge where the guiding center approximation applies.

We first consider the central region of the discharge where "b_b = constant" lines can be approximated, as shown in Fig. 12.5, by "x (or X) = constant" lines corresponding to $X \gtrsim 2$. In this region, b_b is independent of y and decreases exponentially when x (or X) increases (eqn. (11.164)). The adiabaticity condition (eqn. (12.35)) leads to

$$X \ll \log_e \frac{2}{\lambda_{b0}} ,$$

(12.36)

which gives, for the numerical values previously adopted, $X \ll 2.8$ (or $x \ll 0.89$ d). This condition cannot therefore be met in the central region because of a too low magnetic field ($X \geq 2$), leading to a too large Larmor radius. We now turn to regions of high magnetic field intensity.

The adiabaticity condition is always fulfilled in the region which surrounds the saddle point P (see Fig. 12.5): at this point, the scale length of the field inhomogeneity becomes infinite while the Larmor radius r_p of primary electrons is finite ($B_0 = B_{0m}/2$).

Another region of intense magnetic field is that close to the magnets. In the vicinity of the magnets, the "b_b = constant" lines are nearly equidistant as shown in Fig. 12.5, indicating constant magnetic field gradients. Thus, to define the extent of the region close to the magnets where the adiabaticity condition is fulfilled, we may choose for simplicity to study the cusp region (Y = 0 line). According to eqns. (12.13) and (12.14), the reduced magnetic scalar potential $\mu_b(x,y)$ and the reduced magnetic field intensity $b_b(x,y)$ become respectively

$$\mu_b(x,0) = \left(\sinh \frac{\pi x}{d} \right)^{-1} ,$$

(12.37)

$$b_b(x,0) = \frac{\cosh (\pi x/d)}{\sinh^2 (\pi x/d)} .$$

(12.38)

The adiabaticity condition along the "$\lambda_b(x,0) = 0$" line can be written, according to eqns. (12.35), (12.37) and (12.38), as

$$\frac{\lambda_{b0}}{\mu_b} \frac{2\mu_b^2 + 1}{\mu_b^2 + 1} \ll 1.$$

(12.39)

As the second fractional part of eqn. (12.39) takes values between 1 and 2 whatever μ_b, the adiabaticity condition reduces to

$$\mu_b \gg \lambda_{b0}.$$

(12.40)

The lower the value of λ_{b0}, the larger the extent of the region where the approximation of the guiding center can be applied. With the conventional magnets used in most multipolar devices, the adiabaticity condition is generally verified in the whole region of high magnetic field intensity which includes the cusp and the saddle point regions and the region in-between (Fig. 12.5).

Mirror points. The *first adiabatic invariant* [6-8] is the *magnetic moment* $\mathfrak{M}_p$ of a primary electron gyrating around the guiding center. We define it as

$$\mathfrak{M}_p \equiv \frac{m_e v_{pg}^2}{2B_0(x,y)},$$

(12.41)

where v_{pg} is the electron velocity of gyration about the guiding center. Assuming energy conservation, eqn. (12.16) can be rewritten according to

$$v_p^2 = v_{ps}^2 + \frac{2\mathfrak{M}_p\, B_0(x,y)}{m_e},$$

(12.42)

where v_{ps} is the drift velocity of the guiding center. The invariance of $\mathfrak{M}_p$ predicts that the velocity v_{ps}

$$v_{ps} = \pm \left[v_p^2 - \frac{2\mathfrak{M}_p B_0(x,y)}{m_e} \right]^{1/2}$$

(12.43)

will vanish at the magnetic field intensity

$$B_0(x,y) = \frac{m_e v_p^2}{2\mathfrak{M}_p}$$

(12.44)

and that the primary electron will be reflected there. The + sign corresponds to electrons leaving the magnet region and the - sign to electrons approaching the magnets.

In a cusp, the magnetic field $\mathbf{B}_0(x,y)$ is nearly parallel to Ox and is thus practically independent of y. In this region, the coordinates (x,0) of the mirror point can be determined from eqn. (12.44).

Drift velocity of electrons. The drift velocity perpendicular to the magnetic field lines followed by the guiding center can be expressed as [8]

$$\mathbf{v}_{pd} = - \frac{m_e v_{p\parallel}^2}{e} \frac{\mathbf{B}_0 \times (\mathbf{B}_0\cdot\nabla)\mathbf{B}_0}{B_0^4} - \frac{m_e v_{p\perp}^2}{e} \frac{\mathbf{B}_0 \times \nabla B_0^2}{B_0^2},$$

(12.45)

where $\mathbf{v}_{p\parallel}$ and $\mathbf{v}_{p\perp}$ are the primary electron velocities in the direction parallel and perpendicular to the $\mathbf{B}_0$ field respectively. The first term corresponds to the *curvature drift* and the second term to the *B-gradient drift*. If the current carried by primary electrons is assumed not to modify the magnetic field ($\nabla \times \mathbf{B}_0 = 0$), we have $\nabla B_0^2 = 2(\mathbf{B}_0 \cdot \nabla)\mathbf{B}_0$, and the two components of the drift velocity are parallel and can be expressed as[8]

$$\mathbf{v}_{pd} = - \frac{m_e}{e} \frac{\mathbf{R}_c \times \mathbf{B}_0}{R_c^2 B_0^2}\left(v_{p\parallel}^2 + \frac{1}{2} v_{p\perp}^2\right), \tag{12.46}$$

where $\mathbf{R}_c$ is the curvature radius vector of the magnetic field line.

In the cusp region, the curvature of the field lines is small ($R_c \rightarrow \infty$) and ∇B_0 is parallel to $\mathbf{B}_0$, yielding an almost zero drift velocity. In contrast, between two cusps, ∇B_0 is perpendicular to $\mathbf{B}_0$ and the curvature radius of field lines is short, therefore the drift velocity is the highest.

Since the vectors involved in eqns. (12.45) and (12.46) all belong to the (x,y) plane, the drift velocity $\mathbf{v}_{pd}$ is parallel to Oz. A trapped electron also undergoes a drift motion along Oz while oscillating along a "λ_b = constant" line between mirror points in the cusps of two successive magnet rows. Thus, the projection parallel to Oz of the guiding center trajectory in the (x,y,z = 0) plane always follows the same field line, provided the adiabaticity condition is fulfilled along the whole trajectory. Furthermore, considering the axes in Fig. 12.5, the drift velocity is directed towards positive z on the left of a north magnet and towards negative z on the right of it.

Bouncing period of trapped electrons. We consider now the electrons trapped in the multipolar magnetic field between the field lines λ_{bm} and λ_{bM}. As seen in the previous paragraph, the projection of the guiding center in the (x,y) plane always follows the same field line, which links two successive magnets. A given trapped electron oscillates between two well defined mirror points near the magnet poles.

A constant of this motion, the so-called *second or longitudinal adiabatic invariant* is given by[6]

$$\mathbb{K} = \oint v_{ps}(s)\, ds \tag{12.47}$$

where v_{ps} is the velocity of the guiding center and ds is an element of path length along the "λ_b = constant" line followed by the guiding center. The integral is taken over a complete oscillation between mirror points. Since these points on a given field line are perfectly determined (see eqn. (12.44)), the path length $\mathbb{L}$ covered by the guiding center during a complete period $\mathbb{T}_t$ is constant. Thus, according to the longitudinal invariant (eqn. (12.47)), the mean velocity v_{pm} of the guiding center along the field line

$$v_{pm} = \frac{1}{\mathbb{L}} \oint v_{ps}(s)\, ds \qquad\qquad (12.48)$$

is constant over a complete period. The bouncing period of a given trapped electron, which reduces to

$$\mathbb{T}_t = \frac{\mathbb{L}^2}{\mathbb{K}}, \qquad\qquad (12.49)$$

is also constant. Recall that this motion is perfectly periodic only provided that the adiabaticity condition is fulfilled along the whole field line followed by the guiding center.

12.4. Results from numerical calculations

In order to illustrate the results of the previous section, we present trajectories obtained by numerical integration of the above equations of motion.[9] For the sake of simplicity, we use the following dimensionless quantities: the reduced coordinates X and Y, already defined in Sec. 12.3.2, and $Z = \pi z/d$; the reduced velocities $v_{px} = v_{px}/v_p$, $v_{py} = v_{py}/v_p$ and $v_{pz} = v_{pz}/v_p$; the reduced momentum along OZ, $p_z = p_z/m_e v_p$; the reduced magnetic moment $\mathcal{M}_p = 2\mathfrak{M}_p B_{0m}/m_e v_p^2$; and the reduced time $\mathcal{T} = \omega_{cp}/t$ where $\omega_{cp} = eB_{0m}/m_e$ is the cyclotron angular frequency of a primary electron for $B_0 = B_{0m}$. We now examine typical trajectories as a function of initial conditions, starting with trapped electrons under quasi-adiabatic conditions.

12.4.1. Quasi-periodic trapped electrons. Figure 12.8 shows the trajectory of a trapped electron. As expected from the two invariants of motion (Sec. 12.3.3), the electron is confined within a crescent-like region limited by two field lines. In the example presented, $p_z = -3.3$ and $\lambda_{b0} = 0.1212$, and the limiting field lines (eqns. (12.24) and (12.25)) are $\lambda_{bm} = 0.2788$ and $\lambda_{bM} = 0.5212$. In the absence of collisions, the electron oscillates indefinitely between the cusps of two successive magnets. The drift velocity is parallel to OZ and directed towards positive Z (Figs. 12.8b and 12.8c). The trajectories are associated with two periods of the motion: the first one (Larmor period) corresponds to the gyration around the guiding center,

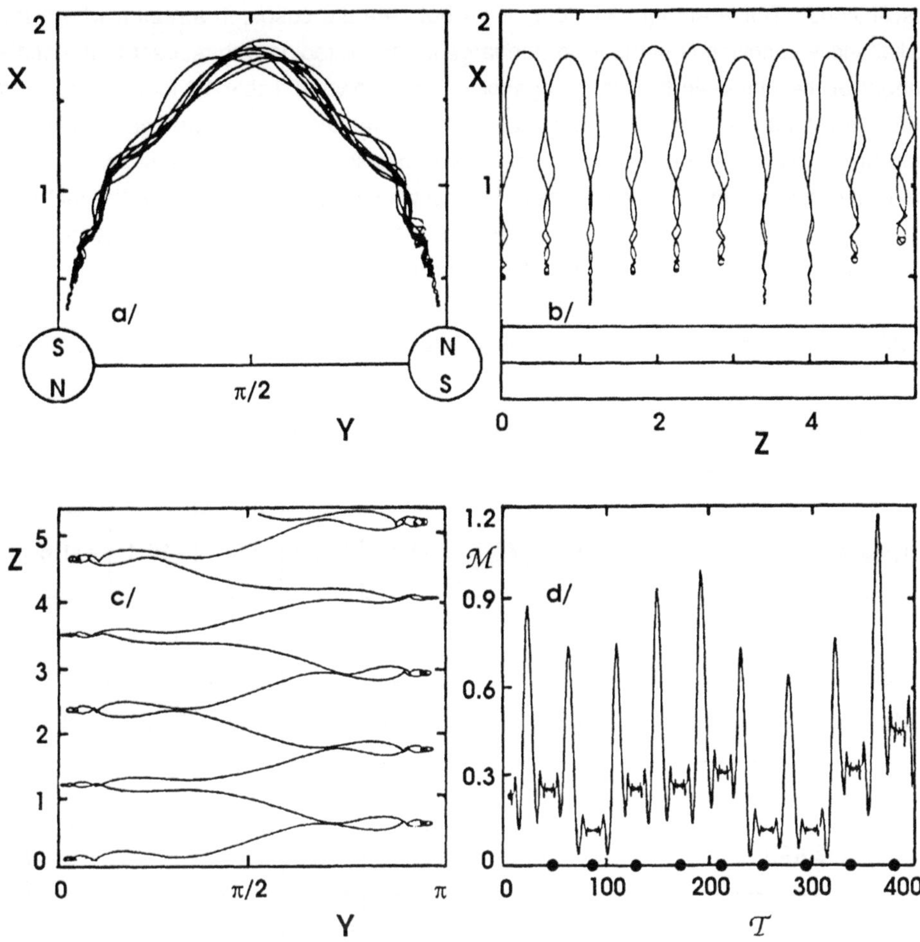

Fig. 12.8. Trajectory of a trapped electron ($p_z = -3.3$). The initial conditions are X = 0.5, Y = 0.1, $v_{px} = 0.0595$, $v_{py} = -0.9982$; a) projection in the (X,Y) plane; b) projection in the (X,Z) plane; c) projection in the (Z,Y) plane; d) time evolution of the magnetic moment; times at which reflections occurs are indicated by dots.

while the second period (bouncing period) corresponds to the reflection of the guiding center at the mirror points.

Table 12.1 gives numerical information about the electron trajectory in Fig. 12.8. The field line $(\lambda_{bm} + \lambda_{bM})/2 = 0.400$ followed by the guiding center practically does not vary. In contrast, we can notice that, between two reflections, the magnetic moment varies considerably. Figure 12.8d shows the detail of this variation as a function of time: the magnetic moment exhibits successive plateaus separated by sharp oscillations. These abrupt variations occur between the cusps, in a region where the adiabaticity condition is not fulfilled (too large a Larmor radius with respect to the scale length ot the magnetic field inhomogeneity). As shown in Table 12.1, each crossing through the non-adiabatic region induces random variations of the magnetic moment, and also a slight shift of the mirror point. Nevertheless, Table 12.1 also shows that the bouncing period remains nearly constant. As a whole, we see that despite the fact that the adiabatic invariants are not fully satisfied, the trajectory of the guiding center and the bouncing period are not significantly affected.

Table 12.1. Numerical data about the primary electron motion in Fig. 12.8 (from ref. 9).

Reflection (number)	1	2	3	4	5	6	7
Time $\mathbb{T}$	1.9	43.7	84.8	126.7	169.1	210.1	251.6
Magnetic moment $\mathfrak{M}_p$	0.230	0.242	0.103	0.255	0.286	0.106	0.252
Mirror points (abscissa X)	0.486	0.491	0.319	0.510	0.541	0.690	0.497
Mirror points (X) (calculated from eqn. (12.44))	0.489	0.501	0.324	0.515	0.546	0.738	0.512
Guiding center (field line λ_b)	0.400	0.399	0.398	0.384	0.392	0.392	0.390
Bouncing period $\mathbb{T}_t$	41.80	41.42	41.88	42.80	41.05	41.45	

12.4.2. Non-periodic trapped electrons. When p_z is greater than, but close to unity (eqn. (12.27)), the region where the primary electrons are trapped ($-2\lambda_{b0} \lesssim \lambda_b \lesssim 0$) extends far from the magnets. In this region, the adiabatic approximation is not valid and the periodicity associated with the longitudinal invariant dissapears. In addition, the magnetic moment exhibits large variations with time and there are no defined mirror points because the position where the electrons turn back varies considerably from one reflection to the other. Figure 12.9 shows an example of such a trajectory with $p_z = 1.1$. The drift velocity along OZ is still present, but the reproducibility of the trajectory in Fig. 12.8 is lost. The transition from aperiodic to

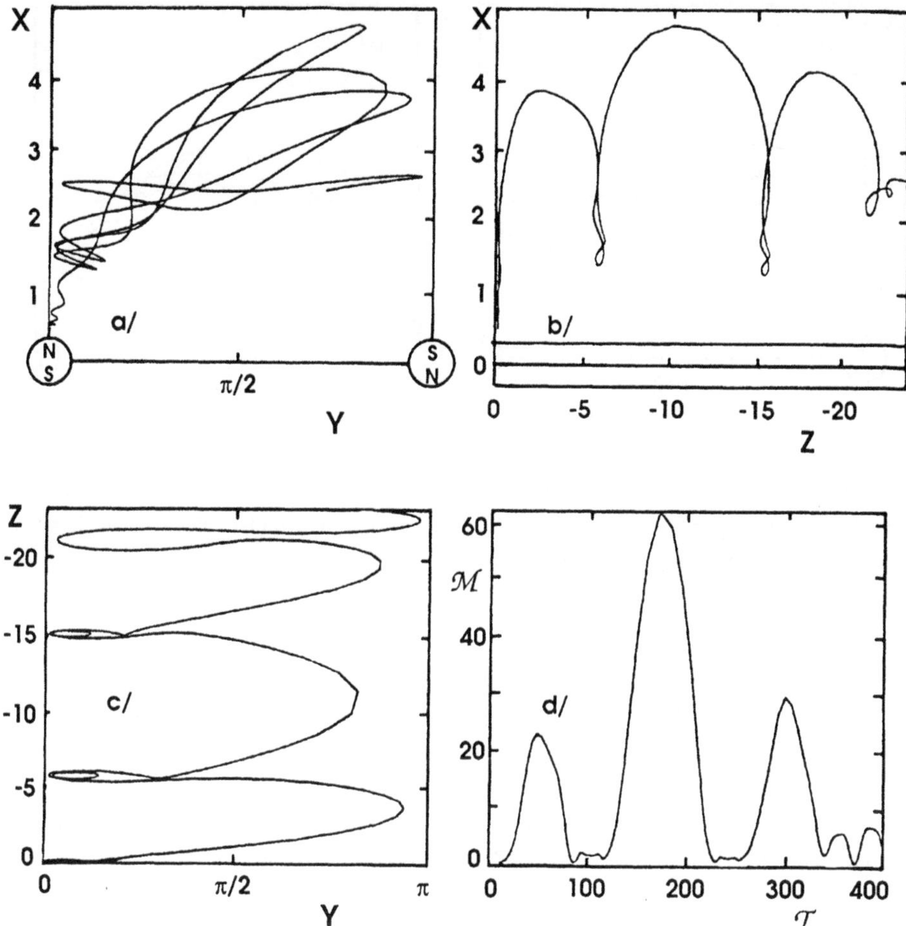

Fig. 12.9. Trajectory of a trapped electron ($p_z = 1.1$). The initial conditions are X = 0.5, Y = 0.032, $v_{px} = 0.0595$, $v_{py} = -0.9982$; a) projection in the (X,Y) plane; b) projection in the (X,Z) plane; c) projection in the (Z,Y) plane; d) time evolution of the magnetic moment.

periodic trajectories occurs progressively when p_z increases as illustrated in Fig. 12.10 by the progressive apparition of two symmetrical mirror points.

12.4.3.   Free electrons.   Figures 12.11a and 12.11b present three typical trajectories of free electrons which are reflected in the cusp region. The electrons come from the same point of the (X,Z) plane and their intial velocity vector lies in the (X,Y) plane. We notice that the electrons which penetrate the most deeply in the cusp are those injected with the smallest angle ϕ with the OX axis. The smaller the angle ϕ, the closer to the magnet is the mirror point (Figs. 12.11a and 12.11b). In the limiting case of

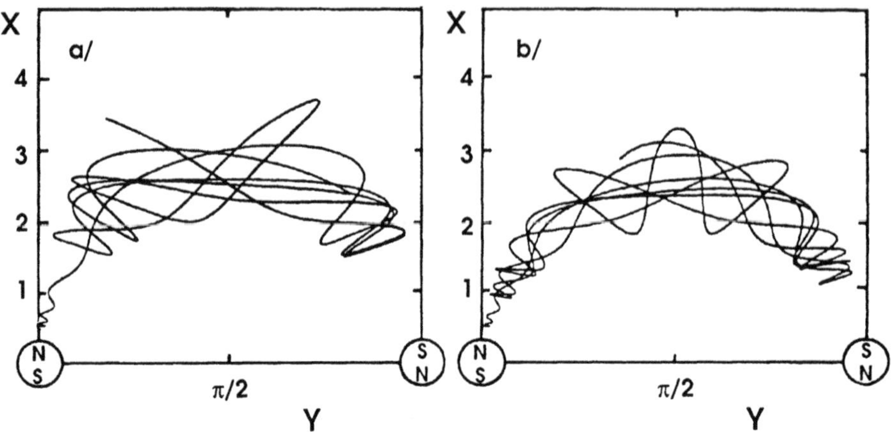

Fig. 12.10. Trajectory of a trapped electron in the (X,Y) plane. a) $p_z = 1.36$. The initial conditions are $X = 0.5$, $Y = 0.04$, $v_{px} = 0.0595$, $v_{py} = -0.9982$; b) $p_z = 1.67$. The initial conditions are $X = 0.5$, $Y = 0.049$, $v_{px} = 0.0595$, $v_{py} = -0.9982$.

an initial velocity vector $\mathbf{v}_p$ along OX ($\phi = 0$), the Lorentz force ($-e\mathbf{v}_p \times \mathbf{B}_0$) is zero. The trajectory is then a straight line and the electron does not turn back. Figures 12.11c and 12.11d also show the variation of the magnetic moment along the trajectory, for two values of the angle ϕ. When the electron penetrates deep enough inside the cusp ($X \leq 2$), the magnetic moment oscillates with small deviations from a mean given value, in agreement with the adiabatic approximation. In addition, as already discussed in Sec. 12.3.4, there is no drift along the OZ axis (Fig. 12.11b).

12.4.4. Trapping of free electrons. In a multipolar discharge, there are two distinct populations of primary electrons: electrons that are free to move in regions of low magnetic field intensity and electrons that are trapped near the magnets. In the multipolar discharges described in Chaps. 10 to 12, the primary electrons are emitted from hot filaments located in the central volume, free from magnetic field. The presence of trapped electrons thus requires the existence of trapping mechanisms. This problem has also been raised in the case of Van Allen radiation belts which are composed of solar wind particles trapped by the dipolar magnetic field of the earth. The trapping of particles (primary electrons or solar wind particles) can only be achieved by violation of the motion invariants. We discuss below the different ways of violating these invariants.

Collisional trapping mechanism. The violation of the motion invariants can result from collisions. The collisions of primary electrons with neutrals lead to an efficient momentum transfer. As an example (see Appendix 11.1), up to 4 elastic collisions occur on the average with argon atoms before the first ionizing collision takes place.

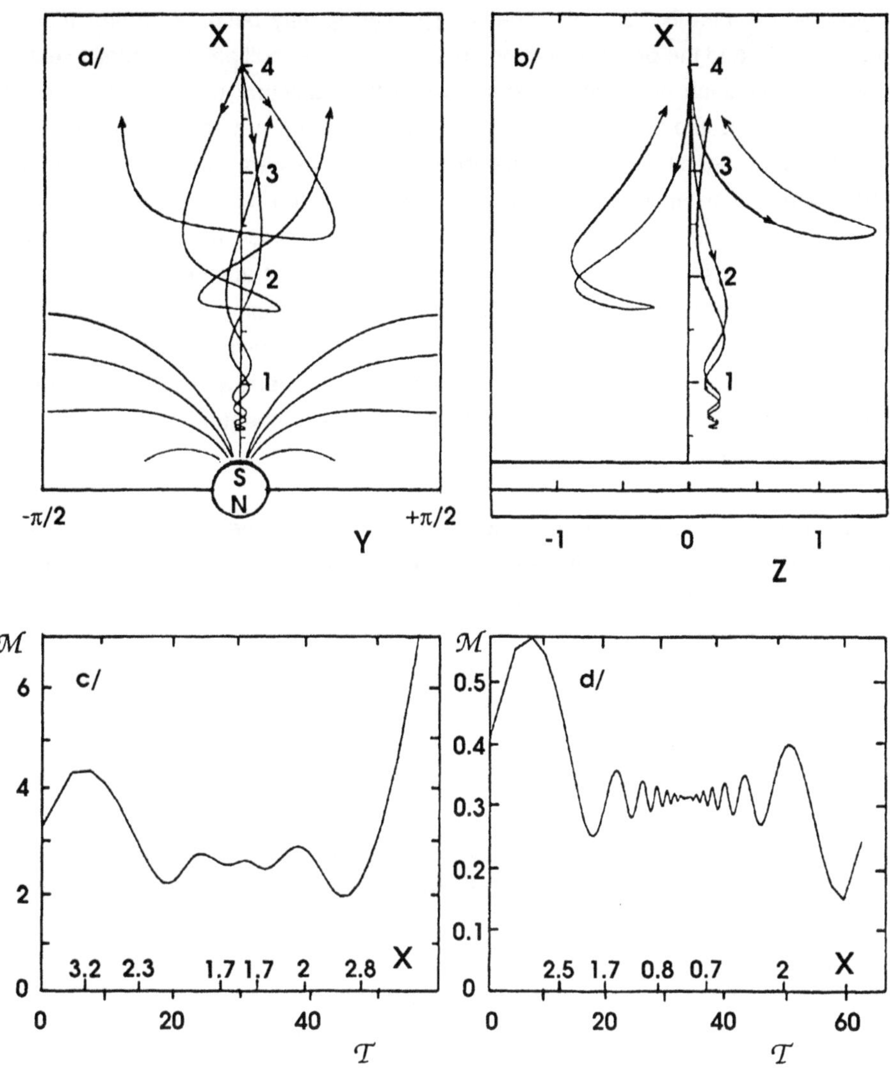

Fig. 12.11. Trajectory of free electrons ($p_z = 0$, $\phi = 7°$, $\phi = -20°$, $\phi = 30°$ with ϕ being the angle between the velocity vector and the OX axis); a) projection in the (X,Y) plane; b) projection in the (X,Z) plane; c) variation of the magnetic moment along the trajectory as a function of time ($\phi = -20°$); d) variation of the magnetic moment along the trajectory as a function of time ($\phi = 7°$).

These elastic collisions can produce a sharp variation in the velocity vector (the magnitude v_p remains practically constant). Thus, all values of v_{pz} between $-v_p$ and v_p are possible (with the same probability), but a maximum variation $|\Delta v_p| = 2m_e v_p$ of the momentum p_z can result on each elastic collision. Thus, in the magnetic field region (see Appendix 12.1), the quantity $|p_z|/m_e v_p$ may become greater than unity (eqn. (12.27)) and the primary electron get trapped in the multipolar magnetic field. The probability for such an event increases when the trajectory of the primary electron passes nearer to the forbidden region. Thus, the capture of free primary electrons by the collisional mechanism occurs mainly in the cusps where the field lines converge and where primary electrons can penetrate deeply in the magnetic field and interact over a long period of time with it. However, the existence of large populations of trapped electrons also implies the irreversibility of the trapping mechanism. The collisional processes, which generally lead to an isotropic distribution of velocity vectors, cannot explain this irreversibility. In fact, as shown in Appendix 12.1, the irreversibility of the trapping mechanism results from the narrowness of the momentum domain for free primary electrons with respect to the momentum domain for trapped electrons (see Sec. 12.3.3).

Electric field trapping mechanism. Another way of violating the invariants in the motion to account for electron trapping is to superimpose to the multipolar magnetic field an electric field, such as the space charge field neglected up to now. The presence of an electric field in the (x,y) plane due to plasma diffusion (Chap. 11) does not affect p_z. Thus free electrons remain free while trapped electron remain trapped, the introduction of the electric field leading only to a drift along Oz, i.e. along the rows of magnets.

Consider now the effect of an electric field assumed to be parallel to Oz and localized near the discharge wall. As a result of this field, the electron will drift towards the initially forbidden region and get eventually trapped. Such a situation is shown in Fig. 12.12. In the first case (Fig. 12.12a), there is no electric field. The electron coming from infinity and gyrating around a field line, gets reflected, returning back to infinity. In the second case (Fig. 12.12b), the same electron coming from infinity meets an electric field directed along Oz and localized into two rectangles, as shown in Fig. 12.12b. The electron kinetic energy $m_e v_p^2/2$ remains unchanged because the work done in the electric field during half of the Larmor period is lost during the other half (since the drift velocity induced by the electric field is perpendicular to the latter, no work is associated with this drift motion). However the direction of the velocity is modified. If the momentum variation Δp_z is large enough for the primary electron to escape the domain of free electrons and, as is the case in Fig. 12.12b, it reaches the domain of trapped electrons ($|p_z| / m_e v_p > 1$).

However, the hypothesis of a DC electric field along Oz is not consistent with the fact that the plasma was assumed invariant by translation along Oz. Another possibility is to consider electric fields produced by instabilities of the plasma.[9,10] In that respect, an ion acoustic instability, attributed to the diamagnetic drift of plasma electrons in the

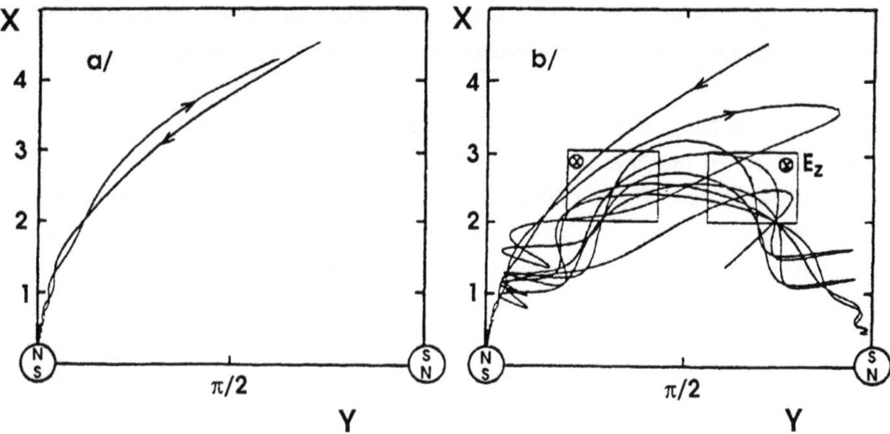

Fig. 12.12. a) Trajectory of a primary electron without electric field; b) trapping of the same electron under the influence of a small E_{0z} electric field ($E_{0z} = 0.03\, v_p B_{0m}$) localized inside the rectangles.

multipolar magnetic sheath, has been invoked to introduce irreversibility in the motion of primary electrons.[10]

12.5. Mechanisms of selective peripheral ionization

We showed in Sec. 12.2 that, in multipolar discharges, plasma homogeneity could not be accounted for either by assuming a constant ionization rate S_i in the discharge volume or by considering a sufficient decrease of the plasma leakage through the multipolar magnetic sheath. We suggested on the basis of a simple model, in accordance with Sec. 9.3.4, that the observed density profiles can be recovered by assuming a sharp increase in the ionization rate near the discharge walls. The arguments for such a selective ionization are developed below.

The study of primary electron trajectories presented in Secs. 12.3 and 12.4 showed in detail how the primary electrons, emitted from filaments in the central volume of the discharge free of magnetic field, interact with the peripheral multipolar magnetic field. We need to discuss how this interaction may lead to a selective peripheral ionization.

We first recall that a preliminary condition for such a peripheral ionization is a mean free path λ_{pi} of primary electrons for ionization of neutrals larger than the dimensions of the discharge chamber. If this was not the case, the electrons would undergo several inelastic collisions and degrade into plasma electrons before reaching the magnetic sheath and interacting with it: then ionization would mainly occur in the central volume of the discharge and plasma homogeneity could not be achieved (Sec. 12.2). In contrast, if λ_{pi} is much larger than the plasma dimensions, the primary electrons can reach the magnetic sheath without inelastic collisions and then interact with it.

Peripheral ionization via reflection in the magnetic field. This first mechanism, which provides selective peripheral ionization, is general as it could be applied to other magnetic field configurations. When a flux of electrons reflects on a magnetic mirror, i.e. in a magnetic field gradient, the velocity of the guiding center of each electron decreases and then vanishes at the mirror point. From the latter fact, two equivalent approaches may be considered to explain the increase in peripheral ionization as compared to volume ionization in the magnetic field free region of the discharge: i) the first approach is to assume that the mean time spent in the sheath is much larger than in the same volume free from magnetic field; ii) the second approach assumes flux conservation

$$n_p v_p = n_{ps} v_{ps} \qquad\qquad (12.50)$$

of the primary electrons that penetrate into the magnetic sheath. In eqn. (12.50), n_p and n_{ps} are the density of primary electrons in the volume free of magnetic field and in the magnetic sheath respectively. When the velocity v_{ps} vanishes at the mirror point, the density n_{ps} becomes very large, as well as the ionization rate which is proportional to the primary electron density.

Peripheral ionization via trapping mechanisms. The second mechanism for peripheral ionization results from the irreversible trapping of primary electrons either via elastic collisions of primary electrons with neutrals or via the interaction with electric fields due to the ion acoustic instability. Then, once trapped, primary electrons oscillate indefinitely between two successive cusps until they suffer inelastic (e.g. ionizing) collisions. The density of trapped primary electrons results from the balance between the different trapping processes and the inelastic collisions with neutrals which slow down the primary electrons and degrade them into thermal plasma electrons. Because of the presence of competitive trapping mechanisms, this balance is difficult to establish.

Finally, a significant increase in the ionization rate S_i' in the magnetic sheath as compared to the ionization rate S_i in the magnetic field free central volume appears as a likely explanation for plasma homogeneity in low pressure multipolar discharges. In reality, in filament-excited discharges, where primary electrons are emitted in the magnetic field free central volume of the chamber, one observes that the ionization rate S_i decreases in the central volume when pressure decreases. When the ionization rate

S_i finally vanishes, perfect plasma homogeneity is obtained in the central volume of the discharge (see eqn. (12.7)).

12.6. Conclusion

The homogeneity of the plasma density observed in low pressure multipolar discharges has been shown to result from the localization of ionization (due to primary electrons) at the periphery of the plasma. The detailed study of the characteristics of motion and of the trajectories of primary electrons led us to distinguish two categories of electrons, those that are free to move in the regions without magnetic field and those that are trapped in the multipolar magnetic field. These two electron populations are characterized by values of the parameter $|p_z|/m_e v_p$ respectively below or above unity (p_z is the momentum along the magnet rows, i.e. in the direction perpendicular to the magnetic field). The variation of p_z due to electric fields parallel to the magnet rows or under the effect of elastic collisions explains how free electrons can become trapped irreversibly in the multipolar magnetic field. It is the interaction of primary electrons with the multipolar magnetic field that accounts for a significant peripheral ionization, as required to explain the plasma homogeneity achieved in these low pressure discharges. The mechanism is as follows. When the mean free path λ_{pi} for ionization of neutrals by primary electrons is less than the discharge size, the ionization rate in the central volume of the plasma leads to density gradients and space charge fields. In contrast, when λ_{pi} is larger than the discharge central size, the ionization rate becomes negligible in the discharge central volume and no density gradients and electric fields develop there. The plasma is then homogeneous, provided volume recombination is small.

The above results apply to multipolar discharges where the electrons are emitted from filaments located in the discharge central volume, which is free of magnetic field. In this case, homogeneity of the plasma is achieved when the interaction of primary electrons with neutrals in the peripheral magnetic sheath becomes predominant as compared to that in the magnetic field free central region of the discharge. However, in such a system, homogeneity degrades rapidly as pressure is increased. An efficient solution to circumvent this difficulty is to produce primary electrons directly in the multipolar magnetic field where they will remain trapped until they make inelastic collisions and degrade into plasma electrons. Such a scheme is currently achieved by setting filaments directly in the magnetic sheath where the energetic electrons are trapped ab initio and cannot escape in the free field central region of the plasma. Another possibility, presented in Chap. 14, is to transfer HF energy to electrons via electron cyclotron resonance within the magnetic field itself.

APPENDIX 12.1

TRAPPING OF PRIMARY ELECTRONS IN A MULTIPOLAR MAGNETIC FIELD: IRREVERSIBILITY OF THE COLLISIONAL MECHANISM

Consider primary electrons of velocity v_p in a multipolar discharge of infinite extent along Oz. The magnetic field configuration is as described in Sec. 12.3.2. Depending on the value of the quantity $|p_z|/m_e v_p$ with respect to unity, where m_e is the mass of electrons and p_z their momentum along Oz, the primary electrons either move freely in the magnetic field free region of the discharge ($|p_z|/m_e v_p < 1$) or are trapped in the multipolar magnetic field (Sec. 12.3.3). The transition from one electron population to the other can only occur by violating the invariants in the primary electron motion, which are the velocity v_p (or the energy $m_e v_p^2/2$) and the momentum p_z (eqn. (12.19)).

Violation of the motion invariants can result from collisions. In particular, elastic collisions of primary electrons with neutrals can produce strong variations in the velocity vector $\mathbf{v}_p$ while the velocity v_p remains practically constant. Since these are random processes, all v_{pz} values between $-v_p$ and $+v_p$ are equally probable. Thus, on each collision, a variation Δp_z of the momentum p_z occurs, with a maximum variation $2 m_e v_p$.

The momentum p_z whose general expression (eqns. (12.19) and (12.22)) is

$$p_z = m_e v_{pz} - m_e v_p \frac{\lambda_b(x,y)}{\lambda_{b0}}, \tag{12.51}$$

reduces to

$$p_z = m_e v_{pz} \tag{12.52}$$

in the magnetic field free region. In this region, the quantity $|p_z|/m_e v_p$ remains less than unity whatever the variation in momentum along Oz ($-v_p < v_{pz} < v_p$). Thus, in the magnetic field free region, elastic collisions are inefficient to violate the motion invariants. In contrast, in the magnetic field region, the Δp_z variation due to elastic collision add algebrically to the momentum value p_z before collision. This may result in a momentum $p_z + \Delta p_z$ such that

$$\frac{|p_z + \Delta p_z|}{m_e v_p} > 1, \tag{12.53}$$

which is the condition for primary electrons to get trapped. Clearly, violation of the motion invariants can occur quite efficiently in the magnetic field region because any variation $m_e \Delta v_{pz}$ induces a correlative variation $\Delta \lambda_b$ of the field line that is followed by the guiding center of the primary electron. However, in principle, the opposite variation $-m_e \Delta v_{pz}$ of the momentum can also lead to de-trapping of an electron. Therefore, an important point to examine is the degree of reversibility of the trapping mechanism through elastic collisions.

To investigate the reversibility of collisional trapping, some simplifying assumptions are in order. First, we only consider elastic collisions of primary electrons with neutrals occurring in the cusps, i.e. along the $\lambda_b \approx 0$ lines in front of magnet poles (see. Fig. 12.5). In the following, n is the number of such successive collisions suffered by an electron. Second, we assume that the variation of momentum along Oz on each collision is $\Delta p_z = \pm m_e v_p$, with equal probabilities for positive and negative variations. The elementary probability after n elastic collisions is thus equal to $(1/2)^n$, as indicated in Table 12.2. The evolution of the distribution of momentum is summarized in Table 12.2 as a function of the number n of successive elastic collisions in the magnetic cusp region. The initial population of primary electrons ($n = 0$) are free electrons with a positive momentum ($0 < p_z < m_e v_p$) along Oz. For reasons of symmetry, the result would be the same for an initial population with a negative momentum ($-m_e v_p < p_z < 0$). Each number in Table 12.2, when normalized to the elementary probability $(1/2)^n$, yields the probability for an electron to exhibit a given momentum value after n elastic collisions. The probability $\mathcal{P}_f$ for an initially free electron to remain free after n successive elastic collisions in the magnetic field is

$$\mathcal{P}_f = \frac{n!}{\left(\dfrac{n}{2}\right)! \ \left(\dfrac{n}{2}\right)! \ 2^n} \tag{12.54}$$

for even n, and

$$\mathcal{P}_f = \frac{n!}{\left(\dfrac{n-1}{2}\right)! \ \left(\dfrac{n+1}{2}\right)! \ 2^n} \tag{12.55}$$

for odd n. The decrease of $\mathcal{P}_f$ with n clearly shows the irreversiblity of the trapping mechanism via elastic collisions. After $n = 5$ collisions, the propability $\mathcal{P}_f$ is less than one third. Using Stirling's formula for large n

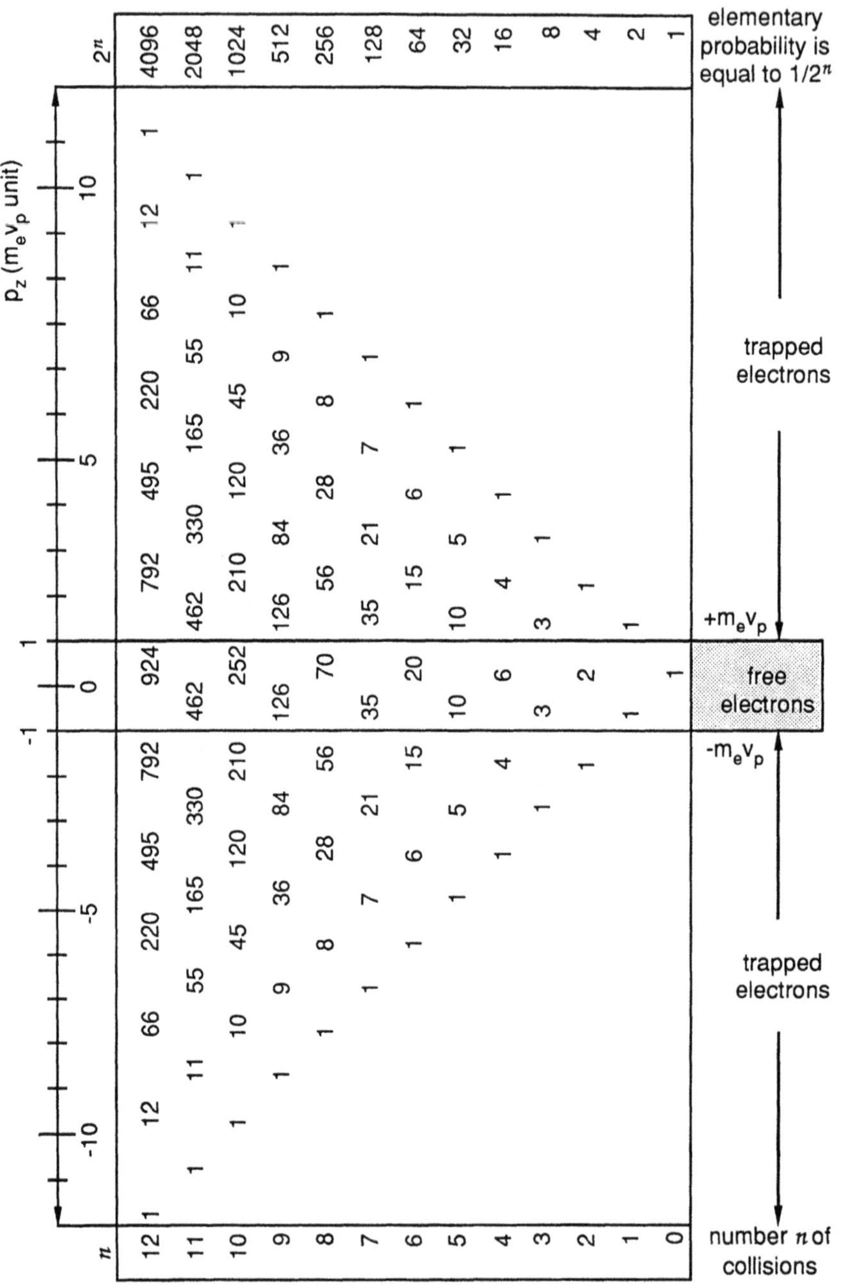

Table 12.2. Evolution of the distribution of the momentum p_z of a free electron with velocity v_p, after n successive elastic collisions in the cusp region. The probability $\mathcal{P}_f$ for a free electron to remain free ($|p_z| / m_e v_p < 1$) after n collisions in the cusps is equal to the number in the central column, normalized to 2^n, the total number of elementary probabilities.

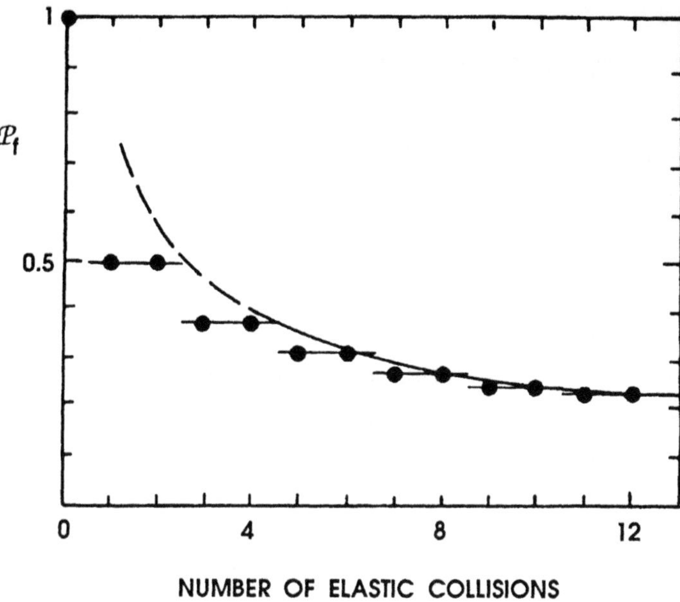

Fig. 12.13. Probability for a primary electron to remain free after n elastic collisions in the magnetic sheath. The data points are the exact calculated values while the full line indicates the approximate value derived from Stirling's formula (valid for large n).

$$n! \approx (2\pi n)^{1/2}\, n^{n}\, \exp(-n), \tag{12.56}$$

the probability $\mathcal{P}_f$ can be approximated to

$$\mathcal{P}_f \approx \left(\frac{2}{\pi n}\right)^{1/2}. \tag{12.57}$$

The exact calculated values of the probability $\mathcal{P}_f$ (eqns. (12.54) and (12.55)) and the corresponding approximate values (eqn. (12.57)) obtained from Stirling's formula are plotted in Fig. 12.13 up to $n = 12$, a value which corresponds to three or four times the free path for ionization of argon (see Appendix 11.1).

References

[1]   M. Sadowski, Phys. Lett. **28A**, 626 (1969).

[2]   J.M. Buzzi, J. Snow and J.L. Hirschfield, Phys. Lett. **A54**, 344 (1975).

[3]   K.N. Leung, N. Hershkowitz and K.R. MacKenzie, Phys. Fluids **19**, 1045 (1976).

[4]   R. Limpaecher and K.R. MacKenzie, Rev. Sci. Instrum. **44**, 726 (1973).

[5]   C. Gauthereau and G. Matthieussent, Phys. Lett. **A102**, 231 (1984).

[6]   T.G. Northrop and E. Teller, Phys. Rev. **117**, 215 (1960).

[7]   T.G. Northrop, Annals of Physics **15**, 79 (1961).

[8]   J.L. Delcroix, *Physique des Plasmas* (Dunod , Paris, 1963) Tome 1, Chap. 3.

[9]   C. Gauthereau and G. Matthieussent, J. Physique **45**, 1113 (1984).

[10] C. Gauthereau and G. Matthieussent, Phys. Lett. **A121**, 342 (1987).

CHAPTER 13

HIGH FREQUENCY SUSTAINED MULTIPOLAR PLASMAS *

13.1. Introduction

We have seen in the previous chapters that plasmas can be produced in a multipolar magnetic confinement structure by means of electron emission from a filament or, for large volumes, from an array of filaments. These electrons, which are given an adequate energy, are known as primary electrons. At moderate neutral pressures (10^{-4} to 10^{-3} torr), this yields uniform plasmas in the 10^9 to 10^{11} cm^{-3} density range. We have shown in Chap. 12 that homogeneity of the plasma in these filament-excited discharges results from the localization of ionization at the periphery of the discharge where primary electrons interact with the multipolar magnetic field.

The production of plasma by primary electrons presents some disadvantages: i) heating of the surfaces exposed to filament radiation; ii) heating of surfaces by energetic electrons; iii) metallic contamination resulting from filament sublimation and sputtering. Moreover, such discharges in reactive gases, owing to chemical erosion of the filament, cannot lead to stable and long lifetime plasmas. For example, Mantei,[1] using CF_4 as the etching gas, found that filament lifetime did not exceed 8 hours. In order to retain the advantages of multipolar confinement structures and eliminate or minimize most of the above drawbacks, different types of discharges have been investigated to sustain plasmas in these systems. We review some of them below, starting with RF discharges.

13.2. Multipolar plasmas sustained by RF discharges

The first HF discharge used to excite plasma in a multipolar magnetic structure is due to Schumacher et al.[2] They applied RF power at 75 MHz to a large-surface grid located at one end of a cylindrical multipolar magnetic structure. Figure 13.1 is a schematic view of this device. They found that the multipolar magnetic confinement increased the density of an argon plasma by two orders of magnitude. The plasma potential and density were measured with a Langmuir probe. Figure 13.2 presents the variation of these parameters as a function of the argon plasma pressure. It shows that a maximum ion density of 10^{10} cm^{-3} is reached at 7×10^{-5} torr, which is also the pressure at which the relative density fluctuation (noise level) is minimum. We note that the plasma

---

* Presented by C. Pomot and J. Pelletier

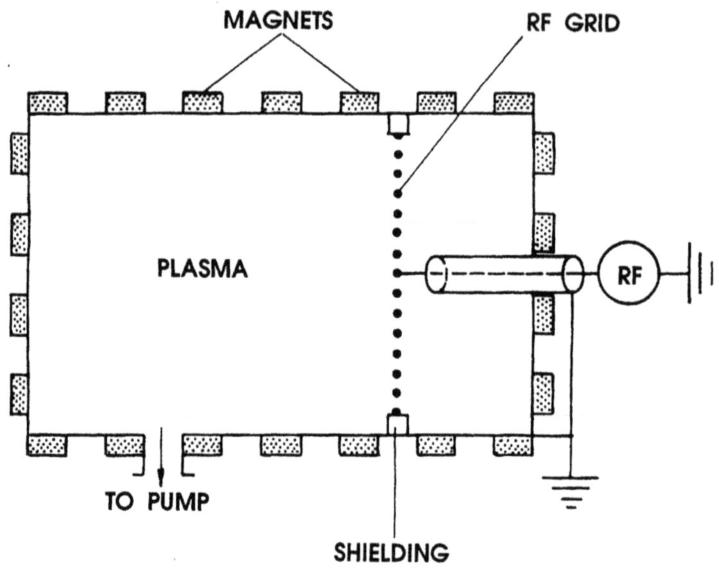

Fig. 13.1. Schematic diagram of the RF-grid plasma device with multipolar magnetic confinement (from ref. 2).

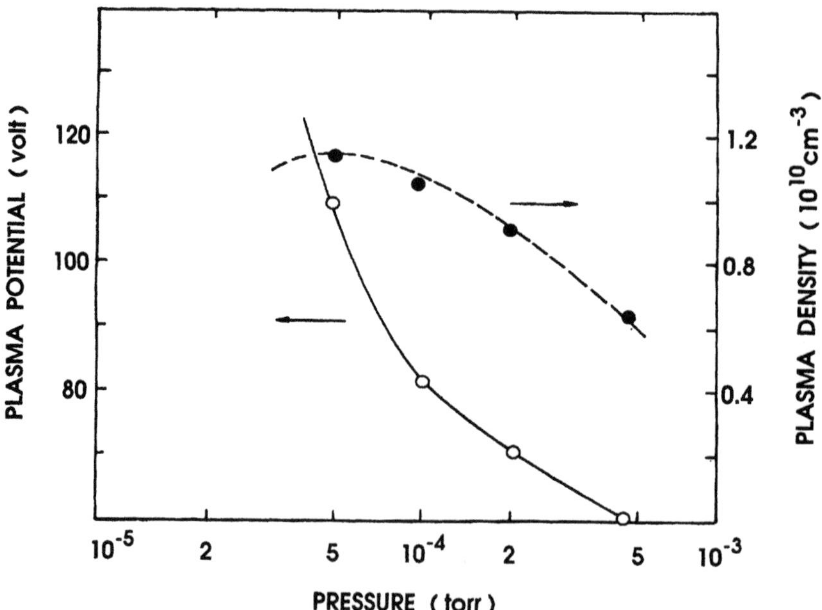

Fig. 13.2 Plasma potential and density as a function of argon pressure in the RF-grid plasma device of Figure 13.1 (from ref. 2).

potential attains high positive values with respect to the wall potential, leading to adverse effects: i) an important ion sputtering of the confinement structure is observed, imposing limitations on the use of this discharge. For example, one should refrain from producing reactive plasmas as it would lead to high contamination rates because of the reactive sputtering occurring on the walls; ii) because the plasma potential is high, low energy ion bombardment cannot be achieved when required. Finally, since the plasma potential depends on the gas pressure and RF power, one cannot control independently the plasma parameters and the ion bombardment energy on the substrate.

Another RF multipolar discharge configuration was investigated by Lejeune et al.,[3] which was intended for ion-assisted materials processing. The schematic diagram of their triode structure is shown in Fig. 13.3. This discharge associates: i) a three-electrode or triode structure to provide independent control of the ion bombardment energy; ii) RF excitation of the glow discharge to ensure long lifetime operation when using reactive gases; iii) multipolar magnetic confinement to provide plasma uniformity over large-diameter beams. Figure 13.4 shows the improvement in beam uniformity

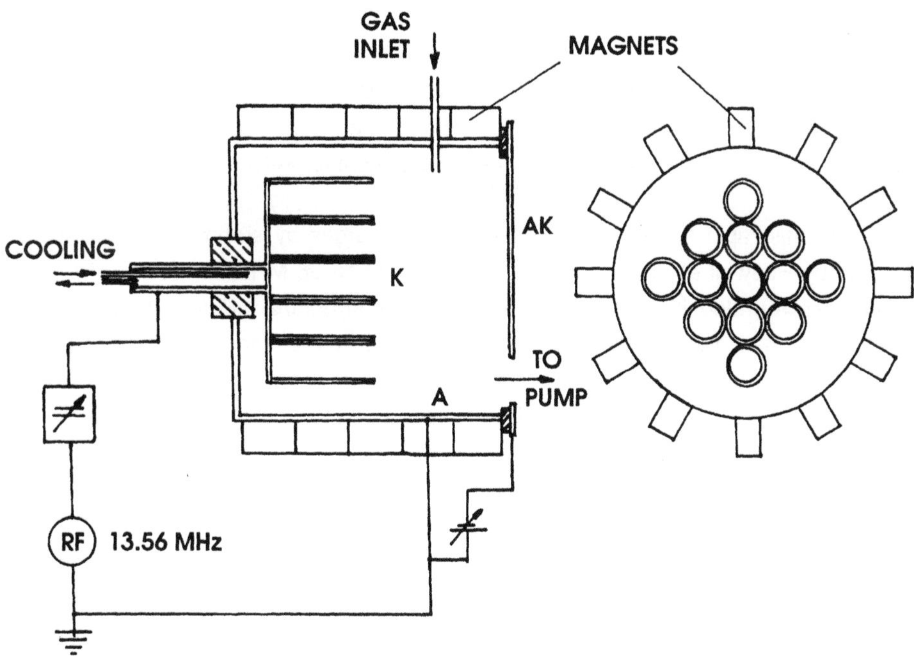

Fig. 13.3. Schematic diagram of an RF triode discharge with multipolar magnetic confinement. K is the hollow cathode, A is the chamber wall acting as the anode, and AK is an independently biased anticathode (from ref. 3).

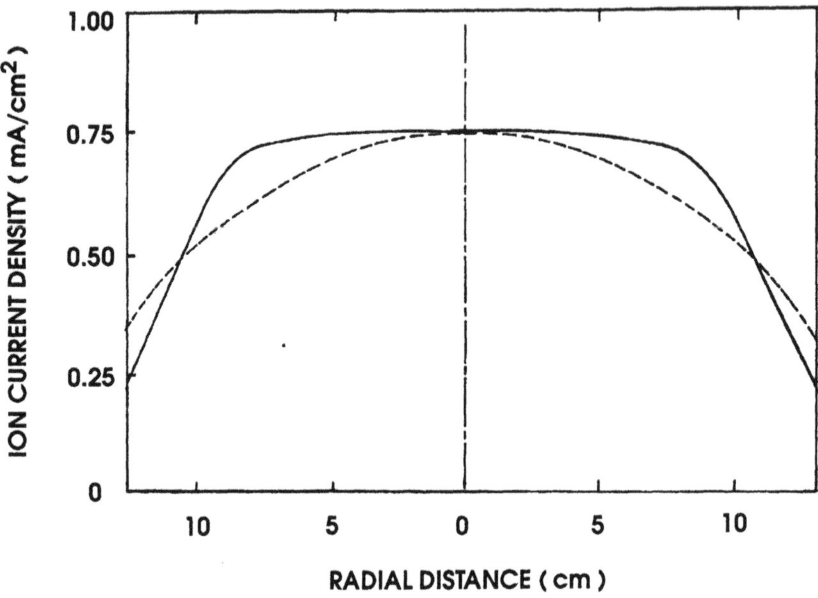

Fig. 13.4. Radial distribution of ion current density in a CF_4 plasma with (full line) and without (dashed line) magnetic multipolar confinement. The gas pressure is 10^{-3} torr (from ref. 3).

obtained with the multipolar magnetic confinement. This device provides current densities of 0.1 to 1.0 mA/cm^2. However, although not reported, contamination arising from the ion bombardment of the hollow cathode system is likely.

We end this rapid survey on RF multipolar discharges by summarizing the results obtained by Boswell et al.[4] with a plasma excited at 13.56 MHz, at the top of a glass chamber surrounded by a multipolar confinement structure. Figure 13.5 shows a schematic of their plasma reactor. It consists of a 30 cm diameter, 40 cm long pyrex tube surrounded by a multipolar structure made up of magnet rows parallel to the tube axis, which yield 600 gauss at the poles. The discharge tube fits in the center of the reactor top plate. The solenoid that encircles the plasma source can produce an axial static magnetic field of up to 1200 gauss. Surrounding the source tube, there is an RF antenna which applies an RF magnetic field perpendicularly to the static magnetic field. This arrangement ensures an efficient coupling with the plasma through the helicon wave mode (Secs. 6.6.2 and 7.4.4). At low pressure, i.e. when the ion mean free path is at least of the order of the chamber size, the plasma from the source diffuses into the multipolar structure. The characteristics of the plasma in the diffusion chamber depend on plasma source parameters such as pressure, RF power, static magnetic field, but are also strongly controlled by the multipolar magnetic field. In the absence of such confinement, the plasma density decreases exponentially with distance from the

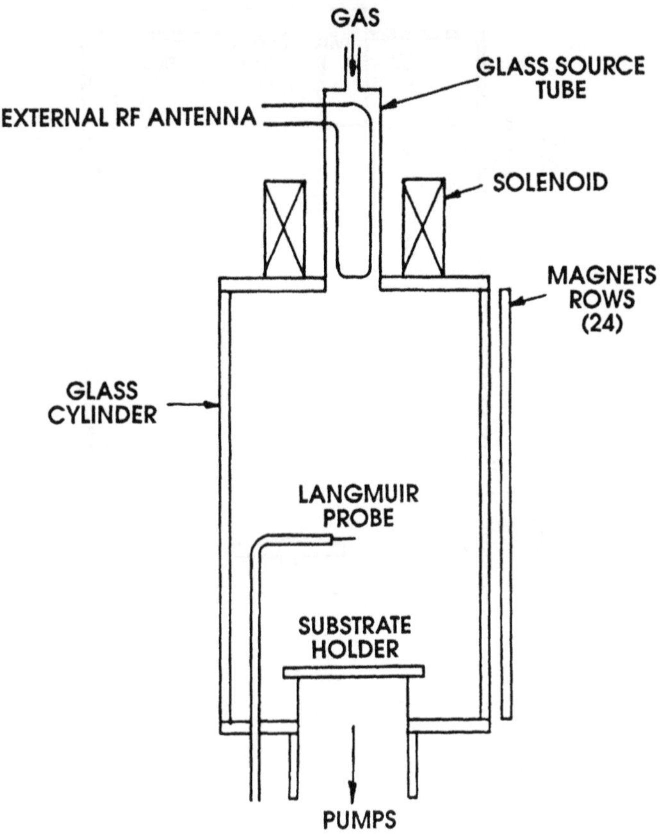

Fig. 13.5. Multipolar plasma reactor using an RF inductively coupled discharge (from ref. 4).

source, as shown in Fig. 9.5. In contrast, with multipolar confinement, the density is practically constant within the confined region. This inductively coupled RF plasma source produces a large homogeneous plasma, with a density exceeding 10^{11} cm^{-3} at 800 W RF power, below 10^{-3} torr of argon. This very efficient plasma source presents some drawbacks, however: i) at high plasma density, the ion plasma frequency can exceed the RF excitation frequency. Then the ions produced in the discharge oscillate in the electric field applied through the dielectric tube. As a result, ion bombardment of the dielectric material by energetic ions is expected to significantly sputter the walls where the RF field is high; ii) independently of the previously mentioned ion bombardment effect, glass or dielectric tubes are eroded by the reactive (fluorine or chlorine-based) gases used in etching processes, thus generating an important dielectric contamination on the substrate.

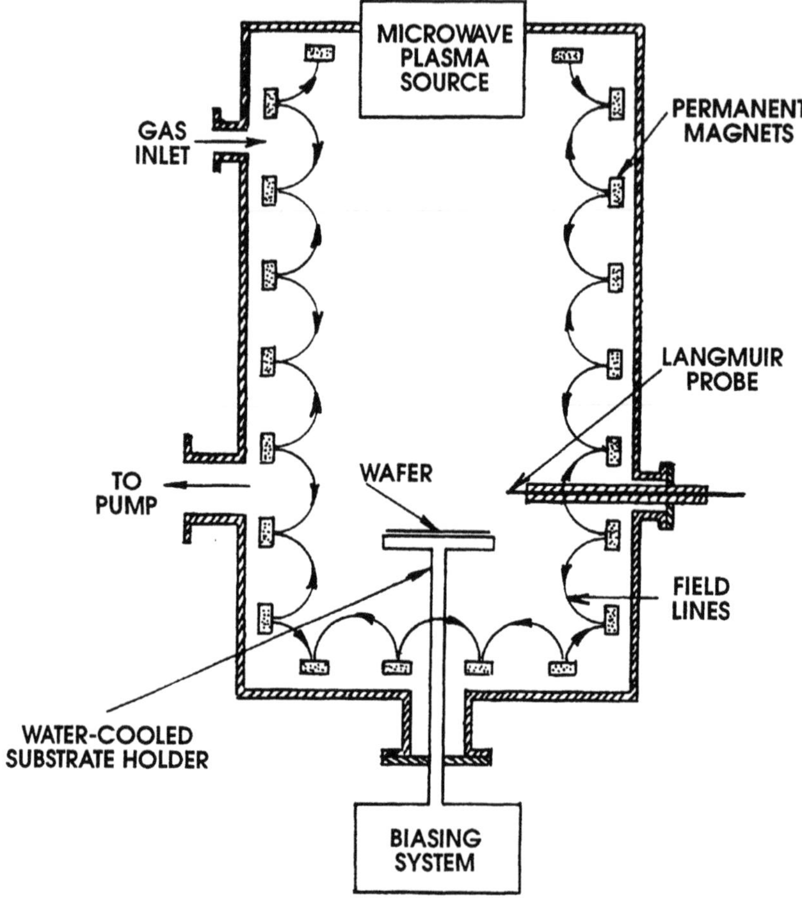

Fig. 13.6   Schematic representation of a plasma reactor combining a localized microwave discharge and multipolar magnetic confinement (from ref. 5).

In concluding this section on RF multipolar discharges, we note that RF discharges, unlike filament-excited discharges, offer the possibility of long lifetime operation in reactive gases. In addition, the use of multipolar magnetic confinement leads to high density plasmas at low gas pressures. However, these plasmas present drawbacks such as: i) sputtering of the electrodes and reactor walls by energetic ions, resulting from the large difference between plasma and floating potentials. This difference is due to the presence of energetic electrons; ii) bombardment of the electrodes and reactor walls by energetic ions responding to the applied electric field at high plasma density, i.e. when the ion plasma frequency is of the order of, or larger than, the excitation frequency; iii) since these RF discharges produce energetic charged particles, substrate biasing does not allow an independent control of the ion bombardment energy in the low energy range.

These shortcomings can be circumvented by increasing the excitation frequency. In particular, this avoids the acceleration of ions in the HF field, which leads to erosion of the electrode or of the dielectric wall when the electrode is located outside the plasma. This suggests using microwave excitation.

13.3. Multipolar plasmas sustained by microwave discharges

13.3.1. Introduction. The ability to create high densities of both excited and charged species have made high-pressure as well as low-pressure microwave discharges an attractive, efficient technology for many plasma processing applications. To obtain an intense and homogeneous plasma, we can associate a microwave discharge with a multipolar magnetic confinement. This scheme requires low operating pressures such that the ion mean free path is at least of the order of the chamber size.

One possibility is to use a localized microwave discharge that diffuses into the multipolar confinement structure,[5] as schematically represented in Fig. 13.6. Two commonly used microwave plasma sources have been utilized for this purpose, as described below.

13.3.2. Microwave discharges sustained by a surface wave. The use of electromagnetic surface waves to sustain a plasma[6-8] has been reviewed in Chap. 5. We saw that surface-wave plasmas may be generated at low gas pressure (mtorr range) in dielectric tubes with diameters up to at least 200 mm. Such a plasma source can be made to diffuse into a multipolar magnetic structure. The schematic diagram of the corresponding multipolar plasma is shown in Fig. 13.7.[9] In this example, the 2.45 GHz microwave discharge system is realized with rudimentary means.[10] The microwave applicator is made from a section of a rectangular waveguide in which a circular opening has been made in one of its wide walls. The closed end of the silica discharge tube penetrates into the guide opening, while the other end opens into the reactor chamber, fitted with a multipolar magnetic field. At low pressure, the plasma sustained by the surface wave inside the silica tube diffuses into the chamber. Figure 13.8 shows the variation of the plasma density along the reactor axis at two different pressures.[11] We note several effects of the multipolar magnetic confinement: i) at 10^{-2} torr, diffusion of the plasma into the chamber is pressure limited. Since the plasma does not reach the chamber wall, the magnetic confinement is ineffective and useless; ii) at both 10^{-2} torr and 2×10^{-3} torr without the magnetic confinement, we observe that the density decreases exponentially as a function of the distance from the

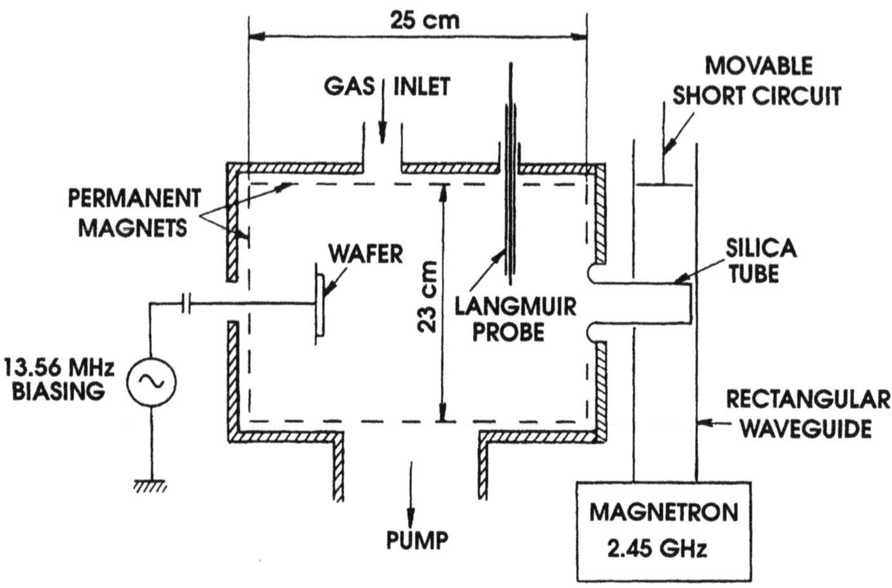

Fig. 13.7   Schematic diagram of a microwave multipolar plasma excited by a surface wave, using a surfaguide-type applicator (from ref. 9).

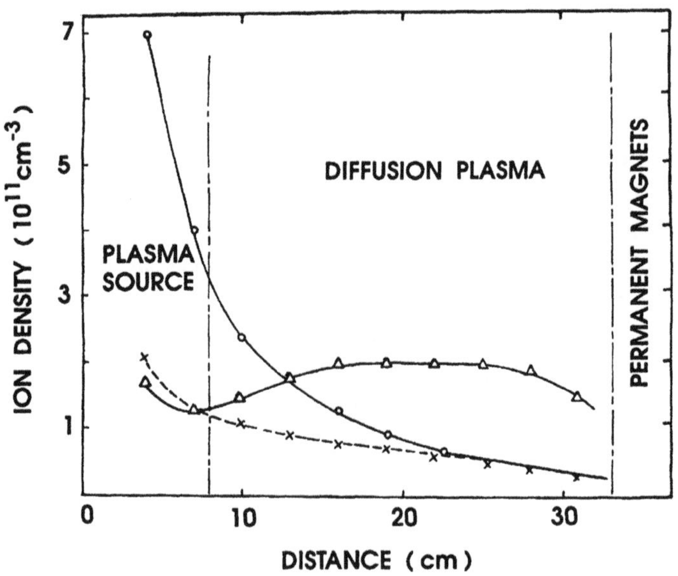

Fig.13.8   Ion density profile as a function of axial position in the system of Fig. 13.7, at two argon pressures. O: 10^{-2} torr; Δ: 2×10^{-3} torr; x: 2×10^{-3} torr (multipolar magnets removed) (from ref. 11)

source.   This decrease is larger at the higher pressure; iii) at 2×10^{-3} torr with multipolar magnetic confinement, the plasma diffuses throughout the chamber.   The density is axially constant over more than 10 cm, and larger by a factor of about 4 than without confinement.   The plasma is homogeneous over a 5 cm radius, which corresponds to the magnetic field free region; iv) as shown in Fig. 13.9, the electron temperature is much lower in the presence of the multipolar magnetic field confinement.[12]

The latter effect can be attributed to the removal of energetic electrons, which can be trapped in the multipolar magnetic field (see Chap. 12).   Assuming this, the above experimental results can be interpreted in the light of Chaps. 9 to 12:   when the pressure is low enough for the plasma to diffuse in the whole chamber, the energetic electrons can interact with the multipolar magnetic field, achieving there what we called peripheral ionization.   Evidence for this additional plasma source is the depression in density (see Fig. 13.8), which may be observed between the surface-wave plasma source (main plasma source) and the multipolar confinement structure: this is because density increases in the multipolar structure as a result of peripheral ionization.   To

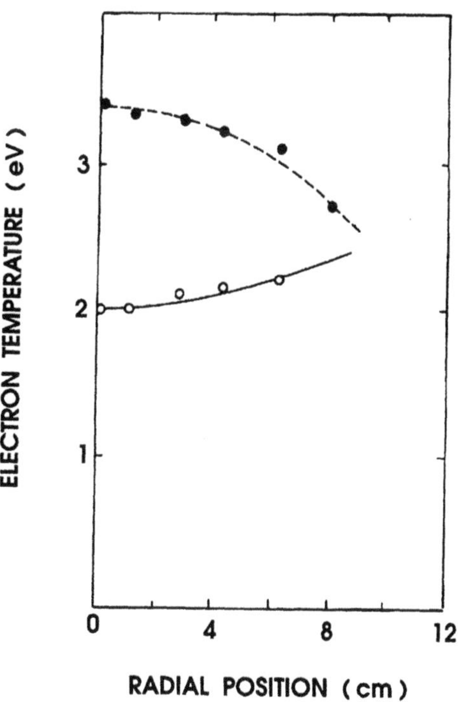

Fig. 13.9. Radial distribution of the electron temperature in a surface-wave-excited diffusion plasma with (o) and without (●) multipolar confinement. The argon pressure is 2×10^{-3} torr (from ref. 12).

summarize, the confinement of the diffusion plasma (leading to an increase in density by a factor of two, as shown in Chaps. 10 and 11) and the additional peripheral ionization are together responsible for the total increase in density (by a factor of about four) and for the homogeneity achieved at low pressure. In contrast, at high pressure ($\approx 10^{-2}$ torr here), where electron-neutral and ion-neutral collisions reduce plasma diffusion and slow down the energetic electrons, the plasma and the energetic electrons produced in the main source cannot enter the magnetic field configuration, so that neither plasma confinement nor additional ionization occurs. Thus, at high pressure, multipolar magnetic confinement is useless.

Figure 13.10 shows the spatial variation along the chamber axis of ion density, electron temperature and plasma potential in an SF_6/Ar mixture at constant microwave power. These results demonstrate that 10 cm downstream of the primary plasma source, i.e. beyond the end of the silica tube (see Fig. 13.7), the plasma characteristics inside the confinement structure are all spatially constant in the field-free volume.

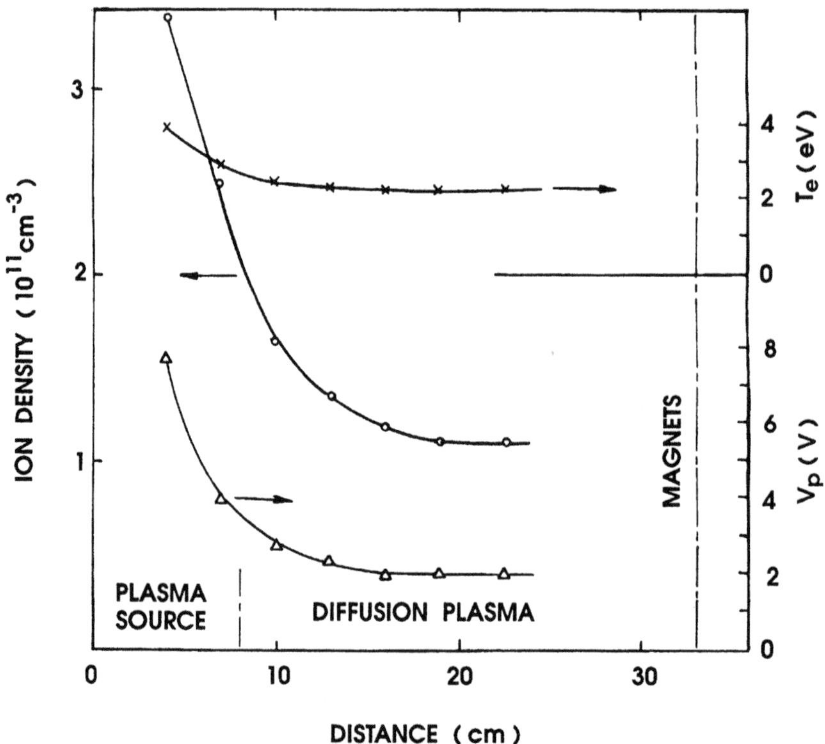

Fig. 13.10. Axial profiles of ion density, electron temperature and plasma potential in an SF_6/Ar mixture at pressures of 4×10^{-4} torr SF_6 and 3×10^{-3} torr Ar (from ref. 11).

In summary, surface-wave-excited multipolar plasmas at low pressure (a few mtorr) can produce homogeneous, high density (10^{11} cm^{-3}) plasmas. Moreover, as the plasma potential is only a few k_BT_e/e higher than the ground potential, ion sputtering of the reactor wall is negligible. Finally, substrate surfaces can be biased with DC or RF potentials, independently of discharge conditions.

Several factors limit the use of surface-wave-excited plasma sources: i) reactive gases may produce chemical erosion of the dielectric silica tube and thus lead to undesirable contamination during surface processing.[†] Deposition from metallic gaseous precursors is also to be ruled out since such deposition on the silica tube would impede wave propagation; ii) the processing of very large areas requires large volumes of homogeneous plasma, which can only be obtained when the ion mean free path is at least of the order of the chamber size. The low pressure limit for surface-wave excitation without magnetic field is in the mtorr range, where the ion mean free path is of the order of a few centimeters. This prevents filling large size reactors with a diffusion plasma. To operate at lower pressures, we turn to magnetized plasma sources such as those sustained by electron cyclotron resonance (ECR), for example.

13.3.3. Microwave discharges sustained by ECR. The microwave excitation of discharges by electron cyclotron resonance originates from works on fusion[13] and ion propulsion.[14] It has been intensively applied to plasma processing by Japanese groups.[15,16] A description of several of these ECR systems adapted to etching and thin film deposition has been presented by Asmussen[17] and in Chap. 7, and their use in microelectronic circuit fabrication is reviewed in Chap. 15.

Electron cyclotron resonance is obtained when the excitation frequency is close to the gyration frequency of the electrons in the static magnetic field. At an excitation frequency of 2.45 GHz, this condition is satisfied for a magnetic field intensity close to 875 gauss (Sec. 6.4). This method of plasma excitation is efficient in the low pressure range where collision rates are low enough not to disrupt the resonance mechanism, as discussed in Sec. 6.4.

Figure 13.11 depicts a microwave multipolar plasma sustained by an ECR discharge. In this device, the microwave power is supplied to the plasma via a circular waveguide through an alumina window. The magnetic field required for ECR is produced by a coil surrounding the window. The current in the coil is adjusted to obtain the resonant magnetic field strength at the entrance port to the multipolar magnetic

---

[†] Using a metallic vessel instead of a silica tube may also lead to undesirable contamination during surface processing.

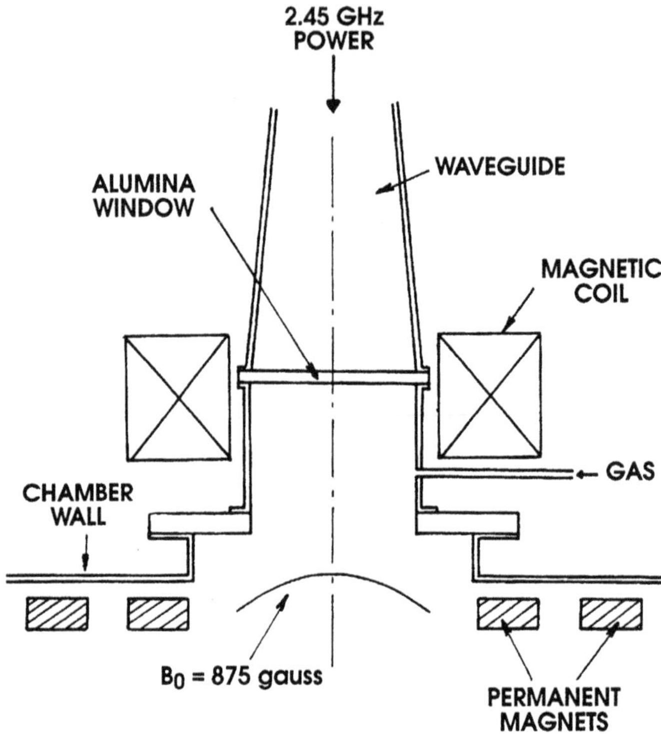

Fig. 13.11. ECR excited microwave discharges diffusing into a multipolar magnetic structure (from refs. 5 and 18).

confinement structure. The ECR plasma can be obtained in a wide pressure range from 10^{-2} down to 10^{-5} torr, but the plasma is homogeneous in the 2 meter long multipolar magnetic structure only below 10^{-4} torr in krypton,[18] i.e. when the plasma produced by the ECR discharge completely fills the multipolar magnetic structure. Similarly to the case of surface-wave sustained multipolar plasmas (Sec. 13.3.2), one can presume that energetic electrons are trapped in the multipolar structure and constitute a secondary source of plasma at the periphery of the chamber.

The above device suffers from the drawbacks of conventional ECR excitation: i) the magnetic field produced by the coil persists over several decimeters, yielding anisotropic plasma parameters; ii) as a result of the diverging magnetic field lines which create magnetic field gradients, the energetic ions originating from the ECR region are propelled throughout the chamber. However, this effect is strongly attenuated when the ECR discharge operates in such a way that a homogeneous plasma is achieved in the multipolar magnetic structure.[18]

13.4. Control of process parameters in microwave multipolar plasmas

13.4.1. Generalities. In most surface treatment applications, the main parameters that characterize the plasma-surface interaction are, besides surface temperature, the flux of incoming reactive species (atoms and/or radicals), and the flux and energy of the ion bombardment. The flux and energy distribution of the electron bombardment, the flux and energy distribution of the photons incident on the surface, as well as the temporal distribution of both the electron and ion bombardment may also be relevant.

The production rates of ions, photons and reactive species increase with microwave power density in the plasma. Since the concentration of reactive species largely depends on ion-molecule reactions occurring in the gas phase, one can control selectively the concentration of reactive species by varying independently the partial pressure of the reactive parent gas. The energy of the charged particles impinging on a substrate can be adjusted by biasing it (bias potential V_0) with respect to the plasma potential V_p. When the substrate is a conductor, it can be biased by simply applying a DC voltage to it. When the substrate is an insulator, it can be biased only by applying a periodic (RF) voltage through a low impedance capacitor, as shown in Fig. 13.12. The substrate surface exposed to the plasma can thus be charged, i.e. electrically polarized, by capacitive coupling.

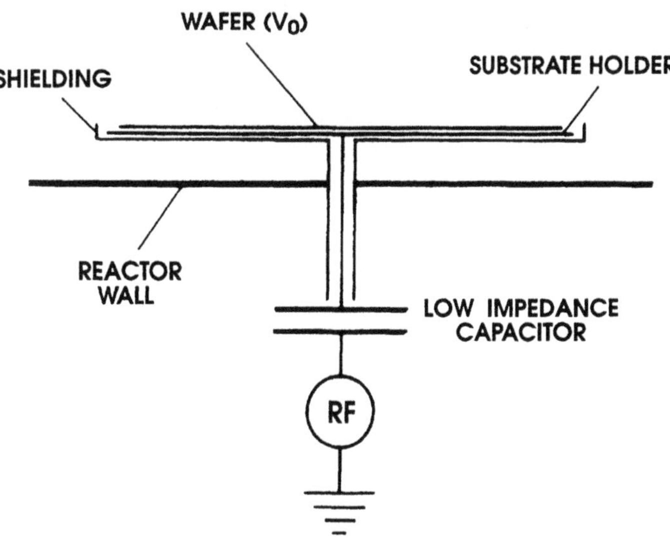

Fig.13.12. Sketch showing the principle of RF biasing.

In microwave multipolar plasmas, the energetic electrons are confined near the wall where the multipolar magnetic field is located, while the diffusion plasma in the central volume of the chamber is free of magnetic and electric fields, isotropic and homogeneous, with a quasi-Maxwellian electron energy distribution function.[18] Since this diffusion plasma is nearly ideal as far as the assumptions made in the theory of electrostatic probes are concerned, one can thus consider the substrates immersed in the plasma as *large plane probes*.† As a result, the biasing voltage with respect to the plasma potential and the corresponding electron and ion bombardment current densities can readily be derived.

13.4.2. Substrate biasing with DC voltages

Plasma free of negative ions. The current-voltage characteristic of a plane probe is sketched in Fig. 13.13 for the case of a plasma free of negative ions. For DC voltages positive with respect to the plasma potential $(V > V_p)$, the positive ions assumed to be cold $(T_i \ll T_e)$ are all repelled, while all the electrons crossing the sheath edge are

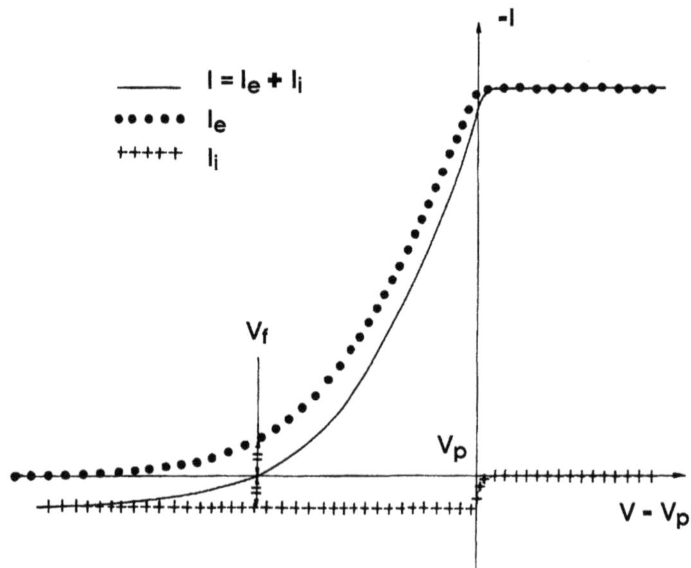

Fig. 13.13. Idealized sketch of the current-voltage characteristic of a plane probe. I_e and I_i are the electron and ion currents respectively, and $I = I_e + I_i$ is the total current. The plasma potential V_p is taken as the origin of the potentials, and V_f is the floating potential $(I = 0)$.

---

† A substrate with a radius of curvature much larger than the Debye length can also be considered as a large plane probe.

collected. The electron current at saturation is thus constant and proportional to the random (thermal) flux of the electrons in the plasma (eqn. (9.27))

$$I_{es} = - e n_e \, \mathcal{S} \left(\frac{k_B T_e}{2 \pi m_e} \right)^{1/2},$$
$$\tag{13.1}$$

where n_e, T_e, e and m_e are the electron density, temperature, charge and mass respectively, k_B is the Boltzmann constant, and $\mathcal{S}$ is the surface of the probe. For DC voltages negative with respect to the plasma potential ($V < V_p$), the saturation ion current density is given by application of the Bohm criterion (eqn. (9.29))

$$I_{is} = e n_i \, \mathcal{S} \left(\frac{k_B T_e}{m_i} \right)^{1/2},$$
$$\tag{13.2}$$

where n_i is the ion density ($n_i = n_e$) and m_i is the ion mass, and all the electrons with thermal energy less than $e(V_p - V)$ are repelled. For $V < V_p$, since we have a Maxwellian distribution, the current of electrons collected is (eqn. (9.28))

$$I_e = - e n_e \, \mathcal{S} \left(\frac{k_B T_e}{2 \pi m_e} \right)^{1/2} \exp \left[\frac{e(V - V_p)}{k_B T_e} \right].$$
$$\tag{13.3}$$

The *floating potential* V_f corresponds by definition to equal electron and ion current intensities

$$I_e + I_i = 0,$$
$$\tag{13.4}$$

and is given by[†]

$$V_f - V_p = - \frac{k_B T_e}{2e} \log_e \frac{m_i}{2 \pi m_e}.$$
$$\tag{13.5}$$

We now turn from a plane probe to a large conducting substrate in the plasma. When it is biased negatively at a DC potential V_0 with respect to the plasma potential, the energy W_i of the ion bombardment is constant in time and equal to

---

[†] Various approximations can be adopted for this expression related to the Bohm criterion. These different approximations all fit $2e \, (V_f\text{-}V_p) = - k_B T_e \log_e (c \, m_i/m_e)$, where c is a coefficient of the order of unity, i.e. they do not differ from each other by more than a fraction of $k_B T_e$. In addition to these analytical approximations, one must also bear in mind that, for a given distribution, say Maxwellian, the floating potential is not a *universal* value, since it depends on the shape of the probe. Because the ion saturation current collected on a cylindrical (or spherical) probe is higher than on a plane probe, the floating potential measured on a cylindrical (or spherical) probe is less negative than the one measured on a plane probe.

$$W_i = e(V_p - V_0). \tag{13.6}$$

When submitting electrons with a Maxwellian distribution to a DC voltage difference $V_p - V_0$, the whole distribution is shifted accordingly and the electron current collected follows eqn. (13.3) for $V = V_0$, as shown in Fig. 13.13. In practice, the electron current can be assumed negligible with respect to the ion current density when

$$V_0 - V_f \le V_f - V_p , \tag{13.7}$$

as can be seen from Fig. 13.13, where $V_p = 0$. When biased positively relatively to the reactor wall, a substrate with an area that is not negligible with respect to the reactor wall leads to a correlative positive shift of the plasma potential (see Sec. 10.5.2). As a result, it is impossible to bias a large substrate positively with respect to the plasma potential.†

Plasma with negative ions. In the more general case of a reactive plasma with negative ions, an additional current component appears in the current-voltage characteristic of a plane Langmuir probe, as shown in Fig. 13.14. For a DC voltage negative with respect to the plasma potential ($V < V_p$), all the negative ions, assumed to be cold (temperature of the negative ions of the order of T_i) are repelled. For a DC voltage positive with respect to the plasma potential ($V > V_p$), the saturation negative ion current is generally negligible compared to the saturation electron current. However, the electron current (eqns. (13.1) and (13.3)) is correlatively reduced according to

$$I_{es} = - e\mathcal{B}n_i \, S \left(\frac{k_B T_e}{2\pi m_e} \right)^{1/2} \tag{13.8}$$

for $V > V_p$, where $\mathcal{B} = n_e/n_i$ is the fraction of electrons in the plasma ($0 < \mathcal{B} \le 1$) and for $V < V_p$, we have

$$I_e = - e\mathcal{B}n_i \, S \left(\frac{k_B T_e}{2\pi m_e} \right)^{1/2} \exp \left[\frac{e(V - V_p)}{k_B T_e} \right], \tag{13.9}$$

while the positive ion saturation current remains unchanged (eqn. (13.2)).

Since large substrates cannot be biased positively with respect to the plasma potential as seen above, negative ions, which are present in most reactive plasmas, cannot be directly involved in plasma surface interaction through ion bombardment.

---

† Such a behavior is also found with double probes.

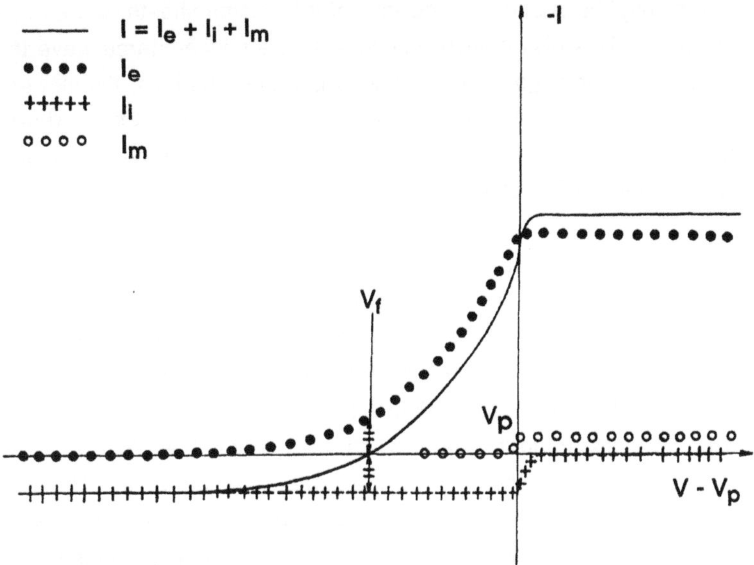

Fig. 13.14. Idealized sketch of the current-voltage characteristic of a plane probe. I_e, I_i and I_m are the electron, positive ion and negative ion currents respectively and $I = I_e + I_i + I_m$ is the total current. The plasma potential V_p is taken as the origin of the potentials, and V_f is the floating potential ($I = 0$).

Actually, their main effect is to reduce the electron density in the plasma and thus, according to eqns. (13.2), (13.4) and (13.9), to decrease the difference between the floating potential and the plasma potential

$$V_f - V_p = -\frac{k_B T_e}{2e}\left(\log_e \frac{m_i}{2\pi m_e} + 2 \log_e \mathcal{B}\right). \tag{13.10}$$

In conclusion, large conducting substrates in plasmas including those with negative ions, can only be biased negatively with respect to the plasma potential by applying DC voltages. The main interest of this biasing mode is that the electron current impinging on the substrate can be eliminated with respect to the positive ion bombardment current, provided that the bias voltage is more negative than $2V_f - V_p$ (eqn. (13.7)). However, due to arcing between substrate and walls, DC biasing is usually limited to approximately -100 V. We will see below that RF biasing is in general more appropriate than DC biasing, since it acts on both conducting and insulating substrates while also permitting one to attain more negative voltages.

13.4.3. Substrate biasing with periodic voltages. When a periodic voltage is applied to an insulating substrate, the variations in the electric charge at the surface of the substrate modify the boundary conditions of the plasma sheath between the plasma and the substrate. This disturbance propagates as a space-charge wave through the sheath with a time scale t_s governed by the dynamics of the ions, the slowest charged species. In the case of a collisionless sheath, t_s is of the order of $2\pi/\omega_{pi}$ where ω_{pi} is the ion plasma angular frequency. As a result, the influence of the bias frequency $\omega_{RF}/2\pi$ will depend on its value with respect to ω_{pi}.

To characterize the effect of the biasing voltage, we distinguish three cases depending on the value of ω_{RF} compared to ω_{pi} and to ω_{pe}, the electron plasma angular frequency: i) $\omega_{RF} \ll \omega_{pi} \ll \omega_{pe}$: at such a low bias frequency, the sheath is at each instant in static equilibrium with the applied periodic voltage; ii) $\omega_{pi} \ll \omega_{RF} \ll \omega_{pe}$: the electrons are still in static equilibrium with the periodic voltage whereas the ions appear frozen with respect to it. The ions are then only influenced by the DC voltage component induced on the surface by the periodic voltage, as discussed below; iii) $\omega_{pi} \ll \omega_{pe} \ll \omega_{RF}$, both electrons and ions appear frozen with respect to the periodic voltage. The sheath is not perturbed and the substrate surface remains at the floating potential, and biasing is thus useless, as would be biasing an insulating substrate with a DC voltage.

On a substrate surface which cannot drain away DC currents,† a DC voltage component appears because of the nonlinearity of the voltage-current characteristic, which is similar to that of a plane probe. This voltage V_0 can be derived by using the fact that the variation of the electric charge on the substrate surface over one period is zero, which we write

$$\overline{I_e + I_i} = 0. \tag{13.11}$$

As a result of the dissymmetry between the ion and electron current densities, the potential of the surface usually remains negative with respect to the plasma potential (see Appendix 13.1). In the case of a plane substrate, which exhibits a constant ion saturation current, the nonlinearity is only due to the electron contribution in the repulsive domain ($V < V_p$). In the general case of a substrate of undefined shape, the nonlinearity of the characteristic is due to both ions and electrons.

Figure 13.15 shows the effect of applying a periodic signal to a plane surface substrate. In the simple case of a square wave signal, the DC bias voltage settles at a

---

† In the case of a conducting substrate, this results from the presence of the low-impedance capacitor inserted in the RF biasing circuit (Fig. 13.12).

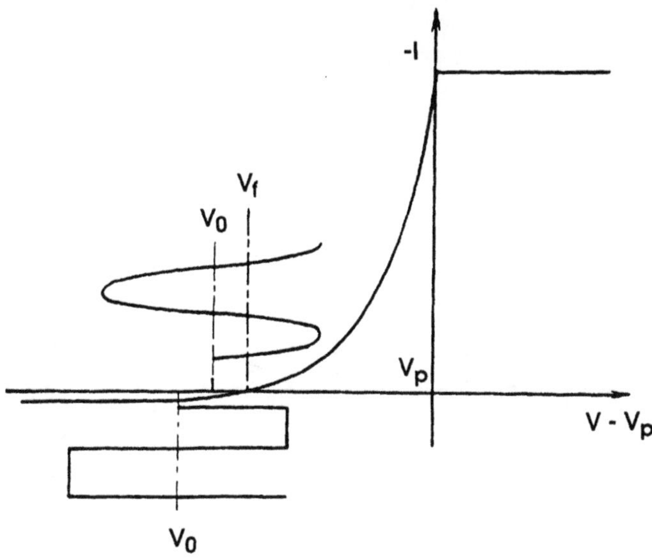

Fig. 13.15. Schematic diagram showing the result of superposing a periodic voltage on the nonlinear I-V characteristic of a plane substrate. The origin of the potentials is V_p, and V_0 is the value of the DC component of the biasing periodic voltage. Note that with a sinusoidal voltage, V_0 is less negative than with a square signal.

value V_0 such that the total current collected during one half-period exactly balances the total current collected during the next half-period.

This figure shows the variation of ion, electron and total currents, and of ion energy as a function of the applied biasing voltage V_{RF}, in the three limiting frequency domains defined above. When $\omega_{pi} \ll \omega_{RF} \ll \omega_{pe}$, the ion energy is constant with time, as with DC biasing but, in contrast to DC biasing, an electron current component is always present.

The derivation of the DC voltage component V_0 for different shapes (shown in Fig. 13.17) of the applied voltage on a plane substrate is presented in Appendix 13.1, assuming a Maxwellian distribution. This appendix also demonstrates that V_0, when referred to the floating potential, is independent of the presence of negative ions in the plasma. In the case of a square wave biasing signal, provided that $eV_{RF} \gg k_BT_e$, the DC component V_0 takes the following asymptotic value[19]

$$V_0 - V_f = - V_{RF} + \frac{k_BT_e}{e} \log_e 2. \tag{13.12}$$

  High frequency sustained multipolar plasmas

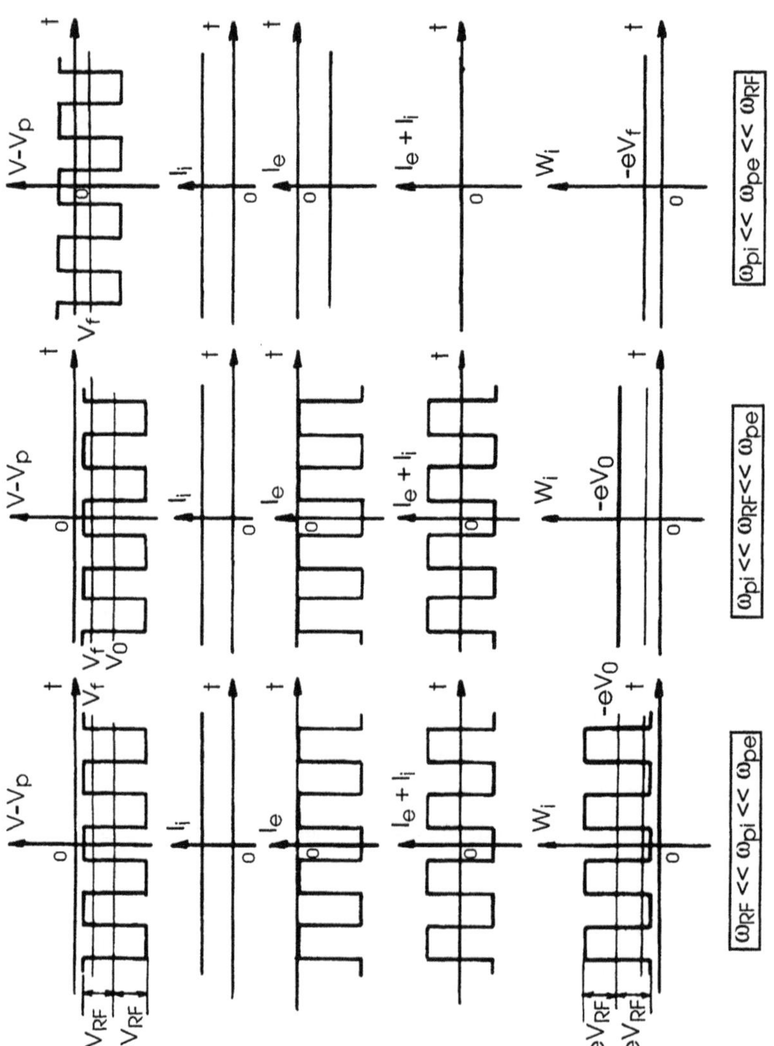

Fig. 13.16. Variation of ion current I_i, electron current I_e and total current I, and of ion energy W_i as a function of the variation of the periodic voltage V_{RF} in three limiting frequency domains where ω_{RF} is the bias angular frequency, and ω_{pe} and ω_{pi} are the electron and ion plasma angular frequencies respectively.

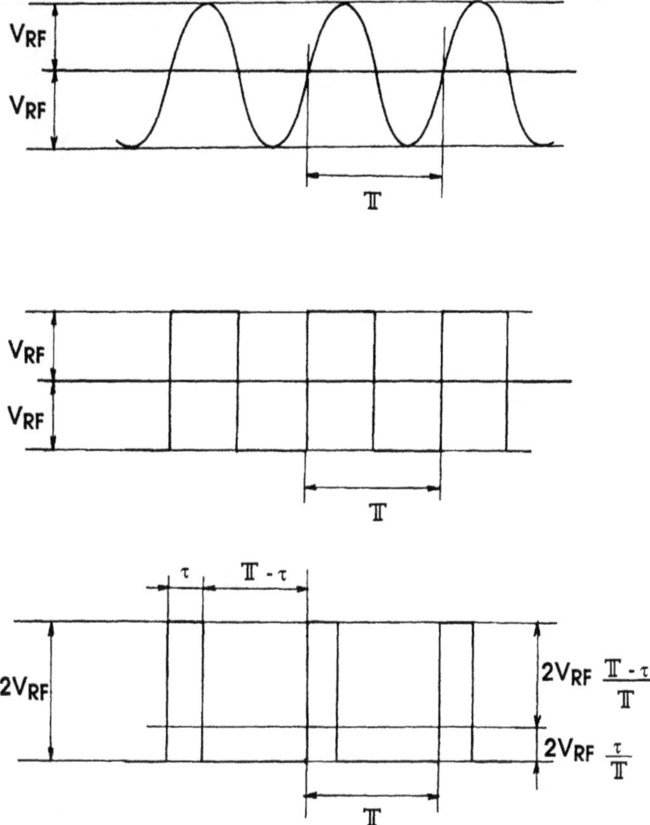

Fig. 13.17. Schematic diagrams showing the bias DC component V_0 for three different shapes of the RF biasing voltage.

The second term in the right-hand side of this equation is a corrective term, independent of the amplitude of the periodic voltage V_{RF}. For a sine wave signal, the asymptotic form becomes[19-21]

$$V_0 - V_f = -V_{RF} + \frac{k_B T_e}{2e} \log_e \left(\frac{2\pi\, e V_{RF}}{k_B T_e} \right), \tag{13.13}$$

where the corrective term now does depend on V_{RF}.

In general, one observes a perfect agreement between the calculated and measured V_0 values for different signal shapes and frequency domains as a function of V_{RF}.[1,21-23] Figure 13.18 shows an example of such a comparison for a 10 kHz square wave signal in the case $\omega_{RF} \ll \omega_{pi}$.[23] Although an accurate bias control of substrates is most often

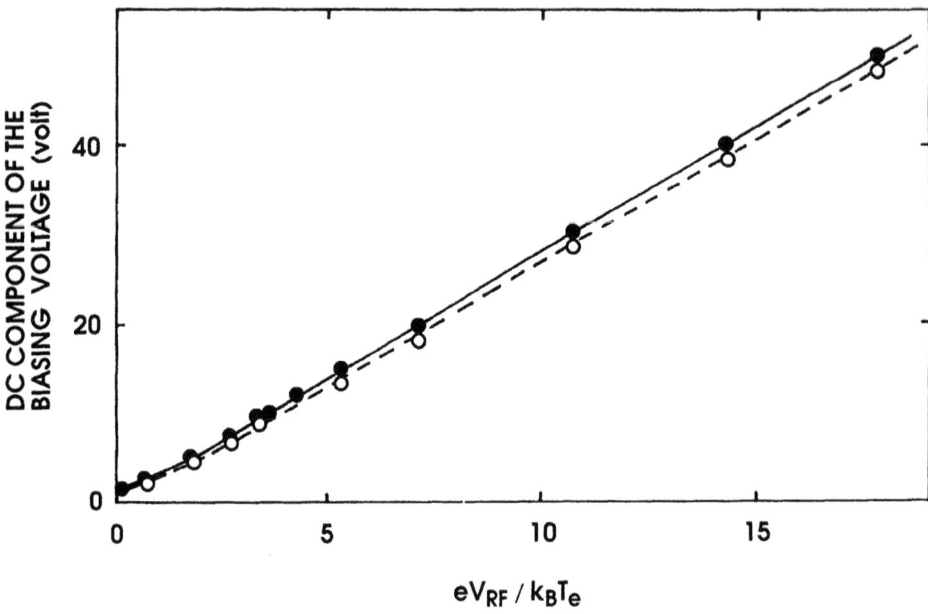

Fig. 13.18. Comparison between theory and experiment of the DC component V_0 resulting from periodic voltage biasing as a function of the biasing amplitude V_{RF}, for a square wave signal at 10 kHz. Plasma parameters: $\omega_{RF} \ll \omega_{pi}$, $k_B T_e = 2.8$ eV (from ref. 23).

possible, problems may occur when attempting to operate in limiting situations, as we will see in the next section. Also, we will take a closer look at the biasing parameters (signal shape, frequency, amplitude) in order to optimize them for a given process.

13.4.4. Limitations in substrate biasing

Range of V_0 with respect to V_p. As discussed in Sec. 13.4.2, one cannot DC bias large substrates positively with respect to the plasma potential. The possibility of biasing positively sufficiently small substrates with DC voltages (Sec. 10.5.2) can be of interest, for example, to understand the role of electrons in surface processes in reactive plasmas.[5]

With periodic biasing, the DC component V_0 is limited to voltages such that $V_0 < V_f$. Hence, to achieve very low ion bombardment energies (eqn. (13.6)), one must succeed in lowering $V_p - V_f$ (eqn. (13.10)). This can be accomplished by reducing the electron temperature or increasing the concentration of negative ions in the plasma.

Control of electron bombardment flux. In the case of DC biasing, the electron contribution to surface bombardment decreases when decreasing the bias voltage towards negative values. In contrast, when using periodic voltage biasing, the electron contribution to surface bombardment is independent of V_0 for plane substrates. The only remaining degree of freedom is the control of the time distribution of the electron current arrival during a period. Applying dissymmetrical rectangular periodic voltages to the surface substrate (see Appendix 13.1) is a way of achieving this.

Influence of the biasing frequency on the ion energy distribution function. A critical parameter for many surface processes is the distribution of the ion energy as a function of the bias frequency. As shown earlier in Fig. 13.16, when $\omega_{pi} \ll \omega_{RF} \ll \omega_{pe}$, the ion energy is time independent. According to eqns. (13.6) and (13.12), the monokinetic ion bombardment energy for an applied square wave signal is then given by

$$W_i = e\left(V_{RF} + V_p - V_f - \frac{k_B T_e}{e} \log_e 2\right). \tag{13.14}$$

In contrast, when $\omega_{RF} \ll \omega_{pi}$, the energy distribution follows the evolution of the periodic biasing voltage so that it is $2eV_{RF}$ wide, with the maximum ion energy given by

$$W_{max} = e\left(2V_{RF} + V_p - V_f - \frac{k_B T_e}{e} \log_e 2\right). \tag{13.15}$$

The control of the ion energy distribution in surface processes is of prime importance when bombardment effects, such as sputtering, do not vary linearly with ion energy. It implies knowing the ion plasma frequency $\omega_{pi}/2\pi$ to choose the best bias frequency $\omega_{RF}/2\pi$ for the application considered. In Fig. 13.19, we have plotted the variation of the ion plasma frequency as a function of ion atomic mass, for different plasma densities. For plasma densities above 10^{11} cm^{-3}, biasing at 40.68 MHz (an ISM authorized frequency) ensures that $\omega_{RF} \gg \omega_{pi}$ for practically all ion masses.

Limitations due to capacitance effects. Another limitation to biasing with periodic voltages can result from the coupling capacitor in the biasing circuit (Fig. 13.12). When the plasma density or the substrate surface is increased, the collected currents and the voltage drop through the capacitor also increase, reducing the effective voltage on the substrate surface. The answer is to decrease the capacitor impedance, by increasing either the capacity value or the frequency of the biasing voltage.

A similar problem arises in the case of a thick dielectric substrate, since it behaves as a high impedance coupling capacitor. The decrease of its intrinsic impedance thus requires a corresponding increase in the bias frequency. This shows the difficulty, if not

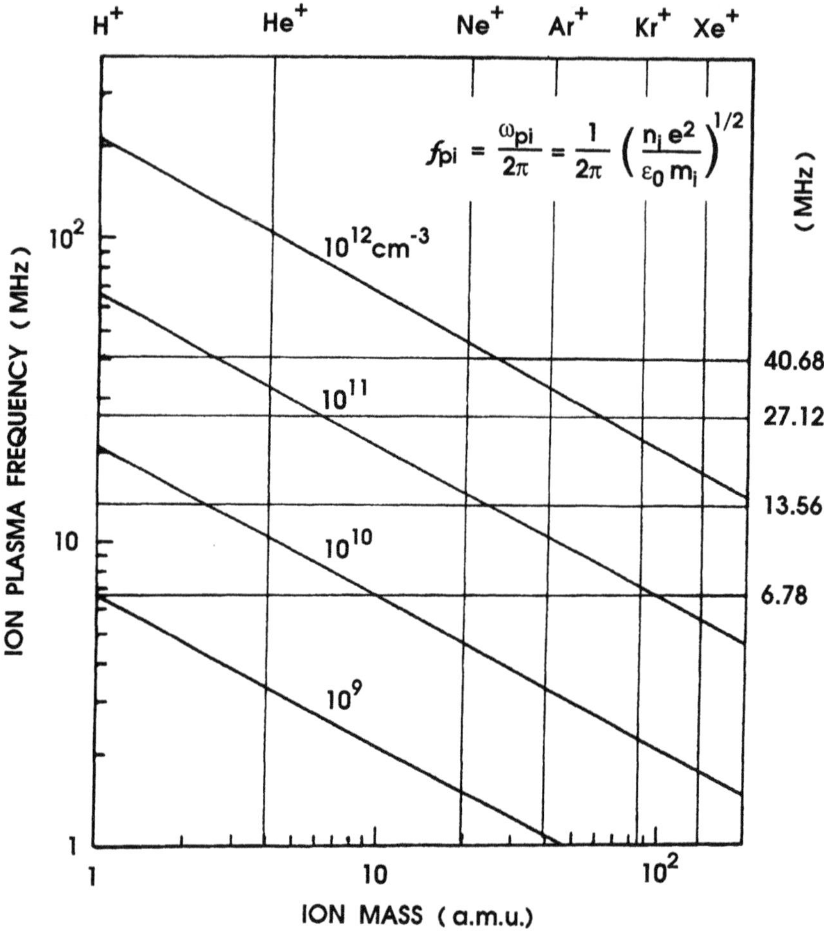

Fig. 13.19. Ion plasma frequency $\omega_{pi}/2\pi$ as a function of ion mass m_i, for different plasma densities. The horizontal full lines correspond to ISM authorized frequencies.

the impossibility, of controlling the bias of too thick dielectric substrates in high density plasmas.

Very high negative voltage biasing. High voltages of up to tens of kV are required for ion implantation on substrates immersed in plasmas.[24-26] When the negative bias voltage V_0 increases, the sheath thickness surrounding the substrate increases. However, if this sheath reaches the reactor wall, arcing occurs between the substrate and the wall. To estimate the sheath thickness $\mathcal{9}$, we consider the ion sheath as a plane diode with a width equal to $\mathcal{9}$. The ion current between the plasma (emitter) and the substrate of area $\mathcal{S}$ (collector) can then be deduced from the Child-Langmuir relationship

$$I_i = \frac{4\,\varepsilon_0\,\mathcal{S}}{9}\left(\frac{2e}{m_i}\right)^{1/2}(V_p - V_0)^{3/2}\frac{1}{g^2}.\tag{13.16}$$

The sheath thickness is obtained by equating the ion currents in eqns. (13.2) and (13.16), yielding

$$g \approx \lambda_D\left[\frac{e\,(V_p - V_0)}{k_B T_e}\right]^{3/4},\tag{13.17}$$

where g is proportional to the electron Debye length λ_D. Typically, for $k_B T_e \approx 1$ eV and $eV_0 \approx 10$ keV, the sheath thickness is of the order of 10^3 Debye lengths, i.e. it is generally much larger than the distance to the reactor wall. In a collisionless sheath, the space charge perturbation induced by the application of the bias voltage propagates at a velocity close to the ion acoustic velocity

$$c_i = \lambda_D\,\omega_{pi} = \left(\frac{k_B T_e}{m_i}\right)^{1/2}.\tag{13.18}$$

This means that the expansion time t_s of the sheath through a cylindrical reactor of radius R is of the order of

$$t_s \simeq R\left(\frac{m_i}{k_B T_e}\right)^{1/2},\tag{13.19}$$

and independent of the plasma density. Taking $m_i = 10$ a.m.u., $k_B T_e = 1$ eV and $R = 0.1$ m, one obtains an expansion time of 30 μs. Consequently, the duration of a single negative high voltage pulse applied to a substrate should not exceed a few tens of μs to avoid arcing between the substrate and the reactor wall. The dose implanted during a single pulse is proportional to the plasma density. The right implantation dose is then obtained by applying a series of elementary pulses.

13.4.5. Concluding remarks. We have shown that in relatively cold Maxwellian plasmas, the lowest available ion energy which can be attained by substrate biasing is equal to a few $k_B T_e$. This ion energy limit is much lower in plasmas free of energetic electrons, such as microwave multipolar plasmas, than in RF plasmas, where the electron energy distribution function generally has a significant tail of energetic electrons.

We have also demonstrated that the bias frequency is a critical parameter for the control of the energy distribution function of the ions that strike the substrate. An important result is that a monokinetic ion bombardment energy can only be obtained when the bias frequency is higher than the ion plasma frequency. In contrast, when the bias frequency becomes lower than the ion plasma frequency, the ion energy can reach a value about twice that of the previous monokinetic energy. As a result, when a plasma is excited at a frequency below the ion plasma frequency, intense ion bombardment of the wall, or of the excitation electrodes, generally occurs, producing significant contamination. This is, for instance, the case of high density 13.56 MHz discharges (Sec. 13.2). At 2.45 GHz, in contrast, the condition of an excitation frequency much higher than the ion plasma frequency is always satisfied, even at very high plasma densities. Thus microwave plasmas leave us free to look for plasmas providing the largest possible plasma density in order to reduce processing time. We thus turn to the physical limitations that prevent from achieving large densities in homogeneous large volume plasmas.

13.5. Density limitations in microwave multipolar plasmas

The treatment of increasingly large substrates requires larger and larger volumes of dense and homogeneous plasmas. As the multipolar magnetic confinement is only efficient when the plasma diffuses into the entire confined volume (see Sec. 13.3), i.e. when the ion mean free path is at least of the order of the reactor size, an increase in chamber dimensions requires a correlative decrease of plasma pressure. As a result, the surface-wave-excited discharges (without a magnetic field) described in Sec. 13.3, whose minimum working pressure is close to one mtorr, must be ruled out in favor of magnetically confined discharges when the plasma dimensions exceed a few decimeters.

Trying to meet the trend towards larger volume plasma sources with multipolar magnetic systems requires lower and lower operating gas pressures, leading to certain drawbacks as far as surface processing is concerned: i) lowering the gas pressure generally yields a corresponding decrease in the concentration of neutral reactive species avalaible in the plasma. As this concentration determines the flux of reactive species on surfaces, the kinetics of surface treatment, usually limited by surface mechanisms in high pressure plasmas, may then become plasma limited; ii) for a given microwave power, the increase in plasma volume leads to a decrease in plasma density,[18] hence to a decrease in the kinetics of ion bombardment induced reactions. To investigate how the plasma density is limited in large plasma volumes confined in a multipolar magnetic structure and sustained by a microwave discharge, we derive below density relations based on simple assumptions.

13.5.1. Plasma density achievable in microwave discharges. We start by considering the minimum plasma density that is obtained in microwave plasma sources. From Secs. 4.3 and 8.5.1, we know that it is usually at least of the order of the *critical density* n_c (Sec. 4.3), which is given by

$$n_c = \frac{\varepsilon_0 m_e \omega^2}{e^2},$$
(13.20)

where ε_0 is the permittivity of free space and ω is the microwave field angular frequency.

At 2.45 GHz, we have $n_c = 7.5 \times 10^{10}$ cm^{-3} but densities of up to a few 10^{11} cm^{-3} are currently reached over discharge diameters from a few centimeters to more than a decimeter.[10,19,27] Since the critical density varies according to the square of the excitation frequency, a large plasma density increase is to be expected from an increase in frequency. At 5.8 GHz (the ISM frequency next to 2.45 GHz), $n_c = 4 \times 10^{11}$ cm^{-3}. Recall that, whatever the frequency, any increase in plasma density always requires a correlative increase in the microwave power absorbed by the plasma. Unfortunately, physical limitations such as, for example, too low a neutral density, and technical limitations, for example, too large a microwave power density level in the field applicator and related circuitry, prevent an unlimited increase in plasma density.

13.5.2. Ion density in a multipolar plasma sustained by a microwave discharge. Taking into account the above plasma density characteristics of microwave plasma sources, we examine the corresponding density in the multipolar structure. We assume that the plasma is only supplied by the microwave discharge (no additional peripheral ionization) and that volume ion recombination is negligible. Under steady-state conditions, the number of ions which cross the source aperture per second and diffuse into the multipolar structure is equal to the number of ions per second that are lost on the wall. According to eqn. (13.2), this balance yields

$$n_d S_d \left(\frac{k_B T_{ed}}{m_i}\right)^{1/2} = n S \left(\frac{k_B T_e}{m_i}\right)^{1/2},$$
(13.21)

where n_d and n are the plasma densities in the source and in the multipolar structure, T_{ed} and T_e are the corresponding electron temperatures, and S_d and S are the areas of the source aperture into the multipolar chamber and of the chamber wall respectively. Since Fig. 13.9 shows a decrease of the electron temperature in the multipolar plasma compared to the temperature in the plasma source, it is not realistic to assume isothermal conditions during plasma expansion from the source into the chamber. Instead, we consider this expansion to be adiabatic, which leads to

$$\frac{T_{ed}}{T_e} = \left(\frac{n_d}{n}\right)^{2/3}. \tag{13.22}$$

Substituting eqn. (13.22) in eqn. (13.21), we find that the density in the multipolar structure is reduced to

$$n = n_d \left(\frac{S_d}{S}\right)^{3/4}. \tag{13.23}$$

This expression summarizes the dependence of the multipolar plasma density on the external plasma source by which it is sustained, and suggests different ways of increasing plasma density, as discussed below.

13.5.3. Discussion and conclusion. The surface S of the chamber wall is a factor which increases homothetically with the plasma volume V in the chamber as

$$S = C\, V^{2/3}, \tag{13.24}$$

where C is a constant which depends on the geometrical configuration of the chamber. Therefore, the plasma density in the chamber varies as a function of the plasma volume according to

$$n = n_d \frac{S_d^{3/4}}{C^{3/4} V^{1/2}}. \tag{13.25}$$

This expression shows that plasma density can be increased: i) by decreasing the surface to volume ratio (constant C), and ii) by reducing the volume V to the lowest possible value compatible with a given process on a given substrate.

As mentioned in Sec. 13.5.1, the plasma density n_d can be increased by increasing the excitation frequency ($n_c \propto \omega^2$). In the case of conventional ECR systems, this imposes a corresponding increase in the intensity of the magnetic field to reach resonance. At the same time, the section of the fundamental mode waveguide decreases (as $1/\omega^2$) so that overdimensioned waveguides have to be used in order to keep S_d constant, a solution that can introduce power losses due to multimode operation. Thus, increasing the frequency in conventional ECR systems leads to more bulky and costly devices.

Finally, considering eqn. (13.23), we see that increasing the area S_d of the plasma source is an important parameter affecting the density in the multipolar plasma. A simplistic solution to increase the surface S_d is to connect a series of microwave discharges to the multipolar confinement structure. This increases the complexity and cost of the system, especially in the case of conventional ECR systems. An elegant solution is to make use of the multipolar magnetic confinement structure to provide the resonant magnetic field necessary for the ECR excitation, i.e. to utilize the so-called distributed ECR (DECR) scheme. The interesting aspect of this excitation technique, presented in the next chapter, is that the area of the plasma source can be made equal to the surface of the chamber walls ($S_d = S$). Hence, from eqn. (13.23), we see that the density n in the multipolar plasma becomes equal to n_d. In addition, in this case, we deal with a peripheral plasma source (see Sec. 9.3.4), which favors plasma homogeneity.

APPENDIX 13.1

SUBSTRATE BIASING WITH PERIODIC SIGNALS: CALCULATION OF THE DC BIAS VOLTAGE

The substrate, considered as a large plane probe, is biased by application of a periodic voltage through a low impedance capacitor (Fig. 13.12). The potential taken by the substrate surface immersed in the plasma is of the form $V_0 + V(t)$ where V_0 is the mean DC component and $V(t)$ the periodic component of the potential (see Fig. 13.17). For a Maxwellian distribution function of electrons, the electron current collected at time t for $V < V_p$, where V_p is the origin of potentials ($V_p = 0$), is given by (see eqn. (13.3))

$$I_e = - en_e \, S \left(\frac{k_B T_e}{2\pi m_e}\right)^{1/2} \exp\left[\frac{e\,[V_0 + V(t)]}{k_B T_e}\right].$$

(13.26)

On a plane substrate, for $V < V_p$, the saturation ion current is time independent. Its value is given by eqn. (13.2).

In a steady-state, assuming a zero electric charge collected on the substrate surface over a period $\mathbb{T}$, (eqn. (13.11)) yields

$$\frac{n_i \mathbb{T}}{(m_i)^{1/2}} = \frac{n_e}{(2\pi m_e)^{1/2}} \, \exp\left(\frac{eV_0}{k_B T_e}\right) \int_0^{\mathbb{T}} \exp\left[\frac{eV(t)}{k_B T_e}\right] dt .$$

(13.27)

The DC bias voltage V_0 thus takes the simple form

$$V_0 = V_f - \frac{k_B T_e}{e} \log_e \mathbb{U} .$$

(13.28)

Here V_f is the floating potential

$$V_f = - \frac{k_B T_e}{2e}\left(\log_e \frac{m_i}{2\pi m_e} + 2 \log_e \mathbb{B}\right),$$

(13.29)

with $\mathbb{B} = n_e/n_i$ ($\mathbb{B} = 1$ in plasmas without negative ions), and

$$\mathbb{U} = \frac{1}{\mathbb{T}} \int_0^{\mathbb{T}} \exp\left[\frac{e\,V(t)}{k_B T_e}\right] dt .$$

(13.30)

The variable $\mho$ depends on the amplitude and shape of the periodic signal. Without RF biasing ($V(t) \equiv 0$), the DC bias (eqns. (13.28) and (13.29)) is equal to the floating potential, of course.

The DC bias voltage is now calculated for three simple cases of signal shape (see Fig. 13.17):

i) sine shape. The time dependence of the signal has the form

$$V(t) = V_{RF} \cos\left(\frac{2\pi t}{T}\right).$$

(13.31)

The corresponding value of $\mho$ is

$$\mho = \frac{1}{T} \int_0^T \exp\left[\frac{eV_{RF}}{k_B T_e} \cos\left(\frac{2\pi t}{T}\right)\right] dt.$$

(13.32)

Integration gives

$$\mho = I_0\left(\frac{eV_{RF}}{k_B T_e}\right),$$

(13.33)

where I_0 is the zero-order modified Bessel function. The resulting DC bias voltage is

$$V_0 = V_f - \frac{k_B T_e}{e} \log_e\left[I_0\left(\frac{eV_{RF}}{k_B T_e}\right)\right].$$

(13.34)

For $eV_{RF} \gg k_B T_e$, the asymptotic value of V_0 becomes

$$V_0 = V_f - V_{RF} + \frac{k_B T_e}{2e} \log_e\left(\frac{2\pi eV_{RF}}{k_B T_e}\right);$$

(13.35)

ii) square shape. The $\mho$ value is given by

$$\mho = \frac{1}{T}\left[\int_0^{T/2} \exp\left(\frac{eV_{RF}}{k_B T_e}\right) dt + \int_{T/2}^{T} \exp\left(-\frac{eV_{RF}}{k_B T_e}\right) dt\right].$$

(13.36)

Integration gives

$$\mho = \cosh\left(\frac{eV_{RF}}{k_B T_e}\right).$$

(13.37)

The bias voltage (eqn. (13.28)) is now

$$V_0 = V_f - \frac{k_B T_e}{e} \log_e \left[\cosh \left(\frac{e V_{RF}}{k_B T_e} \right) \right].$$
(13.38)

For $e V_{RF} \gg k_B T_e$, the asymptotic value for V_0 becomes

$$V_0 = V_f - V_{RF} + \frac{k_B T_e}{e} \log_e 2;$$
(13.39)

iii) rectangular dissymmetrical shape (Fig. 13.17). The $\mathbb{U}$ value is given by

$$\mathbb{U} = \frac{1}{\mathbb{T}} \left[\int_0^{\tau} \exp \left(\frac{2 e V_{RF}}{k_B T_e} \frac{\mathbb{T} - \tau}{\mathbb{T}} \right) dt + \int_{\tau}^{\mathbb{T}} \exp \left(- \frac{2 e V_{RF}}{k_B T_e} \frac{\tau}{\mathbb{T}} \right) dt \right]$$
(13.40)

where τ is the duration of the positive pulse. Integration gives

$$\mathbb{U} = \left[\frac{\tau}{\mathbb{T}} + \frac{\mathbb{T} - \tau}{\mathbb{T}} \exp \left(- \frac{2 e V_{RF}}{k_B T_e} \right) \right] \exp \left(\frac{2 e V_{RF}}{k_B T_e} \frac{\mathbb{T} - \tau}{\mathbb{T}} \right)$$
(13.41)

and

$$V_0 = V_f - 2 V_{RF} \frac{\mathbb{T} - \tau}{\mathbb{T}} - \frac{k_B T_e}{e} \log_e \left[\frac{\tau}{\mathbb{T}} + \frac{\mathbb{T} - \tau}{\mathbb{T}} \exp \left(- \frac{2 e V_{RF}}{k_B T_e} \right) \right].$$
(13.42)

For $e V_{RF} \gg k_B T_e$, the asymptotic value of V_0 is given by

$$V_0 = V_f - 2 V_{RF} \frac{\mathbb{T} - \tau}{\mathbb{T}} - \frac{k_B T_e}{e} \log_e \frac{\tau}{\mathbb{T}}.$$
(13.43)

For a given value of $\tau/\mathbb{T}$, i.e., for a given time distribution of electron bombardement, the bias voltage can be controlled independently via the signal amplitude. For $\tau = \mathbb{T}/2$, it is evident that the result of the square voltage is obtained (eqn. (13.39)).

Finally, an important point to be dealt with is the domain of validity of the above calculations. In particular, the basic hypothesis $V < V_p$ with $V_p = 0$, adopted for the calculation of the collected currents, may not always be valid. In the case of the sine-shaped signal, the most positive potential $V_0 + V_{RF}$ reached during a period takes the value (see eqn. (13.35))

$$V_0 + V_{RF} = V_f + \frac{k_B T_e}{2e} \log_e \left(2\pi \frac{e V_{RF}}{k_B T_e} \right).$$
(13.44)

In the case of the rectangular signal, the most positive potential reached during a period takes the value (eqn. 13.43)

$$V_0 + 2V_{RF}\,\frac{\mathbb{T} - \tau}{\mathbb{T}} = V_f + \frac{k_B T_e}{e}\,\log_e \frac{\mathbb{T}}{\tau}. \tag{13.45}$$

The plasma potential ($V_p = 0$) is reached when the right-hand side of eqn. (13.44) or (13.45) becomes zero. For a plasma free of negative ions, one obtains

$$\frac{eV_{RF}}{k_B T_e} = \frac{m_i}{4\pi^2 m_e} \tag{13.46}$$

for the sine shaped signal and

$$\frac{\mathbb{T}}{\tau} = \left(\frac{m_i}{2\pi m_e}\right)^{1/2} \tag{13.47}$$

for the rectangular signal. For argon (40 a.m.u.), the numerical application gives $eV_{RF} \approx 1.9 \times 10^3\, k_B T_e$ and $\mathbb{T}/\tau \approx 100$, indicating the drastic conditions required to escape the domain of validity for the calculations. Obviously, these conditions are less severe when negative ions are present, but for most surface processes, the assumption $V < 0$ remains valid so that negative ions are generally not involved in the bombardment of surfaces by charged species.

References

[1]   T.D. Mantei and T. Wicker, Appl. Phys. Lett. **43**, 84 (1983).

[2]   R.W. Schumacher, N. Hershkowitz and K.R. Mackenzie, J. Appl. Phys. **47**, 886 (1976).

[3]   C. Lejeune, J.P. Grandchamp, O. Kessi and J.P. Gilles, Vacuum **36**, 837 (1986).

[4]   R.W. Boswell, A.J. Perry and M. Emami, J. Vac. Sci. Technol. **A7**, 3345 (1989).

[5]   B. Petit and J. Pelletier, Jap. J. Appl. Phys. **26**, 825 (1987).

[6]   M. Moisan, Z. Zakrzewski and R. Pantel, J. Phys. **D12**, 219 (1979).

[7]   M. Moisan, Z. Zakrzewski, R. Pantel and P. Leprince, IEEE Trans. Plasma Sci. **PS-12**, 203 (1984).

[8]   M. Moisan, M. Chaker, Z. Zakrzewski and J. Paraszczak, J. Phys. **E20**, 1356 (1987).

[9]   Y. Arnal, J. Pelletier, C. Pomot, B. Petit and A. Durandet, Appl. Phys. Lett. **45**, 132 (1984).

[10] A. Durandet, Y. Arnal, J. Margot-Chaker and M. Moisan, J. Phys. **D22**, 1288 (1989).

[11] C. Pomot, B. Mahi, B. Petit, Y. Arnal and J. Pelletier, J. Vac Sci. Technol. **B4**, 1 (1986).

[12] A. Durandet, Ph.D. Thesis, Université Joseph-Fourier (Grenoble), France (1987) unpublished.

[13] W. B. Ard, M.C. Becker, R.A. Dandl, H.O. Eason, A.C. England and J. R. Kerr, Phys. Rev. Lett. **10**, 89 (1963).

[14] D.B. Miller and E.F. Gibbons, AIAAJ. **2**, 35 (1964).

[15] K. Suzuki, S. Okudaira, N. Sakudo and I. Kanomata, Jap. J. Appl. Phys. **16**, 1979 (1977).

[16] S. Matsuo and Y. Adachi, Jap. J. Appl. Phys. **21**, L4 (1982).

[17] J. Asmussen, J. Vac. Sci. Technol. **A7**, 883 (1989).

[18] L. Pomathiod, R. Debrie, Y. Arnal and J. Pelletier, Phys. Lett. **106A**, 301 (1984).

[19] J. Pelletier, Y. Arnal, B. Petit, C. Pomot and M. Pichot, J. Phys. **D19**, 795 (1986).

[20] A. Boschi and F. Magistrelli, Nuovo Cimento **29**, 487 (1963).

[21] T.D. Mantei, J. Electrochem. Soc. **130**, 1959 (1983).

[22] T.E. Wicker and T.D. Mantei, J. Appl. Phys. **57**, 1638 (1985).

[23] B. Petit, A. Durandet and J. Pelletier, Vacuum **36**, 799 (1986).

[24] J.R. Conrad, J.L. Radtke, R.A. Dodd, F.J. Worzala and N.C. Tran, J. Appl. Phys. **62**, 4591 (1987).

[25] M.A. Lieberman, J. Appl. Phys. **66**, 2926 (1989).

[26] J.T. Schever, M. Shamim and J.R. Conrad, J. Appl. Phys. **67**, 1241 (1990).

[27] Y. Suetsugu and Y. Kawai, Jap. J. Appl. Phys. **23**, 237 (1984).

CHAPTER 14

**DISTRIBUTED ELECTRON CYCLOTRON
RESONANCE (DECR) PLASMAS** *

14.1. Introduction

The difficulty of achieving industrially surface treatments through plasma processes mainly lies in the realization of large, uniform and dense plasmas in which, moreover, it must be possible to adjust the energy of ion bombardment on surfaces independently of the plasma generation. The problem of plasma homogeneity has been investigated in Chaps. 9 to 12, considering first electrostatic and then magnetic confinements. We concluded that homogeneity could be greatly improved by confining the plasma in multipolar magnetic structures. As shown in Chap. 12, plasma homogeneity then results from the localization of ionization at the periphery of the discharge where energetic electrons interact with the multipolar magnetic field. In this way, low pressure, homogeneous plasmas, free of energetic electrons, are obtained in the central volume of the discharge, which meet the above industrial requirements.

The intrinsic advantages of microwave sustained multipolar plasmas have been extensively presented in Sec. 13.3. We have then investigated two types of microwave plasma sources, the surface-wave and ECR sustained discharges, which are both external to the multipolar magnetic structure that they supply. These sources exhibit some drawbacks that result from the deposition of conducting or semiconducting materials on the dielectric tube on which the wave propagates (case of surface-wave discharges) or on the dielectric window transmitting the microwave energy into the discharge (case of common ECR setups). Furthermore, in such configurations, the plasma density is limited by the ratio of the surface area of the source to the volume of the reactor into which it diffuses. Solving these problems was the primary goal that we pursued in developing a new reactor concept based on multipolar magnetic confinement, termed distributed ECR (DECR) discharges.[1]

This new design allows to integrate the microwave field applicator to the multipolar magnetic confinement structure in such a way that ECR conditions (approximately 875 gauss at a 2.45 GHz excitation frequency) are met. As we will see, such a scheme permits one to adjust the plasma source extent to the required plasma size.

---

* **Presented by M. Pichot and J. Pelletier**

We start our presentation by describing the essential features of DECR plasma systems. Then we consider the specific case of a cylindrically shaped device operated at 2.45 GHz and examine its plasma characteristics. Finally, we discuss the possibility of using the principle of DECR to match conical and plane reactor geometries.

14.2. The concept of DECR reactors

The development of this new plasma reactor is based on two ideas: i) the permanent magnet bars destined for multipolar magnetic confinement are further used to provide the magnetic field intensity needed for ECR coupling; ii) the microwave field required for ECR is provided by linear antennas running along and close to the magnet bars. This scheme results in the generation of the plasma at the periphery of the chamber where the so-called energetic electrons (accelerated through ECR) are trapped within the magnetic field cusps until they ionize the gas.

14.2.1. ECR linear antenna applicator. Figure 14.1 is a schematic representation of one of the microwave field applicator in a cylindrically shaped DECR discharge chamber. It consists of a linear conductor of cylindrical cross-section, called the *antenna*, placed a few millimeters above a so-called *ground plane*. In the present configuration permanent magnets contained within casings of rectangular cross-section are resting on the ground plane in a similar way as for classical multipolar magnetic field confinement. These magnets provide the required 875 gauss isomagnetic surface in the vicinity of the antenna along its entire length. The field of

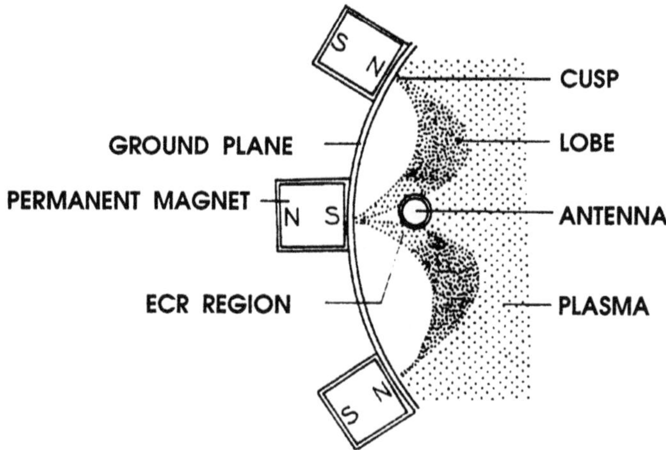

Fig. 14.1. Schematic of one of the linear field applicators in a cylindrically shaped DECR discharge chamber. The conducting wall of the chamber acts as a ground plane for the antenna. The three different regions of the discharge are sketched.[1]

each such magnetic surface is closed by two adjacent magnetized bars as in a conventional multipolar field. A magnetic field configuration is thus created whereby microwave field, applied at 2.45 GHz through a coaxial feedthrough, results in ECR coupling along the full antenna length. Within the pressure range extending from 0.02 to 0.4 Pa ($\simeq$ 0.1 to 3 mtorr), the gas is ionized by the energetic electrons in the *lobes* (see Fig. 14.1) and the plasma produced then diffuses toward the central part of the chamber.

Extension to higher excitation frequencies (for example 5.8 GHz, an ISM worldwide authorized frequency) can be anticipated. Ternary NdFeB or SmCo magnets exhibiting magnetic fields of the required intensity (more than 2070 gauss) are now commercially available. Based on considerations presented in Sec. 8.5.1 concerning ECR, this could lead to plasmas with higher minimum densities than at 2.45 GHz.

14.2.2. Distribution of linear applicators inside the reactor: cylindrical configuration case. Multipolar magnetic confinement structures are well adapted to the setting up of arrays of linear field applicators. As shown in Fig. 14.2 for a cylindrical chamber (with internal magnets this time), the reactor wall (including the magnets) constitutes the ground plane for the set of linear antennas that are symmetrically distributed azimuthally. These carry the microwave power coming through coaxial feedthroughs which ensure vacuum sealing. The plasma volume that is free from magnetic field, the so-called *useful plasma volume*, is delimited by a virtual cylinder that excludes the magnetic field lobes (Fig. 14.1). These lobes constitute the effective

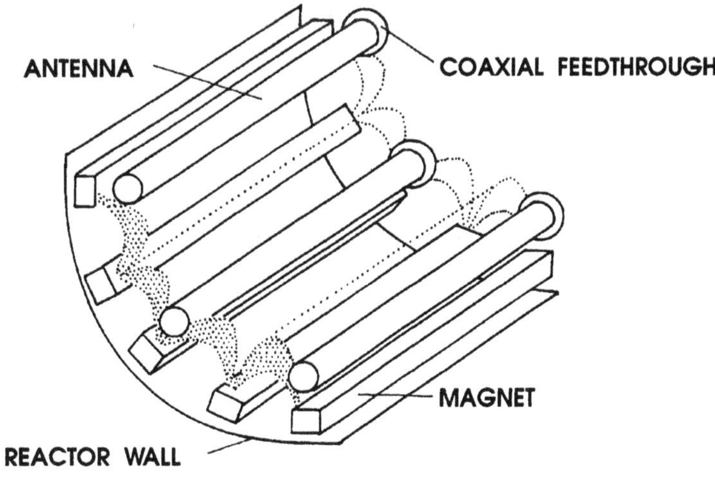

Fig. 14.2. Display of linear antennas inside a cylindrical discharge chamber fitted with a multipolar magnetic field structure.[1]

plasma source from which the plasma diffuses. This diffusion occurs perpendicularly to the magnetic field lines, in contrast with conventional ECR systems where plasma diffusion follows the magnetic field lines. For reasons presented in Sec. 12.3, the energetic electrons created inside the lobes remain confined there until they degrade into thermal electrons after ionizing the gas, so that the plasma diffusing perpendicularly to the magnetic field lines is free from fast electrons.

When using this method of plasma generation, the useful plasma volume can be increased by simultaneously increasing the diameter and length of the vessel and the number and length of antennas. Note that when keeping constant the total microwave power and the plasma volume, increasing the number of linear applicators leads to a decrease of the required power per antenna, hence to safer operation (e.g. lower heating of the feedthroughs). The symmetrical distribution of the plasma sources at the periphery of the reactor, as we have seen in Sec. 9.3.4, naturally leads to better spatial homogeneity, allowing uniform surface treatment over large areas and volumes.

14.3. Experimental setup and current performances

The results presented in this section were obtained in a cylindrical reactor working at 2.45 GHz and designed for microelectronics etching applications.

14.3.1. Microwave feeding circuit. The problem is to supply microwave power to several antennas. This can be realized in two ways.

Independent microwave source for each applicator. Power is fed from the microwave generator through a coaxial cable, followed by a directional coupler used for power measurements and by an impedance matching device (e.g. a triple stub tuner). The latter terminates into a microwave vacuum sealing feedthrough to which the antenna is connected. In our experimental setup, the microwave power on each antenna could be adjusted from 0 to 200 W. Once optimal matching is achieved, no further tuning is required to operate throughout the whole ECR pressure range with the various gases tested, which include Ar, Kr, SF_6, O_2, He, H_2, CH_4, N_2 and mixtures. The fact that the impedance seen at the input of each antenna is found to be very little sensitive to discharge conditions suggests using a multichannel distributor with a single microwave generator.

Single microwave source for all applicators. Here, one single triple stub tuner located between the transmission line (cable or waveguide) and the distributor ensures an adequate impedance matching for all antennas. As a rule, the system can be operated with zero reflected power from the tuner. A possible design for the power distributor is based on a resonant cavity equipped with as many variable depth coupling posts as there are antennas, these coupling posts lying symmetrically with respect to the cavity axis. With this system, we again observe that impedance tuning is practically not affected by varying gas content and pressure as well as microwave power.

14.3.2. Reactor design. The present reactor was intended for plasma processing of 150 mm diameter Si substrates. This diameter needs to fit what we termed the useful plasma region, requiring an overall reactor diameter of 280 mm which includes the antenna and cusp regions. In the present design, eight antennas are used, and as many magnet bars are implemented symmetrically on the reactor outer wall. The

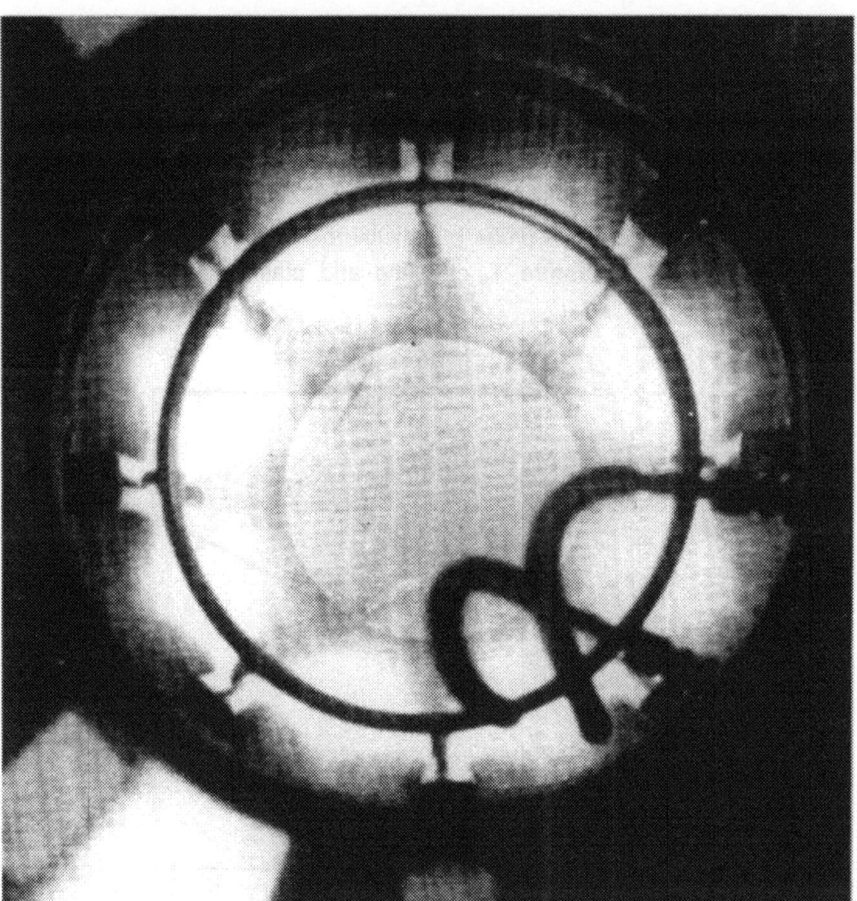

Fig. 14.3. Front view photograph of a DECR plasma in a cylindrical reactor with 8 magnet rows and antennas. The substrate holder (160 mm in diameter) is centered on the chamber axis and perpendicular to it. The plasma is produced in the multipolar magnetic field region at the periphery of the reactor and it diffuses into the field free central region.

strength of the ferrite magnets employed is such that the intensity of the magnetic field at the pole is 1500 gauss. The quasi-exponential decrease of the magnetic field intensity towards the reactor axis is such that the field is really negligible at the substrate location.

The linear applicators are made of 8 mm diameter, non-magnetic stainless steel tubes through which water flows to ensure their cooling (the reactor wall is also water cooled). The length of the antenna and of the magnet bar is 160 mm, generating 3.5 liters of useful plasma. The wafer holder (also water-cooled) is 160 mm in diameter and it is held parallel to the feedthrough flanges at 20 mm from it. It can be biased independently of the plasma generation.

Figure 14.3 is a photograph, taken from the front of the reactor, of the plasma produced. The bright regions at the periphery of the reactor correspond to plasma generation in the magnetic field lobes. The substrate holder is located in the magnetic field free region of the reactor. We have characterized the plasma in this region by means of an electrostatic Langmuir probe made from a tungsten wire (0.5 mm in diameter and 5 mm long). This probe is movable axially and radially. It yields ion density n_i, electron temperature T_e, floating and plasma potentials V_f and V_p respectively.

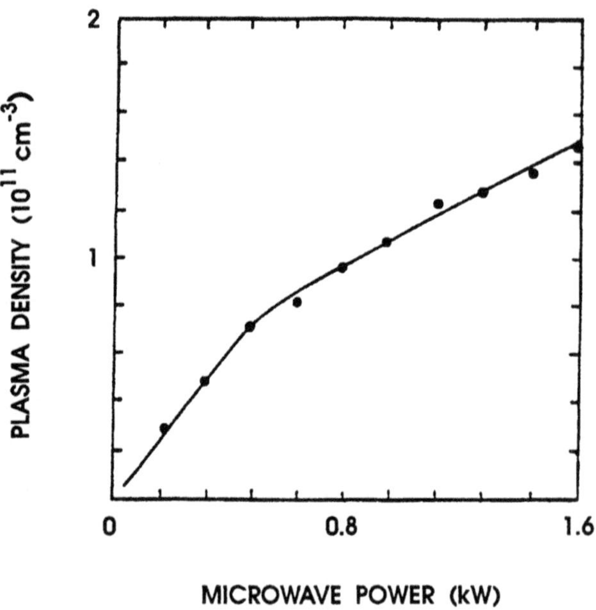

Fig. 14.4. Variation of the ion density as a function of microwave power in the magnetic field free region of a DECR argon discharge, at a pressure of 1 mtorr.[1]

14.3.3. Plasma properties. Figure 14.4 shows the variation of the ion density as a function of the microwave power going through the tuner input (zero reflected power) in an argon reference plasma at 1 mtorr (0.13 Pa). This measurement is made in the magnetic field free region where the plasma is homogeneous as discussed below. In the course of measurements, we noted that a fraction of the microwave power heated the tuner, the power distributor and the feedthroughs. One can thus expect larger plasma densities by reducing power transmission losses and, of course, by increasing microwave power. Figure 14.4 shows a decrease of the density slope passed a certain power value. It occurs at approximately the critical electron density ($n_c = 7.5 \times 10^{10}$ cm^{-3} at 2.45 GHz).

Another important operating parameter is the gas pressure. Figure 14.5 shows that the electron temperature decreases with increasing pressure, as expected, and so does the plasma potential while the floating potential is constant , as can be seen in Fig. 14.6. We find that the difference V_p - V_f is less than the theoretical value expected when assuming a Maxwellian electron distribution function (V_p - V_f = 5.4 $k_B T_e$ for argon). This could be related to a depletion of fast electrons in the distribution, which would be consistent with the fact that the fast electrons are

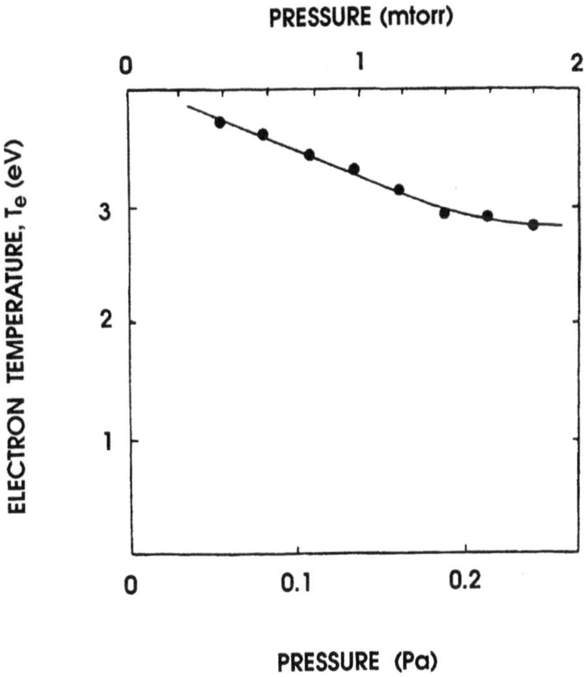

Fig. 14.5. Variation of the electron temperature T_e as a function of pressure in a DECR argon plasma, at a constant microwave power (1600 W).[1]

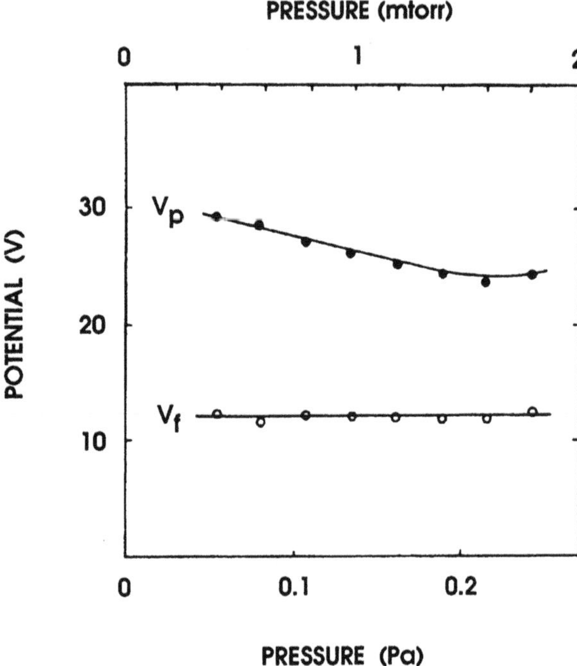

Fig. 14.6. Variation of the floating and plasma potentials V_f and V_p as a function of pressure in a DECR argon plasma, at a constant microwave power (1600 W). These potentials are referenced to the chamber wall potential.[1]

trapped inside the magnetic field cusps. We further note that V_p and V_f are relatively close to the reactor wall potential, yielding negligible sputtering of these walls. Another interesting point is the dependence of plasma density on gas pressure as shown in Fig. 14.7. It remains constant over most of the pressure range of interest in the operation of multipolar devices.

Little radial variation of the plasma parameters (V_f, V_p, n_i, T_e) is observed in the useful plasma region. This is illustrated in Fig. 14.8 for the case of plasma density. However, we found a decrease in plasma density along the axial direction at both extremities of the reactor.[†] Nevertheless, this behavior does not reflect on surface treatment in the present case since the substrate is positioned perpendicularly to the axis.

[†] Here, the ends of the reactor are not magnetically closed. Closing them improves the uniformity (see Sec. 10.3.2).

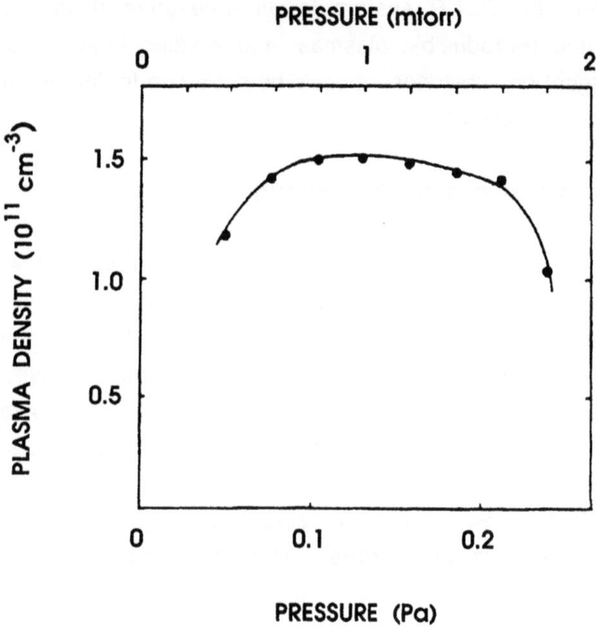

Fig. 14.7. Variation of the ion density as a function of pressure in a DECR argon plasma, at a constant microwave power (1600 W).[1]

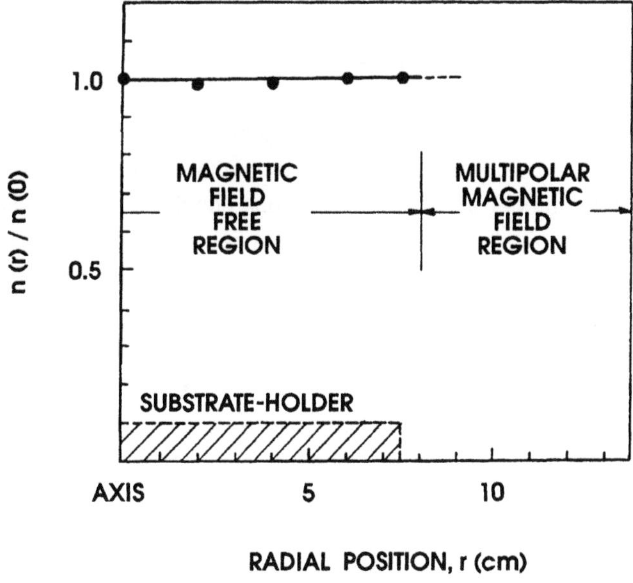

Fig. 14.8. Radial profile of the ion density, n(r)/n(0), in a DECR argon plasma at 1 mtorr and at a constant microwave power (320 W).[1]

In conclusion, the DECR system is an innovative device providing large, homogeneous and reproducible plasmas, and yielding high ion current densities (3 mA/cm^2 in argon) on substrates. It could be scaled up for larger surface processing as we discuss in the next section.

14.4. Reactor scale-up and other configurations

We have seen in Sec. 9.3.4 that localizing plasma generation at the periphery of the reactor is the best approach to the problem of plasma homogeneity. DECR systems provide such a localization as we have shown. Well suited for the processing of small substrate diameters, the cylindrical geometry may become inadequate for uniform plasma processing over large areas. The aim of this section is to determine the limitation of the cylindrical configuration and to propose other configurations.

14.4.1. The homothecy (scaling factor) concept applied to cylindrical configurations, and its limitations. The wafers currently used nowadays in microelectronics are 20 cm in diameter. To avoid edge effects related to the proximity of the magnetic field region when processing them, the diameter of the DECR reactor must be larger than 35 cm. To achieve plasmas with density and homogeneity similar to those found in smaller diameter reactors, the number of antenna-magnet bar pairs must increase proportionally to the reactor diameter in order to keep the plasma production rate constant relatively to the loss rate per unit surface. Following this scale-up rule implies that typically 20 antennas are distributed at the periphery of a cylindrical reactor to process 20 cm diameter wafers.

Another aspect of DECR systems is that they provide perfect plasma homogeneity on condition that volumic losses (by ion recombination) and surface losses at the extremities of the chamber are negligible. These conditions are satisfied with argon plasmas and more generally with atomic gas plasmas (where volume recombination is negligible) when sustained in long cylindrical reactors or in reactors magnetically closed at both extremities (see Sec. 10.3.2). Such a structure can then be used as the basis, for example, of a sputtering deposition system, as shown schematically in Fig. 14.9. The target(s) and substrate(s) can be biased independently with respect to the plasma potential. As compared to conventional sputtering systems, the addition of an independent substrate biasing can improve the characteristics of the deposited layer by providing, for example, better adhesion and purity, and a possible control of grain size and stoichiometry. Furthermore, the uniform bombardment that results from such a biasing, in contrast to magnetron sputtering, leads to a uniform target erosion.

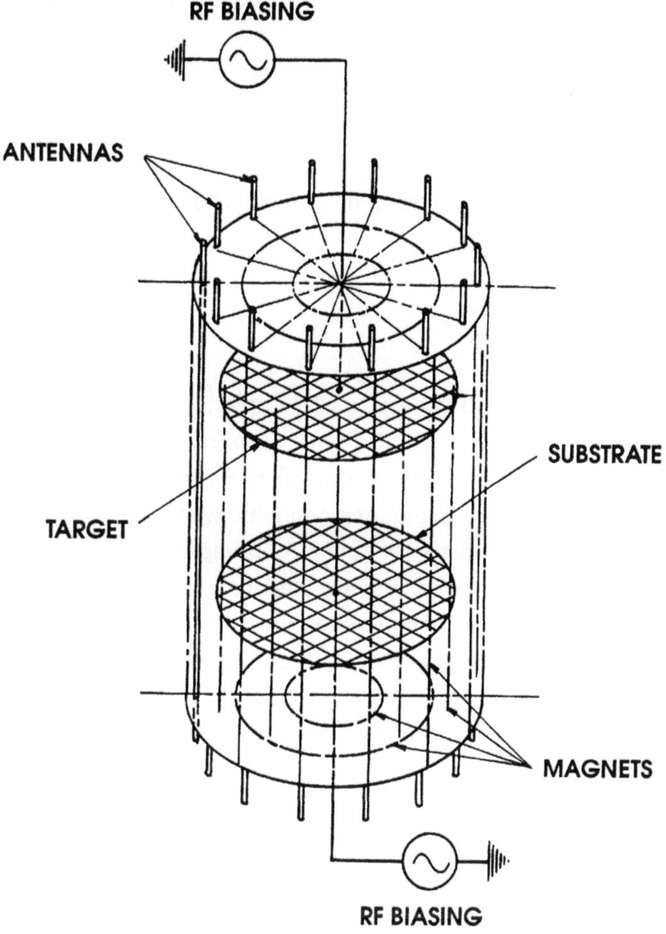

Fig. 14.9. Sketch of a cylindrical DECR sputtering reactor using a fully closed multipolar magnetic confinement configuration, with independent RF biasing of the substrate and target.

In molecular gas plasmas, as shown in Sec. 9.5, ion volume recombination is the major obstacle preventing from achieving homogeneous plasmas. When volume recombination is present, the most favorable situation is clearly that of plasmas where ion-neutral collisions are negligible and where the expansion velocity of plasma can reach the ion-acoustic velocity c_i. Assuming such a plasma to behave as a beam of mean velocity c_i (see Sec. 9.2.1), the variation of plasma density n as a function of coordinate x is then governed by

$$\frac{\partial n}{\partial t} = c_i \frac{\partial n}{\partial x} = -\alpha_r n^2,$$

$$(14.1)$$

where t is the time and α_r is the ion volume recombination coefficient.[†] Integration yields

$$n = n_0 \left[1 + \frac{\alpha_r n_0}{c_i} x \right]^{-1} , \tag{14.2}$$

where n_0 is the plasma density at the origin ($x = 0$). The density reduces by a factor of 4 at

$$x_{1/4} = \frac{3 c_i}{\alpha_r n_0} . \tag{14.3}$$

Assuming $c_i = 6$ km/s and $\alpha_r = 2 \times 10^{-7}$ cm^3/s, one obtains $x_{1/4} = 9$ cm for $n_0 = 10^{12}$ cm^{-3} and 90 cm for $n_0 = 10^{11}$ cm^{-3}. This result clearly shows that plasma homogeneity degrades rapidly with increasing plasma density. Hence, the homothecy concept does not ensure plasma homogeneity when scaling-up molecular gas plasmas at densities exceeding 10^{11} cm^{-3}. Turning to cylindrical configurations, it could be shown that plasma inhomogeneity then increases with diameter, and thus the plasma processing of large diameter wafers in high density molecular gas plasmas requires alternate DECR configurations, as those described below.

14.4.2. The conical configuration: an intermediate solution. Due to its rotational symmetry, the DECR conical configuration follows directly from the cylindrical one. As compared to the cylindrical system, the conical DECR reactor shown in Fig. 14.10 exhibits significant advantages with respect to improving plasma homogeneity, namely: i) a reduced ion volume recombination loss. The plasma sources located on the cone surface are clearly closer to the axis than in the cylindrical configuration. Thus, for a given characteristic ion recombination length $x_{1/4}$ (eqn. (14.3)), the corresponding ion loss between the sources and the center of the wafer is less than in the cylindrical DECR; ii) an increased plasma density at the axis to compensate for ion volume losses. Figure 14.10 shows that the density of microwave power deposited into plasma decreases from the top to the base of the cone. As a result, the plasma density increases towards the axis, thus hopefully compensating for ion volume recombination. Clearly, the degree of compensation depends on the plasma density n_0 and on the ion recombination coefficient α_r.

As far as pumping kinetics and wall contamination are concerned, the conical configuration is more advantageous than the cylindrical one. This comes from the fact

---

[†] A similar derivation can be found in Sec. 9.5 for the case of a collisional plasma.

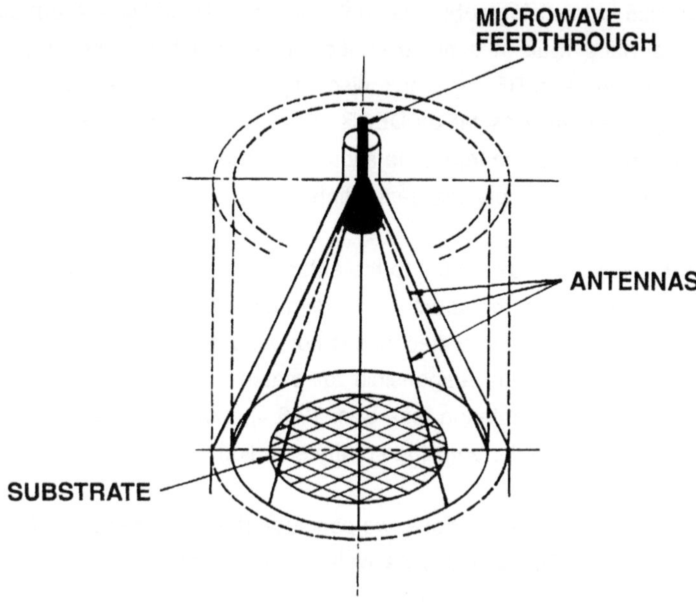

Fig. 14.10. Sketch of a conical DECR reactor. The broken lines outline for comparison the corresponding cylindrical configuration.

that the volume of a cone is 1/3 of that of a cylinder and that its lateral surface is 1/2 of that of a cylinder, assuming equal diameter and height. A further advantage is the possibility of feeding the antennas by using a single feedthrough and a rudimentary microwave power distributor (Fig. 14.10).

In summary, the conical configuration can be helpful in solving some uniformity problems in reactive gases. However, the plane configuration offers more potential in terms of uniform plasma processing, as discussed below.

14.4.3. The plane configuration: an ideal scale-up solution. As far as achieving a uniform plasma-surface interaction is concerned, the infinite plane configuration where plane substrates are placed parallel to a plane uniform plasma source is undeniably an ideal configuration: each point of the substrate surface with respect to the plane source has the same geometrical factor, i.e. identical solid angle and distance. Therefore, volume recombination cannot affect the uniformity of the plasma-surface interaction. Furthermore, since plasma density decreases as a function of distance from the source, obviously higher plasma fluxes are obtained since the plane configuration favors smaller distances from the source. Moreover, beside providing uniformity, the plane configuration leads to the largest substrate to reactor wall surface ratio, minimizing substrate contamination. In addition, the plasma volume to substrate surface ratio in a plane configuration, which can be compared to the interelectrode distance in RF capacitive discharges, provides the highest efficiency in terms of pumping kinetics as far as DECR systems are concerned. Furthermore, in contrast with capacitive discharges, the substrate surface in plane DECR systems can be biased independently of plasma generation.

In most instances, the power sent to the antenna is large enough for the wave to be reflected at the extremity of the structure, yielding a standing wave pattern. This means that microwave power transfer to electrons is not uniform along the antenna. However, since the energetic electrons responsible for plasma ionization undergo a magnetic drift motion parallel to the antenna (see Sec. 12.3.4), plasma production along the axis remains uniform, provided the amplitude of the standing wave pattern is constant.

From the preceding considerations, we conclude that, in a plane DECR system, the ion volume recombination process as well as the eventual existence of stationary waves along the antenna cannot impair plasma uniformity. Thus, this system naturally lends itself to achieving long linear applicators. Figure 14.11 shows such a possible

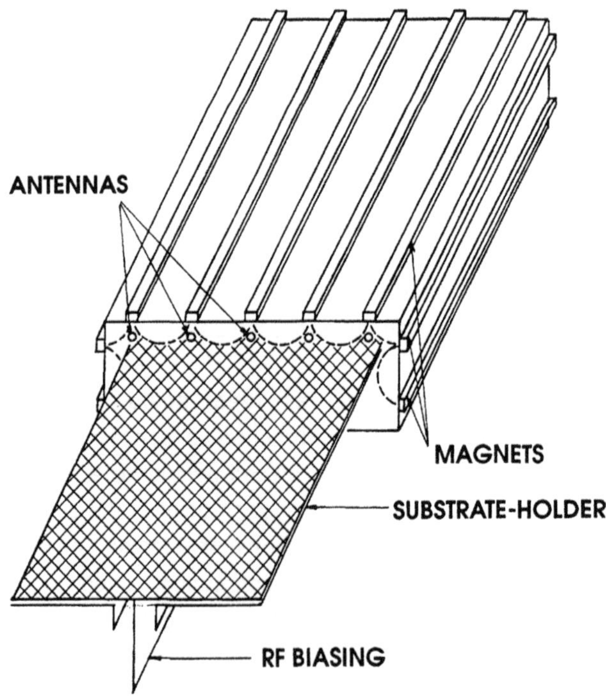

Fig. 14.11. Sketch of a plane DECR reactor based on an array of parallel, linear microwave antennas.

design that uses a set of parallel linear antennas. The possibility of achieving large areas of uniform plasma with the plane DECR system assumes that each single linear applicator can generate a uniform plasma along it.

The corresponding experiments were performed on a 2 meter long single antenna DECR system using the basic arrangement shown in Fig. 14.1. It was found that plasma uniformity largely depends on discharge conditions and microwave power. For example, for a given gas and a given microwave power, uniformity is obtained only at a definite pressure, as shown in Fig. 14.12, whereas, at a given pressure, uniformity is obtained at a definite microwave power.[2] A possible interpretation refers to the standing wave pattern discussed previously where it was suggested that uniformity along the antenna is achieved, provided the amplitude of the standing waves is constant along the antenna. Nonetheless, the aforementioned experiments have shown that a linear (up to 2 meters) plasma with a uniformity better than a few per cent can be achieved. Clearly, many such parallel linear antennas could be associated to form a large plane source delivering a uniform plasma.

Finally, as a general remark applicable to all the DECR systems presented, we stress the fact that linear magnetic field bars can lead to electron leakage at the

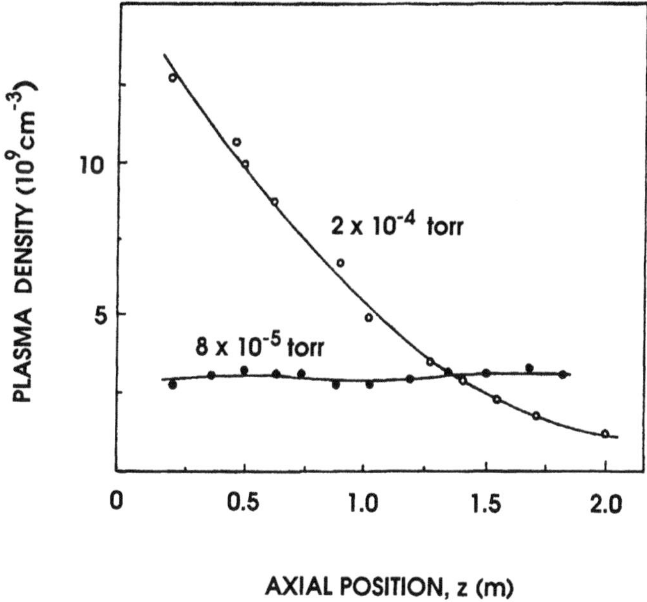

Fig. 14.12. Distribution of the ion density in an argon plasma as a function of axial position along a 2 meter long, single antenna, plane DECR system. The $z = 0$ position corresponds to the power input side (feedthrough). The microwave power is constant at 300 W (2.45 GHz).[2]

extremities of the magnet bars. This problem can be circumvented by implementing magnets at these extremeties as shown in Fig. 10.9, in such a way as to close the magnetic structures on themselves. This leads to a still better uniformity and to an increase of approximately a factor of 100 in density compared to the conditions of Fig. 14.12.

14.5. Conclusion

The concept of localization and uniform distribution of plasma sources at the periphery of a closed volume appears to be a very attractive solution to provide plasma homogeneity for the processing of substrates in various configurations. This can be achieved in DECR plasmas by combining ECR excitation with a multipolar magnetic field intensity localized at the periphery. In an actual DECR system, the excitation zone can be moved away as much as required from the dielectric windows (or feedthroughs) through which the microwave power is introduced into the vacuum chamber, allowing one to deposit conducting materials without perturbing the microwave feeding. However, ion volume recombination, responsible for the lack of homogeneity in molecular gas plasmas, imposes limitations to the uniformity of plasma processing. In the case of large plane substrates, this problem can be evaded by adopting a plane configuration for the plasma reactor. Large plane DECR plasmas can be obtained by associating parallel linear antennas. The possibility of producing such uniform plane plasmas over a distance of more than one meter has been experimentally proven with a single antenna plane structure.

References

[1]   M. Pichot, A. Durandet, J. Pelletier, Y. Arnal and L. Vallier, Rev. Sci. Instrum. **59,** 1072 (1988).

[2]   Y. Arnal, J. Pelletier, L. Pomathiod and L. Vallier, *8 th International Colloquium on Plasma Processes* (CIP 91) (Antibes, France) Le Vide, les Couches Minces, Suppl. **256,** 235 (1991).

CHAPTER 15

**APPLICATIONS OF MICROWAVE PLASMAS
IN MICROCIRCUIT FABRICATION** *

15.1. Introduction

Microwave stimulated plasmas have been known for almost as long as the techniques for generating microwave fields.[1] Their applications include plasma torches,[2] light sources,[3] lasers,[4] sources of atoms for analytical applications,[5-7] hazardous waste disposal[8] and various etching and deposition applications in microcircuit fabrication. This chapter is devoted to the latter application.

Most commercial microwave plasma applications are excited by power sources which operate at 2450 MHz. This is primarily due to the widespread use of this frequency band for microwave ovens. Cheap, noisy, frequency variant magnetrons can be purchased in quantity for less than $30 US, thus providing an inexpensive source of high power ($\simeq$ 1kW) microwave radiation. For the interested reader, a historical account of the development of magnetrons is available in ref. 9. For those inexperienced in the use of microwaves, initial encounters with this technology can be somewhat daunting. At this frequency, the free space wavelength is around 12 cm, thus any attempt to use this radiation becomes a matter of geometrical control of guiding elements such as cables, connectors, waveguides.

At RF frequencies, it is common to use coaxial cables to transmit power between two points. The same cables which are capable of handling 1 kW of power at 13.56 MHz (a commonly used ISM allowed band, see Appendix 1.1) are derated to 40 W at 2450 MHz, due to their rise in attenuation with frequency. Changes in impedance, such as that found at carelessly assembled connectors, give rise to large reflections, which cause heating and, in extreme cases, vapourisation of the connection. Thus, for many microwave applications, it is necessary to use waveguides to transfer power between the source and the point of use (applicator). This geometric constraint is overcome in many commercial applications by locating the microwave source very close to the applicator.

The notion that capacitors and inductors which are separate physical units, used to tune plasmas at lower frequencies, are now elements which are distributed across the entire waveguide or cable, make first forays into this field a little bewildering. Since the

---

* **Presented by J. Paraszczak and J. Heidenreich**

wavelength of microwave radiation is close to the dimensions of the vessel used to contain the plasma, the field distribution in these systems is important and can be affected by the vessel shape and contents.

These effects often lead to considerable frustration on the part of the inexperienced user. Unfortunately, much of the simple and easy to read information in this field is old enough to be out of print although an inexpensive series of reference tables[10] can be very useful, as is an excellent monograph by Veley.[11] Many of the practical aspects of microwave power transfer and control are dealt with in Chaps. 2 and 4 of this monograph and thus will not be addressed in detail here.

As a consequence of these new requirements and demands, most engineers attempting to create plasmas at these frequencies often ask themselves the question "Why should I use microwave plasmas?".

Despite manufacturers and practitioners claims to the contrary, there are no unequivocal answers to this question since some of the processes which are first developed at microwave frequencies, turn out to be just as effective when implemented at RF frequencies. This does not mean that there are not distinct and measurable differences between plasmas excited at different frequencies, some of which are specifically alluded to in Chap. 3. It must be noted that the ultimate result of processing a particular substrate in a variety of plasma systems may not be significantly different, and may be a consequence of a specific set of priorities which may be economic, environmental or based on individual preference. For example, plasmas generated at 13.56 or 2450 MHz are both capable of etching polymers. For some applications, the differences between the two approaches (such as stripping the resist from SiO_2 covered wafers) may be of little import and the choice of technique may be based on economic considerations. In other cases, where selectivity between, say, the removal of the polymer and an underlying material, such as polysilicon, is of prime importance, the use of a particular process may be dictated.

It is the authors' observation that the field of plasma processing undergoes the same vagaries of fashion as any other market. New plasma processing techniques appear, promising to deliver better or faster or cleaner processes. New companies spring up and established companies anxious to maintain their customer base provide equipment based on the new process. It is only years later that it is discovered that the old tool sitting in the corner of the production line or laboratory is being used to generate similar results to that obtained at greater expense with the newer tooling.

Our own interest in the use of microwave frequency plasmas arose as a consequence of Dzioba's[12] work which demonstrated that it was possible to etch polymers at rates exceeding 1 μm/min. This rate was significantly greater than the 0.5 μm/min which was achievable at that time using an RF etching process. The latter not only heated the substrate significantly, but produced sputtering of the target and of chamber fixtures.

Despite the observations and reservations that we expressed, there are many instances where a particular frequency is more effective or a particular plasma process dominates the field. It is thus instructive to review the various plasma based reactors used in microcircuit fabrication and then to investigate a number of microwave plasma based processes.

This chapter is divided into sections which describe the individual reactors used in microwave based plasma processing, including a brief overview of the characteristics of the various systems, and review some of the processes and technological applications of microwave plasmas.

15.2. Plasma deposition and etching systems

There are a wide variety of plasma processing systems in use today, with an extensive range of applications. To understand their use, their characteristics and the processes they induce must be considered.

One classification of plasma processing systems concerns the proximity of the plasma to the substrate. The distinction is made between having the plasma in direct contact with the part being processed, or being located remotely such that only the plasma effluent impinges upon the part. In the latter case, the process is dominated by chemical reactions between excited atomic and molecular species created in the plasma with the part being processed. In contrast, when an object being processed is placed directly into a plasma, it is bombarded with charged and chemically excited species. Such a physical interaction between the plasma and the substrate can modify the resulting chemical reactions, as will be shown below. Many real plasma systems, especially those driven by microwave excitation, represent intermediates between *direct* and *remote* plasmas.

Another classification of processing systems is based on whether the electrodes responsible for the excitation of the plasma are contained within the process chamber or are external to the chamber. This characteristic is important in the case of RF driven systems with internal electrodes which develop a large DC bias (with respect to the plasma) on the electrode. The latter is ultimately responsible for bombardment of the electrodes with directed high energy ions. It is this factor which is primarily responsible for the utility of Reactive Ion Etching (RIE) RF systems, which account for the largest proportion of plasma production equipment in the semiconductor industry.

The classification of a particular system can depend upon the process parameters being used. A system in which the plasma is generated remotely may behave as a direct plasma system when operated at sufficiently low pressure that the plasma extends into the processing chamber and into contact with the part being processed.

Most of these systems belong to one of the following categories

- Barrel systems
- Parallel plate systems
- Downstream plasma systems
- Magnetically enhanced plasma systems
- Ion guns
- Combinations of the above

These systems all work on the same principle that a substrate is placed in a controlled gas atmosphere where a plasma generated in the gas reacts with the substrate. This results in either material removal from the substrate, surface treatment of the substrate or material deposition on the substrate. Schematically, we may write

Substrate + Plasma → Gaseous product (material removed)

or

Substrate + Plasma → Gaseous product (material deposited).

Thus, the basic components of a plasma system include a vacuum chamber, a gas source to provide feedstock for the plasma, an excitation source to create the plasma and a pump to remove the effluent.

15.2.1. Barrel and parallel plate systems. Barrel systems are constructed of quartz (fused silica) or glass cylinders with low dielectric losses, whose diameter can range from a few inches to several feet. A typical system is shown in Fig. 15.1.

The most common RF version uses an external coil electrode which is wrapped around the large diameter of the chamber and transfers energy to a plasma inside the

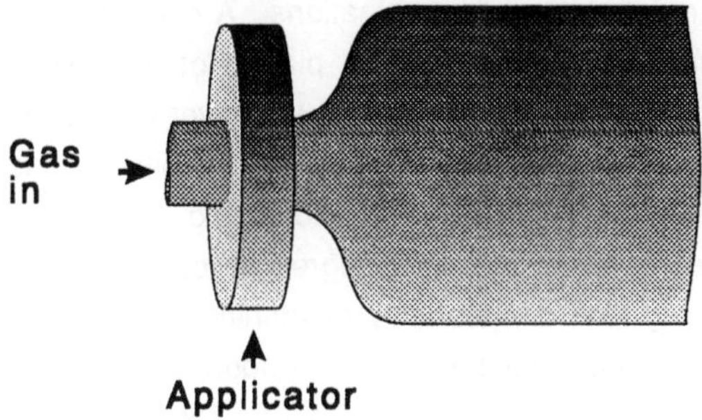

Fig. 15.1. Schematic of the barrel plasma system showing the microwave applicator position.

chamber. Microwave based systems use a remote plasma to generate reactive species which are subsequently pumped into the barrel shaped reaction chamber. Samples to be etched are oriented vertically in boats at some pre-determined spacing and placed in the chamber. This allows a large number of samples to be processed simultaneously. In some instances, RF electrodes are used in various arrangements within the reactor chamber when the chamber body is constructed of metal rather than a dielectric material.

In both the RF and microwave plasma based systems, the flow and diffusion of reactive species can determine the uniformity of processing within the chamber. As a consequence, the uniformity of etching or deposition of a batch of substrates in such systems is poor and these systems are rarely used for exacting work, where high dimensional tolerances need to be maintained. However, where a highly selective process is used, such as the removal of photoresist in the presence of ceramics, these systems can be useful and relatively inexpensive. Nonetheless, it is not unusual to see these systems in use in manufacturing situations, processing only one part at a time. This is due to the discovery that when fully loaded, the system produces unacceptable processing uniformities.

Of all of the different systems that are in use, RF parallel plate reactors are the most common. One example of this type of system is commonly referred to as a Reinberg plasma system, after R. Reinberg of Texas Instruments, who was the first to develop it.[13] As the name implies, the *parallel plate systems* consist of two (normally equal area) plates spaced apart by several centimeters. These systems can be operated in different modes where one or both of these electrodes can be connected to a source of RF. In certain cases, a third electrode can be added, usually placed between the first two plates. Often this third electrode is cylindrical in geometry and excited at a lower (< 1 MHz) frequency. Many different names are ascribed to the modes of operation or excitation of the various electrode combinations. A commonly used term for the case where the part undergoing processing is placed on the RF powered electrode is *Reactive Ion Etching* (RIE); in general, this is the lower electrode for the simple reason that the part can lie prone with no fixturing. The DC bias on this electrode is derived from the difference in the mobilities of the ions and the electrons in a plasma (see Sec. 13.4) and leads to a considerable degree of ion bombardment, which can also improve the uniformity and directionality of etching, particularly in cases where ion bombardment is a rate determining step. A frequently used appelation for the case where the top electrode is powered and the part remains on the lower electrode is *Plasma Etching* which is sufficiently used in other cases as to render the term confusing.

The parallel plate arrangement is the most widely used plasma processing system in the microelectronics industry, because of its versatility, relative simplicity and ease of fabrication from commonly obtainable metal stock parts.

15.2.2. Downstream and pancake reactors

Downstream plasma systems. The systems are based on the principle that a gas passing through a discharge region is excited, and produces chemically reactive species. These species may survive long after the gas has passed through the discharge, and may react with a part being processed at a remote location. In many instances, it is difficult to determine whether the last vestiges of the plasma affect the process and thus whether electrically charged species are involved in the subsequent interaction. For example, a microwave oxygen plasma created in a pancake reactor with a top mounted microwave plasma applicator, such as that mentioned below, will fill a 50 cm diameter by 18 cm high cylinder at 10 mtorr, whereas the same deposited power at 1 torr localises the plasma to a small volume in the applicator, while producing a low but measurable ion current at the substrate. Microwave based downstream plasma systems which attempt to separate the plasma and the substrate are generally used for similar purposes as the barrel systems described above, namely selective yet non-uniform etching or deposition systems. Particular examples of these systems include the previously described barrel etching systems, photoresist strippers[14] and SiO_2 etching systems.

In many of these systems, a key parameter affecting uniformity is the ability to distribute the plasma effluent responsible for the etching or deposition step around the system uniformly.

Pancake reactors. The basic premise of the pancake reactors is that placement of substrates in a horizontal plane allows a higher degree of process uniformity, simply because fixturing which retains the substrates is not required. Fixturing which protrudes from the plane of the substrate can interfere with etching and deposition processes, and many of the reactor electrodes in these and parallel plate systems are designed to keep the substrate front surface coplanar with the electrode surface. This has empirically been shown to result in the highest process uniformity in these reactors. Most of them are circularly symmetrical, with a pumping port which is placed on the axis of the system. Many implementations of the pancake system couple the remote microwave plasma along with an RF biased part platen as shown in Fig. 15.2. This enables independent control of the plasma density (via the microwave plasma) and the energy with which ions strike the part being processed (via the RF electrode). The addition of an RF biased electrode to a pancake system also provides anisotropy to many processes. The fact that the microwave systems make use of the pancake geometry is an outgrowth of the popularity of parallel plate RF systems which have been used for at least 20 years.

15.2.3. ECR and magnetic multipole systems.

Two examples of the use of magnetic enhancement of processing plasmas are ECR (Electron Cyclotron Resonance) and multipolar magnetic confinement systems. A combination of these two

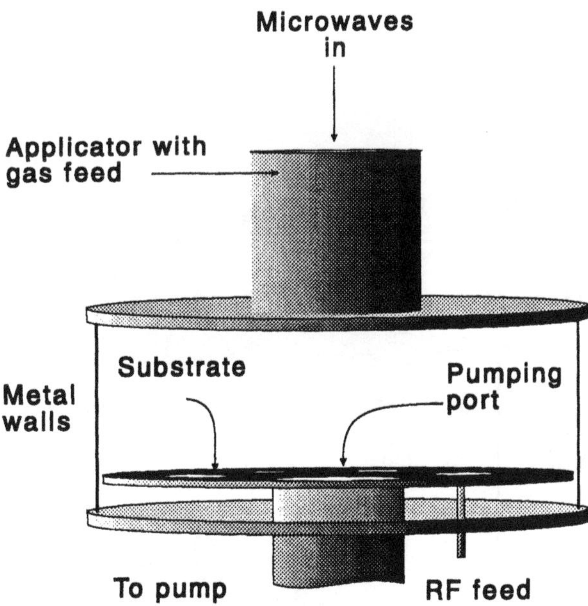

Fig. 15.2. Microwave plasma system based on a pancake reactor. The RF feed provides substrate biassing when required.

results in the DECR (Distributed Electron Cyclotron Resonance[15]) system, which is dealt with in some detail in Chap. 14.

An example of a plasma system based on ECR system is presented in Fig. 15.3. This shows a microwave plasma source attached to the top of a chamber and a set of electromagnets placed around the plasma vessel. This kind of system is used to generate a so-called electron-cyclotron resonance plasma which is generally denser and created at a lower pressure than the popular parallel plate RF plasmas. In some instances, a second set of magnets is placed in the vicinity of the substrate with a view of controlling the divergence of the plasma within the system. The principle of ECR power transfer is described in detail in Chaps. 6 and 8 while ECR plasma sources are discussed in Chap. 7.

For a 10 eV electron, whose translational velocity is about 1.9×10^6 m/s, the radius of cyclotronic motion is around 120 μm for an 875 gauss magnetic field. Once the external oscillating field is applied at the corresponding Larmor frequency (2.45 GHz in the present case) to generate a plasma, energy is absorbed rapidly by the electron, whose radius of motion increases. Since the electron describes a helical path in the presence of the magnetic field, a larger volume of the plasma is swept out than would be the case if the electron were moving along a linear trajectory. This increases the probability that the electron will interact with other species, the majority of which are

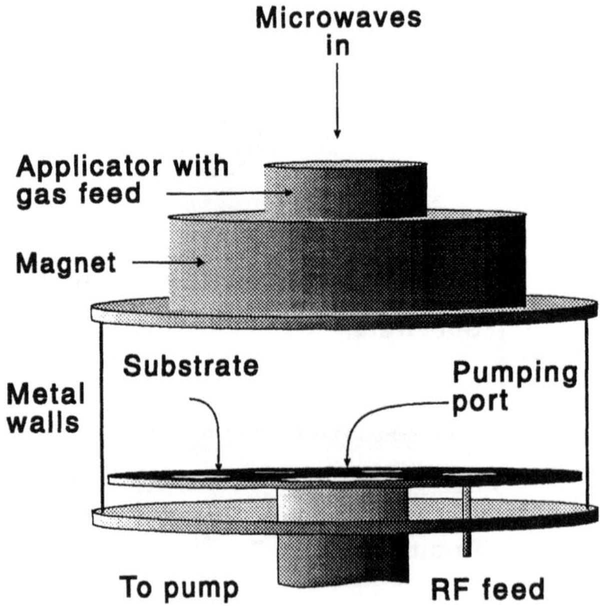

Fig. 15.3. ECR plasma system based on pancake reactor configuration.

neutral atoms at low degrees of ionisation of the plasma. The increase in the ionisation efficacy allows plasmas to be sustained at lower pressures than would be the case without the externally applied magnetic field. The magnetic field also confines the motion of the electrons to these helical paths, reducing their drift to the walls of the plasma chamber, which reduces their loss (see Secs. 6.2 and 8.2.2). Typically, the density of neutral species at 1 mtorr is about 3.5 x 10^{13}/cm^3, while plasma densities at 2.45 GHz may approach 10^{12}/cm^3. This results in a plasma which is much denser that would be the case without this resonance.

Another variant on these ECR microwave plasma systems, shown in Fig. 15.4 for comparison with the typical ECR reactor, is the distributed magnetic multipole (or DECR) plasma system, in which antennae are spread around the plasma chamber and attached to a source of microwaves. The magnetic field which is provided by a series of axially parallel magnets placed around the circumference of the chamber, serves to achieve ECR at the periphery and also reduces plasma diffusion loss to the chamber walls. A few centimeters from the chamber wall, in the vicinity of the substrate, the plasma is free of electric and magnetic fields. This type of system is claimed to have excellent processing uniformity[15] which results from a variety of factors discussed in Chap. 9.

The dominant characteristic of all these systems is their circular symmetry, and the fact that every attempt to provide a uniform gas feed and a homogeneous plasma generation is made to avoid non-uniformities. As a rule of thumb, if the rate determining step of a process is the interaction of the substrate with a neutral specie generated in

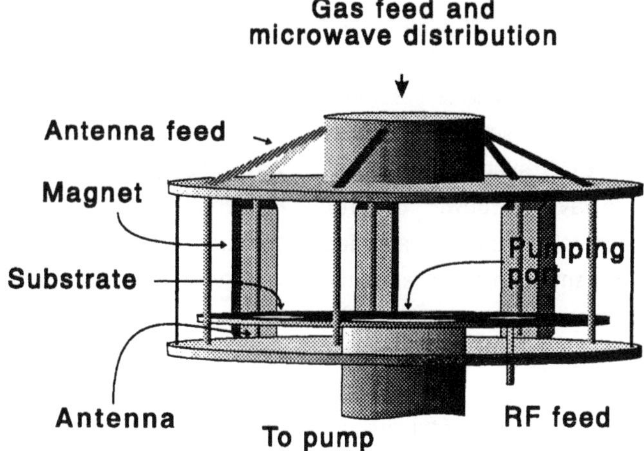

Fig. 15.4. Multipolar DECR plasma system showing magnets and plasma excitation antennae, each antenna is associated with a magnet (not all shown).

the plasma such as an atom or molecule, geometrical effects (non-uniformities) may be observed. The circular symmetry of most reactors generally ensures that these non-uniformities also possess a circular symmetry.

15.2.4. Other microwave based plasma processing systems.

Although most microwave based plasma systems use cavity or other kinds of resonant applicators, yielding localised active zone discharges (Chap. 4), there are also systems using travelling wave applicators, which allow them to couple to larger areas. An example of such a device is the so-called large microwave plasma (LMP) system,[16] where a periodic ladder type of structure is used to couple a wave into a gas volume through a dielectric, such as glass or quartz. Other examples of travelling wave based devices include the "leaky wave structures",[17] travelling wave structures based on an interdigitated arrangement[18] which have been offered commercially for some time, helical resonator,[19] and surface wave produced plasmas, which are dealt with in Chap. 5.

15.3. Materials for use in microwave based plasma systems

Many materials which were useful and could be applied simply in plasmas at 13.56 MHz require careful consideration when used in reactors at microwave frequencies. Dielectrics, such as borosilicate glass, which were slightly lossy at 13.56 MHz, dissipate sufficient energy at 2.45 GHz to generate localised heating and can lead to catastrophic failure.[20] A simple copper or aluminium mesh, used as a ground plane or shield at 13.56 MHz, can become an efficient antenna at microwave frequencies, unless a good return path is provided. For this and other reasons, great care has to be used to ensure that geometrical discontinuities are kept to a minimum, and that the plasma generating apparatus is properly shielded.

In many cases, the plasma is created inside a dielectric vessel such as quartz or pyrex, and its effluent transmitted to a chamber as previously described. An alternative, often adopted in ECR systems, is to use a waveguide connected to the chamber through a microwave window. Unfortunately, under certain conditions, mainly governed by pressure and the care afforded to minimise standing waves at the window, a high intensity plasma can be created in the direct vicinity of the window. Since this window can consist of slots cut in a circular metal plate containing an insulating dielectric used as a vacuum seal between the low pressure of the chamber and the atmospheric pressure of the waveguide, great care must be taken to maintain its integrity. Some embodiments of this vacuum seal use a full quartz window.[21] In some other instances, a short metal rod attached to a feedthrough is used to connect the microwaves between the waveguide and the processing chamber. This acts as an antenna inside the chamber, whose power has to be redistributed to the plasma.

There is much to be said for plasma systems which are designed to be well matched to the microwave source and which use good design to surmount problems of heating caused by impedance mismatches. In many cases, some of these problems are obviated in poor designs by the expedient of providing water cooling to these critical parts and a high power source to overcome poor microwave matching at these points.

15.4. Microwave field applicators

Like many first time users of microwave equipment, the first applicators which we used were simple Evenson[22] cavities, applicators which were designed to couple a maximum of 100 watts or so into a 1/2 inch (12.7 mm) diameter quartz tube. These cavities, which are typically used for atomic emission analysis, can provide excellent coupling at these low powers, but are prone to arcing, external microwave field generation and substantial heating under conditions useful for large area (> 500 cm^2) etching or deposition. Using an all quartz system, similar to the barrel equipment described earlier, with a maximum diameter of 10 cm, and a neck diameter of 2.5 cm (Fig. 15.1), we were able to etch polyimides at rates above 10 µm/min, using O_2/CF_4 plasmas.

Although this system was adequate for etching demonstrations, a much larger system was required for processing real substrates which have diameters as large as 20 cm. Thus, the system shown in Fig. 15.2 was used. The chamber is made from stainless steel and has a diameter of 53 cm. Unlike the figure, the pumping port was placed perpendicular to the axis of the chamber. An automated throttle valve was employed to control pressure, which was measured with a capacitance manometer in the chamber. Microwave applicators were attached to the chamber as shown. They included

- Surfaguide-type waveguide applicators (Fig. 5.13)
- Cavity applicators[23]
- Surfatron applicators (Fig. 5.13)
- Surfacan applicator (Fig. 15.5)
- Travelling wave applicators.[16]

Fused silica containment vessels with different geometrical shapes were also constructed. Figure 15.6 shows a quartz plate attached to a tube which terminated in a

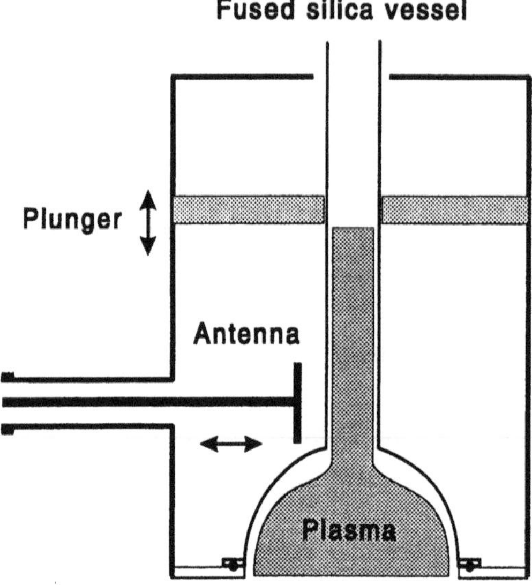

Fig. 15.5. Schematic of surfacan applicator showing the microwave power input via the capacitive coupler (antenna) and the movable plunger used for tuning.

cone. A similar arrangement was also constructed using a flat plate termination as shown in the same figure.

15.5. Pancake microwave plasma reactors and plasma processing

15.5.1. Polymer etching. Some of the earliest results of the use of microwave plasma based reactors are summarised in a review by Musil[24] which describes many processes which employ microwave plasmas and focuses on the technological aspects

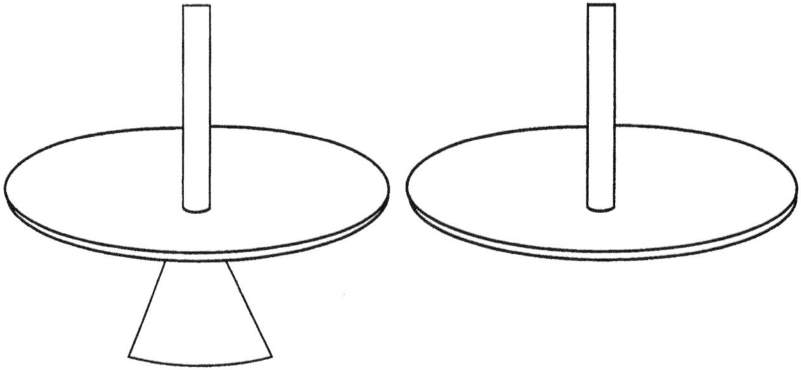

Fig. 15.6. Schematic of applicator quartz termination: (left) cone; (right) flat plate.

of these processes. Suhr[25] reviewed the reaction of a variety of organic and inorganic materials in plasmas and provided a chemist's view of plasma processing. The first data concerning the etching of polymers by microwave plasmas was described by Shibagaki et al.[26] who also appear to be the first researchers using a microwave plasma for the processing of materials in semiconductor manufacturing. Subsequently groups headed by Jinno[27] and Dzioba[12] began their investigations into the etching of polymers in microwave plasmas. In the latter work, Dzioba et al.[12] demonstrated the etching of polymers at high rates approaching 20 μm/min. It was believed at that time that oxygen atoms were responsible for the high rate of removal.

Francis[28] studied the generation of oxygen atom in a microwave O_2 plasma contained in a quartz tube using Electron Paramagnetic Resonance (EPR). It was found that oxygen atom decay was first order in atom concentration and occurred via a homogeneous recombination reaction. Recombination on the quartz tube surface was found to be important only at low gas velocities (6-60 cm/s). As pressure decreased from approximately 9 torr to 1 torr, the rate of decay of atom concentration with distance from the plasma decreased. The primary pathway for oxygen atom removal was found to be due to the following sequence of reactions

$O_2 + O + O \rightarrow O_2 + O_2$ Direct recombination
$O + O_2 + O_2 \rightarrow O_3 + O_2$ Ozone process

followed by

$O_3 + O \rightarrow 2\, O_2.$

This mechanism dominates under conditions where the surface recombination coefficient is low. Greaves and Linnett[29] showed that the surface recombination coefficient of oxygen atoms is highly dependent on the nature of the material. The most active surfaces were metallic oxides, whereas the least active was pyrex. These investigations showed the importance of the material constituting the reactor surfaces in the manufacture of microwave plasma based reactors. Brown[30] investigated the effect of impurities upon the production of oxygen atoms in microwave discharges and found that just 1% of added hydrogen, water or nitrogen could increase the yield of oxygen atoms by a factor of 9, 8 and 5 respectively. A chain mechanism was suggested to be responsible for this effect, although no details were provided.

Subsequent EPR determination of oxygen atom yield in these plasmas by Cook and Benson[31] showed a substantial concentration of oxygen atoms but despite their presence, there was no apparent etching of photoresist placed downstream of the plasma. It turned out that thermal activation of the process was required and an activation energy for etching photoresist of 0.5 eV was determined. Our own determinations of the activation energy of polyimide etching in oxygen microwave plasmas with added CF_4 showed that it was highly dependent upon the halocarbon content.[32] Robinson and Shivashankar[33] also determined the energy of activation of this process. Thus the rapid etching of photoresist noted by Dzioba[12] in pure oxygen was probably due to either substrate heating or fluorine released from the quartzware which had previously been cleaned in aqueous HF. In either case, the etch rate of polymers is significantly increased.

Bell and Kwong[34] developed a model of the kinetics of dissociation of oxygen in a microwave discharge and found that the efficiency of atomic oxygen production depended upon the reactor geometry and upon the applicator design. Brake et al.[35,36] studied the dissociation and recombination of oxygen atoms created in a microwave discharge using NO_2 titration. They found that the degree of dissociation of the oxygen plasma increased with decreasing pressure. They also found that the oxygen atom concentration decreased with distance from the source of the plasma. A kinetic model of this plasma produced results which closely followed the experimental observations and provided initial estimates of the gas temperature and resultant gas velocities.

More recently, Cvetanovic[37] compiled a list of kinetic data relating the reaction of ground state $O(^3P)$ atoms with unsaturated hydrocarbons with a view to creating a comprehensive database of reaction kinetic data for use in computer models. Commercial applications of these oxygen plasmas have included downstream plasma strippers where thermal activation of the etching process has been used (at around 200°C),[38] thus providing extremely high selectivities for polymers etching relative to most inorganic materials.

Effect of chamber geometry and plasma applicator upon the etching of polyimide. Battey[39] studied the stripping (etching) of photoresist in barrel type systems and found that reactant diffusion effects due to system geometry dominated the uniformity of etching. Nagy[40] performed similar measurements on the etch rate of polysilicon and SiO_2 in a geometrically similar reactor. However, the system used in his experiments applied RF directly to a planar electrode placed parallel to the long axis of the cylindrical chamber. He found that the uniformity of etching was affected by local reactant gradients caused by discontinuities of materials, and geometrical discontinuities which produced localised field "hot spots".

Our own investigations initially centred around the etching of polymers in oxygen plasmas in the system previously described in Fig. 15.2. They were intended to determine the relationship of the plasma dimensions, as denoted by the luminous region, to such characteristics as etch rate and etch rate uniformity of polymers. Silicon wafers, 82 mm in diameter, were coated with DuPont (PMDA-ODA) polymide and cured according to the manufacturer's specification. The wafer was then bonded to the graphite covered electrode using colloidal graphite and subjected to a microwave plasma, while monitoring the etching rate in the centre of the wafer with a laser interferometer. Etching gases included O_2 with controlled addition of CF_4 or SF_6. The thickness of the polyimide was measured across the wafer before and after etching in locations which were 2.54 mm apart. The difference between the two measurements was determined and the percentage etch rate variation[41] was then defined as $100\ (\mathcal{E}_c - \mathcal{E}_e)/\mathcal{E}_c$ where $\mathcal{E}_c$ is the etch rate in the centre of the wafer and $\mathcal{E}_e$ is the etch rate at a radius of 36 mm. Several important observations were obtained from this work: i) the most striking and counter intuitive was that the area of plasma, whether generated inside a 50 mm diameter tube or 25 mm diameter tube in the applicator zone (Fig. 15.2), did not affect the etching uniformity in an O_2 or O_2 / additive gas plasma; ii) two different frequencies of plasma excitation (2450 MHz and 915 MHz) were used with various applicators and although differences in the etch uniformities and etch rates were observed, no clear relationship between the plasma excitation frequency, the etch uniformity, and the plasma containment vessel could be established; iii) for most of the different combinations of applicators, excitation frequency, pressure and containment vessel, a minimum in the etch rate variation across the wafer occurred at a specific addition of the CF_4 or SF_6. A typical example is presented in Fig. 15.7 where the etch rate in the centre of the wafer at 200 mtorr is shown together with the etch rate variation as a function of CF_4 addition. The particular applicator used was of the surfaguide type through which a 25.4 mm quartz tube passed, attached to a flat plate (Fig. 15.6); iv) although the maximum etch rate coincided with the minimum etch rate variation (see Fig. 15.7 and Table 15.1) of polyimide when CF_4 was added to the oxygen plasma, this was not the case for the SF_6 additive.

Although we were looking for a simple trend in the etch rate variation as a function of the size of the generated plasma, we were unable to see any clear indication of this phenomenon. We also observed an almost exponential decrease of the etch rate at the centre of the wafer with distance from the applicator exit, under a variety of conditions as shown in Fig. 15.8. This indicates that the *rate of decrease* of the polyimide etch rate appears to be dependent more on the gas pressure than on the gas flow rate. This was explained on the basis of more rapid gas phase recombination reactions of oxygen and fluorine atoms at the higher pressure.[41]

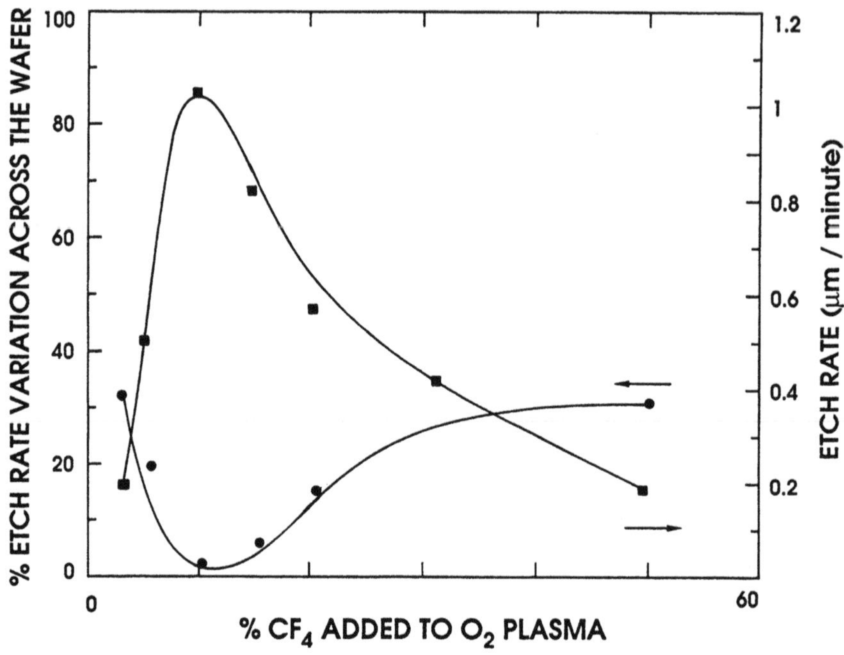

Fig. 15.7. Etch rate and uniformity of polyimide in an O$_2$/CF$_4$ plasma produced in a 25.4 mm diameter tube passing through a WR-284 surfaguide-type applicator and terminating in a flate plate at 200 mtorr (ref. 41).

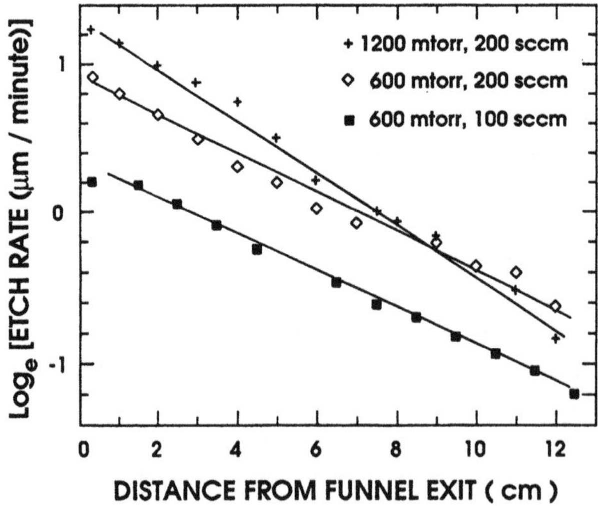

Fig. 15.8. Effect of plasma-wafer separation upon the etch rate of polyimide with 10% CF$_4$ in an O$_2$/CF$_4$ plasma. A surfaguide applicator at 2450 MHz was used. Various parameters were changed to show their effect upon polymer etch rate at the centre of the wafer.

Table 15.1. A comparison of the etching characteristics of polyimide using various applicators, excitation frequencies and percentages of fluorinated gases in various O_2/gas mixtures. The added gas used is CF_4, except where marked.[41]

Device	Pressure (mtorr)	Incident power (watts)	Maximum etch rate (μm/min)	Halogen at maximum etch rate (%)	Minimum etch rate variation (%)	Halogen at minimum variation (%)
25.4 mm tube terminated in a flat plate. WR-284 surfaguide at 2450 MHz	200	450	1.1	12	2	10
	400		2.5	18	18	25
	800		1.2	13	18	13
50 mm tube terminated in a flat plate. WR-340 surfaguide at 2450 MHz	200	1000	0.5	12.5	22	18
	400		0.6	11	35	40
Surfacan at 2450 MHz	200	400	0.55	11	7	30
	400		0.81	10	15	10
	800		1.1	17	22	24
Surfacan at 915 MHz	200	300	0.42	18	8	10
	400		0.7	18	8	20
	800		0.9	25	5	45
Surfatron at 915 MHz. Exit in 100 mm cone	200	300	0.37	16	7	26
	400		0.66	17	10	15-20
	800		0.63	10	25	7
Surfacan at 915 MHz. SF_6 added	400	300	0.57	5	12	16
Surfaguide at 2450 MHz. Plasma exit from 100 mm cone	200	800	0.68	10	15	10
	400		1.0	15	8	20
	800		1.5	20	12	35

Figure 15.9 shows that the etch rate variation of polyimide decreased with distance from the applicator and beyond a distance of approximately 8 cm from the exit of the funnel, etched more rapidly on the outside edge of the wafer than in the centre. We also noticed that it was possible to obtain extremely high etch rates in the centre of the wafer if very high gas flow rates were used. Indeed, when using flow rates of several standard litres/min, it was possible to etch polyimide at 5 to 10 µm/min. Other authors have also seen similar effects,[33,42] and also show that the uniformity of etching degrades. These results were explained on the basis[41] of a difference between the diffusion rates of reactants and the velocity of the gas plug within the plasma tube and that within the much larger chamber.

The empirical data from our set of experiments is summarised in Table 15.1, which shows the effect of applicator type, excitation frequency and gas composition upon the etch rate and etch rate uniformity of polyimide.

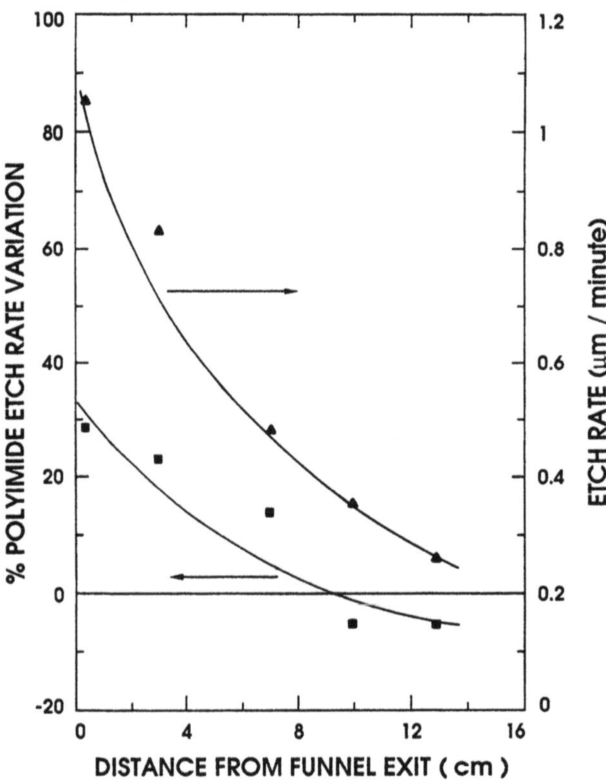

Fig. 15.9. Effect of plasma-wafer separation upon etch rate and etch rate uniformity of polyimide, measured at 600 mtorr, 800 W and 10% CF_4 in an O_2/CF_4 plasma. A surfaguide applicator at 2450 MHz terminated in a 100 mm diameter cone was used.

For a complete analysis of the factors affecting etching uniformity, the identity of the reactants must be established, and their precise distribution within the reactor must be determined. This depends upon the gas flow dynamics, the gas phase reaction of various species, the reactivity of the etched material, and various heterogeneous recombination processes and not just the volumetric flow of a gas plug. Furthermore, kinetics of reaction of these gas phase species with the polyimide must be ascertained along with the electrical effects of the plasma and the effects produced by this plasma which include energetic bombardment by a variety of ions and photons.

This set of experiments into the geometrical limitations of microwave plasma based etching systems suggested that, for downstream systems, the primary factors affecting the etch rate and etch uniformity were geometrical and hydrodynamical in nature. However, these results were complicated by the recombination of reactants, believed to be fluorine and oxygen atoms. This recombination via homogeneous and heterogeneous processes can affect the uniformity and etch rate of the polyimide.

The high polymer etch rates which we had initially seen in our first small reactor were not achievable in this larger system. The smaller reactor had much higher gas flow velocities within the chamber, and more importantly, heated the substrate albeit in an uncontrolled manner, resulting in the high etching rates of polyimide. Subsequent measurements showed that the etch rate of polyimide[43,44] in oxygen microwave plasmas was dependent upon ion bombardment energy and plasma density via the externally controlled parameters of pressure and absorbed microwave power. Other authors have also studied the interaction of oxygen plasmas with polymers, with a particular view[32,45,46] of determining the profiles obtained in organic films after their etching with an in situ mask.

Although it is not the intent of this chapter to dwell upon the specific issues involved in the etching of polyimide in oxygen based plasmas,[47-59] the experiments described above do serve to demonstrate effects which are common to many of the microwave induced plasma etching systems in which the plasma effluent is directed towards a target. These effects include variation of the etch rate with position from the plasma source, location on the substrate, gas composition, temperature and excitation frequency.

Directionality of etching in microwave plasmas. For all of the experiments involving the etching of polyimide in the various reactor configurations described above, the directionality of the etching process is unity, where directionality is defined as the ratio of the vertical to lateral etch rates. This is similar to a case where a material is etched in a chemical solution and there is no external electric field imposed to drive etching in any particular direction with regard to the masking layer.

A profile of polyimide etched in an O_2 microwave plasma with no external field applied to the substrate is shown in Fig. 15.10a together with that etched in the same plasma at three values of the RF field applied to the planar electrode of the system shown in Fig. 15.2. The fact that the imposition of an RF field to the electrode changes the etching directionality is a key issue in most of the downstream or microwave based reactors. Without this external RF bias, most of the microwave plasma based systems cannot provide a completely directional etching process. This issue is discussed in detail in ref. 45 and stems from the difference in the relative mobility of electrons and ions. Initially, electrons which leave the edge of the plasma and are accelerated across the plasma sheath, strike a target such as an electrode connected to the source of RF. A charge imbalance results as a consequence of the difference in the mobility of these species and leads to the formation of a negative bias on the electrode. This bias can be exacerbated by a difference in the grounded and driven electrode areas. Similarly, a capacitor used in series with this electrode as a tuning element results in storage of charge and the development of negative potentials on the electrode (see Secs. 13.4.1 and 13.4.3). This negative potential which is built up on the electrode is limited by a balance between the ions arriving at the electrodes and those electrons with sufficient energy to surmount the potential barrier. The electrode potential also varies with excitation frequency and is dependent upon the impedances involved in the electrical circuit.[60-62] The net result is that ions bombard the electrode with energies that are dependent upon their exit energies from the plasma ($\simeq$ plasma potential) and the potential at the electrode which attracts them.

As mentioned above, the use of a controllable RF bias is very important for the generation of anisotropic profiles and is often employed in situations where the energies from ions leaving the plasma are lower than a critical value required to stimulate etching by ion assisted physical or chemical processes.[63-65]

Although the kinetics and mechanism of silicon etching in CF_4 plasmas are very different from those of polymer etching in oxygen plasmas, the imposition of an RF bias to the planar electrode, while maintaining a constant microwave plasma above, has similar effects upon the etched profiles of silicon and polyimide. In both cases, the directionality of etching changes as shown in Fig. 15.10a for polyimide etching and Fig. 15.11 for silicon etching.

Improving selectivity with dual frequency systems. Another advantage of using an independent RF bias with microwave plasmas is the increase of selectivity obtainable

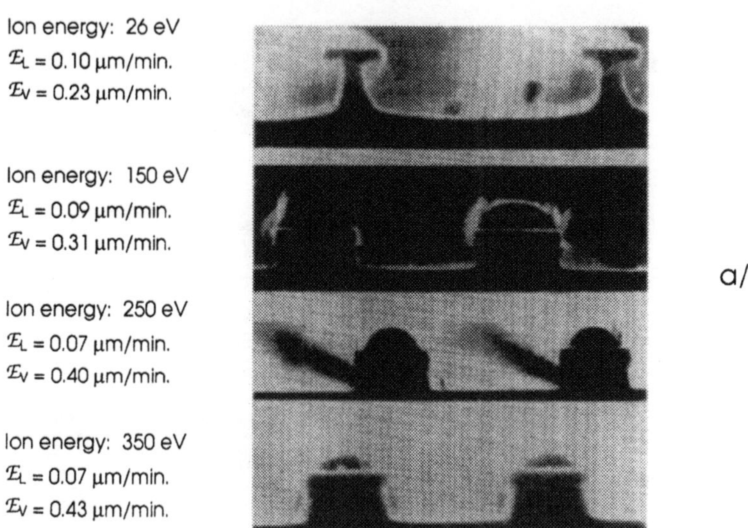

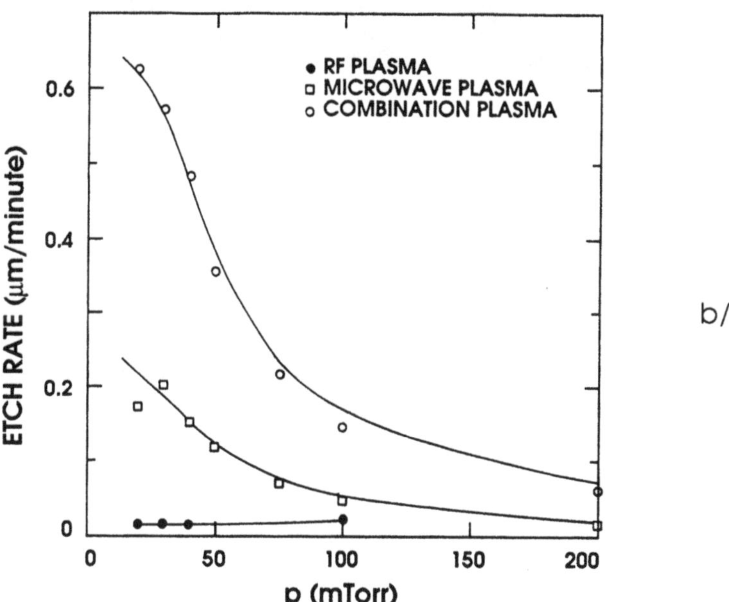

Fig. 15.10. (a) Scanning electron micrographs showing effect of application of RF at 13.56 MHz to an oxygen microwave plasma etched polyimide layer. Lateral ($\mathcal{E}_L$) and vertical ($\mathcal{E}_V$) etch rates are indicated with respect to the ion energy resulting from RF biassing. (b) Variation of the etch rate of polyimide as a function of gas pressure for a microwave plasma at 915 MHz alone (substrate at floating potential), for RF alone applied to the substrate (no microwave plasma), and for the combination of microwave plasma and RF biassing.[44]

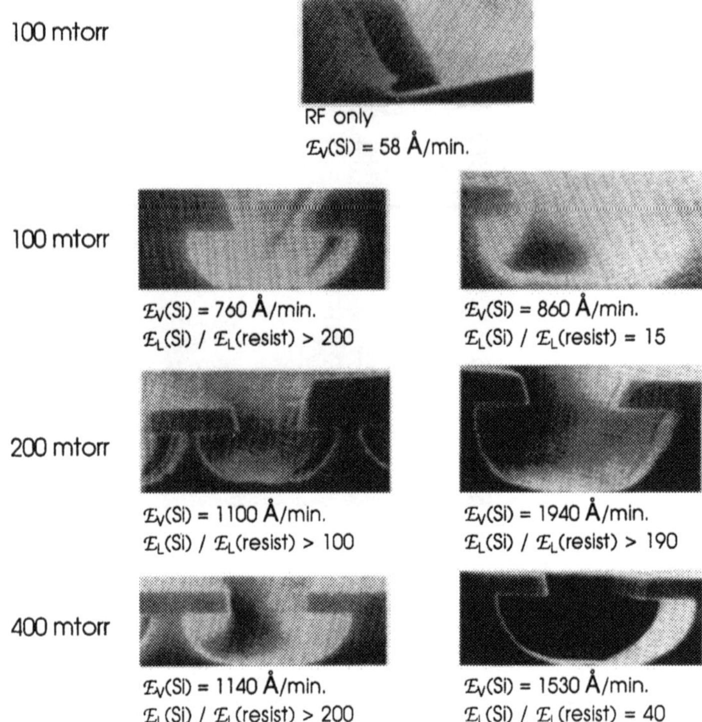

Fig. 15.11. Scanning electron micrographs of etched silicon using a photoresist mask, showing effect of application of RF bias at 13.56 MHz to a CF_4 microwave plasma at 915 MHz upon directionality of etching.[41] Photos on the left side show results obtained with microwave only while those on the right side are for RF + microwave.

over RF parallel plate plasmas. In the latter case, chemical selectivity is reduced by ion bombardment which removes material indiscriminately. Indeed, DC voltages measured on RF parallel electrodes are often above several hundred volts which results in very little selectivity between the etching of a mask and the exposed material. Thus, the ion bombardment stimulated etching mechanism is dominant under these conditions. In contrast, the energy of ions impinging upon a substrate subjected to a microwave plasma is much lower, typically 10-20 eV, whereas the density of neutral and excited species in these plasmas is higher than in the RF case. Hence, the chemical reaction occurring between these species and the substrate dominates and the process is much more selective for the microwave plasma. Application of a small RF bias results in little difference in the plasma density, yet a large range of DC bias can be achieved and hence the energies of ions striking the surface can be controlled. This should result in an increased anisotropy for processes which depend on both a chemical and ion stimulated mechanism of etching. Hence, the combination of RF bias and microwave plasmas can result in increased selectivity over the RF plasma case and provides an improved anisotropy over the microwave plasma.

An example of this is shown in Fig. 15.11: (i) for the case where RF power alone was applied to the planar electrode at 100 mtorr, the etch rate of the silicon was 58 Å/min. The etch rate of the photoresist mask was 40 Å/min. When the microwave plasma alone was applied, the etch rate of silicon increased to 760 Å/min and the ratio of the etch rate of the silicon to that of the photoresist mask was greater than 200 (measured interferometrically). Application of RF bias to the system increased the etch rate of silicon to 860 Å/min, while decreasing the etch rate ratio of silicon to photo resist to 15. At this pressure, the anisotropy of silicon etching did not seem to change; (ii) at 200 mtorr, the etch rate of silicon rose to 1100 Å/min using the microwave plasma alone, while the etch rate ratio between silicon and photoresist was found to be above 100. Application of RF bias to the plasma increased the etch rate of silicon to 1940 Å/min in the vertical direction, while the lateral etch rate remained at 1100 Å/min. The silicon to photoresist etch rate ratio increased to above 190. Simultaneously, the directionality of etching changed due to the application of the RF bias; (iii) at 400 mtorr, the etch rate of silicon in a pure microwave CF_4 plasma did not change significantly from that observed at 200 mtorr. However, application of an RF bias produced a higher vertical etch rate of 1530 Å/min, while decreasing the etch rate ratio to 40 or so. The three cases examined show that the optimisation of etching anisotropy and selectivity depends on both the RF bias and gas pressure.

Heindenreich et al.[43] studied the etch rate of polyimide under various combinations of RF and microwave oxygen plasmas and concluded that the etching of polyimide in these systems was driven by an ion assisted process: the rate limiting step appeared to be the removal of etch product by ion bombardment from the polyimide surface. Figure 15.10b is taken from this work. It shows the large differences in the etch rate of polyimide using microwave oxygen plasmas with and without RF bias. Note that the etch rate of the combined RF and microwave plasmas far outweighs the algebraic sum of the two indivual etch rates (RF and microwave). They also found that for polyimide etching, using an RF biassed electrode at 13.56 MHz and plasma excitation of a surfacan at 915 MHz, the onset of directionality was below an average ion energy of 26 eV. Measurements of the Ar ion beam etching of SiO_2, Si^{66} and Cu^{67} show sputtering thresholds of around 25 eV, which is similar to the value obtained, much less rigorously for the ion stimulated etching of polyimide in oxygen plasmas mentioned above.

Frequency dependence of the etch rate of polyimide in O_2/CF_4 plasmas. Various attempts to characterise the effect of the plasma excitation frequency upon the results of a given plasma process[68,69] have proven the difficulty of doing so unequivocally. Although many electrical, physical and chemical parameters can be determined in a given plasma, they provide little insight into the result of the process. In part, this is due to the spatial variation of the parameters within the process chamber and to the difficulty of relating the measured parameters to the final outcome of the process. However, it is possible to investigate the dependence of a result upon the trend of a few select parameters. Most of the literature data which relates results such as etch rate to plasma density, or deposition rate to electron temperature, are examples of such work.

Rapeaux and Turban[70] studied the etching of a variety of polymers in a parallel plate based etching system in the frequency range of 15 kHz to 13.56 MHz in O_2/SF_6 and O_2/CF_4 plasmas. They found that for a 30% admixture of CF_4 in an oxygen plasma, there was a distinct difference of the frequency at which the etch rate of an epoxy polymer and polyimide reached a maximum: the maximum etch rate of a polyimide (kapton [R]) film occurred at around 0.5 MHz whereas it still had not reached its maximum at 13.56 MHz for an epoxy film. Similarly, they found that at a fixed frequency of 13.56 MHz, the etch rate maxima of polyimide, acrylic and epoxy films occurred at 20, 30 and 40% added CF_4 to a 0.2 torr O_2 plasma. This was later investigated by Cain et al.[56] who used molecular orbital calculations to explain the chemical bonding in these polymers and the resulting differences of etching in oxygen plasmas.

In an attempt to determine the effects of the plasma excitation frequency upon etch rate, Sauvé et al.[71] assembled equipment which would provide the same geometric constraints to the plasma and effluent, while enabling the excitation of the plasma over a wide frequency range from 13 MHz to 2450 MHz. The apparatus utilised surface wave launchers (see Sec. 5.6.2) to create plasmas at these different frequencies. The etching of polyimide in O_2/CF_4 plasmas as a function of gas composition and microwave power was used to probe the effect of the plasma excitation frequency. A simple laser interferometer allowed measurement of the polyimide etch rate while the plasma was applied. With surface wave plasmas, the minimum plasma density (density required to sustain the plasma) is that which occurs at the end of the plasma column when the plasma is operating in a travelling wave mode. When HF power to the launcher is increased so that the wave reaches the end of the plasma column, the wave is reflected. For the experiments described here, reflection occurs from a grounded metal plane and the power reaching the plane is either absorbed in the plasma during the return to the launcher, is radiated out from the apparatus or is dumped into a dummy load. If absorbed, the power increases the plasma density,† hence increasing

---

† Under travelling wave conditions, the plasma density decreases away from the launcher, generally at a constant axial rate. When reflection occurs at the end of the column, the power absorbed in the plasma during the return of the wave to the launcher reduces the initial axial gradient of electron density, which eventually disappears at large enough reflected powers.

the etch rate. Thus, for the data shown in Fig. 15.12, the absorbed power axis is normalised to the power required to just fill the volume[†] (no wave reflection).

Assuming that the attempts to cool the substrate by direct contact of the polyimide coated silicon wafer to the water cooled electrode were successful[72] and that irradiation of the polyimide surface by the luminescence from the plasma was unimportant, the results reproduced below in Fig. 15.12, show some very interesting characteristics. They include the observation of a very rapid variation of the etch rate with both concentration of added halogen and absorbed power at 2450 MHz and a significantly different behaviour as the frequency is decreased. The concentration of added CF_4 at which the etch rate peaks, remains approximately constant at all the absorbed powers at a fixed frequency and increases to higher values with decreasing frequency.

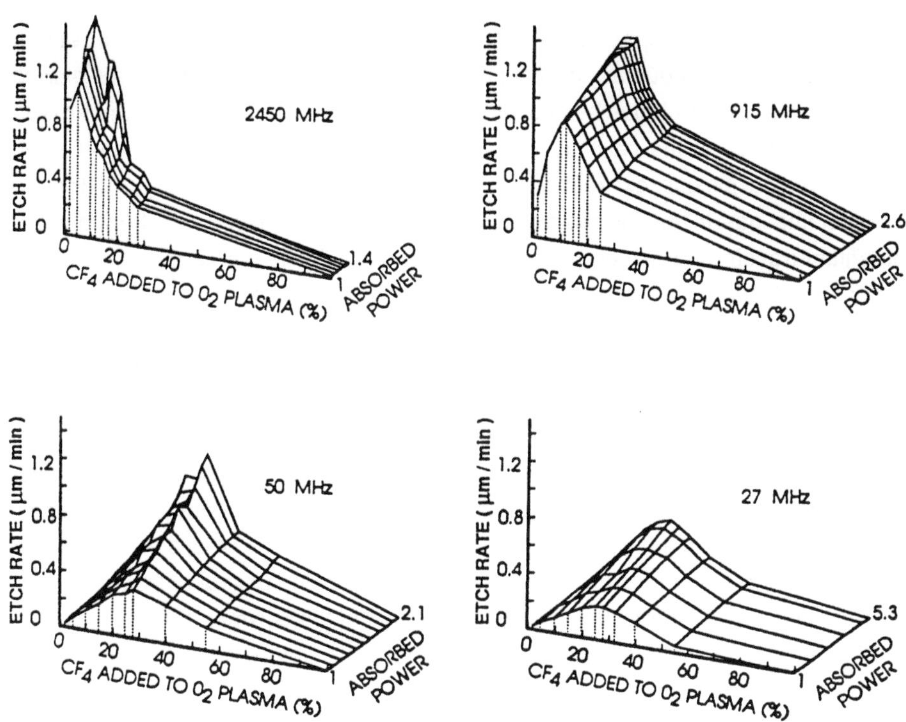

Fig. 15.12. Effect of gas composition, excitation frequency and microwave power upon the etch rate of polyimide in O_2/CF_4 plasmas.[71]

[†] In the present setup, these are 24 W (13.56 MHz), 21 W (27.12 MHz), 37 W (50.5 MHz and 200 MHz), 160 W (915 MHz) and 900 W (2450 MHz).

Simultaneously, the peak etch rate decreases, partially due to the fact that the absolute powers which are available at each frequency of operation are different. However, the correlation between the etch rate and CF_4 concentration are undeniably influenced by frequency. Later measurements[73] indicated that the variation of the atomic oxygen concentration in the plasma, as measured by actinometry, showed peaks in atomic oxygen yield which varied with excitation frequency. At 200 mtorr, 100 sccm O_2, the maximum atomic oxygen yield at an excitation frequency of 915 MHz, occurred at 12% added CF_4 and 20% CF_4 at 200 MHz. This corresponds well with the equivalent etch rates, whose maximum value occurred at 11% added CF_4 at 915 MHz and 22% (not shown) at 200 MHz. Although the correlation between the physical characteristics (e.g. electron energy distribution function) of the plasma and the etching process is not directly established by these experiments, the etch rate of polyimide does show a clear dependence on plasma excitation frequency.

It is interesting to speculate that the observed differences of the etching characteristics with frequency in the above experiment are related to the electron energy distribution, which has been shown[43] not to vary with absorbed power in oxygen plasmas. Indeed, as the calculations of Sec. 3.2 indicate, there should be a strong influence of the plasma excitation frequency upon the electron energy distribution whenever the gas in the discharge has an electron-neutral collision frequency for momentum transfer that varies with electron energy. This is the case in O_2 plasmas (Sec. 3.2.2). Variations of the electron distribution change the rate of production of fluorine atoms and oxygen atoms, both of which are known to effect the etching of polymers.[46-59] There appears to be a high correlation of the etching rate of polymers with the concentration of oxygen atoms in the generated plasma.[28,30,31,34-37,51-54] Some investigators[55-56] have used extended Huckel molecular orbital calculations[74] to determine the process of degradation of a range of polymers in oxygen based plasmas. From these calculations, they concluded that etching should proceed through a saturated intermediate which can only be formed via addition reactions in saturated polymers. Oxygen atom insertion into the backbone could subsequently degrade the backbone and cause fragmentation and volatilisation of the polymer. For polyimides in O_2/CF_4 plasmas, this occurs through saturation of the phenolic ring by fluorine atoms and insertion of the oxygen atom into the aliphatic backbone.[50] Thus, the removal of polymers with these plasmas should be affected by the polymeric structure, by the composition of the plasma effluent and by excitation of the polymer surface by thermal means or with energetic ions or photons.[75]

15.5.2. Diamond and diamond-like carbon coatings. Many organic and inorganic films have been deposited from microwave plasmas using a variety of reactors, and plasma systems. Microwave plasmas have been prominent in this field since they can provide high density plasmas which are capable of heating the substrate. The dense plasma obtained by microwave stimulation and a high (> 500°C) substrate temperature are essential for the formation of diamond films. The inertness, durability, thermal conductivity and chemical resistance of this material make it very attractive for many microelectronic applications. The main distinction between diamond and diamond-like carbon is that the latter is usually amorphous in structure. The term "diamond-like coating" seems to cover a wide range of material properties ranging from amorphous to microcrystalline which according to Aisenberg et al. (see Messier et al.[76], their reference 10) can contain anywhere from 0 to 25 atomic weight % hydrogen. As such, its electrical and physical properties may be substantially different from that of diamond but nevertheless harder than most other films. The hardness of this material, its good adhesion and the lower temperature requirement for its deposition make it particularly attractive for various protective coatings.

The quest for the transmutation of cheap starting materials such as marsh gas (methane) into diamond combines the romance of the search for the philosopher's stone with many practical applications. As such, this field has seen a variety of talents devoted to it including literary as well as scientific. Unfortunately, to date, it has been impossible to artificially produce diamonds of the calibre of the Jonker diamond, which had a weight of some 726 carats, thus removing some of the more romantic (and speculative) aspects of its fabrication from the popular realm.

The notion that diamonds could be made from carbon was first popularised by H.G. Wells in a short story,[77] which was borne out by work at the General Electric Company, giving credence to the notion that diamond could be synthetically produced.[78] Today, synthetic diamonds are produced in quantity for a wide variety of cutting, polishing, lapping and grinding processes, among others. A lucid review concerning diamond synthesis can be found in papers by Messier et al.[76] and by Zhu et al.[79]

Diamond films. In general, the use of plasmas for the deposition of diamond films has been driven by the fact that they contain active species not normally found in other environments. In particular, plasmas have been used which contain atomic hydrogen species, which "dissolve" or etch graphitic carbon formed during a typical deposition process, while leaving the diamond allotrope behind. However, high substrate temperatures (> 500°C) are required for this process.

Typically, a graphite susceptor[79] will be placed inside a region of high microwave fields, which both heat the susceptor and create a plasma based on some carbon containing gas such as methane or acetone with a hydrogen carrier. In general, silicon substrates upon which the film is deposited are pretreated by scratching with diamond powder to encourage growth. The main methods of determination of the structure of these diamond coatings include scanning electron microscopy, Raman spectroscopy, which is generally used to identify the diamond phase and the graphitic content of the film, and various electron and X-ray diffraction techniques for crystalline structure determination. Transmission electron energy loss spectroscopy (EELS) has also been used with some efficacy to distinguish between amorphous carbon, graphite and diamond. One of the underlying precepts in much of the literature concerning the deposition of diamond from the vapour phase is the removal of graphite from the depositing material using a variety of gases including water vapour, oxygen and hydrogen.

Matsumoto et al.[80] described the deposition of diamond particles from methane using thermal CVD from a mixture of methane and hydrogen and deposited on silicon and molybdenum substrates. They found that it was possible to grow microcrystallites of diamond which they surmised were strained, based on a shift from 1331 cm^{-1} to 1334.5 cm^{-1} of the diamond Raman peak, and its increased width. A wide variety of plasma techniques have been used to fabricate diamond from the gas phase. These have been described by several authors, including Whitmel and Williamson,[81] Holland and Ojha,[82] Sokolowski and Sokolowska[83] and Vora and Moravec.[84] Saito et al.[85] described the deposition of diamond films from a microwave plasma of methane and hydrogen in a silica tube 18 mm in diameter at 2 torr. Electron and X-ray analysis showed that the crystallinity of the diamond depended very strongly on the deposition temperature, while silicon carbide was found on the silica tube walls. Optical emission spectra showed that CH_2 and CH radicals were formed in the plasma together with excited hydrogen atoms. The authors stated, without experimental substantiation,† that graphite deposited from the methane was removed by hydrogen atoms. Subsequently, they[87] attempted synthesis of these diamond films from mixtures of methane, hydrogen and water vapour in a microwave plasma on silicon substrates. Presumably, this combination was chosen since acetone and hydrogen had been found to be very effective in producing diamond using a hot filament method[88], suggesting the investigation of the addition of oxygen via water vapour. Fom X-ray diffraction and Raman spectra, it appeared that the addition of water did indeed foster the growth of high purity diamond films. Addition of increasing amounts of methane also appeared to increase growth rate, while producing some amount of diamond-like carbon and graphite.

---

† Rosner has shown that atomic hydrogen and other radicals and even diatomic molecules could be used to attack graphite.[86]

A year later, the same authors extended this work with water vapour to include parametric data concerning film growth[89] and physical characteristics of the diamond films. They also showed that the rate of removal of diamond in a water/hydrogen plasma was much less than that of graphite (selectivity of etching).[†] A pure hydrogen plasma removed graphite faster than diamond, but not as fast as the water/hydrogen plasma. High growth rates (comparatively speaking) of 10 μm/h were achievable. Hardnesses of the diamond films approaching 6000 kg mm^{-2} were obtained (cf. natural diamond 10,000 kg mm^{-2}). This was a significant advance in the state of the art and supported the graphite removal hypothesis of diamond growth. Despite the possibility of the involvement of atomic carbon in diamond film growth, there has been little spectroscopic attention paid to it.[91]

Watanabe and Sugata[92] compared Raman and electron spin resonance (EPR) spectroscopy as methods of analysis of the structure of diamond films. Since only defects in the diamond film would be expected to generate EPR spectra, the authors felt that the quality of the diamond films could be characterised by the EPR technique. Diamond films were prepared from methane, methyl alcohol and ethyl alcohol in a microwave plasma at substrate temperatures of 850 to 950° C and pressures ranging from 5 to 20 torr. No clear relationship between the spin density and various features in the Raman spectrum was established. However, the authors found that as the diamond content in the film increased, as evidenced by the increase of the 1331 cm^{-1} Raman transition intensity, the peak-to-peak EPR linewidth decreased.

Liou et al.[93] were able to use much lower substrate temperatures of 400° C in a bell jar microwave system to form diamond films. They used MgO, silicon, fused silica and soda lime which were scratched with 0.5 μm diamond powder as substrates for deposition of the diamond. Again, they turned to microwave radiation to heat the graphite holder. They also used a range of concentrations of methane in hydrogen, and methane with oxygen as feed gases and found that the gas mixtures containing oxygen actually increased the growth rate, since it allows the use of higher C containing concentration for a given diamond purity.

Watanabe and Sugata[94] synthesised diamond films from the same precursors as above, and added acetone to the list. The apparatus again consisted of a quartz reaction tube, 36 mm in internal diameter. Fused silica which had been roughened by diamond powder was used as a substrate. The optical phonon in the Raman spectrum of diamond at 1331 cm^{-1} was seen for all films except that prepared from acetone.

---

† A more extensive work on that subject has been recently published by Vietzke et al.[90]

Ethyl alcohol appeared to be the material which provided the sharpest diamond peak with least graphitic background, at least when formed at 20 torr.

A method of improving on the rate of deposition or the quality of the diamond films deposited from the plasma is to use different precursors. Ito et al.[95] attempted the deposition of diamond from hydrogen/carbon monoxide feed gas at a substrate temperature of 900° C in a microwave plasma on a scratched silicon substrate. The authors used the microwave field to heat the substrate and to generate the plasma. From Raman spectra of diamond films deposited at CO concentrations from 5 to 100%, they showed that diamond was formed at concentrations up to 80%. It appeared that a concentration of 5% yielded a spectrum with little graphitic carbon. Despite the fact that Raman spectra showed evidence of a graphitic nature of the diamond at CO concentrations above 5%, X-ray diffraction data showed no evidence of this allotrope (see further below).

Hoshino et al.[96] studied the mechanical properties of diamond-like carbon films deposited from DC plasmas, and although this is a review of microwave plasma assisted processes, their article is quite instructive, particularly as it describes mechanical characteristics such as Young's modulus and hardness, together with electrical characteristics such as dielectric constant and a measurement of the electrical resistivity of these films. They also mention that disk wear tests using 2×10^4 starts/stops did not damage the surface of a disk coated with 160 Å of a diamond-like film layer.

Diamond growth and coating diagnostics. Badzian et al.[97] also studied the growth mechanisms of diamond on different substrates including diamond, graphite, silicon, βSiC, SiO_2 and nickel. Using methane/hydrogen/argon mixtures, they found that diamond films would deposit directly on most of the substrates except for SiO_2. They also claimed that silicon had a catalytic effect on diamond nucleation. They found that in cases where Raman spectra do not show deposits to have a diamond character, electron diffraction patterns do show these characteristics. They ascribed this to the low sensitivity of Raman spectroscopy to the ordered diamond structure and the low sensitivity of the electron diffraction to the disordered amorphous structure. Thus the two techniques used to analyse the sample can lead to erroneous conclusions.

Harris and Weiner[98] used in situ mass spectroscopy measurements of gas composition to investigate the effect of oxygen addition to the diamond forming gases. They used a tungsten filament to decompose the gases, as opposed to a microwave plasma. After studying the flammability limits of the H_2-CH_4-O_2 gas mixture, they found that addition of oxygen to the feed gas resulted in the reduction of methane, in good agreement with their kinetic model. They also concluded that the OH radicals formed in this environment removed the non-diamond carbon, thus allowing good quality diamond to be grown at lower temperatures.

Kaplan et al.[99] used solid state ^{13}C nuclear magnetic resonance (NMR) to study the nature of glow deposited amorphous carbon. They also used proton NMR and combustion to determine the hydrogen content of these films. Film hardness, density and the absorption spectrum in the UV-visible range were also obtained as a function of the method of preparation. Although the plasmas used to produce these deposits were created with DC and RF fields, the results are applicable to microwave frequencies. They tried to correlate the method of formation (ion beam, RF, DC) of the amorphous carbon with the sp^2/sp^3 hybridisation ratio,† the film hardness and the band gap. They deduced that hydrogen both passivates dangling bonds, and reduces the graphitic states at the band edge. They also found that the increasing sp^2/sp^3 ratio corresponded with a higher film hardness and a decreasing band gap due to a reduction of H content.

Most diamond films show Raman spectra which are indicative of some amount of a non-diamond containing material. In an extensive analysis of diamond growth, Zhu et al.[100] investigated the formation of graphite during film growth. Using transmission electron microscopy and Raman spectroscopy, they studied the diamond films formed in a microwave plasma from mixtures of methane and hydrogen. At 0.5% concentrations of methane, with a substrate temperature of 975° C, no graphite was detected. As the concentration of methane was increased to 2% and higher, both Raman spectra and electron diffraction patterns showed the presence of graphite, which appeared to be small crystallites some 50 to 100 Å in size and distributed in domains which aligned with different diamond faces. At a 5% concentration of methane, the Raman spectra showed a greater non-diamond contribution which electron diffraction patterns showed to be randomly distributed graphite. Increasing the temperature above 1000° C increased the non-diamond concentration.

Similar work using Auger electron spectroscopy (AES), secondary ion mass spectroscopy (SIMS), electron and X-ray diffraction, electron energy loss spectroscopy (EELS), transmission electron and scanning electron microscopy was undertaken by Williams and Glass.[101] Microwave plasma deposition from a methane/hydrogen plasma at 800° C substrate temperature produced films which were mainly C with

---

† The *hybridisation ratio* sp^2/sp^3 is indicative of the nature of the carbon bond; in diamond, all bonds are expected to be of the sp^3 type, thus the sp^2/sp^3 ratio would be zero. Where mixed bonding occurs, such as in diamond-like hydrogenated carbon, the hybridisation ratio sp^2/sp^3 would not be expected to be zero.

some H, O, Al, Si, Ca and Cu impurities. The SiO and Al impurities were shown to be segregated in regions of up to several μm in diameter in a matrix of diamond. To ensure that the observed material did not include amorphous carbon, the authors used AES and EELS to check that the carbon bonding had sp^3 hybridisation. They showed that the majority of the films they deposited were diamond whose main defect was twinning. It also appeared that ion beam bombardment of the diamond surfaces converted the topmost layers to graphite. They concluded that the very regular fourfold symmetrical diamond crystals formed in the center of their silicon substrates yielded to cauliflower (a common term in this research) textured material on the outer edges because of temperature variations. They found that the dominant diamond crystal facets changed from {111} threefold symmetrical faces to {100} fourfold symmetry in going from 0.3% methane to 2% methane.

Kieser and Sellschop[102] used an LMP[16] large area microwave reactor to deposit polymers on glass slides, probably with the intent of coating compact disks. From their introduction, it appears that they were trying to create a diamond-like material. They used acetylene, ethylene and ethane feed gases for their room temperature deposition experiments. From pressure changes occurring during ignition of the plasma, they concluded that the plasma stimulated oligomer growth in the vapour phase. Moisture permeation measurements showed the films to have localised pinholes.

The hardness of tungsten carbide is the reason that Kamo et al.[103] undertook an investigation of the coating of single crystals of tungsten carbide with diamond from a mixture of methane and hydrogen. After a period of 2 hours at 830° C, diamond nuclei ranging from 0.04 to 0.3 μm were visible by scanning electron microscopic examination. After 6 hours, the nucleated centres coalesced to form a continuous film. The authors found a difference in the nucleation and growth rates on the different faces of the WC crystal. They also found a significant difference in crystal structure and growth rate with deposition temperature, as has been shown by various other groups.

Diamond and diamond-like carbon deposition in ECR systems. The search for a high quality diamond film deposited from the plasma, originally attempted at microwave frequencies using simple, magnetic field free plasmas, was extended to ECR plasmas by Suzuki et al.,[104] who used a mixture of 5% CO in hydrogen. An independently heated substrate was immersed in the ECR plasma chamber and operated between 600 and 900° C. Surprisingly, material was deposited in this system at 0.1 and 10 torr, both pressures at which any electron cyclotron resonance effects are damped by neutral-electron collisions, though it is expected that some magnetic confinement of the plasma still occurs. It should also be noted that the major part of the plasma chamber in which the substrate was located was placed in a magnetic field which was much higher than that required for electron cyclotron resonance.

At pressures between 10 and 50 torr, without bias and at substrate temperatures of 800-900° C, diamonds with {111} and {100} facets were generated. The same products were obtained in the pressure range of 0.5 to 1 torr and at the same temperatures with a positive DC bias. At the lower 0.1 torr pressure, it was possible to form graphite, SiC or diamond, depending on the combination of substrate temperature and DC bias. By using a substrate temperature of 600° C and a positive DC bias, diamond crystals with a {111} facet were obtained.

A much more extensive review of diamond deposition using ECR plasma reactors was undertaken by the same authors.[105] In this case, both CO and CH_4 feed gases were used with a positive substrate bias of 30-100 volts. The authors found that it was possible to form SiC using CH_4/H_2 feed gas at a pressure of 0.5 to 1 torr and substrate temperatures of 750-900° C and no DC bias, whereas application of a positive DC bias resulted in diamond formation. Although many of the results are similar to those described in their earlier work, the authors present new data that shows that decrease of the carbon bearing feed gas changes the dominant crystal facet from {100} to {111}. In general, it was found that a 5% CO/H_2 mixture produced a 3 times higher deposition rate than that obtained with 2% CH_4/H_2 feed gas mixtures. Nunotani et al.[106] deposited diamond films from CO and H_2 gas using an ECR system. They found a considerable effect of oxygen in the plasma upon the grain structure of the polycrystalline diamond films. Furthermore, increasing the amount of added oxygen resulted in an increase of the fraction of the film deposited as diamond.

Ion bombardment effects were studied by Wei et al.[107] in an ECR deposition system. They showed that the deposition rate of diamond films was strongly affected by the potential difference between the plasma and the substrate.

It is not unusual to find that fairly complex polymers can be formed by plasma assisted deposition when attempting to form diamond films. Fujita and Matsumoto[108] produced what appeared to be a π bonded polyethylene-like polymer which was deposited on the walls of a silica tube placed inside an ECR system at 0.04 Pa and 50 watts. At higher powers (200 watts), a graphitic polymer was obtained. Electrical measurements showed that the polymer demonstrated both semiconducting and insulating properties.

15.5.3. Silicon. Generally, most of the plasma processes used to generate silicon from either amorphous or polycrystalline silicon, depend upon the temperature of deposition. A common problem in these systems is the fact that the silicon deposits in the regions where the plasma is most intense, viz. the applicator. This results in a semi-conductive coating which attenuates the passage of the microwave field and the plasma extinguishes. Many methods to circumvent this problem have been reported, but the most common is to use a relatively inert carrier gas, such as argon, as the source of plasma and to feed this effluent into a chamber, where it is reacted with a source of silicon. However, in some cases, the source of silicon, commonly silane, is admixed to a carrier at a low level and fed through the applicator. Surfatron stimulated silane/hydrogen plasmas were used by Biter[109] to deposit amorphous silicon at room temperature at various stimulating frequencies. Although the stoichiometry of the deposits were not investigated, the deposition rate was approximately dependent on the square of the excitation frequency. When glass slides heated to 200° C where used, a much stronger dependence of the deposition rate upon distance from the plasma source was observed. Kato et al.[110] used a microwave plasma generated at 2.45 GHz to decompose a mixture of argon and silane and produce amorphous silicon in a downstream chamber. However, they used the ingenious expedient of passing the carrier argon and the silane/argon gas mixture through a set of co-axial tubes which were placed within the microwave applicator. The inner tube, which contained the silane/argon mixture, was made of stainless steel, thus preventing the formation of an inner plasma, while the outer tube, made of quartz, contained the argon plasma. They also found that using a quartz inner tube instead of steel allowed them to maintain the outer plasma since, they argued, the field penetration into the central tube was too small to generate an inner plasma. They found that the uniformity of deposition varied inversely with the distance of a 10 cm diameter wafer from the exit of the co-axial tubes, whose inside and outside diameters were 2.2 and 2.5, and 0.4 and 0.6 cm.

Masson et al.[111] studied the deposition of amorphous silicon in an LMP(R) reactor and found considerable contamination of the film from atmospheric water and CO_2. Manory et al.[112] used mass spectrometry to analyze microwave induced plasmas of mixtures of $SiCl_4$ and H_2 or Ar in a quartz tube based system. They discovered that a variety of Si_2 Si_3, Si_4, Si_5 and Si_6 species were generated in the plasma. They found that the highest deposition rate of the silicon containing material was upstream of the plasma, while the Cl contamination was the lowest. They also found that there was an optimal power for maximum deposition rate, which they argued was due to the fact that recombination within the plasma occurred at higher powers. Addition of Ar to the plasma greatly increased reactant dissociation. This was believed to be due to an enhancement of ion-molecule reactions in the presence of the excited argon.

Silicon oxides and nitrides. Silicon oxide has been deposited from the vapour phase, using various silicon containing precursors including silane, silicon tetrachloride and tetraethoxysilane (TEOS). Oxygen is admitted in some form, whether as N_2O or O_2, with added buffer gases such as H_2 or He. The main reason for the use of plasma enhanced chemical vapor deposition (PECVD) as opposed to other oxide forming processes such as the thermal method is the ability to form the oxide at lower temperatures (as low as 250° C) and the ability to create very thin films (< 50 Å) in a controllable manner.

Pai and Chang[113] turned to TEOS as a source of oxide in the growth of oxide films using a downstream plasma deposition system, where a He/N_2O or O_2 gas mixture was used as the source of active oxygen atoms in the plasma, and TEOS or silane were added downstream of the plasma, just above a silicon substrate. The deposition rate varied from 75 to over 150 Å/min over a temperature range of 250 to 400° C. For TEOS, a negative activation energy of 0.12 eV was found whereas for the SiH_4 based process, the activation energy was essentially equal to zero. The stress in these films was found to be dependent on the concentration of oxygen with a high oxygen concentration providing the lowest stress films. Rutherford backscattering, HF etch rates and IR spectroscopy were used to measure the quality of the films. Ray et al.[114] used TEOS mixed with N_2 in a microwave downstream O_2 plasma to deposit silicon dioxide on an unheated susceptor. During deposition, they noted surface temperatures of 150° C. MOS capacitors were fabricated which showed that breakdown voltages ranged from 4 to 7 MV/cm.

Warren et al.[115] used a downstream microwave plasma to deposit thin oxide films. They used capacitance/voltage and electron spin resonance measurement techniques to determine the differences in the electrical characteristics of plasma deposited and thermally grown oxides. One of the most important issues limiting the use of either PECVD oxides or plasma oxidation in the semiconductor industry for devices, is the electrical quality of the films. Often, researchers in this field make use of characterization techniques which provide information on the chemical nature of the films (e.g. infrared spectroscopy, Rutherford backscattering, X-ray photoelectron spectroscopy, secondary ion mass spectroscopy). This information while important, is far less sensitive than the electrical characterization of devices produced with these materials. A number of authors have produced MOS capacitors, and measured their breakdown voltages, density of interface traps, trapped charge, etc., and these measurements constitute a much more useful benchmark of material quality for electrical devices.

PECVD nitride formation. Silicon nitride is another important dielectric with many applications in the microelectronics industry. The plasma enhanced chemical vapor deposition (PECVD) of nitride has been practiced for some time and is typically accomplished by reacting N_2 or some other nitrogen containing species with silane, with one or both of the reactants chemically excited by the plasma. This material is amorphous[116] as opposed to the stoichiometric Si_3N_4 that is deposited at much higher temperatures using thermal CVD. The key measures of the success for this process include,[117] i) the silicon-to-nitrogen ratio of the nitride, which effects the resistivity of the film, ii) the hydrogen content of the film, since excess hydrogen can produce adhesion problems with thin films that would subsequently be deposited over the nitride, and electrical instabilities in devices produced with the nitride, iii) the film density, which is indicative of the susceptibility of the film to cracking,[118] and iv) the refractive index, an indication of the quality and reproducibility of the film. There is also a set of electrical characteristics which indicates the utility of these films[117] including resistivity, breakdown strength, dielectric constant, interface charge, etch rate, stress and step coverage. As with oxides, it is these latter properties which measure the utility of these films for particular applications.

Bardos et al.[119] showed evidence that nitrogen atoms produced in the plasma were a key precursor to the nitride. This result, taken in conjunction with the fact that microwave plasmas typically produce relatively high densities of reactive atomic species, is consistent with the demonstration by Tessier et al.[120] that microwave plasma driven PECVD is capable of deposition rates that are 10-25 times greater than those using lower frequency plasma excitation. An additional advantage of the remote microwave plasma technique is that nitride films are not in direct contact with the plasma, and thus avoiding damage by energetic particle bombardment;[121] this technique has been widely applied.[121-124]

15.5.4. Plasma oxidation. Many studies of the oxidation of silicon in microwave plasmas base their model of oxidation on that which has been used in thermal oxidation for several tens of years. Thus the literature contains many examples of fitting of the rate of formation of an oxide in a plasma to some function of time, whose functional form varies, depending on the model.

In the late 50's, Greaves and Linnett[29] studied the recombination rates of oxygen atoms both in the gas phase and on various surfaces. From their work, they were able to surmise that oxygen atoms created in a microwave discharge not only reacted with various surfaces, but in the case of SiO_2, atom exchange with the substrate material occurred. Ligenza[125] showed that the growth of silicon oxide in an oxygen microwave plasma followed the general form $X^2 = ct$, where X is the thickness, t the time of oxidation and c a constant, provided that a positive bias was applied to the silicon electrode on which the samples rested. He concluded that negative ions were "stuffed" into the silicon layer by the externally applied DC field. Without such a field, the oxide thickness reached a limiting value of around 4400 Å. With the application of a field, it was thought that negative ions could be driven through the thin surface layer where they were subsequently able to diffuse within the film. Kraitchman's studies[126] confirmed some of these conclusions and showed that the silicon oxidation rate was dependent upon the positive potential applied to the sample electrode. A silicon rod served as the counterelectrode. Only positive potentials (with respect to the counterelectrode) were capable of creating an oxide layer atop the silicon samples. Surface temperatures of less than 700° C were observed during plasma oxidation of the silicon, whose thickness was dependent upon an expression which included a parabolic growth and a linear sputtering removal rate. The parabolic term was ascribed to the diffusion of some active specie (presumed to be an oxygen atom) through the silicon, whereas the linear term limited the final thickness. Surprisingly good quality films were grown, which were compared to thermally grown films using standard criteria such as refractive index. It was believed that the bombardment of the oxide surface by positive ions and electrons both drove the formation of diffusing oxygen species and promoted removal of the surface layers by positive ion stimulated sputtering. No evidence for the interaction of negative ions with the surface of the oxide was found. Kiermasz et al.[127] further studied the kinetics of silicon oxidation in microwave oxygen plasmas and found an absence of the sputtering (linear) term which Kraitchman[126] ascribed to be responsible for limiting the oxide growth.

Bardos et al.[128] also reported that it was possible to oxidise silicon in a microwave oxygen plasma. Similarly, Weinreich[129] showed that it was also possible to oxidise gallium arsenide in a microwave oxygen plasma at 70° C. Amorphous films as thick as 3850 Å were grown in 100 minutes. Musil et al.[130] discovered that in a magnetically confined plasma, the key parameter which controlled the formation of the silicon oxide was the floating potential. When this became more negative than a certain critical value, silicon oxides could not be grown, whereas above this value, the rate of formation would also be dependent upon the plasma density.

Most analyses of SiO_2 use measurement of the refractive index, IR absorption spectra and techniques such as determination of the etch rate in specific etchants, to characterise the oxide formed in this way. Ruzyllo et al.[131] used current-voltage and voltage breakdown techniques to characterise the quality of oxide films grown in oxygen based plasmas. Using a variety of plasmas as stimulants for a downstream gas mixture, they were able to show that oxide layers could be grown using this method. Measurements of breakdown strengths and interface trap densities, which degrade MOS device performance, were obtained and showed that the use of hydrogen in the plasma and in the downstream excited gas provided oxides with the highest breakdown strengths and the lowest density of fixed charge. Kimura et al.[132] directly compared the performance of MOSFET's whose gates were fabricated using both conventional and plasma oxidised SiO_2. They found that the field effect mobility of the FET's generated using plasma oxidation were somewhat lower than those from the normal thermal oxide. However, the difference was only about 12%. Surface state densities of the plasma grown oxide were about 4 times greater than those of the thermally grown oxide after a 450° C anneal.

15.5.5. Plasma surface treatment. One lesser known, but quite important application of plasmas in the microelectronics industry, is the modification of surfaces so that they are more amenable to subsequent processing. There is much folk-lore surrounding this subject but only a few studies have been reported. Wertheimer and Schreiber[133] used a large microwave plasma (LMP) system with a variety of plasmas designed to react, modify or activate the surface of multifilament cloth formed from polyimide. They showed a range of effects including changes in adhesion strength of the polyimide and changes in its mechanical properties. Mutel et al.[134] found that exposure of a polypropylene surface to a microwave plasma composed of nitrogen and fluorine-containing-species, made it possible to promote adhesion of a subsequent epoxy resin layer. In this case, the plasma treatment resulted in significant inclusion of nitrogen in the polymer surface.

Promoting adhesion of subsequent films is quite a common result of microwave plasma processing, whether the goal of the process is etching, deposition or specifically to pre-treat a surface. An important concern with respect to the chemical and physical condition of a surface which has been exposed to a plasma, is the extent to which the surface was in direct contact with the plasma, or more specifically the bombardment by energetic ions, electrons or photons from the plasma. Ruzyllo et al.[135] have shown that direct ion bombardment of silicon surfaces can induce damage and degrade the performance of devices built on that surface. This demonstrates an important advantage of microwave plasma processing relative to other plasmas, in that it is possible to independently control the relative contribution of the chemical component (exposure to chemically excited species produced in the plasma) of the process and the physical component (energetic ions, electrons and photons), by using a combination of remote microwave excitation and in situ RF biasing.

Adhesion promotion is not the only application of plasma surface treatment. Atomic hydrogen from a crossed-beam microwave plasma source[136] has been used to passivate the surface of GaAs which resulted in superior n-GaAs Schottky diode properties. In this case, the fact that direct ion bombardment of the surface was avoided, was key to the success of the application. Grot et al.[137] have shown that a hydrogen plasma surface treatment of doped diamond films can produce a conductive layer which has ohmic characteristics, making it possible to produce excellent Au-gate Schottky diodes.

15.6. ECR systems

Although Electron Cyclotron Resonance (ECR) plasmas have been used in the microelectronics processing arena for at least 2 decades, most of the recent activity in this type of microwave stimulated plasma has been driven by a large effort from Japanese plasma system manufacturers. Though the first record of the use of ECR systems for such a purpose dates to 1972,[138] and as early as 1964 for high current ion sources[139] and for space propulsion,[140] few groups were actively involved in this particular approach to plasma processing. In part, this may be because of the complexity of these systems and requirements for large electromagnets coupled with the perceived difficulty of working at microwave frequencies. The field gained considerable momentum when it was believed that these systems could provide high resolution etching of silicon to better than 0.5 μm dimensions. The impetus for this belief seemed to be correlated with the entry of a wide range of commercially available ECR systems from vendors such a Hitachi, Sumitomo, Toshiba, NEC, Anelva and Alcatel. Whether this belief was well placed remains to be seen. There also appears to be a large degree of confusion in regard to the promised high anisotropy of etching, high selectivity and low electrical damage of these systems. The effects of the distribution of magnetic fields, the efficiency of these sources and their potential uses are just beginning to be investigated with sufficient rigour to allow sensible conclusions to be made, at least as far as manufacturing applications are concerned.

For these low pressure and relatively high power (1 kW) ECR plasmas, the degree of ionisation is usually high compared to normal RF plasmas of similar volume, and thus they are very suitable for use as ion sources.[141-144] In particular, they are excellent for sources of ions which are created from a corrosive gas such as chlorine, or an oxidant such as oxygen, which rapidly erode the more common filament excited ion sources. The latter survive for periods of only several minutes in these plasmas unless they are coated with protective materials.[145] Asmussen has reviewed the salient issues associated with Electron Cyclotron Resonance plasmas,[146] with particular reference to their use as plasma processing sources, and compared the different approaches that various groups have adopted in implementing the ECR plasma for sundry purposes.

In the following sections, some of the uses and operating characteristics of these various plasma systems are discussed and compared. In order to do so, the basic configurations of the ECR systems are described below.

15.6.1. System configuration

Plasma stream systems. In the basic ECR plasma stream configuration, shown in Fig. 15.13, the plasma is created by means of microwave stimulation through a quartz window which provides a vacuum seal and a path for the entry of microwave power into the plasma chamber. In some cases, the window may become hot, due to the fact that it lies at a geometrical discontinuity in the path of the microwave field. In particular, O-rings have to be chosen and placed with care. Some materials which are highly chemically resistant may be lossy, resulting in their rapid thermal degradation in the high microwave fields. In many instances, these same discontinuities create high localised electric fields which cause the formation of a plasma close to the window seal. In such a case, there may be no other expedient than to cool the sealing area. An alternative approach[147], which isolates the ECR plasma from the window, is to place a soft iron yoke surrounding the waveguide which delivers microwave power to the chamber. The soft iron yoke "chokes off" the magnetic field close to the window, without which a plasma is very difficult to maintain at these low pressures.

Electromagnets placed around the source provide a magnetic field which can diverge, in which case electrons drift along these field lines to the edges of the chamber with their attendant ions. Alternatively, a second field created by magnets placed in the

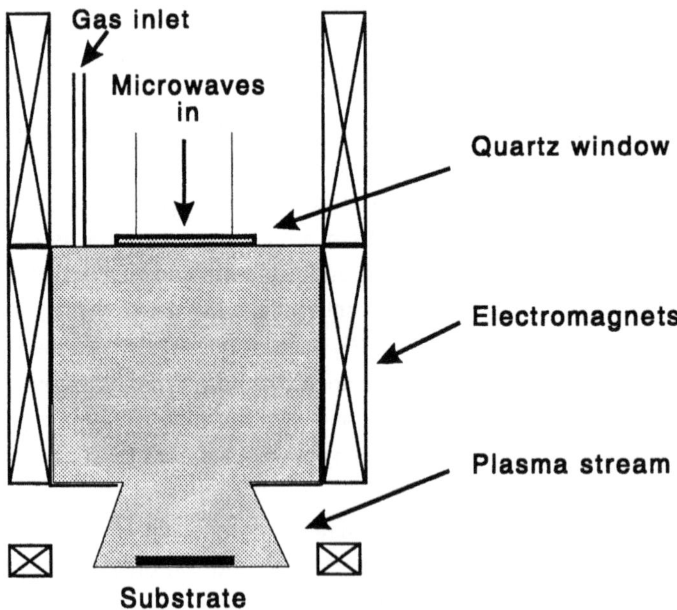

Fig. 15.13. Schematic of ECR stream reactor.

region of the substrates, can be excited in such a way to either oppose the field, hence creating a cusp, or to create a mirror (bottle), which re-focuses the plasma to the substrate from its early divergence. As will be seen below, these configurations of magnetic field correspond to radically different plasma characteristics. In all of these cases, the plasma stream systems disgorge the plasma from the plasma chamber into the process chamber.

ECR based ion guns and sources. In some cases, a set of grids is placed across the exit of the plasma chamber, allowing ions to be extracted from the plasma in a directed manner. However, the energy of these ions as they strike an unbiased substrate is just the difference between the plasma potential and the floating potential of the substrate. At low ion energies, the divergence of the ion beam can be large and, as a consequence of the low energy of the ions relative to their surroundings, the beam can be significantly affected by external fields. To gain higher ion energies at the substrate, some authors have applied a DC bias to it, which is ineffective for insulators. A solution is to bias the metal housing containing the microwave plasma at a fixed potential V, relative to ground, together with the first or screen grid. The subsequent grids can then be used to extract and focus the ion beam, and a secondary electron source placed across the exit (e.g. hollow cathode) to neutralise the beam, thus allowing it to be used to process insulators.

Tokiguchi et al.[148] used a cylindrical-sector energy analyser to study the ion energy distribution of an argon plasma created in an ECR source. After calibrating the system with a thermal Na^+ source, they found that the energy spread of the Ar^+ ions ranged from around 5 eV at 1 mtorr to 4 eV at 20 mtorr. They claimed that a change of microwave power from 200 W to 800 W resulted in a slight (1 eV) increase in the energy width from 5 eV. This was tentatively ascribed to the relatively large energy spread to the thermal energies of the ions gained from the magnetic field and from fluctuations of the plasma potential.

A similar ion source was characterised[149] at Michigan State University (East Lansing), at an argon pressure of 7.5×10^{-4} torr, and showed that currents of 4-500 mA over a 140 mm diameter circle were achievable at uniformities of $\pm 5\%$. Oxygen ECR plasmas were also used as sources for this ion beam system. Significant ion density increases were seen as ECR was approached. In further studies,[150] they found that with an applied microwave power of 150 watts, currents of 100 mA across a 9.4 cm diameter disk could be obtained. A later improvement of the design with a 3.2 cm diameter ion beam disk[151] showed that current densities from this small source were above 10 mA/cm^2, for argon. It was also found that the efficiency of the source was dependent upon the gas flow. Note that permanent magnets have been used[142] instead of electromagnets in many of the designs from this group.

Yafarov et al.[144] described an ion source which used a 100 mm diameter ECR source whose output was controlled with a pair of grids. They found that they could obtain currents of several tens of mA with a variety of plasmas including CCl_4, Ar and air. They also found that typical filament lifetimes for HCl and CCl_4 gas plasmas were around 300 hrs, whereas the fluorine based gases, namely CF_4 and CF_2Cl_2, degraded the tungsten filaments within 5-6 hours.

15.6.2. ECR plasma stream etching. The relative simplicity of plasma stream ECR system results in their use in many applications. Popov[152] has characterised several of these sources using probe measurements. Sources from 2.5 to 15 cm in diameter and 10 to 16 cm in length were built for purposes of determining whether they could be excited below cut-off of the TE and TM cavity modes and how the efficiencies and electrical characteristics of these sources were affected by externally variable parameters. It was found that excitation of the source below cut-off could be achieved, and that plasma densities above 2×10^{11} cm^{-3} could be obtained. The plasma potentials and electron temperatures were also studied and found to vary with geometry, magnetic field, and absorbed power. Similar work by the prodigious team of Matsuoka and Ono[153] showed that the plasma stream systems have ion energy distributions which are very strongly dependent upon the magnetic field gradients. For the case where the magnetic field is divergent from the plasma source to the substrate, the energy distribution of ions (at 0.5 mtorr of an undefined gas) was shown to be broader and centered at higher energy than that obtained with a convergent magnetic field (mirror field). This has significant implication for the uniformity and directionality of etching as will be seen below.

One of the first attempts to use ECR plasmas for etching was made in 1977[154] by a group at Hitachi, who used this system to etch silicon with CF_4/O_2 gas mixtures at a variety of pressures. They found that the shape of the etched profile was strongly dependent upon the gas pressure, and that typical etch rates were around 1000 Å/min. It was argued that since the difference between the plasma potential and the floating potential in this reactor was around 20 V, and the threshold sputtering energy of silicon is 19 eV, very little damage would be encountered by the silicon surface. This argument has persisted, although little solid evidence for it has appeared in the literature.

Anisotropy control. As many subsequent authors have discovered, the control of the etched profile is strongly dependent upon many parameters, chief of which are the ion energy and trajectory (see Sec. 7.5), and the distribution of atomic species within the reactor. If ions have any role in etching of a substrate, then it should be expected that their trajectory will be reflected in the etched profile of the material. Hence if a mask is placed atop a silicon layer and ions impinge normally to the substrate, any enhancement of etching due to their impact should result in a profile which shows anisotropy, i.e. etching in a preferred direction. This is similar to the microwave-RF plasma etching of polymers discussed above (Sec. 15.5.1).

Suzuki et al.[155] used electrically biassed grids, placed atop a silicon wafer masked with a patterned aluminium layer, to assess the effects of ion bombardment upon the etched profile of silicon. They found that if all ion current was suppressed using this grid arrangement, the etch rate of silicon dropped to zero below 10^{-3} torr. Above about 7 mtorr, the suppressed and unsuppressed etch rates were identical. This suggests that either ion bombardment is not important for etching above this pressure, or that other effects such as charge exchange play a significant role in providing a directional beam which is not charged. They attempted to cast light on this issue by tilting the substrate relative to the magnetic field axis. In doing so, they found little change in the inclination of the etched trench in the silicon relative to the plane of the silicon wafer.

Ono et al.[21] also used a plasma stream system to etch molybdenum and polysilicon in Cl_2 and SF_6 plasmas and, in contrast to the former work, showed that it was possible to incline the etched silicon profile relative to the substrate surface. Using emissive probes, they also discover that the plasma potential in the stream varied from 50 or so volts to 5 volts from 7×10^{-2} to 1.5 mtorr. Since the plasma sheath potential was invariant with pressure, and amounted to 10 V, they were able to estimate the ion impact energy on the substrate. By inclining the substrate to the plasma at 60°, they were able to show trenches etched at an angle which was not quite 60°. They accounted for this difference by assuming that the ions travel along the plasma stream and are refracted at the plasma sheath generated between the substrate and the plasma. Since the ions already have energy due to their acceleration in the stream, they are refracted at the sheath boundary, before they impinge on the substrate. Thus the distribution of the magnetic fields and the imposition of external electric fields can change the etched profile in a manner which is not obvious. Using this apparatus, typical etch rates of silicon, photoresist and molybdenum were found to be 1200, 200 and 600 Å/min respectively in Cl_2/SF_6 plasmas of varying composition.

In an attempt to control the etch rate ratio between Si and SiO_2, Suzuki et al.[156] used a DC biassed carbon tube placed in the ECR plasma stream atop a silicon wafer, whose potential could also be changed relative to the grounded chamber walls. It was argued that since the etching of silicon in a plasma depends strongly on the C to F ratio, providing a secondary source of carbon (through sputtering) would alter the etch rate of silicon, while hardly affecting the etch rate of SiO_2. This turned out to be the case. At a negative bias voltage of 200 V on the carbon plate, the etch rate of Si dropped, from around 200 Å/min without the bias, to zero, whereas the etch rate of SiO_2 in the pure CF_4 plasma was unaffected. By accelerating ions towards the silicon wafers by biassing the substrate with a negative DC voltage, with the carbon tube unbiassed, etch rates of silicon in the same plasma increased from 200 Å/min to 480 Å/min. In a subsequent publication,[157] the same group also applied RF bias to the substrate, since this technique allows insulators to be etched under the ion bombardment conditions applied in the DC case. They investigated the etch rate of Si and SiO_2 from 100 kHz to 1 MHz and found that the etch rate of Si was not affected by the RF frequency, whereas the etch rate of the SiO_2 in the same C_3F_8 ECR plasma, increased from about 200 Å/min to 400 Å/min when the RF excitation frequency was increased from 1 to 200 kHz. Above 200 kHz, there was no apparent change in the etch rate for SiO_2.

Hopwood et al.[158] studied the conditions required to etch silicon uniformly and anisotropically in an SF_6 ECR plasma, which was allowed to diffuse from a plasma volume contained by a grounded electrical mesh. Small samples of silicon were placed on an electrode which was biassed negatively with respect to the grounded mesh. These samples had patterned aluminium masks deposited on them. They investigated the lateral etching rate of silicon as a function of the concentration of SF_6 in Ar and found that lateral etching appeared above 5% SF_6. They concluded that there was a critical ratio of neutral fluorine atom flux to ion flux, which determined the occurrence of the isotropic etching. Their data also indicated that although related, the ion density distribution above the substrate and the etching rate distribution were not co-incident. This is similar to the situation in RF plasmas where vertical etching of the silicon dominates as a consequence of the effect of ion bombardment[159-161] when the density of fluorine atoms, which induce spontaneous and omnidirectional etching, is low.

A three layer resist profile[46] was used[162] to determine the nature of directional etching in oxygen ECR plasmas by a joint group from Sumitomo and Matsushita. The etch rates of the polymer were determined as a function of gas pressure from 0.1 to 2 mtorr in oxygen and argon. Using an argon plasma, they found that the polymer etch rate would drop from around 100 Å/min at 0.1 mtorr to 20 Å/min at 0.5 mtorr, whereas the use of an oxygen plasma would yield an etch rate of around 500 Å/min at similar microwave powers at both pressures. A structure of 3 μm of photoresist covered with a patterned mask of SiO_2 was used as a monitor of the directionality (anisotropy) of etching. The resulting scanning electron micrographs showed that a rapid change of directionality occurred in going from 0.1 mtorr to 10 mtorr. At the low pressure, the etched profile was anisotropic, while at the higher pressure, the profile showed considerable lateral etching. In some cases, the etched profile was determined to be re-entrant, with the foot of the etched profile being etched more in the lateral direction than the top of the profile, located just beneath the mask. There was some evidence that the ion trajectory was affected by the presence of the charged insulating film of polymer. Again the energy gained by the ions in the divergent magnetic field appears to be affecting the directionality of etching, just as was found for the case of the silicon earlier. Furthermore, in contrast to the case where ions pass across the dark space of an RF plasma and strike the substrate normal to its plane, the spatial variation of the magnetic field creates a situation where ions strike the substrate at angles which are dependent upon the integrated interaction of the magnetic field with the ion along its trajectory. Not only is the etch rate expected to vary with the magnetic field, but so is the shape of the etched profile.

Influence of the ion energy distribution. Forster and Holber[163] characterised an ECR oxygen plasma from 0.2 to 1.6 mtorr and found that electron densities of 4×10^{11} cm^{-3} could be obtained. They also showed a very strong correlation of the etch rate of photoresist in this plasma with the ion current incident on the substrate. Subsequent work by Holber and Forster[164] in Cl_2 plasmas studied the ion energy and distribution using an electrostatic sector energy analyser. They determined that the width[†] of the ion energy distribution was broadened by a low frequency RF bias of the substrate. When the RF frequency was increased, the width of the ion energy distribution decreased, since the transit time for an ion to pass from the edge of the plasma sheath to the substrate became smaller. The ion energy distribution width was also measured as a function of the divergence of the magnetic field. In general, the width was found to be greater for the divergent than the convergent field. Thus the magnetic field gradient in ECR plasmas has the effect of accelerating or decelerating ions[165], which drift together with electrons in this field. In the absence of the magnetic field, ions would emerge from the edge of the plasma in random directions. The electric field across the plasma sheath serves to re-direct their trajectories, whereas they already have a defined trajectory due to the applied magnetic field. Increasing the applied electric field across the plasma sheath results in a decrease of the ion energy distribution width. Ironically, under pressure conditions (no ion collisions) that ensure a highly directional etching in RF parallel plate systems, the application of the magnetic field in this microwave plasma system can result in an isotropic profile.

---

† The width on the ion distribution is important since it affects the directionality of etching. A larger width implies a wide range of ion incidence angles on the substrate which results in an *isotropic* profile.

McKillop et al.[166] found that there was a large correlation between the ion saturation current and the optical emission from the Ar lines at 7504 Å, thus yielding a simple tool for the measurement of the relative ion densities in the plasma. Recent optical emission studies by O'Neil et al.[167] measured the Doppler lineshape profiles of Ar+ in an Argon ECR plasma. They found that the profile observed along the plasma system axis was significantly broadened compared to the emission lineshape observed radially. They observed this asymmetry in plasmas created at pressures from 0.1 to 2 mtorr and ascribed it to the difference of ion velocity in the radial and axial directions. This directed and divergent ion motion was assumed to be dominated by the electric field gradient set up by electrons diffusing out of the source region; the ions drifted along with them. However, the differences in radial and axial energy of the Ar ions measured in this way was found to be 0.2 to 0.3 eV, much lower than that measured with electrostatic methods earlier. They concluded that although the divergent magnetic fields used in this work could contribute to the out-diffusion of the ions, this could not account for the measured *axial* Doppler broadening. Rather, they assumed that this broadening was due to changes in the electrical properties of the plasma. Their work points out the importance of understanding how ions gain energy from plasmas, since the mechanism influences the etching anisotropy of many processes. This will determine whether ECR plasma systems will be accepted by the manufacturing community, who in turn decides the ultimate sales potential of any new plasma system.

Other measurements and applications in plasma stream reactors. An interesting version of the standard plasma stream ECR applicator is the so-called Lisitano coil, an axially periodic structure, described by Yonetsu et al.,[168] which uses a 40 cm diameter, 9 cm high metal tube, which has 102 slots cut into it. The slots are arranged radially, so that the slot length along the magnetic field plus that perpendicular to the magnetic field are equal to 1/2 microwave wavelength. This provided a uniform microwave plasma, which was generated by a pulsed source, and measured using standard probe techniques.

Bardos and Musil[169] measured the axial dependence of the floating potential and plasma potential in a 2.35 GHz ECR plasmas. They used a microwave interferometer to determine that the electron density could be represented by the saturated ion current drawn to a probe. These probe measurements showed that at points along the axis, where the plasma reached ECR, the floating potential dropped drastically. They also found that the material of construction of a plane probe determined its electrical response.

Similar probe techniques were used by Lee et al.[170] to measure various electrical characteristics of O_2, CF_4 and Ar ECR plasmas generated using a plasma stream reactor. Measurements of plasma density, plasma potentials and electron energy distributions were made: i) the authors found that a rapid increase in plasma density with applied magnetic field could be obtained in argon and CF_4 plasmas for pressures of up to 40 and 10 mtorr respectively; ii) the electron energy distributions measured in an O_2 plasma at 40 mtorr showed an average energy of around 1.58 eV in a region where the magnetic field was around 100 G. This constrasts with earlier measurements made at 50 mtorr in a microwave O_2 plasma where no magnetic confinement was attempted, which yielded average energies of around 2.9 eV; iii) at 1 mtorr, the electron energy distribution measured in an argon plasma directly in the ECR region of the plasma was extremely noisy. However, there are sufficient data points to allow the determination of an average electron energy of 7.4 eV. For all three measurements, the electron energy distribution could be fitted to the equation $c_1 u^{1/2} \exp(-c_2 u^{1.3})$, where c_1 and c_2 are adjustable parameters and u is the energy. For a Maxwellian energy distribution, the exponent of u is 1. A similar set of probe based reactor characterisation experiments were undertaken by Keqiang et al.,[171] who measured electric field breakdown strengths, ion densities and electron temperatures in argon and nitrogen plasmas. They also compared barrel, parallel plate and ECR reactors on the basis of their electrical characteristics.

Although it is usual to determine concentrations of charged species in these plasmas, concentrations of neutral species, generated by dissociation are rarely monitored. Meikle and Hatanaka[172] determined the efficiency of generation of atomic species in a nitrogen ECR plasma. They did so by passing nitrogen through a 33 mm pyrex tube crossing a TE_{10} cavity operated at 2.45 GHz across which a magnetic field was applied. The concentration [N] of atomic nitrogen was measured downstream, by titrating the atomic nitrogen with NO. They found that the efficiency ($[N]/[N_2]$) of N atom production peaked at 6 mtorr and reached values of around 12%.

15.6.3. ECR plasma stream deposition and surface treatment. One of the advantages of the ECR based plasmas is the high ion densities ($> 10^{10} cm^{-3}$) which are generated. For many of these low pressure plasmas, the degree of ionisation approaches values of around 10% or more which can be very useful for processes which involve sputtering. Indeed, some of the apparent isotropy (see Sec. 15.6.2) of processing due to ions with trajectories which are not perpendicular to a substrate, could prove to be advantageous, especially for filling deep structures. Indeed, Machida and Okowa[173] found that it was possible to planarise aluminium lines of 1.5 μm in height by using a plasma stream of O_2, N_2 and Ar and a secondary injected SiH_4 gas stream. The reaction of the silane gas with the plasma effluent in the ECR plasma stream generated SiO_2. This material was sputtered and redeposited by means of an RF plasma created at the substrate by connection to a 13.56 MHz generator. They found that the application of the RF bias reduced the deposition rate, implying that the sputtering of the deposited material was occurring. They also determined the effective lateral etch rate of the SiO_2 deposited on the aluminium line sidewalls. They showed quite remarkable scanning electron micrographs of filled structures which appeared to be almost perfectly planarised.

Metal and dielectric deposition. Several types of ECR sputtering systems have been developed. The simplest, the so-called diode-type sputtering system, relies on the difference in energies between the ions in the plasma and the negative potential applied to the conductive target to provide sputtering. The ejected material is conveyed along the plasma stream. Matsuoka and Ono[147] showed that Fe deposition rates of around 50 Å/min could be obtained with such a system.

A logical development of this simple approach is the ECR based *magnetron sputtering system*, which relies on magnetic confinement of electrons around the target. It appears that the earliest such system was described by Ono et al.[174] who used a doughnut shaped target placed at the exit of the ECR plasma chamber. The tantalum or aluminium targets were surrounded by soft iron yokes, which served to provide local magnetic field confinement and trapping of electrons around the targets. This increased the plasma density locally and provided an increased density of ions to bombard the targets. The latter were accelerated to the target by means of an applied negative (with respect to the chamber and yoke) potential. The same authors[175] later expanded on this work by comparing the target currents, with and without magnetic confinement around the target. Oxygen was added to the argon plasma using aluminium as a target, thus creating a reactive sputtering system, where aluminium is sputtered and reacts with the oxygen containing plasma before the oxide deposits on a substrate. Deposition rates of aluminium oxide of 800 Å/min were obtained with this system. The third type of sputtering system which has been used combines the previous two arrangements. Here both targets have the same potentials applied to them, and secondary electrons are ejected from each of the targets. These secondary, or γ electrons, spiral along magnetic field lines between the two targets gaining energy, which can approach that of the target potential. A beam of high energy electrons thus develops between these two targets and increases the plasma density in this region.[176] The authors found that Al, Mo, Cu and Fe films could be deposited at rates above 2000 Å/min. However, high DC currents were required to do so. Measurements of the ion energy distribution in these plasmas were obtained earlier[177] and showed that the distribution width of the ions extracted from the plasma, increased with target current and decreased when a magnetic mirror field was applied. Deposition uniformities, target erosion profiles and other practical data related to the use of ECR sputtering plasmas were also obtained.[178] If these so-called beam electrons actually exist, they should interact with the plasma and stimulate oscillations which should be measurable. These plasma oscillations were observed by the authors, who found that a certain critical beam current was required before this oscillation could be detected. In a later paper,[179] these same authors prepared zinc oxide films for optoacoustic devices by ECR sputtering Zn in an oxygen atmosphere. Their results indicated significant differences in the crystal structure due to substrate temperature and injection location of the oxygen gas. They also found a significant difference of crystallite orientation upon gas flow rate. These observation were related to the degree of dissociation of the oxygen.

It should be noted that the high rate sputter deposition systems rely on the conductivity of the targets to provide the ion bombardment and beam electron trapping. One of the few ways in which insulators can be produced with this method is to use a metal and inject the appropriate reactive gas, which creates the compound insulator such as ZnO, Si_3N_4, SiO_2. Another method would be to apply RF to these targets, much in the manner of the RF magnetron sputtering systems. Such a system was used by Goto et al.[181] (see below) for the deposition of a high T_c superconductor.

High T_c superconductor deposition. One of the difficulties in fabricating superconducting thin film structures with a high critical temperature T_c lies in the deposition of thin films from targets of a sintered superconductor. The main problem is to deposit the superconductors in a stoichiometric form. ECR sputtering plasmas have been found useful in this endeavour.

An important parameter which determines the superconductivity of oxide containing high T_c superconductors is their degree of oxidation. For the case of $La_1Ba_2Cu_2O_y$, superconductivity does not occur until $y > 6.8$. Ceramics were prepared by mixing the native oxides and subjecting them to thermal treatment; after cooling to 800° C, they were immersed in liquid nitrogen, hence creating oxygen deficient materials. Without any plasma treatment, samples showed a superconducting transition at 54 K. After oxygen ECR plasma treatment at 400° C at an unspecified pressure, Minomo et al.[180] found that the superconducting transition temperature increased to 93 K after 60 minutes of a -40 V DC bias or 120 minutes of a +30 V bias.

Goto et al.[181] used an RF magnetron ECR based sputtering system to deposit the high T_c superconductor $YBa_2Cu_3O_{7-x}$ (123 for short) on crystalline MgO substrates. A mirror magnetic field was applied to the ECR system, such that at the substrate, the magnetic field was 450 gauss. The 123 was deposited at 650° C using IR heating below the substrate (sample B). After deposition, the 123 was cooled to 400° C in 1 atmosphere of oxygen. Films of 123 were also prepared without substrate heating (sample A); these were subsequently annealed in oxygen at 950° C (sample C). It was found that A was amorphous, mirror-like and insulating; subsequent post annealing to produce C gave a T_c of 73 K, whereas the T_c for the sample deposited during heating was found to be 70.5 K. Deposition rates in the unidentified plasma were around 3-4 Ås^{-1}.

The high T_c superconductor BiPbSrCaCuO, having two different phases and thus two different superconducting temperatures, was prepared by ECR sputtering by Nishimori et al.[182] In this set of experiments, the intent was to optimise the in situ oxidation of the superconductor, during cooling of the substrate in the ECR O_2 plasma, rather than the deposition rate. In fact, the latter was quite low, approximately 60 Å/min (this is probably because the DC plasma between the target and substrate appears to be unconnected to the ECR discharge). Unfortunately, the authors were not able to generate the superconducting phase having a T_c of 110 K. However, they were able to show that cooling of the deposited superconductor from 710° C in an oxygen ECR plasma produced a superconductor with a T_c of 71 K.

Oxide formation in ECR plasmas. As in the case of the pure microwave plasma oxidation of silicon (Sec. 15.5.4), one of the key technological issues in the use of ECR systems in the formation of silicon and other oxides is the rate at which the oxide layer forms. As a consequence, much of the research centres around the mechanism of oxidation and relates to earlier work done in microwave oxygen plasmas by Ligenza[125] and others (see above). Often comparison is made to the thermal mechanisms of oxidation to try to establish the effect of the plasma. The participation of O$^-$ ions in the oxidation process still appears to be undecided.

Miyake et al.[183] used an ECR oxygen plasma stream system to form silicon dioxide layers on a silicon substrate heated to 700° C. After an hour of treatment, a 230 Å thick film was grown. Use of a CO_2 pulsed laser capable of delivering 3J/pulse was found to increase the oxidation rate by 30%. Oxide films formed by this route were said to have the same CV characteristics as thermally grown oxides. A year later, Kimura et al.[184] used a similar apparatus to study the oxidation kinetics of silicon subjected to an oxygen ECR plasma. They were able to fit their thickness vs. time curves to a function $X^c = ct$, where X is the oxide thickness, t the time of oxidation and c a constant. They found that the exponent c ranged from 3.06 at a substrate temperature of 350° C to 2.60 at 640° C. They concluded that the Cabrera-Mott (CM) mechanism was operative in the oxidation of silicon. This mechanism supposes that substrate atoms diffuse towards an already oxidised layer in contrast to the Deal and Grove (DG) model, where oxygen species diffuse into the substrate. In the CM model, the electric field from an oxygen ion which arrives on the metal surface (in this case silicon) decreases the activation energy for ion movement. Since the plasma drives the surface potential on the oxide to the floating potential, the authors believed that this increased field across the oxide reduced the activation energy for the diffusion of the substrate atoms to the oxide. Their measurements showed that the activation energy for oxidation in the ECR plasma was 0.4 eV as compared to the 1.24 eV obtained by dry thermal oxidation. However, the quality of the oxide, as measured by IR absorption and P-etch[†] measurement, were different from that of thermal oxide. Disruption of the surface by ion sputtering or removal of the oxide by the plasma was not considered. For a 50 nm film grown at 640° C in this system, 2 hours of oxidation were required.

Matsumura[185] observed that the oxidation of aluminium in a DC plasma (plasma anodization) was probably driven by O^- ions generated on the surface of the aluminium by electron attachment of oxygen atoms, rather than by O^- ions produced in the plasma. The density of the latter was judged to be too low to contribute directly to the process. Thus, the electron density determined not only the generation rate of oxygen atoms, but also that of the O^- ions. Kimura et al.[186] used an oxygen ECR plasma to study the oxidation of silicon in a cusped magnetic field. By studying the electrical characteristics of the plasma using a Langmuir probe, they showed that oxygen ions from the plasma played no role in the oxidation kinetics of silicon. However, they inferred that O^- ions could still be involved in the oxidation process, since electron attachment on the oxide surface by oxygen atoms could occur. In a parallel paper, Kimura et al.[187] investigated the oxidation of silicon in their ECR system employing $^{18}O_2$ as a tracer. Using various combinations of thermal and plasma oxidation with different oxygen isotopes, they concluded that plasma oxidation drives the replacement and consequent "scrambling" of oxygen atoms in the oxide layer. This agrees with the sugestion originally inferred by Greaves and Linnett[29] some 30 years prior.

Although most attempts to create oxides centre around the oxidation of silicon, Matsumura[188] used an ECR oxygen plasma to anodize aluminium, this time, the substrate being connected to a source of DC current and biassed positive with respect to the grounded chamber. The sample used consisted of an optical flat, coated with sputtered aluminium, which was biassed to the plasma potential. This procedure was adopted so that the electrical environment surrounding the sample closely mimicked that in the bulk plasma, thus allowing the current due to the electrons in the plasma to

---

† P-etch: a mixture of acids with a specific composition,[60] commonly used to determine the quality of an oxide by measuring its etch rate. It is called P-etch in honour of its inventor, William (Bill) Pliskin, a colourful moustachio'd character who retired from IBM.

be determined. As anodization progressed, the potential applied to the sample was adjusted to maintain a constant current at the substrate.

The use of lasers for the oxidation of various surfaces coupled with a plasma was also reported by Matsumura[189] who used the same plasma system as in the previous work and coupled it with an Argon ion laser. At a system pressure of 3.3 Pa, the maximum increase in the anodization efficiency of an aluminium sample due to the laser irradiation was around 20%. It was believed that the function of the laser irradiation was to dissociate oxygen molecules in the plasma sheath, near the substrate, although no evidence for this mechanism was provided.

Surface treatment in ECR plasmas. Because of the low ion energies associated with ECR systems, they have been employed in the surface treatment of various materials. In particular they have been used in the cleaning of GaAs surfaces by hydrogen ECR plasmas. Kondo and Namishi[190] used in situ reflection high energy electron diffraction (RHEED) to study the effect of this plasma treatment. They found that it was quite effective in removing carbon, but surprisingly not quite as effective in removing oxygen. Lu et al.[191] also used hydrogen based ECR plasmas to clean GaAs surfaces. They used both x-ray photoelectron spectroscopy (XPS) and low energy electron diffraction to investigate the removal of the surface oxides by H_2 ECR plasmas. They showed that the removal of the arsenic oxides was more rapid than that of the gallium oxides, which they concluded was due to the lower thermodynamic stability of the arsenic based compounds. After a comparatively lengthy processing in the plasma, they did note that it was possible to remove the gallium oxide and to restore a well ordered GaAs surface. They also found that heating the surface greatly increased the rate of oxide removal, as Kondo and Namishi[190] as well as others[192] had also determined. Despite the efficacy of this process, the authors discovered that thermal treatment to 200° C was necessary to completely eliminate all traces of hydrides formed by the plasma.

A companion work by Lu et al.[193] used oxygen based ECR plasmas to purposely create oxides on the GaAs surface. By applying negative DC bias to the substrate, which was believed to exclude electron and negative ion bombardment of the surface, they found a slightly more rapid gowth rate of oxides than with a positive bias. They also determined that the efficiency of oxidation by the ECR plasma was higher than that obtained by thermal or laser oxidation. Oxidation times of 5 minutes were sufficient to produce 22 Å thick oxide layers.

15.6.4. ECR assisted chemical vapor deposition (CVD). A variety of different materials have been deposited from ECR plasmas. The following section reflects the use of an ECR plasma to decompose a reactive gas to yield a layer of film.

Silicon. In contrast to the usual axial magnetic field approach, K. and I. Kato[194] used a magnetic field transverse to the chamber axis to generate an ECR plasma. This required that the magnet polepieces be placed inside the plasma chamber. The motivation for this approach was to reduce substrate heating by ions, which occurs in the externally applied axial magnetic field systems.

Meija et al.[195] used a WR 284 waveguide section as a discharge chamber to contain an ECR plasma with a mirror magnetic field. Silane and hydrogen were used as feed gases to deposit amorphous silicon (a-Si:H) films whose electrical properties were comparable to the best RF deposited a-Si:H obtainable at that time. The authors found some correlation between the microcrystallite size and deposition rate, although this was not definitive. Similar work was undertaken by Watanabe et al.[196] who deposited a-Si:H from silane in a plasma stream ECR reactor operating in the whistler mode. Argon was diverted into the plasma chamber whose effluent mixed with a shower of silane placed outside the chamber and above the silicon substrate. A large number of parametric measurements were taken and showed that the property of the deposited a-Si:H depended upon a wide range of parameters including pressure, substrate temperature, gas flow rate. They found that the best films, as judged by their optical gap, photoconductivity, film density, refractive index and IR absorption, could only be formed over a narrow range of pressure. In subsequent work, Watanabe et al.[197] studied the properties of a-Si:H deposited from either a plasma stream interacting with a shower of silane or direct plasma excitation of silane in an ECR reactor. They used Corning #7059 glass or FZ silicon wafers as substrates upon which 7000 Å films were deposited. They found that both direct excitation of silane and remote excitation, using H_2 as a carrier, were capable of generating a-Si:H on these substrates. In contrast to the previous data obtained using Ar as the carrier gas, they found that the photoconductive properties of the a-Si:H were strongly affected by the plasma conditions.

Boron and phosphorus doped amorphous silicon were prepared by Kitagawa et al.[198] using a plasma stream reactor. Silane, phosphine and diborane were injected below the exit of the plasma stream and deposited on a substrate which was held at 60° C. Although the as-deposited conductivities were below 2×10^{-7} and 3×10^{-8} (ohm cm)$^{-1}$ for the P and B doped samples respectively, annealing at 300° C increased these values to around 10^{-4} (ohm cm)$^{-1}$. This conductivity is approximately the same as that obtained for RF plasma deposition at the same doping level for the B doped material, whereas it is about 3 orders of magnitude too low for the P doped sample. They concluded that the band gap of their material (1.9 eV) would allow it to be used for windows for solar cells.

Fujita et al.[199] prepared a-Si:H by deposition from a 20% SiH_4 ECR plasma. They estimated that the substrate temperature ranged from 50-100° C, although they neither controlled nor measured it. Scanning electron micrographs showed a difference in morphology with placement relative to the intense plasma created at the microwave inlet: those closer to the plasma were much grainier than those deposited further from the source. They also found that oxygen and nitrogen were incorporated into their films.

Amorphous Si-Ge alloys were generated in an ECR plasma by Watanabe et al.[200] using undiluted silane and germane as source gases. They compared the use of hydrogen and argon plasmas as excitation sources for the SiH_4/GeH_4 gases with direct excitation of a 7:3 mixture of $SiH_4:GeH_4$. The photoconductive properties of the alloy were found to be dependent upon a large number of parameters including excitation method, microwave power and pressure. These alloys are very important in the fabrication of high speed semiconductor devices.

Silicon dioxide and silicon nitride. Silicon dioxide and silicon nitride were deposited from silane/oxygen and silane/nitrogen mixtures by Matsuo and Kluchi.[201] They found both SiO_2 and Si_3N_4 could be deposited at temperatures from 50 to 150° C with characteristics which matched those of the thermally grown counterparts. In fact, one of the most controversial issues involved in the deposition of SiO_2 is whether ion bombardment is important for the formation of high quality SiO_2 films.

In an investigation of the mechanism of SiO_2 formation, Murakami et al.[202] considered a combination of in situ electron beam silicon evaporation and simultaneous oxidation using an O_2 ECR plasma. Film thicknesses of around 75 nm were deposited on a 76 mm silicon wafer by simultaneous evaporation and direction of the ECR plasma stream through a 25 mm quartz slit. From comparison of IR and ellipsometric data from SiO_2 grown by deposition in an oxygen atmosphere and in an oxygen ECR plasma, the authors concluded that peroxy structures are generated during deposition, which convert to SiO_2 after annealing. The effect of the ECR plasma on the formation of SiO_2 appeared to be minimal.

Fukuda et al.[203] determined the role of the magnetic field in an ECR system, upon the quality of the SiO_2 film grown by deposition from SiH_4 and O_2. Both divergent and convergent magnetic fields were used to study their effect on the film quality, which was measured by IR absorption, P-etch and buffered HF etch rates, and its refractive index. They found that the quality of the SiO_2 film improved markedly as the magnetic field at the deposition region approached the ECR condition. By applying a DC bias to the substrate and measuring the quality of the deposited film, they found no effect of the DC field upon the film quality. They concluded that ion bombardment has little effect upon the quality of the deposited film, despite the fact that SiO_2 is an insulator and thus would be expected to pass little current when DC potential was applied. Instead, they proposed that excited ions were responsible in an undefined manner for the quality of the films, which approached that of thermal SiO_2. In a subsequent publication,[204] the authors expanded on their earlier work to include optical emission measurements. The results of these experiments showed some correlation between the emission from O_2^+ ions and the film quality, although they were not conclusive.

Van Nguyen and Albaugh[205] used ellipsometry, Auger, Fourier transform infra red (FTIR) and photoelectron spectrocopies together with transmission electron microscopy, HF acid etching and nuclear reaction analysis to determine the physical and bonding properties of silicon nitride and silicon oxide films deposited in various undescribed "commercial" ECR deposition systems. These systems were also equipped with RF substrate excitation. A comparison of the films deposited from these ECR systems with those obtained using i) direct high temperature processing and ii) standard RF plasma deposition showed that the hydrogen content of the nitride and oxide films in ECR systems was much lower than that obtained using standard RF plasma deposition. FTIR spectra showed that the bonding structure of ECR films was similar to that of thermally grown films, even though the former were created at room temperature. Without the RF bias, the ECR grown films showed a greater and more non-uniform distribution of hydrogen, especially for silicon nitride. Higher substrate temperatures produced a hydrogen content of the oxide which approached that of the thermally grown material. Scanning electron micrograph examination of step coverage of deep (7-8 μm, 1 μm wide) trenches showed that the relationship between RF bias, substrate temperature and resultant porosity of the film was complex.

A mixture of 10 % SiH_4 in Ar was used by Chau et al.[206] to deposit SiO_2 films in a waveguide based ECR system. Breakdown strengths of 5-6 MV/cm and CV characteristics of these films, deposited at rates of 20 Å/min, were determined.

15.7 Concluding remarks

From the preceding chapter, it is clear that although the term "microwave plasma system" is widely used, it really encompasses a large number of combinations of electrical, geometrical and chemical elements. Since the interaction between these components is poorly understood and not likely to be resolved in the next few years, it is probable that a plethora of different systems and processes will pervade the market. As with most systems, the particular application is dominated by the process requirements, rather than by one particular component of the system such as the plasma excitation frequency.

Nonetheless, it is true to say that the use of different frequencies appears to change the electron energy distribution in the plasma, which can have profound effects upon the subsequent chemical and physical processes. This is evidenced by experiments such as that used to measure the etch rate variation of polyimide as a function of frequency. As a greater understanding of this and other effects of excitation frequency develops, this parameter will be used to full advantage.

In practical terms, microwave plasmas have been used in instances where high density plasmas are required which do not produce energetic ion bombardment. These include such applications as polymer removal and atom generation from molecular gases. A side effect of the use of microwave plasma excitation which has been used to great advantage in the creation of diamond films from the plasma has been to use the microwave energy to heat the substrate directly.

As in all applications, clear, insightful study of the fundamental aspects of microwave plasma technology brings the most rapid progress. A through understanding of the effect of plasma excitation frequency upon electron energy distribution function would be very useful in this regard, as would a notion of the ion trajectory in these plasmas. Unfortunately, more efforts appear to be expended on technological evolution, than on an understanding of the seemingly abstract issues.

References

[1]   A. Herlin, Phys. Rev. **74**, 291 (1948).

[2]   P. Fauchais, E. Boudrin, J. Coudert and R. McPherson, High pressure plasmas and their application to ceramic technology, in *Topics in Current Chemistry, Plasma Chemistry IV* (Springer Verlag, 1983).

[3]   D. Beauchemin, J. Hubert and M. Moisan, Appl. Spectros. **40**, 379 (1986).

[4]   C. Moutoulas, M. Moisan, L. Bertrand, J. Hubert, J.L. Lachambre and A. Ricard, Appl. Phys. Lett. **46**, 323 (1985).

[5]   S. Greenfield, H. McD. McGeachin and P.B. Smith, Part I in Talanta **22**, 1 (1975), Part II in Talanta **22**, 553 (1975) and Part III in Talanta **23**, 1 (1976).

[6]   J. Sneddon, R. Browner, P. Keliher, J. Winefordner, D.J. Butcher and R. Michel, Prog. Analyt. Spectrosc. **12**, 369 (1989).

[7]   J.A.C. Broekaert, Analytica Chimica Acta **196**, 1 (1987).

[8]   *Development of Microwave Plasma Detoxification Process for Hazardous Wastes, Phase I*, PB-268 526, Lockheed Missiles and Space Co. Inc. (Palo Alto, Calif.) (April 1977).

[9]   G.R. Kilgore, IEEE Trans. on Electron Devices **ED-31**, 1593 (1984).

[10] T. Saad, R.C. Hansen and G. Wheeler, *Microwave Engineer's Handbook* (Artech House, 1971) Vols. 1 and 2.

[11] V.F. Veley, *Modern Microwave Technology* (Prentice Hall, 1987).

[12] S. Dzioba, G. Este and H.M. Naguib, J. Electrochem. Soc. **129**, 2437 (1982).

[13] A.R. Reinberg, *Radial Flow Reactor*, U.S. Patent 3 757 733 (1973).

[14] See for example systems made by GaAsonics or Plasma Technology (USA).

[15] M. Pichot and the Interaction Plasma Surface Department, Microelectronic Eng. **3**, 411 (1985).

[16] R.G. Bosisio, C.F. Weissfloch and M.R. Wertheimer, J. Microwave Power **7**, 325 (1972).

[17] A.J. Mendelsohn, R. Normandin, S.E. Harris and J.F. Young, Appl. Phys. Lett. **38**, 603 (1981).

[18] Gerling Laboratories, 1132 Doker Drive, Modesto, CA 95351 (USA).

[19] J.M. Cook, D.E. Ibbotson, P.D. Foo and D.L. Flamm, J. Vac. Sci. Technol. **A8**, 1820 (1990).

[20] J. Hubert, M. Moisan and Z. Zakrzewski, Spectrochim. Acta **41B**, 205 (1986).

[21] T. Ono, M. Oda, C. Takahashi and S. Matsuo, J. Vac. Sci. Technol. **B4**, 696 (1986).

[22] F.C. Fehsenfeld, K.M. Evenson and H.P. Broida, Rev. Sci. Instrum. **36**, 294 (1965).

[23] S. Goode and K. Baugh, Electrochem. Soc. **131**, 1875 (1977).

[24] J. Musil, Vacuum **36**, 169 (1986).

[25] H. Suhr, Plasma Chem. and Plasma Proces. **3**, 1 (1983).

[26] M. Shibagaki, Y. Horiike and T. Yamazaki, Jap. J. Appl. Phys. **17**, 215 (1978).

[27] K. Jinno, Jap. J. Appl. Phys. **17**, 1283 (1978).

[28] P.D. Francis, Brit. J. Appl. Phys. (J. Phys. D) **2**, 1717 (1969).

[29] J.C. Greaves and J.W. Linnett, Trans. Farad. Soc. **55**, 1338, 1346 and 1355 (1959).

[30] R.L. Brown, J. Phys. Chem. **71**, 2492 (1967).

[31] J.M. Cook and B.W. Benson, J. Electrochem. Soc. **130**, 2459 (1983).

[32] J. Paraszczak, M. Hatzakis, E. Babich, J. Shaw, E. Arthur, B. Grenon and M. DePaul, Proc. Microcircuit Eng. **2**, 517 (1984).

[33] B. Robinson and S. Shivashankar, *Proc. 5th Symp. Plasma Proces.*, G.S. Schwartz and G. Smolinsky eds., Electrochem. Soc. **85-1**, 206 (1984).

[34] A.T. Bell and K. Kwong, Ind. Eng. Chem. Fundam. **12**, 90 (1973).

[35] M. Brake, J. Hinkle, J. Asmussen, M. Hawley and R. Kerber, Plasma Chem. and Plasma Proces. **3**, 63 (1983).

[36] M.L. Brake and R.L. Kerber, Plasma Chem. and Plasma Proces. **3**, 79 (1983).

[37] R.J. Cvetanovic, J. Phys. Chem. Ref. Data **16**, 261 (1987).

[38] J.E. Spencer, R.A. Borel, A. Hoff, J. Electrochem. Soc. **133**, 1922 (1986).

[39] J.F. Battey, J. Electrochem. Soc. **124**, 147 (1977).

[40] A.G. Nagy, J. Electrochem. Soc. **131**, 1871 (1984).

[41] J. Paraszczak, J. Heidenreich, M. Hatzakis and M. Moisan, Microelec. Eng. **3**, 397 (1985).

[42] T.H. Lin, M. Belser and Y. Tzeng, IEEE Trans. Plasma Sci. **16**, 631 (1988).

[43] J. Heidenreich, J. Paraszczak, M. Moisan and G. Sauvé, J. Vac. Sci. Technol. **B6**, 288 (1988).

[44] J. Heidenreich, J. Paraszczak, M. Moisan and G. Sauvé, Microelec. Eng. **5**, 363 (1986).

[45] C. Jurgensen, A. Novembre and E. Shaqfeh, *Proc. of the SPIE Advances in Resist Technology and Processing VII* **1262**, 94 (1990).

[46] J. Paraszczak, E. Babich, J. Heidenreich, R. McGouey, L. Ferreiro, N. Chou and M. Hatzakis, *Proc. of the SPIE Advances in Resist Technology and Processing V*, **920**, 243 (1988).

[47] N.J. Chou, J. Paraszczak, E. Babich, Y.S. Chaug and R. Goldblatt, J. Vac. Sci. Technol. **A5**, 1321 (1987).

[48] F.D. Egitto, V. Vukanovic, F. Emmi and R.S. Horwarth, J. Vac. Sci. Technol. **B3**, 893 (1985).

[49] J. Dedinas, M.M. Feldman, M.G. Mason and L.J. Gerenser, *Proc. 1st Annual International Conference of Plasma Chemistry and Technology* (San Diego, CA) 119 (1982).

[50] M. Tsuda, S. Oikawa, S. Ohnogi and A. Suzuki, Proc. Microcircuit Eng., p. 553 (1980).

[51] J.F. Battey, IEEE Trans. Electron Devices, **ED-24**, 140 (1977).

[52] G.N. Taylor, T.M. Wolf and J.M. Moran, J. Vac. Sci. Technol. **19**, 872 (1981).

[53] R.H. Hansen, J.V. Pascale, T. de Benedictis and P.M. Rentzepis, J. Polym. Sci. **A3**, 2205 (1965).

[54] S.J. Moss, Polymer Degradation and Stability **17**, 205 (1987).

[55] S.J. Moss, A.M. Jolly and B.J. Tighe, Plasma Chem. and Plasma Proces. **6**, 401 (1986).

[56] S.R. Cain, F.D. Egitto and F. Emmi, J. Vac. Sci. Technol. **A5**, 1578 (1987).

[57] F.D. Egitto, Pure and App. Chem. **62**, 1699 (1990).

[58] H. Beachell and L. Smiley, J. Polymer Sci. A-1, **5**, 1635 (1967).

[59] J. Pelletier, Y. Arnal and O. Joubert, Appl. Phys. Lett. **53**, 1914 (1988).

[60] J.S. Logan, IBM J. Res. Develop. **14**, 172 (1970).

[61] D.M. Manos and H.F. Dylla, Diagnostics of Plasmas for Material Processing, in *Plasma Etching, an Introduction* , D.M. Manos and D.L. Flamm eds. (Academic, 1989) pp. 300-307.

[62] R. Gottscho and M. Mandich, J. Vac. Sci. Technol. **A3**, 617 (1985).

[63] J.W. Coburn and H.F. Winters, J. Appl. Phys. **50**, 3189 (1979).

[64] A.W. Kolfschoten, R.A. Haring, A. Haring and A.E. deVries, J. Appl. Phys. **55**, 3813 (1984).

[65] Y. Arnal, J. Pelletier, C. Pomot, B. Petit and A. Durandet, Appl. Phys. Lett. **45**, 132 (1984).

[66] J.M.E. Harper, J.J. Cuomo, P.A. Leary, G.M. Summa, H.R. Kaufman and F.J. Bresnock, J. Electrochem. Soc. **128**, 1077 (1981).

[67] T.M. Mayer, J.M.E. Harper and J.J. Cuomo, J. Vac. Sci. Technol. **A3**, 1779 (1985).

[68] M.R. Wertheimer and M. Moisan, J. Vac. Sci. Technol. **A3**, 2643 (1985).

[69] C.M. Ferreira and J. Loureiro, J. Phys. D: Appl. Phys. **17**, 1175 (1984).

[70] M. Rapeaux and G. Turban, J. Electrochem. Soc. **130**, 2231 (1983).

[71] G. Sauvé, M. Moisan, J. Paraszczak and J. Heindreich, Appl. Phys. Lett. **53**, 470 (1988).

[72] J.J. Hannon and J.M. Cook, J. Electrochem. Soc. **131**, 1164 (1984).

[73] G. Sauvé, M. Moisan, J. Paraszczak, Y. Arnal, J. Heidenreich and A. Ricard, Microelec. Eng. **9**, 471 (1989).

[74] R. Hoffman and W.N. Lipscomb, J. Chem. Phys. **36**, 2179 (1962).

[75] J. Friedrich, Plasma Physik **21**, 261 (1981).

[76] R. Messier, K. Spear, A. Badzian and R. Roy, J. Metals **39**, 8 (1987).

[77] H.G. Wells, *The diamond maker* from the complete short stories by H.G. Wells (E. Benn, London, 1971).

[78] F.P. Bundy, H.T. Hall, H.M. Strong and R.H. Wentorf, Nature **176**, 51 (1955).

[79] W. Zhu, B.R. Stoner, B.E. Williams and J.T. Glass, Proc. IEEE **79**, 621 (1991).

[80] S. Matsumoto, Y. Sato, M. Kamo and N. Setaka, Jap. J. Appl. Phys. **21**, L183 (1982).

[81] D.S. Whitmell and R. Williamson, Thin Solid films **58**, 255 (1976).

[82] L. Holland and S.M. Ojha, Thin Solid Films **58**, 107 (1979).

[83] M. Sokolowski and A. Sokolowska, J. Crystal Growth **57**, 185 (1982).

[84] H. Vora and J. Moravec, J. Appl. Phys. **52**, 6151 (1971).

[85] Y. Saito, S. Matsuda and S. Nogita, J. Mat. Sci. Lett. **5**, 565 (1986).

[86] D.E. Rosner, Annual Rev. of Materials Sci. **2**, 573 (1972).

[87] Y. Saito, K. Sato, H. Tanaka, K. Fujita and S. Matuda, J. Mat. Sci. **23**, 842 (1988).

[88] Y. Hirose and T. Terasawa, Jap. J. Appl. Phys. **25**, L519 (1986).

[89] Y. Saito, K. Sato, H. Tanaka and H. Miyadera, Jap. Mat. Sci. **24**, 293 (1989).

[90] E. Vietzke, V. Philipps, R. Flaskamp, J. Winter, S. Veprek and P. Koidl, *IUPAC Proc. ISPC-10*, U. Ehleman, H. Lergon and K. Wiesemann eds. (Bochum, 1991) p. 3.1.1.

[91] E.Y.Y. Lam, P. Gaspar and A. Wolf, J. Phys. Chem. **75**, 445 (1971).

[92] I. Watanabe and K. Sugata, Jap. J. Appl. Phys. **27**, 1808 (1988).

[93] Y. Liou, A. Inspektor, R. Weimer and R. Messier, Appl. Phys. Lett. **55**, 631 (1989).

[94] I. Watanabe and K. Sugata, Jap. J. Appl. Phys. **27**, 1397 (1988).

[95] K. Ito, T. Ito and I. Hosoya, Chem. Let. **4**, 589 (1988).

[96] S. Hoshino, K. Fujii, N. Shohata, H. Yamaguchi, Y. Tsukamoto and M. Yanagisawa, J. Appl. Phys. **65**, 1918 (1989).

[97] A.R. Badzian, T. Badzian, R. Roy, R. Messier and K.E. Spear, Mater. Res. Bull. **23**, 531 (1988).

[98] S.J. Harris and A.M. Weiner, Appl. Phys. Lett. **55**, 2179 (1989).

[99] S. Kaplan, F. Jansen and M. Machonkin, Appl. Phys. Lett. **47**, 751 (1985).

[100] W. Zhu, C.A. Randall, A.R. Bazdian and R. Messier, J. Vac. Sci. Technol. **A7**, 2315 (1989).

[101] B.E. Williams and J.T. Glass, J. Mater. Res. **4**, 373 (1989).

[102]   J. Kieser and M. Sellschop, J. Vac. Sci. Technol. **A4**, 222 (1986).

[103]   M. Kamo, Y. Sato, T. Tanaka, N. Setaka and H. Chawanya, Material Science and Engineering A - Structural Materials Properties Microstructure and Processing **106**, 535 (1988).

[104]   J. Suzuki, H. Kawarada, K.S. Mar, J. Wei, Y. Yokota and A. Hiraki, Jap. J. Appl. Phys. **28**, L281 (1989).

[105]   A. Hiraki, H. Kawarada, K.S. Mar, Y. Yokoto, J. Wei and J.I. Suzuki, Nucl. Inst. and Methods in Phys. Res. **B37/38**, 799 (1989).

[106]   M. Nunotani, M. Komori, M. Yamasawa, Y. Fujiwara, K. Sakuta, T. Kobayashi, S. Nakamisha, S. Monomo, M. Taniguchi and M. Sugiyo, Jap. J. Appl. Phys. **30**, L1190 (1991).

[107]   J. Wei, H. Kawarada, J. Suzuki, J. Ma and A. Hiraki, Jap. J. Appl. Phys. **30**, 1279 (1991).

[108]   T. Fujita and O. Matsumoto, J. Electrochem. Soc. **136**, 2625 (1989).

[109]   W. Biter, Thin Solid Films **188**, 231 (1984).

[110]   I. Kato, S. Wakana and S. Hara, Jap. J. Appl. Phys. **22**, L40, (1983).

[111]   D. Masson, E. Sacher and A. Yelon, Phys. Rev. **B35**, 1260 (1987).

[112]   R. Manory, A. Grill, U. Carmi and R. Avni, Plasma Chem. Plasma Proces. **3**, 235 (1983).

[113]   C.S. Pai and C.P. Chang, J. Appl. Phys. **68**, 793 (1990).

[114]   S.K. Ray, C.K. Maiti, S.K. Lahiri, N.B. Chak and R.A. Borti, Semicon. Sci. Techn. **5**, 361 (1990).

[115]   W. Warren, J. Stathis, P. Lenahan and B. Robinson, Appl. Phys. Lett. **53**, 482 (1988).

[116]   A.K. Sinha, H.J. Levinstein, T.E. Smith, G. Quintana and S.E. Haszko, J. Electrochem. Soc. **125**, 601 (1978).

[117]   A.C. Adams, in *Plasma Deposited Thin Films*, J. Mort and F. Jansen eds. (CRC Press, Boca Raton, Florida, 1986).

[118]   A.K. Sinha, H.J. Levinstein and T.E. Smith, J. Appl. Phys. **49**, 2432 (1978).

[119]   L. Bardos, J. Musil and P. Taras, J. Phys. D: Appl. Phys. **15**, L79 (1982).

[120]   Y. Tessier, J.E. Klemberg-Sapieha, S. Poulin-Dandurand, M.R. Wertheimer and S. Gujrathi, Can. J. Phys. **65**, 859 (1987).

[121]   I. Kato, Y. Kohyama and K. Noguchi, Electron. Comm. Japan, Part 2 **69**, 39 (1986).

[122]   I. Kato, K. Naguchi and K. Numada, J. Appl. Phys. **62**, 492 (1987).

[123]   I. Kato, K. Numada and Y. Kiyota, Jap. J. Appl. Phys. **27**, 1401 (1988).

[124]   K. Yasui, M. Nasu, K. Komaki and S. Kaneda, Jap. J. Appl. Phys. **29**, 918 (1990).

[125]   J.R. Ligenza, J. Appl. Phys. **36**, 2703 (1965).

[126]   J. Kraitchman, J. Appl. Phys. **38**, 4323 (1967).

[127]   A. Kiermasz, W. Eccleston and J.L. Moruzzi, Sol. State Electron. **26**, 1167 (1983).

[128]   L. Bardos, G. Loncar, I. Stoll, J. Musil and F. Zacek, J. Phys. D: Appl. Phys. **8**, L195 (1975).

[129]   O.A. Weinreich, J. Appl. Phys. **38**, 4323 (1967).

[130]   J. Musil, F. Zacek, L. Bardos, G. Locar and R. Dragila, J. Phys. D: Appl. Phys. **12**, L61 (1979).

[131] J. Ruzyllo, A. Hoff and G. Ruggles, J. Electron. Mat. **16**, 373 (1987).

[132] S. Kimura, E. Murakani, T. Warabisako, H. Sunami and T. Tokuyama, IEEE Electron. Dev. Lett. **ED-7**, 38 (1986).

[133] M.R. Wertheimer and H.P. Schreiber, J. Appl. Polymers Sci. **26**, 2087 (1981).

[134] B. Mutel, O. Dessaux, P. Goudmand, J. Grimblot, A. Carpentier and S. Szarzynski, Revue Phys. Appl. **234**, 1253 (1988).

[135] J. Ruzyllo, A.M. Hoff, D.C. Frystak and S.D. Hossain, J. Electrochem. Soc. **136**, 1474 (1989).

[136] A.A. Balmashnov, K.S. Golovanisvsky, E.M. Omeljanovsky, A.V. Pakhomov and A.Y. Polyakov, Semicond. Sci. Technol. **5**, 242 (1990).

[137] S.A. Grot, G.Sh. Gildenblat, C.W. Hatfield, C.R. Wronski, A.R. Badzian, T. Badzian and R. Messier, IEEE Electron. Dev. Lett. **ED-11**, 100 (1990).

[138] Y. Okamoto and H. Tamagawa, Rev. Sci. Instrum. **43**, 1193 (1972).

[139] R. Bardet, T. Consoli and R. Geller, Nucl. Fusion **4**, 481 (1964).

[140] D.B. Miller and E.F. Gibbon, A.I.A.A. **2**, 35 (1964).

[141] R. Geller, IEEE Trans. Nucl. Sci. **2**, 2120 (1979).

[142] J. Asmussen and J. Root, Appl. Phys. Lett. **44**, 396 (1984).

[143] E. Ghanbari, I. Trigor and T. Nguyen, J. Vac. Sci. Technol. **A7**, 918 (1989).

[144] R.K. Yafarov, S.A. Terentev, A.P. Telitsyn and A.O. Balakin, Instrum. and Experim. Techn. (USSR) **32**, 651 (1989).

[145] R. Guarnieri, private communication.

[146] J. Asmussen, J. Vac. Sci. Technol. **A7**, 883 (1989).

[147] M. Matsuoka and K. Ono, Jap. J. Appl. Phys. **28**, 1503 (1989).

[148] K. Tokiguchi, N. Sakudo and H. Koike, J. Vac. Sci. Technol. **A2**, 29 (1984).

[149] T. Roppel, D.K. Dienhard and J. Asmussen, J. Vac. Sci. Technol. **B4**, 295 (1986).

[150] M. Dahimene and J. Asmussen, J. Vac. Sci. Technol. **B4**, 126 (1986).

[151] L. Mahoney, M. Dahimene and J. Asmussen, Rev. Sci. Instrum. **59**, 448 (1988).

[152] O. Popov, J. Vac. Sci. Technol. **A7**, 894 (1989).

[153] M. Matsuoka and K. Ono, Appl. Phys. Lett. **50**, 1864 (1987).

[154] K. Suzuki, S. Okudaira, N. Sakudo and I. Kanomata, Jap. J. Appl. Phys. **16**, 1979 (1977).

[155] K. Suzuki, S. Okudaira and I. Kanomata, J. Electrochem. Soc. **126**, 1024 (1979).

[156] K. Suzuki, S. Okudaira, S. Nishimatsu, K. Usami and I. Kanomata, J. Electrochem. Soc. **129**, 2764 (1982).

[157] K. Suzuki, K. Ninomiya, S. Nishimatsu and S. Okudaira, J. Vac. Sci. Technol. **B3**, 1025 (1985).

[158] J. Hopwood, P.K. Reinhard and J. Asmussen, J. Vac. Sci. Technol. **B6**, 1896 (1988).

[159] J.W. Coburn and H.F. Winters, Ann. Rev. Mater. Sci. **13**, 91 (1983).

[160] J. Paraszczak and M. Hatzakis, J. Vac. Sci. Technol. **19**, 1412 (1981).

[161] D.L. Flamm, Introduction to plasma chemistry in *Plasma etching: an introduction*, D.M. Manos and D.L. Flamm eds. (Academic, New York, 1988) chap. 2.

[162] Y. Tobinaga, N. Hayashi, H. Araki, S. Nakayama and S. Kudoh, J. Vac. Sci. Technol. **B6**, 272 (1988).

[163] J.M. Forster and W. Holber, J. Vac. Sci. Technol. **A7**, 899 (1989).

[164] W. Holber and J.M. Forster, J. Vac. Sci. Technol. **A8**, 3720 (1990).

[165] M. Matsuoka and K. Ono, J. Vac. Sci. Technol. **A6**, 25 (1988).

[166] J.S. McKillop, J.M. Forster and W. Holber, J. Vac. Sci. Technol. **A7**, 908 (1989).

[167] J. O'Neill, W. Holber and J. Caughmann, *Proc. of the 1990 SPIE Symposium on Real-Time Monitoring and Control* (Santa Clara, 1990).

[168] A. Yonesu, Y. Takeuchi, A. Komori and Y. Kawai, Jap. J. Appl. Phys. **27**, 1748 (1988).

[169] L. Bardos and J. Musil, J. Phys. D. **21**, 1459 (1988).

[170] Y.H. Lee, J.E. Heidenreich and G. Fortuno, J. Vac. Sci. Technol. **A7**, 903 (1988).

[171] C. Keqiang, Z. Erli, W. Jinfa, G. Zuoya and Z. Bangwei, J. Vac. Sci. Technol. **A4**, 828 (1986).

[172] S. Meikle and Y. Hatanaka, App. Phys. Lett. **54**, 1648 (1989).

[173] K. Machida and H. Okowa, J. Vac. Sci. Technol. **B4**, 818 (1986).

[174] T. Ono, C. Takahashi and S. Matsuo, Jap. J. Appl. Phys. **23**, L534 (1984).

[175] C. Takahashi, M. Kiuchi, T. Ono and S. Matsuo, J. Vac. Sci. Technol. **A6**, 234 (1988).

[176] M. Matsuoka and K. Ono, Appl. Phys. Lett. **53**, 202 (1988).

[177] M. Matsuoka and K. Ono, J. Appl. Phys. **64**, 5179 (1988).

[178] M. Matsuoka and K. Ono, J. Appl. Phys. **65**, 4403 (1989).

[179] M. Matsuoka and K. Ono, J. Vac. Sci. Technol. **A7**, 2976 (1989).

[180] S. Minomo, M. Taniguchi, M. Sugiyo, T. Takahashi, M. Tonouchi, S. Kita and T. Kobayashi, Jap. J. Appl. Phys. **27**, L411 (1988).

[181] T. Goto, H. Matsumoto and T. Hirai, Jap. J. Appl. Phys. **28**, L88 (1989).

[182] Y. Nishimori, S. Minamo, M. Taniguchi, M. Sugiyo, Y. Fukumoto, Y. Fujiwari and T. Kobayashi, Jap. J. Appl. Phys. **28**, L1220 (1984).

[183] K. Miyake, S. Kimura, T. Warabisako, H. Sunami and T. Tokuyama, J. Vac. Sci. Technol. **A2**, 496 (1984).

[184] S. Kimura, E. Murakami, K. Miyake, T. Warabisako, H. Sunami and T. Tokuyama, J. Electrochem. Soc. **132**, 1460 (1985).

[185] K. Matsumura, J. Appl. Phys. **65**, 1866 (1989).

[186] S. Kimura, T. Warabisako, H. Sunami and E. Murakami, J. Electrochem. Soc. **135**, 2009 (1988).

[187] S. Kimura, E. Murakami, T. Warabisako, E. Mitani and H. Sunami, J. Appl. Phys. **63**, 4655 (1988).

[188] K. Matsumura, J. Appl. Phys. **65**, 5101 (1989).

[189] K. Matsumura, J. Appl. Phys. **66**, 1103 (1989).

[190] N. Kondo and Y. Namishi, Jap. J. Appl. Phys. **28**, L7 (1989).

[191] Z. Lu, M.T. Schmidt, D. Chen, R.M. Osgood Jr., W.M. Holber, D.V. Podlesnik and J. Forster, Appl. Phys. Lett. **58**, 1143 (1991).

[192] A. Takamori, S. Sugata, K. Asakawa, E. Miyauchi and H. Hashimoto, Jap. J. Appl. Phys. **26**, L142 (1987).

[193] Z. Lu, M.T. Schmidt, R.M. Osgood Jr., W.M. Holber and D.V. Podlesnik, J. Vac. Sci. Technol. **A9**, 1040 (1991).

[194] K. Kato and I. Kato, Jap. J. Appl. Phys. **3**, L343 (1989).

[195] S.R. Mejia, R.D. McLeod, K.C. Kao and H.C. Card, Rev. Sci. Instrum. **57**, 494 (1986).

[196] T. Watanabe, K. Azuma, M. Nakatani, K. Suzuki, T. Sonobe and T. Shimada, Jap. J. Appl. Phys. **25**, 1805 (1986).

[197] T. Watanabe, M. Tanake, K. Azuma, M. Nakatani, T. Sonobe and T. Shimada, Jap. J. Appl. Phys. **26**, 1215 (1987).

[198] M. Kitagawa, K. Setsune, Y. Manabe and T. Hirao, J. Appl. Phys. **61**, 2084 (1987).

[199] H. Fujita, H. Handa, M. Nagano and H. Matsuo, Jap. J. Appl. Phys. **26**, 1112 (1987).

[200] T. Watanabe, K. Azuma, M. Tanaka, M. Nakatani, T. Sonobe and T. Shimada, Jap. J. Appl. Phys. **27**, 1126 (1988).

[201] S. Matsuo and M. Kluchi, Jap. J. Appl. Phys. **22**, L210 (1983).

[202] E. Murakami, S. Kimura, K. Kiyoshi and T. Warabisako, J. Appl. Phys. **62**, 3063 (1987).

[203] T. Fukuda, K. Suzuki, S. Takahashi, Y. Mochizuki, K. Ohue, N. Momma and T. Sonobe, Jap. J. Appl. Phys. **27**, L1962 (1988).

[204] T. Fukuda, M. Ohue, N. Momma, K. Suzuki and T. Sonobe, Jap. J. Appl. Phys. **28**, 1035 (1989).

[205] S. Van Nguyen and K. Albaugh, J. Electrochem. Soc. **138**, 2835 (1989).

[206] T.T. Chau, S.R. Mejia and K.C. Kao, Electron. Lett. **25**, 1088 (1989).

INDEX OF SYMBOLS

Vector quantities are designated by boldface letters while tensors are represented by letters underlined with a tilde. With harmonic functions of time, we use italic letters to designate their real values and upright letters for their complex amplitudes.

Latin letters

a	Radius of an elementary magnet
a	Coefficient
a_w	Width of a waveguide
A	Vector potential
$A(\exp j\omega t)$	Harmonic function of time: complex value
$A(\exp j\omega t)$	Harmonic function of time: real value
$A(\bar{n})$	Power absorbed from the field by the plasma column per unit length
b	Normalized susceptance
b	Coefficient
$\mathbf{b_b}$	Reduced magnetic flux density: $\mathbf{B_0} = B_{0m}\mathbf{b_b}$; "$b_b$ = constant" defines lines of constant magnetic field intensity
b_w	Height of a waveguide $(a_w > b_w)$
$\mathcal{B}$	Susceptance (imaginary part of $\mathcal{Y}$) (eqn. (2.77))
$\mathcal{B}$	Ratio of n_e to n_i in plasmas containing negative ions: $\mathcal{B} = n_e/n_i$
$\mathcal{B}_D$	Imaginary part of $\mathcal{Y}_D$ (eqn. (2.80))
$\mathbf{B_0}$	Static magnetic flux density
$\mathcal{B}_0$	Input susceptance of a field applicator in absence of the discharge (eqn. (2.79))
B_{0m}	Twice the value of the magnetic field intensity at the saddle point
B_{0s}	Magnetic induction at the surface of a magnet
c	Speed of light in free space $(c = 2.998 \times 10^8$ meter/second)
c	Coefficient
C	Exponent
c_i	Ion-acoustic velocity (eqn. (9.16))
C	Capacitance
$\mathcal{C}$	Constant
C_i	Ionization rate coefficient by electron impact
C_{pq}	Rate coefficient for atom (molecule) excitation from level p to level q by electron impact.
d	Distance between two points

$d\Omega$	Element of solid angle: $d\Omega = \sin \chi \, d\chi \, d\phi$
D	Diffusion coefficient (eqn. (2.58))
$\mathcal{D}$	Determinant; $\mathcal{D} = 0$ is the determinental equation (dispersion relation)
$\mathbb{D}$	Vessel dimension
$\underset{\sim}{D}$	Diffusion tensor
D_a	Ambipolar diffusion coefficient (eqn. (2.36))
$D_{a\parallel}$	Ambipolar diffusion coefficient in the direction parallel to the magnetic field (eqn. (6.11))
$D_{a\perp}$	Ambipolar diffusion coefficient in the direction perpendicular to the magnetic field (eqn. (6.12))
D_e	Free diffusion coefficient for electrons (eqn. (2.34))
$D_{e\parallel}$	Free diffusion coefficient for electrons in the direction parallel to the magnetic field
$D_{e\perp}$	Free diffusion coefficient for electrons in the direction perpendicular to the magnetic field
D_i	Free diffusion coefficient for ions
$D_{i\parallel}$	Free diffusion coefficient for ions in the direction parallel to the magnetic field
$D_{i\perp}$	Free diffusion coefficient for ions in the direction perpendicular to the magnetic field
D_s	Effective electron diffusion coefficient
D_{se}	Effective diffusion coefficient related to Self and Ewald's theory (eqn. (2.69))
$D_{M\perp}$	Diffusion coefficient from Monroe's model (eqn. (6.14))
$D_{S\perp}$	Diffusion coefficient from Simon's theory (eqn. (6.13))
$D_\parallel$	Diffusion coefficient in the direction parallel to the magnetic field
$D_\perp$	Diffusion coefficient in the direction perpendicular to the magnetic field
e	Absolute value of the electron charge ($e = 1.602 \times 10^{-19}$ coulomb)
$\mathbf{E}$	Amplitude of the complex electric field vector ($\mathbf{E} \exp(j\omega t)$)
E	Real part of $\mathbf{E}$
$\mathbf{E}^*$	Complex conjugate of $\mathbf{E}$
$\mathcal{E}$	Etch rate
$\mathcal{E}_c$	Etch rate in the center of the wafer
E_e	Effective field intensity (eqn. (2.53))
$\mathcal{E}_e$	Etch rate at a given radius of the wafer
E_p	Sum of the squared field intensities along three orthogonal axes
E_{rms}	Root mean square of the total intensity: $E_{rms} = E_p/\sqrt{2}$

E_{DC}	Electric field in a direct current discharge
$E_\perp$	Total electric field component in the direction perpendicular to $\mathbf{B}_0$. In cylindrical coordinates, $\|E_\perp\|^2 = \|E_r\|^2 + \|E_\phi\|^2$
f	Frequency
f_{pi}	Ion plasma frequency
$f(\mathbf{w})$	Electron velocity distribution function (eqn. (2.29))
$f_0(\mathbf{w})$	Isotropic component of $f(\mathbf{w})$ (eqn. (2.29))
$f_1(\mathbf{w})$	Anisotropic component of $f(\mathbf{w})$ (eqn. (2.29))
$\mathcal{F}(x)$	Function of x
$F_0(u)$	Electron energy distribution function (EEDF) (eqn. (2.32))
g	Normalized conductance
$\mathcal{g}$	Coefficient
$\mathcal{G}$	Sheath thickness
g_n	Escape factor for resonance radiation
$\mathcal{G}$	Conductance (real part of $\mathcal{Y}$) (eqn. (2.77))
$\mathcal{G}_D$	Real part of $\mathcal{Y}_D$ (eqn. (2.80))
h	Complex axial wavenumber: $h \equiv \beta - j\alpha$
h	Planck constant: 6.6262×10^{-34} Js
$h_\perp$	Transverse wavenumber: $h_\perp = (\mathcal{N}\sin\Psi)\,\omega/c$ (Sec. 6.5.5)
$\mathbf{H}$	Amplitude of the complex magnetic field vector ($\mathbf{H}\exp(j\omega t)$)
$\mathbf{H}^*$	Complex conjugate of $\mathbf{H}$
$\mathcal{H}_p$	$\mathcal{H}_p^2 \equiv \beta^2 - \beta_0^2\varepsilon_p$
I	Current amplitude
I	Flux of electrons
$\underset{\sim}{1}$	Identity tensor
I_d	Discharge current
I_e	Current due to plasma electrons
I_{es}	Saturation current of the plasma electrons
I_i	Current due to plasma ions
I_{is}	Saturation current of the plasma ions
I_m	Current due to negative ions
j	Imaginary operator: $j = \sqrt{-1}$
$\mathbf{J}$	Conduction current density (eqn. (2.11))
$\mathbb{J}$	Magnetization intensity: $\mathbb{J} = \mu_0\mathbf{M}$
J_e	Electron current density
J_{es}	Electron saturation current density
J_i	Ion current density
J_{is}	Ion saturation current density
k	Integer number
k_B	Boltzmann constant: 1.381×10^{-23} joule/kelvin

$\mathbb{K}$	Second or longitudinal adiabatic invariant (eqn. (12.47))
K_p	Rate coefficient for Penning reactions
ℓ	Length
L	Power lost in the plasma column per unit length
$\mathcal{L}$	Inductance
$\mathbb{L}$	Total path length back and forth between two mirrors
m	Azimuthal wave number (Sec. 5.3.1)
m	Integer number
m_c	Turns ratio of a transformer
m_e	Electron mass: $m_e = 9.109 \times 10^{-31}$ kg
m_i	Ion mass
$\mathcal{M}_p$	Magnetic moment of primary electrons (first adiabatic invariant)
M	Mass of a heavy particle (atom or molecule)
$\mathbb{M}$	Magnetization density (Appendix 11.4)
$\mathcal{M}$	Reduced magnetic moment of the primary electrons: $\mathcal{M} = 2\mathcal{M}_p\, B_{0m}/m_e v_p^2$
n	Plasma density or electron density
n	Integer number
n	Used when factoring electron density as $n(r,t) = n(r)\,\mathbb{N}(t)$ (eqn. (2.59))
$\mathfrak{n}$	Reduced plasma density (eqn. (11.49))
$\bar{n}$	Cross-section average of the plasma or electron density
n_c	Critical electron density ($\omega_{pe} = \omega$)
n_d	Density in the plasma source
n_e	Electron density to distinguish from n_i when necessary
n_h	Plasma density at some given point
n_i	Ion density
n_p	Electron density at which the plasma conductance equals the characteristic admittance Z_0^{-1} of the waveguide (eqn. (4.19))
n_s	Plasma density at the sheath edge
n_A	Electron density at the tube axis
n_D	Density at resonance ($\beta \to \infty$) for a surface wave. Also, an approximate value for the minimum density in a low pressure, surface-wave sustained plasma column.
n_0	Electron density at $r = z = 0$ or given value
N	Gas density or density of heavy particles
$\mathcal{N}$	Plasma refraction index (Sec. 6.5.1)
$\mathbb{N}$	Used when factoring electron density as $n(r,t) = n(r)\,\mathbb{N}(t)$ (eqn. (2.59))
p	Gas pressure

p	Integer number
$\mathcal{P}_c$	Critical impact parameter
p_z	momentum along OZ of the primary electron
p_0	Reduced gas pressure (referred to 0° C and 1 torr)
$\mathbb{P}$	Power density
$P(z)$	Power flux at z
$\mathcal{P}_f$	Probability for a free electron to remain free after n elastic collisions
P_p	Minimum power required to sustain the discharge (eqn. (4.20))
$\mathcal{P}_s$	Total probability of collision per unit path length
P_A	Power absorbed by the plasma (eqns. (2.45) and (2.52))
P_D	Power flux at $n(z) = 2n_D$
P_I	Incident HF power at the source input
P_L	Total power lost by the electrons in collisions (eqn. (2.73))
P_R	HF reflected power at the source input (Sec. 2.3.2)
P'_S	HF power lost within a matching network (Sec. 2.3.2)
P''_S	HF power lost inside the applicator (Sec. 2.3.2)
P'''_S	HF power radiated into ambient (Sec. 2.3.2)
P_0	Total wave power available at the beginning of a TWD
q_{el}	Differential cross-section for elastic collisions (eqn. (2.25))
q_s	Differential cross-section for scattering (eqn. (2.20))
$\mathcal{Q}$	Quality factor of a resonant cavity
$\mathbb{Q}$	Volumic electric charge
$Q_c(w)$	Electron-heavy particle collision cross-section for momentum transfer
Q_{el}	Cross-section for elastic collisions
Q_{en}	Cross-section for collision of electrons on neutrals
Q_i	Ionization cross-section (eqn. (2.27))
Q_s	Total collision cross-section (eqn. (2.20))
r	Radius in cylindrical or spherical coordinates
$\mathbf{r}$	Displacement vector (eqn. (2.9))
r	Normalized resistance: $r = \mathcal{R}/Z_0$ (eqn. (2.84)); real part of the normalized input impedance
r_e	Electron gyroradius
r_h	Hybrid gyroradius: $r_h = (r_e r_i)^{1/2}$
r_i	Ion gyroradius
r_p	Gyroradius of primary electrons
R	Plasma radius
$\mathcal{R}$	Resistance; real part of Z (eqn. (2.84))

R_b	Integration limit for the numerical solution of the plasma diffusion equation
$\mathbb{R}_b$	Radius of the circle of radius R_b in the (r,φ) plane once transformed in the (λ_b,μ_b) plane
$\mathbf{R}_c$	Curvature radius vector of a field line
R_m	Distance of each magnet row to the axis
R_s	Radius of a spherical substrate
$\mathcal{R}_w$	Characteristic impedance of the plasma column as seen by the surface wave
R_1	Internal radius of a discharge tube
R_2	External radius of a discharge tube
$\mathbb{S}$	Surface
S_i	Ionization rate
$\mathbb{S}_d$	Area of the plasma source aperture in the multipolar magnetic confinement chamber
S_p	Net rate at which the charged particles are created (difference between the rate of creation and loss in the volume) (eqn. (2.32))
S_r	Volume recombination rate
$S_0(F_0)$	Collisional term of the homogeneous Boltzmann equation
t	Time
t_s	Time of propagation of a disturbance across the plasma sheath
T	Gas temperature
$\mathcal{T}$	Reduced time: $\mathcal{T} = \omega_{cp}/t$
$\mathcal{J}$	Power factor in $v_i \propto E^{2\mathcal{J}}$
$\mathbb{T}$	Period of an oscillating field
T_e	Electron temperature (1 eV $\approx$ 11560° K)
T_{ed}	Electron temperature in the plasma source
T_i	Ion temperature
$\mathbb{T}_t$	Bouncing period of trapped electrons
u	Kinetic energy
$\bar{u}$	Mean kinetic energy over a period of oscillation
u_c	Energy transferred on the average (over one period) per collision from the HF field to electrons (eqn. (2.33))
u_k	Electron characteristic energy in eV: $u_k \equiv D_e/\mu_e$
U	Voltage amplitude (Sec. 2.3.3)
$\mathcal{U}$	Magnetic scalar potential (Appendix 11.4)
$\mathbb{U}$	$\mathbb{U} = \dfrac{1}{\mathbb{T}} \displaystyle\int_0^{\mathbb{T}} \exp\left[\dfrac{eV_{RF}}{k_B T_e} \cos\left(\dfrac{2\pi t}{\mathbb{T}}\right)\right] dt$ (eqn. (13.32))
v	Complex velocity particle (single trajectory description); also complex average (over EEDF) velocity (hydrodynamic description)

v	Real part of complex velocity $\mathbf{v}$
υ	Reduced velocity: $\upsilon = v/v_p$
v_p	Velocity of primary electrons
v_{ps}	Velocity of the guiding center of primary electrons
v_{th}	Thermal velocity of electrons (most probable velocity): $(2\,k_BT_e/m_e)^{1/2}$
v_{Te}	Average thermal velocity of electrons: $(8\,k_BT_e/\pi m_e)^{1/2}$
v_{Ti}	Thermal velocity of ions
V	Electric scalar potential
$\mathbb{V}$	Volume
V_d	Discharge voltage
V_f	Floating potential (eqn. (13.5))
V_i	Threshold potential for ionization
V_k	Threshold potential for atom or molecule excitation to level k
V_p	Plasma potential
V_s	Threshold potential for an inelastic process
V_{RF}	Radio frequency biasing voltage
V_0	Extraction voltage for ions; also DC component of biasing
w	Individual (or microscopic) velocity of particles
w'	Individual (or microscopic) velocity of particles after interaction with another particle
W	Total average electromagnetic energy stored in a volume (eqn. (2.51)): $W \equiv W_M + W_E + W_K$
W_i	Ion energy
W_{pe}	Kinetic energy of primary electrons
W_E	Average energy stored in the electric field within free space (eqn. (2.48))
W_K	Average kinetic energy related to the ordered motion of electrons under the influence of the HF electric field (eqn. (2.49))
W_M	Average energy stored in the magnetic field (eqn. (2.46))
W_P	Average energy stored in the electric field in a dielectric medium of permittivity $\varepsilon_r\varepsilon_0$ ($W_P = W_E - W_K$); represents the combined effect of the average energy of the electric field stored within free space and by electrons (eqn. (2.47))
x	Rectangular coordinate
χ	Normalized reactance (eqn. (2.84)): $\chi = X/Z_0$
x_m	m - th zero of $J_0(x)$ ($x_0 = 2.405$)
X	Reduced coordinate
$\mathcal{X}$	Reactance; imaginary part of $\mathcal{Z}$ (eqn. (2.84))
$\mathbf{X}$	Thickness of deposition

y	Rectangular coordinate
y	Normalized admittance: $y = Z_0/Z$
Y	Reduced coordinate
$\mathcal{Y}$	Admittance: $\mathcal{Y} = \mathcal{G} + j\,\mathcal{B},\ \mathcal{Y} = \mathcal{Z}^{-1}$
$\mathcal{Y}_D$	Admittance of a discharge (eqn. (2.80)): $\mathcal{Y}_D = \mathcal{G}_D + j\mathcal{B}_D$
z	Rectangular or cylindrical coordinate
z	Normalized impedance with respect to Z_0 (eqn. (2.84)): $z = Z/Z_0 = r + j\chi$
z_L	Normalized input impedance of a source (eqn. (2.83)): $z_L = r_z + j\chi_L$
Z	Reduced coordinate or charge number of ions
$\mathcal{Z}$	Impedance: $\mathcal{Z} = \mathcal{R} + j\mathcal{X}$
$\mathbb{Z}$	Plasma dispersion function
$\mathcal{Z}_p$	Impedance of the plasma in an equivalent circuit (eqn. (4.1))
$\mathcal{Z}_L$	Input impedance of a source: $\mathcal{Z}_L = \mathcal{R}_L + j\mathcal{X}_L$
$\mathcal{Z}_0$	Characteristic impedance of a feed line (real quantity)

Greek Letters

α	Attenuation coefficient of a wave: $\gamma = \alpha + j\,\beta$ (eqn. (5.8))
α_r	Volume recombination coefficient (eqn. (2.28))
α_D	Attenuation coefficient at $n(z) = 2n_D$
β	Axial wavenumber
β_{th}	$\beta_0\,(\xi^2\tau\,c/v_{th})^{1/2}$ (Sec. 6.5.5)
β_0	Free-space wavenumber: $\beta_0 = \omega/c$
γ	$\gamma = \alpha + j\beta$
γ_c	$\gamma_c = e^2/m_e\,m_i\,v_{in}\,v_e$ (eqn. (11.9))
Γ	Particle flux
$\Gamma_L,\ \Gamma_G$	Reflection coefficients in a transmission line (Fig. 2.8)
δ	Ratio v_{eff}/ω
ε_g	Relative permittivity of the discharge tube material
ε_i	Imaginary part of ε_p
ε_p	Complex plasma permittivity relatively to ε_0 ($\varepsilon_p = \varepsilon_r + j\varepsilon_i$; dimensionless) (eqn. (2.2))
$\underline{\varepsilon}_p$	Plasma permittivity tensor
ε_r	Real part of ε_p
ε_L	$1 - \xi^2/(1 + \tau)$ (Sec. 6.5.1)
ε_R	$1 - \xi^2/(1 - \tau)$ (Sec. 6.5.1)
ε_0	Permittivity of free space: $\varepsilon_0 = 8.854 \times 10^{-12}$ farad/meter

ζ	Axial electron density profile: $\zeta(2z/\ell) = n(r,z)/n_0 J_0(x_0 r/R)$ (Sec. 4.4.2)
η	Overall efficiency of a plasma source (eqn. (2.74))
η_c	Coupling efficiency
η_L	Launching efficiency
θ	Power absorbed or lost, per electron, since under steady-state conditions, $\theta_A = \theta_L$
θ_A	Power absorbed per electron from the electromagnetic field (eqn. (2.18))
θ_L	Power lost per electron in collisions of all kinds (eqn. (2.71))
Θ	Polarization angle of the electric field in the direction perpendicular to $\mathbf{B}_0$ (eqn. (6.23))
κ	Shape factor
κ_α	$\kappa_\alpha = \omega_{c\alpha}/\nu_{eff}$, where the subscript α stands for i in the case of ions and e for electrons (Sec. 6.2.1.1)
λ	Wavelength: $\lambda = 2\pi/\beta$
λ_a	Ambipolar mean free path
λ_b	Reduced magnetic vector potential (Appendix 11.4); "λ_b = constant" defines a field line
λ_e	Electron mean free path (eqn. (11.10))
λ_{en}	Electron mean free path for collisions with neutrals
λ_g	Wavelength in a waveguide
λ_{in}	Ion mean free path for collisions with neutrals: $\lambda_{in} = v_{Ti} v_{in}^{-1}$
λ_{pi}	Mean free path of primary electrons for ionization
λ_{pn}	Mean free path of primary electrons for momentum transfer with neutrals
λ_s	Electron mean free path for the collision cross-section Q_s (eqn. (2.23))
λ_D	Debye length: $\lambda_D^2 = \varepsilon_0 k_B T_e/ne^2$
λ_0	Free-space wavelength
Λ	Characteristic diffusion length
Λ_c	Coulomb screening factor (eqn. (6.27))
Λ_e	Effective diffusion length (eqn. (5.17))
Λ_r	Diffusion length in the radial direction of a cylinder
Λ_z	Diffusion length in the axial direction of a cylinder
Λ_D	Effective diffusion length (eqn. (4.10))
Λ_∞	Diffusion length for long cylinders, i.e. $\ell^2 \gg R_1^2$: $\Lambda_\infty = R_1/2.405$
μ	Mobility
$\tilde{\mu}$	Mobility tensor
μ_a	Ambipolar mobility

σ_i	Imaginary part of σ
σ_r	Real part of σ
τ	ω_{ce}/ω or time duration of a square wave signal
ϕ or φ	Azimuthal angle
Φ	Reduced electric potential: $\Phi = eV/k_B T_e$
χ	Polar angle; also scattering angle
χ_k	Elements of the tensor $\underset{\sim}{\chi}$ used in defining the diffusion tensor $\underline{D}$ and the mobility tensor $\underset{\sim}{\mu}$ (eqn. (6.2))
ψ	Phase angle
Ψ	Angle between the wave vector $\boldsymbol{\beta}$ and the static magnetic field $\mathbf{B}_0$
ω	Angular frequency of the field: $\omega = 2\pi f$
ω_{ce}	Electron cyclotron angular frequency
ω_{ci}	Ion cyclotron angular frequency
ω_{cp}	Cyclotron angular frequency of a primary electron
ω_h	Upper hybrid angular frequency: $\omega_h = \sqrt{\omega_{pe}^2 + \omega_{ce}^2}$
ω_{pe}	Electron plasma angular frequency
ω_{pi}	Ion plasma angular frequency
ω_L	Cutoff frequency of the right-hand plasma wave (Sec. 6.5.2)
ω_R	Cutoff frequency of the left-hand plasma wave (Sec. 6.5.2)
ω_{RF}	Angular frequency of the RF biasing voltage
Ω	Solid angle

Subscripts

e	Stands for electrons
h	Indicates the value of a quantity at a given position in space
i	Generally stands for ions (except with ε and σ where it designates the imaginary part of these complex quantities)
n	Stands for neutral atoms (molecules)
p	Stands for primary electrons
r	Stands for the real part of a quantity
rms	Root mean square of the quantity
s, t	Identify coordinate axes in cylindrical geometry $s, t = \mathrm{r},\phi,\mathrm{z}$
$\parallel$	Identify the direction parallel to the static magnetic field $\mathbf{B}_0$
$\perp$	Identify the direction perpendicular to the static magnetic field $\mathbf{B}_0$

Mathematics symbols

∞	Infinity
$\approx$	Approximately equal to

$\equiv$	Identically equal to
$< \, >$	Denotes averaging over the EEDF
$\overline{A}(t)$	Mean value of A(t) over a period of oscillation
∇	Gradient
$\nabla \cdot$	Divergence
$\nabla \times$	Curl
∇^2	Laplacian
$\overline{A}$	Cross-section average value of A
$\underset{\approx}{A}$	Tensor
$\mathbf{A}$	Vector (complex)
$\boldsymbol{A}$	Vector (real)
$\mathbf{A}^*$	Complex conjugate of a complex vector $\mathbf{A}$
$\mathbf{A} \cdot \mathbf{A}$	Scalar product of $\mathbf{A}$ ($\mathbf{A} \cdot \mathbf{A} = A^2$, where A^2 is a positive real scalar)
$\mathbf{A} \times \mathbf{A}$	Vector product of $\mathbf{A}$ with $\mathbf{A}$
$|z|$	Modulus of a complex number z: $|z| = \sqrt{zz^*}$
$|\mathbf{A}|$	Magnitude of a vector quantity
Re(A)	Real part of A
Im(A)	Imaginary part of A
dA	Differential element of the quantity A
∂A	Partial differential element of A

Abreviations

AES	Auger electron spectroscopy
CVD	Chemical vapor deposition
DC	Direct current
DECR	Distributed electron cyclotron resonance
ECR	Electron cyclotron resonance
EEDF	Electron energy distribution function
EELS	Electron energy loss spectroscopy
EM	Electromagnetic
EPR	Electron paramagnetic (spin) resonance
EVDF	Electron velocity distribution function
FTIR	Fourier transform infra red
HF	High frequency
IR	Infra red
ISM	Industrial, scientific and medical
LIF	Laser induced fluorescence
LMP[R]	Large volume microwave plasma

MIP	Microwave induced plasma
MW	Microwave
NMR	Nuclear magnetic resonance
PECVD	Plasma enhanced chemical vapor deposition
RF	Radio frequency
RHCP	Right-hand circular polarized
RHEED	Reflection high energy electron diffraction
RIE	Reactive ion etching
SCCM	Standard cubic centimer per second
SIMS	Secondary ion mass spectroscopy
SWD	Surface wave discharge
TE	Transverse electric field
TEM	Transverse electromagnetic field
TM	Transverse magnetic field
TWD	Traveling wave discharge
ULSI	Ultra large scale integration
UV	Ultra violet
VSWR	Voltage standing wave ratio
XPS	X-ray photoelectron spectroscopy

INDEX

9 780444 888150